FINITE ELEMENT ANALYSIS AND DESIGN OF STEEL AND STEEL–CONCRETE COMPOSITE BRIDGES

by

EHAB ELLOBODY

Professor, Department of Structural Engineering,
Faculty of Engineering, Tanta University, Egypt

ELSEVIER

AMSTERDAM • BOSTON • HEIDELBERG • LONDON
NEW YORK • OXFORD • PARIS • SAN DIEGO
SAN FRANCISCO • SINGAPORE • SYDNEY • TOKYO

Butterworth-Heinemann is an imprint of Elsevier

Butterworth-Heinemann is an imprint of Elsevier
225, Wyman Street, Waltham, MA 01803, USA
The Boulevard, Langford Lane, Kidlington, Oxford OX5 1 GB, UK

First edition **2014**

Library of Congress Cataloging-in-Publication Data
Ellobody, Ehab, author.
   Finite element analysis and design of steel and steel-concrete composite bridges / by Ehab
Ellobody. – First edition.
      pages cm
   Includes bibliographical references and index.
   ISBN 978-0-12-417247-0
   1. Iron and steel bridges–Design and construction. 2. Concrete bridges–Design and
construction. 3. Finite element method. I. Title.
   TG380.E45 2015
   624.2′5–dc23

                                                                                2014011942

British Library Cataloguing in Publication Data
A catalogue record for this book is available from the British Library

For information on all **Butterworth-Heinemann** publications
visit our web site at store.elsevier.com

Transferred to Digital Printing in 2014

ISBN: 978-0-12-417247-0

FINITE ELEMENT ANALYSIS AND DESIGN OF STEEL AND STEEL–CONCRETE COMPOSITE BRIDGES

CONTENTS

CHAPTER 1

Introduction

1.1 GENERAL REMARKS

Steel and steel–concrete composite bridges are commonly used all over the world, owing to the fact that they combine both magnificent aesthetic appearance and efficient structural competence. Their construction in a country not only resembles the vision and inspiration of their architects but also represents the country's existing development and dream of a better future. Compared to traditional reinforced concrete (RC) bridges, steel bridges offer many advantages, comprising high strength-to-self weight ratio, speed of construction, flexibility of construction, flexibility to modify, repair and recycle, durability, and artistic appearance. The high strength-to-self weight ratio of steel bridges minimizes dead loads of the bridges, which is particularly beneficial in poor ground conditions. Also, the high strength-to-self weight ratio of steel bridges makes it easy to transport, handle, and erect the bridge components. In addition, it facilitates very shallow construction depths, which overcome problems with headroom and flood clearances, and minimizes the length of approach ramps. Furthermore, high strength-to-self weight ratio of steel bridges permits the erection of large components, and in special circumstances, complete bridges may be installed in quite short periods. The speed of construction of steel bridges is attributed to the fact that most of the bridge components can be prefabricated and transported to the construction field, which reduces working time in hostile environments. The speed of construction of steel bridges also reduces the durations of road closures, which minimizes disruption around the area of construction. Flexibility of construction of steel bridges is attributed to the fact that the bridges can be constructed and installed using different methods and techniques. Installation may be conducted by cranes, launching, slide-in techniques, or transporters. Steel bridges give contractors the flexibility in terms of erection sequence and program. The bridge components can be sized to suit access restrictions at the site, and once erected, the steel girders provide a platform for subsequent operations. Flexibility to modify, repair, and recycle steel bridges is a result of the ability to modify the current status of the bridges such as widening the bridges to accommodate more lanes of traffic. Also, steel bridges can be repaired or strengthened by adding steel plates or advanced composite laminates to carry more traffic loads. In addition, if for any reason, such as end of their life of use or change of environment around the area, steel bridges can be recycled. Steel bridges are durable bridges, provided that they are well designed, properly maintained, and carefully protected against corrosion. Finally, steel bridges can fit most of the complex architecture designs, which in some cases are impossible to accommodate using traditional RC bridges.

Highway bridges made of RC slabs on top of the steel beams can be efficiently designed as composite bridges to get the most benefit from both the steel beams and concrete slabs. Steel-concrete composite bridges offer additional advantages to the aforementioned advantages of steel bridges. Compared to steel bridges, composite bridges provide higher strength, higher stiffness, higher ductility, higher resistance to seismic loadings, full usage of materials, and particularly higher fire resistance. However, these advantages are maintained, provided that the steel beams and concrete slabs are connected via shear connectors to transmit shear forces at the interface between the two components. This will ensure that the two components act together in resisting applied traffic loads on the bridges, which will result in significant increases in the allowable vehicular weight limitations, ability to transport heavy industrial and construction equipment, and possibility to issue overload permits for specialized overweight and oversized vehicles. One of the main advantages of having steel beams acting together with concrete slabs in composite bridges is that premature possible failures of the two separate components are eliminated. For example, one of the primary modes of failure for concrete bridges is cracking of the concrete slabs and beams in tension, while for the steel bridges, the possible modes of failure are the formation of plastic hinges and the buckling of webs or flanges. By having the steel beams work together with the concrete slab, the whole slab will be mainly subjected to compressive forces, which reduces the possibility of tensile cracking. On the other hand, the presence of the concrete slab on top of the steel beams eliminates the buckling of the top flange of the steel beams. Efficient design of steel-concrete composite bridges can ensure that both the steel beams and concrete slabs work together in resisting applied traffic loads until failure occurs in both components, preferably at the same time, to get the maximum benefit from both components.

Numerous books were found in the literature highlighting different aspects of design for steel and steel-concrete composite bridges; for examples, see [1.1–1.11]. The books highlighted the problems associated with the planning, design, inspection, construction, and maintenance of steel and steel-concrete composite bridges. Overall, the books discussed the basic concepts and design approaches of the bridges, design loads on the bridges from either natural or traffic-induced forces, and design of different components of the bridges. On the other hand, numerous finite element books are found in the literature; for examples, see [1.12–1.18], explaining finite element method as a widely used numerical technique for solving problems in engineering and mathematical physics. The books [1.12–1.18] were written to provide basic learning tools for students mainly in civil and mechanical engineering classes. The books [1.12–1.18] highlighted the general principles of finite element method and the application of this method to solve practical problems. However, limited investigations, with examples detailed in [1.19, 1.20], are found in the literature in which researchers used finite element method in analyzing case studies related to steel and steel-concrete composite bridges. Recently, with continuing developments of computers and solving and modeling techniques, researchers started to detail the use of finite element method to analyze steel and steel-concrete composite bridges, with examples presented in [1.21, 1.22]. Also, extensive experimental and numerical research papers were found in the literature highlighting finite element analysis of steel and steel-concrete composite bridges, which will be detailed in Section 1.3. However, up-to-date, there are no detailed books found in the literature addressing both finite element analysis and design of steel and steel-concrete composite bridges, which is credited to this book. The current book will present, for the first time, explanation of the latest finite element modeling approaches specifically as a complete piece work on steel and steel-concrete composite bridges. This finite element modeling of the bridges will be accompanied by design examples for steel and steel-concrete composite bridges calculated using current codes of practice.

There are many problems and issues associated with finite element modeling of steel and steel-concrete composite bridges in the literature that students, researchers, designers, practitioners, and academics need to address. Incorporating nonlinear material properties of the bridge components in finite element analyses has expanded tremendously over the last decades. In addition, computing techniques are now widely available to manipulate complicated analyses involving material nonlinearities of the bridge components. This book will highlight the latest techniques of modeling nonlinear material properties of the bridge components. Also, simplified analytic solutions were derived to predict the distribution of forces and stresses in different bridge components based on many assumptions and limitations. However, accurate analyses require knowledge of the actual distribution of forces and stresses in the component members, which is the target of the nonlinear finite element modeling approach detailed in this book. In addition, in case of steel-concrete composite bridges, if the slab cracks under heavy traffic loads or the steel beam yields or buckles, it becomes extremely important to know the location of failure, the postfailure strength of the component that has failed, and the manner in which the forces and stresses will redistribute themselves owing to the failure. Once again, traditional simplified analyses cannot account for these complex failure modes because no interaction between bridge components was considered. The finite element modeling approach aimed in this book will capture all possible failure modes associated with steel-concrete composite bridges. It should also be noted that while simplified design methods have been developed to predict the ultimate capacity of steel bridges or their components, none of these methods adequately predicts the structural response of the bridge in the region between design load levels and ultimate capacity load levels. Therefore, the proposed finite element modeling approach will reliably predict both the elastic and inelastic responses of a bridge superstructure as well as the structural response in the region between the design limit and the ultimate capacity. Another complex issue is the slip at the steel-concrete interface in composite bridges that occurs owing to the deformation of shear connectors under heavy traffic loads. This parameter also cannot be considered using simplified design methods and can be accurately incorporated using finite element modeling. The aforementioned issues are only examples of the problems associated with modeling of steel and steel-concrete composite bridges. Overall, this book provides a collective material, for the first time, for the use of finite element method in understanding the actual behavior and correct structural performance of steel and steel-concrete composite bridges.

Full-scale tests on steel and steel-concrete composite bridges are quite costly and time-consuming, which resulted in a scarce in test data for different types of bridges. The dearth in the test data is also attributed to the continuing developments, over the last decades, in the cross sections of the bridges and their components, material strengths of the bridge components, and applied loads on the bridges. Therefore, design rules specified in current codes of practice for steel and steel-concrete composite bridges are mainly based on small-scale tests on the bridges and full-scale tests on the bridge components. In addition, design rules specified in the American Specifications [1.23–1.25], British Standards [1.26], and Eurocode [1.27, 1.28] are based on many assumptions, limitations, and empirical equations. An example of the shortcomings in current codes of practice for steel-concrete composite bridges is that, up-to-date, there are no design provisions to consider the actual load-slip characteristic curve of the shear connectors used in the bridges, which results in partial degree of composite action behavior. This book will detail, for the first time, how to consider the correct and actual slip occurring at the steel-concrete interface in composite bridges through finite element modeling. This book will highlight the latest numerical investigations performed in the literature to generate more data, fill in the gaps, and compensate the lack of data for steel and steel-concrete composite bridges. This book also highlights the use of finite element modeling to improve and propose more accurate design guides for steel and steel-concrete composite bridges, which are rarely found in the literature. In addition, this book contains examples for finite element models developed for different steel and steel-concrete composite bridges as well as worked design examples for the bridges. The author hopes that this book will provide the necessary materials for all interested researchers in the field of steel and steel-concrete composite bridges. Furthermore, the book can also act as a useful teaching tool and help beginners in the field of finite element analysis and design of steel and steel-concrete composite bridges. The book can provide a robust approach for finite element analysis of steel and steel-concrete composite bridges that can be understood by undergraduate and postgraduate students.

The book consists of seven well-designed chapters covering necessary topics related to finite element analysis and design of steel and steel-concrete composite bridges. This chapter provides a general background for the types of steel and steel-composite bridge and explains the classification of bridges. The chapter also presents a brief review for the components of the bridges and how the loads are transmitted by the bridge to the ground. The chapter also gives an up-to-date review for the latest available investigations carried out on steel and steel-concrete composite bridges. The chapter focuses on main issues and problems associated with the bridges and how they are handled in the literature. The chapter also introduces the role of finite element modeling to provide a better understanding of the behavior of bridges. Finally, this chapter highlights the main current codes of practice used for designing steel and steel-concrete composite bridges.

Chapter 2 focuses on the nonlinear material behavior of the main components of steel and steel-concrete composite bridges comprising steel, concrete, reinforcement bars, shear connectors, etc. The chapter presents the stress-strain curves of the different materials used in the bridges and defines the important parameters that must be measured experimentally and incorporated in the finite element modeling. The definitions of yield stresses, ultimate stresses, maximum strains at failure, initial stiffness, and proportional limit stresses are presented in the chapter. The chapter enables beginners to understand the fundamental behavior of the materials in order to correctly insert them in the finite element analyses. Covering the behavior of shear connectors in this chapter is important to understand how the shear forces are transmitted at the steel–concrete slab interfaces in composite bridges. In addition, the chapter presents how the different materials are treated in current codes of practice.

Chapter 3 presents the different loads acting on steel and steel–concrete composite bridges and the stability of the bridges when subjected to these loads. The chapter starts by showing the dead loads of steel and steel–concrete composite bridges that are initially estimated for the design of bridges. Then, the chapter moves to explain how the live loads from traffic were calculated. After that, the chapter presents the calculation of wind loads on the bridges and highlights different other loads that may act on the bridges such as centrifugal forces, seismic loading, and temperature effects. When highlighting the loads in this chapter, it is aimed to explain both of the loads acting on railway and highway bridges. The calculations of the loads are based on the standard loads specified in current codes of practice. In addition, the chapter also presents, as examples, the main issues related to the stability of steel and steel–concrete composite plate girder and truss bridges, which enable readers to understand the stability of any other type of bridges.

Chapter 4 presents detailed design examples of the components of steel and steel–concrete composite bridges. The design examples are calculated based on current codes of practice. The design examples are shown for the stringers (longitudinal beams of the bridges), cross girders, plate girders, trusses, bracing systems, bearings, and other secondary members of the bridges. Also, design examples are presented for steel–concrete composite bridges. It should be noted that the aim of this book is to provide all the necessary information and background related to the design of different bridges using different codes of practice. Therefore, the design examples presented are hand calculations performed by the author. The chapter explains how the cross sections are initially assumed, how the straining actions are calculated, and how the stresses are checked and assessed against current codes of practice.

Chapter 5 focuses on finite element analysis of steel and steel–concrete composite bridges. The chapter presents the more commonly used finite elements in bridges and the choice of correct finite element types and mesh size that can accurately simulate the complicated behavior of the different components of steel and steel-concrete composite bridges. The chapter highlights the linear and nonlinear analyses required to study the stability of the bridges and bridge components. Also, the chapter details how the nonlinear material behavior can be efficiently modeled and incorporated in the finite element analyses. In addition, Chapter 5 details modeling of shear connection for steel-concrete composite bridges. Furthermore, the chapter presents the application of different loads and boundary conditions on the bridges. The chapter focuses on the finite element modeling using any software or finite element package, for example, in this book, the use of ABAQUS [1.29] software in finite element modeling.

Chapters 6 and 7 present illustrative examples of finite element models developed to understand the structural behavior of steel and steel-concrete composite bridges, respectively. The chapters start with a brief introduction of the presented examples as well as a detailed review of previous investigations related to the presented examples. The chapters detail how the models were developed and the results obtained. The presented examples show the effectiveness of finite element models in providing detailed data that complement experimental data in the field. The results are discussed to show the significance of the finite element models in predicting the structural response of the different bridges investigated. Overall, they aim to show that finite element analysis not only can assess the accuracy of the design rules specified in current codes of practice but also can improve and propose more accurate design rules. Once again, it should be noted that in order to cover all the latest information regarding the finite element modeling of different bridges, the presented finite element models are developed by the author as well as by other researchers and previously reported in the literature.

1.2 TYPES OF STEEL AND STEEL-CONCRETE COMPOSITE BRIDGES

Steel bridges can be classified according to the type of traffic carried to mainly *highway* (roadway) bridges, which carry cars, trucks, motorbikes, etc. with an example shown in Figure 1.1; *railway* bridges, which carry trains, with an example shown in Figure 1.2; or *combined highway-railway* bridges, which carry combinations of the aforementioned traffic as shown in Figure 1.3. There are also steel bridges carrying pipelines (Figure 1.4), cranes (Figure 1.5), and pedestrian bridges (Figure 1.6), which are also secondary

Figure 1.1 A highway arch steel bridge (bikethehoan.com). (For the color version of this figure, the reader is referred to the online version of this chapter.)

Figure 1.2 A railway arch steel bridge (highestbridges.com). (For the color version of this figure, the reader is referred to the online version of this chapter.)

Figure 1.3 A combined highway-railway truss steel bridge (chinatravelguide.com). (For the color version of this figure, the reader is referred to the online version of this chapter.)

Figure 1.4 An arch steel bridge carrying pipelines (civilenginphotos.blogspot.com). (For the color version of this figure, the reader is referred to the online version of this chapter.)

Figure 1.5 A crane truss steel bridge (paperstreet.iobb.net). (For the color version of this figure, the reader is referred to the online version of this chapter.)

Figure 1.6 A pedestrian arch steel bridge (photos.uc.wisc.edu). (For the color version of this figure, the reader is referred to the online version of this chapter.)

types of this classification. Railway bridges may be constructed such that the rails rest on sleepers, which rest on the longitudinal beams of the bridge. In this case, the bridges are called *open-timber floor railway bridge* and commonly used outside towns as shown in Figure 1.7. Alternatively, railway bridges

Figure 1.7 An open-timber floor bridge (123rf.com). (For the color version of this figure, the reader is referred to the online version of this chapter.)

Figure 1.8 A ballasted floor bridge (hothamvalleyrailway.com). (For the color version of this figure, the reader is referred to the online version of this chapter.)

may be constructed such that the rails rest on sleepers, which rest on compact aggregates confined by a RC box transmitting the load straightaway to the main structural system. In this case, the bridges are called *ballasted floor railway bridges* and commonly used in towns as shown in Figure 1.8. Railway bridges with no concrete slabs on top of the carrying steel beams are called *railway steel bridges* (Figure 1.2). On the other hand, highway bridges constructed such that the concrete slabs are connected to the steel beams underneath via shear connectors ensuring that the two components act together in resisting traffic loads are called *highway steel-concrete composite bridges* as shown in Figure 1.9. Figure 1.9 shows a steel–concrete box girder composite bridge under construction where headed stud shear connectors are used to connect both the concrete slab and the steel box girder section.

Steel and steel-concrete composite bridges (highway or railway) can be classified according to the type of the main structural system considered in the design of the bridges to *plate girder bridges, box girder bridges, rigid-frame bridges, truss bridges, arch bridges, cable-stayed bridges, suspension bridges,* and

Figure 1.9 A steel-concrete composite box girder bridge under construction (mto.gov. on.ca). (For the color version of this figure, the reader is referred to the online version of this chapter.)

orthotropic floor bridges. Plate girder bridges are the bridges having their main carrying structural system made of plate I-shaped girders, which are suitable for simply supported spans up to 40 m. For normal bridge cross-section widths (less than or equal 10 m), twin plate girder bridges may be used. Otherwise, multiple plate girders can be used as main structural systems transmitting different loads to foundations, as shown in Figure 1.10. Box girder bridges (see Figure 1.11) are the bridges having their main structural system made of box-shaped girders, which are suitable for continuous spans up to 300 m. Rigid frame bridges (see Figure 1.12) are the bridges having their main structural system made of rigid frames, which are suitable for continuous spans up to 200 m. Truss bridges (see Figure 1.3) are the bridges having their main structural system made of trusses, which are suitable for simple and continuous spans from 40 to 400 m. Arch bridges (see Figures 1.1, 1.2, 1.4, and 1.6) are the bridges having their main structural system made of arches, which are suitable for simple and continuous spans from 200 to 500 m. Cable-stayed bridges (see Figure 1.13) are the bridges having their main structural system made of cables hung from one or more towers, which are economical when the spans are in the range of 200 to 800 m. Suspension bridges (see Figure 1.14) are the bridges having their main structural system made of decks suspended by cables stretched over the bridge span, anchored to the ground at two ends, and passed over towers at or near the edges of the bridge, which are, similar to cable-stayed bridges, economical when the spans are in the range of 200 to 1000 m. Finally, orthotropic floor bridges (see Figure 1.15) are the bridges having their main structural system made of structural steel deck plate stiffened either

Figure 1.10 A multiplate girder bridge (haks.net). (For the color version of this figure, the reader is referred to the online version of this chapter.)

Figure 1.11 A box girder bridge (alviassociates.com). (For the color version of this figure, the reader is referred to the online version of this chapter.)

Figure 1.12 A rigid-frame bridge (en.structurae.de). (For the color version of this figure, the reader is referred to the online version of this chapter.)

Figure 1.13 A cable-stayed bridge (bridgemeister.com). (For the color version of this figure, the reader is referred to the online version of this chapter.)

Figure 1.14 A suspension bridge (ikbrunel.org.uk). (For the color version of this figure, the reader is referred to the online version of this chapter.)

longitudinally or transversely, or in both directions. The orthotropic deck may be supported straightaway on the main structural system such as plate girder and truss or supported on a cross girder transmitting the load to the main structural system.

Steel and steel-concrete composite bridges can also be classified according to the position of the carriageway relative to the main structural system to *deck bridges*, *through bridge*, *semi-through bridge*, and *pony bridge*. Deck bridges are the bridges having their carriageway (highway or railway) resting on top of the main structural system as shown in Figures 1.1 and 1.2 and the highway bridge in Figure 1.3. Through bridges are the bridges having their carriageway resting on the bottom level of the main structural system and the top level of the main structural system is above the carriage as shown for the railway bridge in Figure 1.3. In this case, a top-bracing system can be installed at the top level of the main structural system. Semi-through bridges

Figure 1.15 An orthotropic steel floor truss bridge (steelconstruction.info). (For the color version of this figure, the reader is referred to the online version of this chapter.)

Figure 1.16 A semi-through truss bridge under construction (steel-trussbridge.com). (For the color version of this figure, the reader is referred to the online version of this chapter.)

are the bridges having their carriageway resting between the bottom and top levels of the main structural system and the top level of the main structural system is below the carriage, with an example shown in Figure 1.16. In this case, a top-bracing system cannot be installed at the top level of the main structural system. Finally, pony bridges are semi-through bridges having their carriageway resting on the bottom level of the main structural system and the top level of the main structural system is below the carriage as shown in Figure 1.17. In this case, similar to semi-through bridges, a top-bracing system cannot be installed at the top level of the main structural system. It should be noted that most of modern bridges are fabricated in workshops and transferred to the construction field. Also, most of modern bridges are fabricated such that the main structural system components are

Figure 1.17 A pony truss bridge (bphod.com). (For the color version of this figure, the reader is referred to the online version of this chapter.)

Figure 1.18 An old-fashioned riveted truss bridge (pbase.com). (For the color version of this figure, the reader is referred to the online version of this chapter.)

connected by welding to replace the old-fashioned riveted bridge shown in Figure 1.18. However, in case of continuous bridges and long-span bridges, it is more convenient to divide the bridge into separate welded parts that are connected to the construction field by bolted connections.

Let us now look in more detail to the structural components of a traditional railway bridge. Figure 1.19 shows the general layout of a double-track open-timber floor plate girder railway steel bridge. A train track of this railway bridge consists of a pair of rails resting on timber sleepers. For a single track, the sleepers are supported by two longitudinal steel beams known as *stringers*. The stringers are spaced at specified distances (a_3), given by the national code of practice in the country of construction, depending on the spacing between rails and the spacing between center-lines of trains (a_2), in case of more than a single track. The stringers are supported on cross steel beams known as *cross girders*. The cross girders

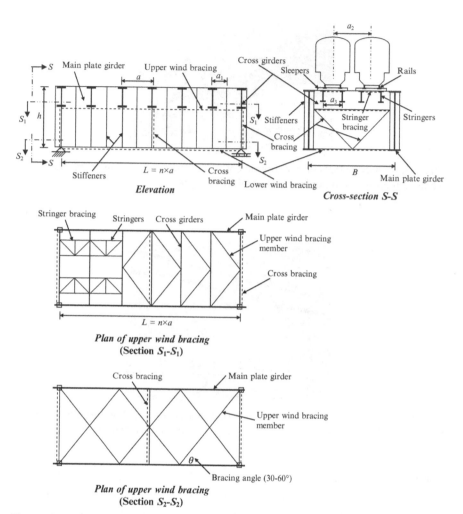

Figure 1.19 General layout of a double-track open-timber floor plate girder railway steel bridge.

are supported by two, in this case of bridges (Figure 1.19), longitudinal main steel beams known as *main plate girders*, which are the main structural system for this type of bridges. The main plate girders are supported on supports called "bearings" such as the *hinged and roller bearings* shown in Figure 1.19, which rest on foundations or piers, in case bridges are constructed over obstacles such as rivers, roads, and seas. The main girders are spaced at a distance (*B*), which is the width of the bridge. The moving train loads are transmitted from the rails to the sleepers, from the sleepers to the stringers, from the stringers to the cross girders, from the cross girders to main plate girders, from the main plate girders to the bearings, from the bearings to the foundations or piers, and finally from foundations or piers to the ground. Wind and lateral loads acting on the bridge can be transmitted by systems of horizontal (*upper and lower wind bracings*) and vertical (*cross wind bracings*) bracing systems, which carry out wind loads safely to the bearings. Also, the stringers can be attached to horizontal systems of bracings called *stringer bracing* or *lateral shock* (*nosing force*) *bracing*, which transmit lateral shock (nosing) forces resulting from the moving train safely to cross girders where it causes additional small axial force on the cross girders. The web of the main I-shaped plate girder bridge is very sensitive to buckling since it has a thin thickness compared to its depth. Therefore, the web of the plate girder is strengthened by vertical and horizontal *stiffeners*. The spacing between the vertical stiffeners should be reasonably assumed (1.5-2 m) not to increase the thickness of the web. Hence, the spacing between cross girders (*a*) is dependent on the number of vertical stiffeners used between two adjacent cross girders. Finally, the length of the bridge (*L*) is equal to the number of (*a*).

The structural components of a traditional highway bridge can be reviewed as shown in Figure 1.20. The figure shows the general layout of a through truss highway steel bridge. The bridge has a RC floor supported by a number of stringers. The stringers are spaced at designed distances (*a*$_3$) reasonably assumed between 2 and 3 m. Similar to railway bridges, the stringers of this type of bridges are supported by cross girders. The cross girders are supported by two longitudinal trusses, which are the main structural system for this type of bridges. The main trusses are supported on hinged and roller bearings, which rest on foundations or piers. The truck and car loads are transmitted from the RC floor to the stringers, from the stringers to the cross girders, from the cross girders to the main trusses, from the main trusses to the bearings, from the bearings to the foundations or piers, and finally from the foundations or piers to the ground. Wind loads acting on the bridge can be transmitted by systems of horizontal upper, since this bridge is a through bridge with enough height to contain traffic in addition to overhead clearance, and end portal frames, since cross bracing will close the bridges, which carry out wind loads safely to the bearings. The bracing systems are also important to define the buckling lengths of compression members of the main trusses. However, the stringers do not need a bracing since the RC concrete floor takes care of any lateral and

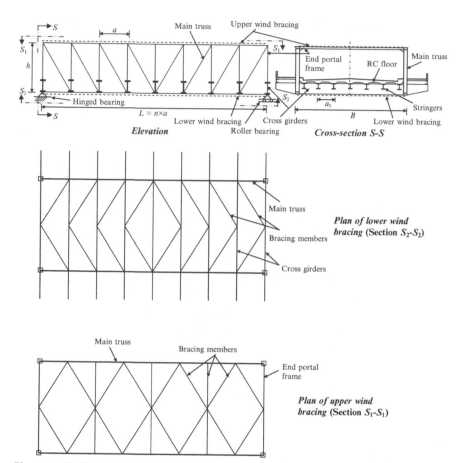

Figure 1.20 General layout of a through truss highway steel bridge.

longitudinal loads associated with moving traffic. Cross girders must be aligned with vertical members to avoid adding bending moments to truss members. Hence, the spacing between cross girders (a) is the spacing between vertical truss members. The spacing between vertical truss members is dependent on the angle of inclined truss members, which is defined by the height of the vertical members (h) that is dependent on the length of the bridge (L). The length of the bridge (L) is equal to the number of spacing between cross girders or vertical truss members (a).

Let us now look at the structural components and general layout of a steel–concrete composite highway bridge shown in Figure 1.21. The bridge has a RC floor supported by a number of main I-shaped plate girders.

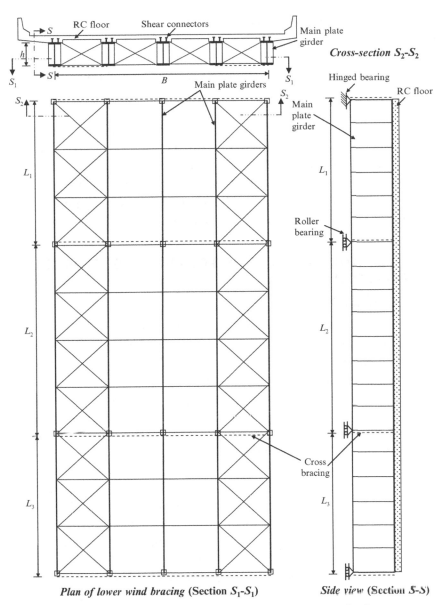

RC floor   Shear connectors   Main plate girder

Cross-section S_2-S_2

h

S_1 B S_1

Main plate girders

Hinged bearing

RC floor

S_2 Main plate girder

L_1 L_1

Roller bearing

L_2 L_2

Cross bracing

L_3 L_3

Plan of lower wind bracing (Section S_1-S_1)   *Side view (Section S-S)*

Figure 1.21 General layout of a highway steel-concrete composite bridge.

Headed stud shear connectors were used to transmit shear forces at the steel-concrete interface and to ensure that both components work together in resisting applied traffic loads. The main plate girders are supported on hinged and roller bearings, which rest on foundations or piers. The traffic loads are carried out by the composite action between the steel plate girders and the RC concrete floor transmitting the loads to the hinged and roller bearings attached to the steel plate girders. Wind loads acting on the bridge can be transmitted by lower bracing systems and cross bracings. However, systems without upper bracing are used since the RC concrete floor carries out all lateral and longitudinal loads associated with moving traffic. It should be noted that for this continuous-span steel-concrete composite plate girder bridge, there are sagging and hogging bending moments. The composite action relies on that the concrete slab must be in the compression zone. Therefore, parts of the composite plate girder where the concrete slab is in the tension zone are designed without considering the composite action between the steel plate girder and the concrete slab.

1.3 LITERATURE REVIEW OF STEEL AND STEEL-CONCRETE COMPOSITE BRIDGES

1.3.1 General Remarks

Steel and steel-concrete composite bridges have been the subject of extensive investigations, reported in the literature, highlighting the design and structural behavior of the bridges. The investigations were mainly research papers presenting small-scale laboratory tests on the bridges and their components, limited full-scale tests on the bridge components, and numerous numerical and analytic investigations of the bridges and their components. The investigations covered different types of bridges subjected to different loads and designed according to rules specified in current codes of practice. The main objective of the investigations was to satisfy safety and serviceability requirements imposed by current design codes of practice as well as to fulfill other requirements set by the public such as cost, self-weight, and aesthetic appearance. However, the investigations were hindered by the high-cost and time-consuming full-scale tests on this form of construction. Numerous books were found in the literature, with examples given in [1.1–1.11], addressing different parameters related to the design, construction, inspection, and maintenance of the bridges. The aforementioned books contained literature reviews and historical developments of steel and steel-concrete composite girders. These reviews will not be repeated in this study since the main objective of this book is to present the latest and current investigations related to the design and finite element modeling of the bridges investigated. This section presents recent experimental and numerical investigations on the bridges and their components. The author aims that the presented material can update the information related to steel and steel-concrete composite bridges and act as basis for future investigations.

1.3.2 Recent Investigations on Steel Bridges

Curved steel I-shaped plate girder bridges have been the subject of experimental and analytic studies presented by Zureick *et al.* [1.30]. The authors have shown that due to the need to augment traffic capacity in urban highways and the constraints of existing constructions, there has been a steady increase in the use of curved bridges. This is attributed to the advantages of curved steel girders comprising simplicity of fabrication and construction, speed of erection, and serviceability performance. The study [1.30] described a full-scale experimental and analytic program to develop new design guidelines for horizontally curved steel bridges. The authors have shown that although horizontally curved steel bridges constitute around one-third of all steel bridges being erected today, their structural behavior is not fully understood. The study was divided into six stages starting with a review of previous research and followed by an investigation of construction issues, determination of straining actions, connection details, serviceability considerations, and determination of the levels of analysis required for horizontally curved girders. Based on, mainly, the comprehensive bibliography on curved steel girders, containing over 200 references, presented by McManus *et al.* [1.31], the state-of-the-art review performed by the ASCE-AASHTO Committee on Flexural Members [1.32] and the book published by Nakai and Yoo [1.33], the authors have performed an extensive literature review comprising around 900 references reported in [1.34], which showed that approximate analytic methods for curved steel I-shaped plate girder bridges have shortcomings since they do not consider the bracing effect in the plane of the bottom flange and their reliability depends on the selection of the proper live-load distribution factors. Thus, approximate methods are only recommended for preliminary analyses. Also, the authors [1.30] concluded that compared to different analytic methods (finite strip, finite difference, closed-form solutions to differential equations, and slope-deflection method), the finite element method can act as a general and comprehensive technique to perform static/dynamic and elastic/inelastic analyses with different mechanical and thermal loadings. The other analytic methods can be as good as the finite element method but are limited to certain configurations and boundary conditions. In addition, the authors concluded that the geometrically and/or materially nonlinear behavior of horizontally curved ridges was not fully understood. The study has also outlined the shortcomings in the previously published experimental investigations comprising stability issues related to curved box and I-girder bridges during construction; effects of ties, bracing, and web stiffeners on the distortional behavior of the bridges during construction; field experimental programs to measure internal forces and deformations in the main girders and the bracing during construction; experiments demonstrating local and lateral-torsional buckling; experiments demonstrating the limit states in a transversely and/or longitudinally stiffened web; experiments addressing the effective width of the concrete slab in both curved box and I-girders; and cost-effective construction methods and erection guidelines that incorporate the experience of steel fabricators and erectors.

Padgett and DesRoches [1.35] performed a nonlinear 3D time history analysis for typical multispan simply supported and multispan continuous steel girder bridges to evaluate the effectiveness of various retrofit strategies. The influence of using restrainer cables, steel jackets, shear keys, and elastomeric isolation bearings on the variability and peak longitudinal and transverse responses of critical components in the bridges was investigated by the authors. The authors concluded that different retrofit measures may be more effective for each class of bridges. The restrainer cables are effective for the multispan simply supported bridge, shear keys improve the transverse bearing response in the multispan continuous bridge, and elastomeric bearings improve the response of the vulnerable columns in both bridges. The study [1.35] has also shown that while a retrofit may have a positive influence on the targeted component, other critical components may be unaffected or negatively impacted. Shoukry et al. [1.36] investigated the long-term sensor-based monitoring of the Star City Bridge in Morgantown, WV, USA, which was a steel girder bridge designed according to Load and Resistance Factor Design (LRFD) of the American Association of State Highway and Transportation Officials (AASHTO) [1.37]. The bridge had a length of 306 m over four spans. Overall, the study aimed to demonstrate the long-term performance of existing lightweight bridge decks. The bridge was heavily instrumented with over 700 sensors that recorded the response of the main superstructure elements to various loading parameters. The authors have recorded data to monitor and evaluate the performance of the bridge since construction over a 4-year period. The authors have shown that the expansion and contraction of the superstructure at one end contributed to the relief of environmentally induced internal stresses in the longitudinal direction. It was also found that bearing movement constraints on the other end introduced normal forces in the steel girders that were not considered in deck designs. In addition, the study has shown that a nonlinear gradient across the bridge width was developed, which resulted in additional stresses found on diaphragm members at the outside girders.

Cheng and Li [1.38] performed a reliability analysis for a long-span steel arch bridge against wind-induced stability failure during construction. An algorithm was developed based on stochastic finite element method to evaluate the reliability analysis. The study has incorporated uncertainties in static wind load-related parameters. The proposed algorithm integrated the finite element method and the first-order reliability method. The authors performed the analysis as an example on a long-span steel arch bridge with a main span length of 550 m built in China. The reliability analysis was performed in two different construction stages. The first construction stage involved the construction process before closure of main arch ribs. On the other hand, in the second construction stage, all the remaining parts of the bridge have been completed except the stiffening girder of the main span. Three components of wind loads (drag force, lift force, and pitch moment) acting on both steel girder and arch ribs were considered in the study [1.38]. The authors have concluded that the steel arch bridge during the second construction stage was more vulnerable to wind-induced stability failure than that during the first construction stage. The authors have performed a parametric study to investigate the effects of the variations of wind speed with height, drag force of wind loads, design wind speed at the bridge site, and static aerodynamic coefficients on the probability of wind-induced stability failure during the construction stages for the steel arch bridge. Yoo and Choi [1.39] proposed an iterative system buckling analysis to determine the effective lengths of girder and tower members of cable-stayed bridges. The proposed technique included a fictitious axial force that was added to the axial force of each member in the geometric stiffness matrix to represent an additional force for the individual buckling limit of the member. The proposed method was initially used to analyze a three-story plane frame under two different load cases. After that, it was applied to cable-stayed bridge examples with several center span lengths and girder depths. The effective lengths of the individual members in these example bridges were computed using the proposed method and compared with those found using system buckling analysis. The study has shown that the critical load expression in combination with system buckling analysis yields excessively large effective length for members subjected to small axial forces. Also, it was shown that the proposed method reasonably estimated the individual buckling limit of each member by introducing a fictitious axial force in the geometric stiffness matrix during the iterative system buckling analysis.

The optimum design of steel truss arch bridges was investigated by Cheng [1.40] using a hybrid genetic algorithm. In the study, the weight of the steel truss arch bridge was used as the objective function, and the design criteria of strength (stress) and serviceability (deflection) were used as the constraint conditions. All design variables were treated as continuous/discrete variables. The author considered different methods, analysis types, and formulations and their effects on the final designs were studied. It was shown that the proposed algorithm integrated the concepts of the genetic algorithm and the finite element method. Also, the proposed algorithm was compared with the first-order method and proved to perform better than the first-order method. In addition, it was concluded that when the proposed optimum design was used for a steel truss arch bridge, the weights can be considerably reduced compared with those of the traditional design. Finally, it was concluded that the geometric nonlinearity is not significant for the investigated application. Hamidi and Danshjoo [1.41] studied the effects of various parameters comprising velocity, train axle distance, the number of axles, and span lengths on dynamic responses of railway steel bridges and impact factor values. The study replaced the traditional method specified in current codes of practice, which considered traffic load as a static load increased by an impact factor. In the traditional methods, impact factor was represented as a function of bridge length or the first vibration frequency of the bridge. The authors investigated dynamic responses and impact factors for four bridges with 10, 15, 20, and 25 m lengths under trains with 100–400 km/h velocity and axle distances between 13 and 24 m. It was shown that, in most cases, the calculated impact factor values are higher than those recommended by the relevant codes. It was also shown that the train velocity affected the impact factor, so that the value of impact factor has risen considerably with the train velocity. In addition, it was shown that the ratio of train axle distance to the bridge span length affects the impact factor value such that the impact factor value varies for the ratio below and above unity. Finally, it was concluded that the train number of axles only affected the impact factor under resonance conditions. The authors have proposed some relations for the impact factor considering train velocity, train axle distance, and the bridge length.

The performance of high-strength bolted friction grip joints commonly used in steel bridges was investigated by Huang et al. [1.42]. The experimental and numerical study aimed to study the mechanical behavior including load–slip relationship, load transfer factors, stress state, and friction stress distribution of this type of joints. The study has shown that the loads resisted by bolts in the edge rows are larger than the loads resisted by bolts in the middle rows. It was also shown that the stress distributions in the connected plate and cover plate were wavelike with large stress. The authors concluded that the numerical simulation method of the HSFG joints is recommended for connection design. Guo and Chen [1.43] discussed the field stress and displacement measurements in controlled load tests and long-term monitoring of retrofitted steel bridge details. The retrofitted details were used to alleviate the cracking problems of the existing steel bridge. The authors compared the displacements of the retrofitted details with that of the nonretrofitted details. Based on the field-monitored data and the AASHTO specifications, a time-dependent fatigue reliability assessment was performed. The effective stress ranges derived from daily stress range histograms and lognormal probability density functions were used to model the uncertainties in the effective stress range. The study has shown that the stress ranges in the instrumented details were below the corresponding constant amplitude fatigue limits. It was concluded that the study can provide references to bridges with similar fatigue cracking problems. Kim et al. [1.44] investigated experimentally structural details of steel girder-abutment joints in integral bridges. Integral bridges are the bridges that maintain the rigid behavior of their joints. The study proposed structural details of girder-abutment joints in integral steel bridges to enhance rigid behavior, load-resisting, and crack-resisting capacity. The authors suggested various joints that apply shear connectors to existing empirically constructed girder-abutment joints. The performance of the proposed steel girder-abutment joints was observed through experimental loading tests. The study also performed nonlinear finite element analyses, which applied contact interaction of the interface at the steel-concrete composite joints. It was shown that all joints investigated had sufficient rigidity and crack-resisting capacity. It was also concluded that the proposed joints had good structural performance.

Miyachi et al. [1.45] investigated progressive collapse of three continuous steel truss bridge models with a total length of 230.0 m using large deformation and elastoplastic analysis. The study aimed to clarify how the live-load intensity and distribution affected the ultimate strength and ductility of two steel truss bridge models having different span ratios. Sizes and steel grades of the truss members were determined such that they were within the allowable stress for the design dead and live loads. After the design load was applied, the live load was increased until the bridge model collapses. It was shown that the collapse process differed depending on live-load distribution and span length ratio. It was concluded that the investigation clarified the collapse process, buckling strength, and influences of live-load distribution and the span ratio on the investigated steel truss bridges. Ye et al. [1.46] carried out an experimental investigation to determine the stress concentration factor and its stochastic characteristics for a typical welded steel bridge T-joint. The study reported a test on a full-scale segment model, which had the same profile as an existing railway beam section of the suspension Tsing Ma Bridge. The test had also the geometric dimension and material property as well as weld details. Strain gauges were fitted on the web and flanges and the hot spot strain at the weld toe is determined by a linear regression method. The stress concentration factor was calculated as the ratio between the hot spot strain and the nominal strain. The tests were carried out under different moving load conditions. It was shown that the stress concentration factor for the welded steel bridge T-joint conformed to a normal distribution.

A current research topic on steel bridges is the investigation of the structural behavior of the bridges under different fire conditions. Zaforteza and Garlock [1.47] investigated numerically the fire response of steel girder bridges by developing a 3D numerical model for a typical bridge of 12.20 m span length. A parametric study was performed considering different axial restraints of the bridge deck, different types of structural steel for the girders, different constitutive models for carbon steel, different live loads, and different fire loads. The numerical study showed that restraint to deck expansion coming from an adjacent span or abutment should be considered in numerical models. Also, times to collapse were very small when the bridge girders are built with carbon steel (between 8.5 and 18 min), but they can almost double if stainless steel is used for the girders. The authors recommended that stainless steel be used as a construction material for girder bridges in a high fire-risk situation. It was also concluded that the methodology developed in the study and the results obtained can be useful for researchers and practitioners interested in developing and applying a performance-based approach for the design of bridges against fire.

Structural health monitoring of steel bridges is a recent research topic for evaluating bridge condition and safety. Measured strain data from a structural health monitoring system can be used to assess the status of fatigue of steel bridges, which is a common form of damage in this form of construction. Ye et al. [1.48] proposed a standard daily stress spectrum method for fatigue life assessment of steel bridges using structural health monitoring data. The authors applied the proposed method to assess the fatigue status of critical welded details on the instrumented Tsing Ma Bridge carrying both highway and railway traffic. It was shown that the proposed method makes it convenient to simultaneously consider the effects of different loads (highway traffic, railway traffic, monsoon, and typhoon) with the use of a single standard stress spectrum. It was shown that, in applying the proposed method, it was unnecessary to separate the temperature-induced ingredient and slow-varying drift from the raw measurement data. The authors concluded that a standard daily stress spectrum accounting for highway traffic, railway traffic, and typhoon effects can be formulated from the long-term monitoring data by combining the standard traffic-stress spectrum and standard typhoon-stress spectrum proportionally. It was also concluded that, for the Tsing Ma Bridge, the predicted fatigue life was varying slightly when using more than 10 daily strain data and keeps almost the same when using more than 20 daily strain data. In addition, it was concluded that the proposed method provided a feasible approach for fatigue life assessment of welded details on steel bridges by using field monitoring data from a structural health monitoring system. Investigating the mechanical properties of new materials used in steel bridges is also a current research area. Mo et al. [1.49] investigated the mechanical properties of thin epoxy polymer overlay materials upon steel bridge decks. Overall, the authors highlighted the epoxy binder steel bonding behavior, dynamic response of epoxy binder, and response and fatigue behavior of epoxy polymer concrete. The test data obtained indicated that epoxy binder-steel bonding exhibited a strong temperature dependency. Also, fatigue models on epoxy binder-steel bonding and epoxy polymer concrete were developed using the power-law equation. It was shown that response models for epoxy binder and their concrete can be properly established. The developed material response models can be served for finite element simulations on thin epoxy polymer concrete overlay upon steel bridge decks.

Proposing approximate methods for estimating collapse loads of steel bridges with complex geometry is a current research area. Although nonlinear inelastic analysis is commonly used to determine the collapse loads of these bridges and accounts for all geometric and material nonlinear aspects of the bridge system, approximate methods can be useful in the preliminary design stages. Yoo et al. [1.50] proposed a simple alternative for complex nonlinear inelastic analysis to estimate the collapse load of a steel cable-stayed bridge. The fundamental theories of nonlinear inelastic analysis and inelastic buckling analysis were briefly reviewed. The authors proposed a new criterion for a beam–column member based on the axial-flexural interaction equation in combination with the classical tangent modulus theory and the column-strength curve. The study has shown that the proposed criterion for a beam–column is appropriate to determine the collapse loads of steel cable-stayed bridges. It was also concluded that the inelastic buckling analysis using the proposed criterion for a beam–column can substitute complex nonlinear inelastic analysis in estimating the collapse load of a steel cable-stayed bridge. Shifferaw and Fanous [1.51] investigated fatigue-crack formation in the web-gap region of multigirder steel bridges. The authors have shown that the region has been a common occurrence of fatigue-crack formation due to differential deflections between girders resulting in diaphragm forces that subject the web-gap to out-of-plane distortion. The study investigated the behavior of web-gap distortion of a skewed multigirder steel bridge through field testing and finite element analyses. The study also investigated different retrofit methods that include the provision of a connection plate between the stiffener and the girder top flange, loosening of the bolts connecting the cross bracing to the stiffener, and supplementing a stiffener plate opposite to the original stiffener side. The study has shown that the connection plate addition and loosening of bolts alternatives were effective in reducing induced strains and stresses in the web-gap region. An inverse relationship between web-gap height and induced strains and stresses with the shortest web-gap height resulting in the highest strains due to increased bending by diaphragm forces in the web was also shown. The authors concluded that expressions developed to relate vertical stresses and relative out-of-plane displacements combined with measurements of out-of-plane displacements by transducers can be utilized for prediction of induced stresses at other critical web-gap regions of the bridge and at critical locations in the web gaps of similar bridges.

Postrehabilitation assessment of existing bridges is a current research topic. Cavadas *et al.* [1.52] presented the postrehabilitation assessment of the Eiffel Bridge, which is a centenary double-deck bridge located in Portugal. Recently, the bridge was rehabilitated involving the replacement of the top deck, the strengthening of the top chords, and the replacement of the support bearings. After the rehabilitation, a load test and an environmental test were carried out in order to analyze the bridge behavior and the live-load distribution for the new structural conditions, to evaluate the effectiveness of the chords' strengthening, and to establish the new baseline condition for future structural assessments. The field results were augmented by an appropriate numerical model of the bridge. The authors concluded that the deflection of the main girder chords results from the overlapping of both global and local behaviors. Also, the rotations do not reproduce the global behavior of the structure. Therefore, the fundamentals about rotations in a beam and the derived methods to estimate the bridge deflection using rotations are not applicable to this structure. In addition, the external prestress system has a significant influence on the response of the structure to the temperature changes. Furthermore, observing the effectiveness of the top chords' strengthening, the internal forces estimated using a multiple linear regression model agree well with the internal forces obtained with the numerical model. Overall, it was concluded that the study provided valuable information regarding the installation of a permanent monitoring system for the surveillance of the bridge.

The effect of local damage on the behavior of steel bridges is also a current research area. Brunell and Kim [1.53] investigated the effects of local damage in steel trusses on the overall behavior of the bridge. The study comprised a combined experimental and numerical investigation. The experimental results of a scaled model bridge were used to validate the developed numerical model. The numerical model was used to perform analyses investigating the relationship between damage and bridge failure. The behavior of 16 damage scenarios was compared with that of a control truss. A static analysis was carried out which utilized a damage index to quantify the level of damage present in the bridge, to examine the load transfer relationship between truss members, and to quantify the failure load for various scenarios. In addition, a dynamic analysis was carried out to highlight the effect of damage on mode frequency and changes in mode shape. The authors mentioned that, since the dynamic behavior of the test specimen was not measured in the laboratory, the findings reported could be experimentally verified in future research. A simple reliability analysis was conducted to assess the safety of the truss systems. The authors showed that the results and conclusions of the study were based on laboratory-scale research, and thus a size effect might exist when implemented in practice. Also, the technical findings reported could be conservative to a certain extent because the contribution of a RC deck was not included in the analyses. The authors concluded that the presence of local damage in the truss system significantly influences the serviceability of the system. Also, the current AASHTO load rating method was reasonably applicable to the truss bridge system. From a dynamic analysis perspective, a higher mode shape and corresponding frequency were useful to detect the presence of local damage in the truss systems. Finally, it was concluded that the stress of the damaged truss member was not effectively redistributed to other members, except for those adjacent to the damage.

Recently, Cheng *et al.* [1.54, 1.55] performed a numerical study on steel truss bridges with welded box section members and bowknot integral joints. The investigation [1.54] highlighted the linear and nonlinear mechanical behaviors of the bridges. The truss studied was simply supported at two end nodes of bottom chords with two concentrated dead and live loads being applied at each unsupported bottom chord node. The finite element method was employed to analyze the elastic and elastoplastic behaviors of trusses with bowknot/conventional integral joints. Based on the study, it was concluded that the axial forces of members were very insignificantly enhanced by the section shrinking of the member ends. It was also concluded that the secondary moments at the member ends and the sectional maximum stresses of the unshrunken segments of the truss are significantly reduced by the section shrinking of the member ends but the vertical stiffness and elastic stability of the bowknot truss are deteriorated compared to the conventional one. Finally, it was shown that when the steel strength of the shrunken segments has been moderately enhanced, the ultimate bearing capacities of axially compressed shrunken members and of Warren trusses with bowknot integral joints are as high as those of uniform members and of conventional trusses, respectively. The study [1.55] presented a minimum weight optimization based on the provisions of current design codes for both conventional and bowknot trusses. The optimization investigated was illustrated through analytic derivation of minimum weight optimization of a single member. The results of the numerical study indicated that the member weight reduction increased as the primary stress to secondary stress ratio, or the end moment reduction, was increased. The authors extended the minimum weight optimization of the truss on the basis of linear finite element analysis of the same truss that was discussed in [1.54], by the use of first-order optimization method. It was concluded that the cost rise due to steel strength enhancement of shrunken segments was taken into account in the nominal weight of whole truss. Also, a series of requirements related to truss vertical stiffness, member strength, member stability, and truss stability were set as constraint functions.

1.3.3 Recent Investigations on Steel-Concrete Composite Bridges

The short-term and long-term behaviors of composite steel box girder bridges have been investigated theoretically by Kwak *et al.* [1.56]. The study discussed the effect of the slab deck casting sequence on the short-term and long-term behaviors of the bridges. Three cases of sequential casting and one case of continuous casting were investigated. The study showed that the effect of slab casting sequence can be neglected for both short-term and long-term behaviors as well as the resulting moments of the bridges. The authors concluded that continuous casting for closed box sections can be used as an easy and fast construction without any danger of increasing transverse cracking. The authors recommend that continuous casting can be applicable to open box sections used widely in many countries. However, the study showed that the effect of drying shrinkage was most critical for the long-term behavior of a bridge and transverse cracking, which can result in crack development at interior supports. Steel-concrete composite girders are analyzed using beam bending theory by utilizing the effective flange width concept to evaluate deflections, stresses, strengths, etc. Shear lag effects can be considered by replacing the actual slab width by an appropriate reduced effective width. Nonlinear finite element analyses can effectively consider the shear lag effects. However, current codes of practice provide simplified empirical equations to evaluate effective flange width, which differ from country to country. Ahn *et al.* [1.57] investigated effective flange width provisions of several countries including America, Britain, Canada, Japan, and Europe. The provisions were compared qualitatively and quantitatively by the authors. It was shown that each specification shares common organization in describing the effective flange width comprising basic formulation and effective span length. In the basic formulation part, the effective flange width of a simply supported span can be specified. For continuous girders, the lengths of independent spans to which basic formulation can be applied were specified in the effective span length part. It was also shown that the way to describe the basic formulation differed from one provision to the other following the underlying philosophy that drove the development of each specification. AASHTO and Eurocode 4 provisions used a list of descriptions. However, Canadian and Japanese provisions used equations; BS 5400 used a table format. Through a numerical example of simply supported spans, it is observed that BS 5400 (interior) develops the largest effective flange width. Eurocode 4 delivered the largest effective flange width. Effective flange widths from AASHTO varied considerably compared with the others. Without the thickness limitation, AASHTO provisions were similar to the values from Eurocode 4. It was concluded that the interrelation between effective flange width, loading effects on the bridges, and design of concrete deck (especially crack control) should be consistent.

The behavior of curved composite steel I-shaped plate girder bridges was investigated by Chang and White [1.58]. The authors considered different parameters affecting the composite bridge modeling including girder web distortion, cross frames, support and load height, and displacement compatibility between the girders and slab. In the study, web distortion effects were investigated and an approximate approach using open-section thin-walled beam theory for the steel I-girders was proposed. Different analysis approaches including line girder analysis, V-load method, grid methods, and general finite element methods for analysis of curved I-girder bridge structural systems were highlighted. The investigated plate girders were horizontally curved bridges subjected to coupled torsion and bending. It was shown that the plate girder behavior in these bridges involved significant web distortion, which caused additional lateral displacements and lateral bending stresses at the girder bottom flanges. The study showed that a general 3D analysis using shell elements for the slab and for the girder webs and 3D grid models using open-section thin-walled beam theory for the plate girders were recommended for efficient modeling of the bridges. However, the study suggested approximate approaches for simulating the composite I-shaped plate girder web distortion effects using 3D grid methods. It was found that when using an open-section thin-walled beam element for the bridge plate girder and either shell elements or a beam grid system for the slab, a rotational release must be placed between the slab and the top flange of the I-girders in order to compensate the web distortion effects. It was also found that when using an open-section thin-walled beam element for the combined slab and steel plate girder via an equivalent composite plate girder cross-sectional model, the contribution from the slab to the St. Venant torsional constant J was suggested to be neglected. In addition, the lack of consideration of the web distortion effects results in a significant underestimation of the girder bottom flange lateral bending stresses. The authors have compared results from a full-scale composite I-shaped plate girder bridge against the results of the 3D grid models.

Structural performance of bridge decks with high load resistance capacity as well as high fatigue strength is a current research topic. Ahn et al. [1.59] conducted tests to investigate the fatigue behavior of a new type of steel-concrete composite bridge deck. The proposed composite bridge deck consisted of corrugated steel plate, welded steel ribs, headed stud shear connectors, and RC filler. Fatigue tests were conducted under a four-point bending test with four different stress ranges in constant amplitude. A total of eleven fatigue specimens were subjected to cyclic loading to evaluate the stress category and fatigue behavior of the proposed composite bridge deck. In order to determine the influence of the concrete filling, the authors conducted fatigue tests on partial steel specimens with plain corrugated steel plates. The partial steel specimens and the steel–concrete composite deck specimens showed that fatigue failure occurred in the tension part. The research concluded that the fatigue behavior of the proposed steel–concrete composite decks under sagging moment can be estimated based on the classical S-N approach, focusing on steel components. The structural behavior and ultimate strength of steel–concrete composite bridge deck slab with profiled sheeting were investigated by Kim and Jeong [1.60, 1.61]. The study [1.60] presented an experimental investigation on a steel–concrete composite bridge deck slab with profiled sheeting and perfobond shear connectors. Two full-scale deck slab specimens cast onto three concrete blocks were fabricated and tested under static loading to examine the ultimate load-carrying capacity of the proposed deck slab system under sagging and hogging bending actions. The ultimate behavior of the full-scale deck slab specimens was compared with that of simply supported deck specimens under hogging bending only. In addition, the load-deflection behavior of the proposed deck system was compared with that of a RC deck slab. The test results indicated that the ultimate load-carrying capacity of the proposed deck system was at least 220% greater than that of the RC deck system and that the deck weighs about 23% less than the RC deck system. The study [1.61] investigated experimentally the ultimate behavior of steel–concrete composite deck slab system with profiled steel sheeting and perfobond rib shear connectors. The experimental investigation aimed to develop a composite deck slab for girder bridges that spans longer but weighs less than the conventional RC deck slab. Eight deck specimens were tested with different shear span lengths to evaluate the horizontal shear capacity of the proposed deck system by using the empirical m-k method. The study also presented the results of two full-scale deck slab specimens supported by a set of steel box blocks to examine the ultimate load-carrying capacity of the proposed deck slab system under sagging and hogging bending actions. It was found that the ultimate strength and initial concrete cracking load of the proposed deck system under hogging bending action were approximately 2.5 and 7.1 times greater than those of an RC deck, respectively, while the deck weighed about 25% less than RC deck systems.

Steel–concrete composite cable-stayed bridges were investigated numerically by Pedro and Reis [1.62]. The composite deck and the concrete towers were modeled by three node steel and concrete frame elements having seven degrees of freedom. Shear connection of the deck was modeled using continuous spring elements. The numerical investigation considered geometric and material nonlinear behaviors of both steel and concrete materials. Cable's sag and time-dependent effects due to load history, creep, shrinkage, and aging of concrete were also included in the analysis. The cable-stayed bridge investigated a 420m main span composite cable-stayed bridge under service conditions. The failure load and the failure mechanism were also analyzed, both at the end of construction and at long term. The influence on the structural behavior of deck load pattern, time-dependent effects, cables' yielding, existence of intermediate piers at the lateral spans, effective slab width, and flexibility of the shear connection was investigated in the study. The structural performance of orthotropic steel bridge decks renovated using advanced composite bonded systems was the subject of experimental and analytic investigations reported by Freitas et al. [1.63]. The proposed renovation solution for orthotropic steel bridge decks studied consisted of bonding a second steel plate to the existing steel deck in order to reduce the stresses and increase the life span of the orthotropic bridge deck. The authors performed a parametric study on the flexural behavior of beams representing the renovation solution. The influences of different thickness, temperatures, and spans were investigated. The results obtained for the stress reduction factor showed that it was independent of temperature. Also, more efficient solutions can be achieved by minimizing the second steel plate thickness and maximizing the adhesive layer thickness reducing the weight and increasing the stiffness of the composite structure. Both elastic behavior and yield load of the composite beams were dominated by the steel plate properties and therefore were not affected significantly by temperature. However, the ultimate failure of the beams occurred by shear of the adhesive layer.

The shear connection of the unfilled composite steel grid deck was experimentally investigated by Kim and Choi [1.64]. A total of 14 pushout specimens having different number of holes, areas of reinforcements through holes, and reinforcement diameters were fabricated and conducted to evaluate the load-slip behavior and the shear strength of the connection. The study highlighted the effects on the shear resistance of the connection owing to the friction force between the steel beam and the concrete, the concrete dowel force, and the shear force due to reinforcement bars. An analytic expression was developed based on an existing formula to predict the shear resistance of the connection. Based on the limited test specimens for the shear connector of the unfilled composite steel grid deck, it was concluded that the failure of the specimen occurred in the concrete slab and the steel beam was intact. The crack patterns showed longitudinal splitting on the concrete slab close to the steel beam, spreading outward to the bottom of the specimen and falling off the concrete face near the reinforcing bars. It was also concluded that the shear resistance of the connection was influenced by the friction force between the steel beam and the concrete slab, the number of holes, and the amount of reinforcing bars passing through the shear holes. In these factors, the shear capacity was almost directly dependent on the area of reinforcing bars. In addition, it was shown that the shear strength of the connection was predicted by the sum of the friction force, the concrete dowel force, and the shear force due to reinforcement bars. The authors recommended that in a further study, more tests should be required to highlight the size effects of shear holes, the effect of multiple holes, and the concrete strength. Machacek and Cudejko [1.65] investigated numerically distribution of longitudinal shear along an interface between steel and concrete parts of various composite truss bridges from elastic phase up to plastic collapse. The study was based on previous experimental research reported by the authors. The numerical analysis and the Eurocode approach highlighted distribution of the longitudinal shear flow. Overall, the study considered elastic and elastoplastic distribution of the flow corresponding to the design level of bridge loading and plastic collapse. The analysis covered both the common elastic frame 2D modeling of the shear connection used by designers and the 3D geometrically and materially nonlinear analysis using. The results of the numerical models were compared against design rules specified in Eurocode 4 for composite bridges. It was shown that the nonlinear distribution of the longitudinal shear, required for correct design of shear connection of composite steel and concrete bridges (in both ultimate limit state including fatigue and serviceability limit state), depended on rigidity of the shear connection and densification of the shear connectors above truss nodes. Parametric studies were performed by the authors and recommendations for practical design were proposed.

The fatigue of steel and composite highway bridges in terms of the structural system service life was analyzed by Leitão [1.66]. A steel–concrete composite bridge with a 12.50 m roadway width and 0.2 m concrete deck thickness, spanning 40.0 m by 13.5 m, was investigated in the study. The computational model, developed for the composite bridge dynamic analysis, adopted the usual mesh refinement techniques present in finite element method simulations. The proposed analysis methodology and the procedures presented in the design codes were initially assessed to evaluate the bridge fatigue response in terms of its structural service life. The study has shown that the composite bridge service life results corroborated the importance of considering the roughness of the pavement surface and other design parameters such as floor thickness, structural damping, and beam cross-sectional geometric properties in the bridge dynamic and fatigue analyses. The analysis methodology considered a vehicle structure mathematical model, which included the interaction between their dynamic properties. It was shown that the proposed methodology can be general and can be used as a solution strategy on other highway bridge types such as multigirder bridges, continuous multigirder bridges, cable-stayed bridges, and rigid-frame bridges. The authors showed that the fatigue problem was much more complicated and was influenced by several highway bridge types. It was concluded that the investigated composite (steel-concrete) highway bridge can perform safely with an acceptable probability that indicated that failure by fatigue cracking can be eliminated. It was also shown that when the dynamic actions related to the vehicles moving on the bridge lateral track path and two simultaneous lateral track paths were applied on the bridge structure, it was observed that the service life values proposed by current design codes were exceeded. Okamoto and Nakamura [1.67] proposed and applied a new type of hybrid high tower to a multispan cable-stayed bridge. The proposed type was a sandwich-type structure and consisted of a steel double box section filled with concrete. The filled concrete increased its strength due to the confined effect, and the steel plates increased the resistance against local buckling because the deformation was restricted by the filled concrete. The study showed that the hybrid tower can have high bending and compressive strength and also a good ductile property. In the study, static analysis was conducted for different live-load intensity and distribution. The live loads distributed in alternate spans gave larger bending moment of the towers than the live loads distributed in full spans. The authors checked the safety of the tower using limit state design method. Serviceability was not a major problem for the hybrid tower. Following the static analysis, seismic analysis was performed for a multispan cable-stayed bridge subjected to the medium and ultra-strong seismic waves. Three support conditions of the girder at the tower cross beams were considered, which were movable, connection with linear springs and bilinear springs. The study showed that bilinear springs were very effective in reducing the dynamic displacements and bending moments of the towers. The study also showed that a new steel-concrete hybrid tower was feasible for multispan cable-stayed bridges and most effective for seismic forces when the girder was connected with bilinear springs.

Ji *et al.* [1.68] presented static and fatigue performance of composite sandwich bridge decks with hybrid glass fiber-reinforced polymer-steel core. The composite sandwich bridge deck system consists of wrapped hybrid core of glass fiber-reinforced polymer grid and multiple steel box cells with upper and lower glass fiber-reinforced polymer facings. The study investigated the structural performance under static loading and fatigue loading with a nominal frequency of 5 Hz was evaluated. The responses from laboratory testing were compared with the finite element predictions. The study showed that the failure mode of the proposed composite sandwich bridge deck was more favorable because of the yielding of the steel tube when compared with that of glass fiber-reinforced polymer decks. It was also shown that the ultimate failure of the composite sandwich deck panels occurred by shear of the bonded joints between glass fiber-reinforced polymer facings and steel box cells. In addition, results from fatigue load test indicated that no loss in stiffness, no signs of debonding, and no visible signs of deterioration up to 2 million load cycles were observed. The authors recommended that the thickness of the composite sandwich deck retaining the similar stiffness be decreased to some extent when compared with the glass fiber-reinforced polymer deck. Furthermore, the study presented design of a connection between composite sandwich deck and steel girder. Turer and Shahrooz [1.69] investigated different parameters related to structural identification, calibrated model-based load rating, and sensitivity of rating to the analytic model, along with experimental studies conducted on an existing concrete-deck-on-steel-stringer bridge. The proposed model-updating procedure used collected dynamic data comprising mode shapes, modal frequencies, and order of modes as well as static deformed shape information. The authors developed 2D grid models to simulate the transverse load transfer mechanisms between girders, torsional flexibility, and effects of skewed bridge architecture. It was shown that the rating results obtained from the 2D grid models were close to 3D finite element method-based evaluation, while simplified 1D bar models had serious shortcomings. It was shown that grouping the parameters of the analytic model at different stages of model calibration enhanced the speed and convergence success of the objective function. It was also shown that although cross braces were considered as nonstructural members, they were found to be the most critical members of the selected bridge during rating studies.

Mechanical behavior of composite joints for connecting existing concrete bridges and steel-concrete composite beams was the subject of recent investigation by Nie *et al.* [1.70]. The authors showed that in a technique of widening existing concrete bridges with steel-concrete composite beams, the old existing concrete bridge and new composite beam were connected by a composite cross beam with a composite joint. Six specimens were tested to compensate the lack in experimental studies on the mechanical behavior of composite joints. The study showed that, based on the existing methods, the shear strength of the interface of the old and new concrete was calculated. It was shown that the shear failure of the interface between the old and new concrete was the failure mode of the composite joint and the interface between the steel plate and new concrete was always in good condition. It was also shown that there was approximately no slip between the old and new concrete before the bonding failure of the interface. The interface between the old and new concrete had good ductility and high strength. Based on the constitutive law of the materials, a simplified three-stage mechanical model was proposed and the load-slip relationship was predicted. The study showed that the ultimate shear strength of the interface was determined by the strength of the concrete, roughness degree, and friction coefficient of the interface and the normal stress could increase the ultimate shear strength. In addition, the residual shear strength of the interface can be determined by the embedded bars, and the ratio and yield strength of the embedded bars can be the main influence factors. Based on the tests results, a practical design method was proposed. Finally, studies on fiber-reinforced polymer deck-on-steel girder systems are current research topics. Davalos et al. [1.71] investigated the performance of the fiber-reinforced polymer deck-on-steel girder system, which depends substantially on the connectors used. The authors proposed a prototype shear connector and showed its advantages through experimental studies and field applications. The effectiveness of the shear connector at bridge system level, including the static and fatigue performance of the shear connector and the bridge system, the degree of composite action of the system, and the influence of the partial degree of composite action on load distribution factor and effective flange width were investigated. The authors tested a 1:3 scaled fiber-reinforced polymer deck bridge model, with a fiber-reinforced polymer sandwich honeycomb deck connected to steel girders using the prototype shear connector. The experimental investigation comprised static and fatigue load tests on the scaled bridge model. The experimental investigation was accompanied by a numerical investigation using finite element method. The study showed that the shear connection was able to provide partial composite action of about 25% and sustain a cyclic fatigue loading equivalent to 75-year bridge service life span. It was also shown that AASHTO specifications can still be applicable for load distribution factor, while the effective flange width needs to be redefined for a bridge system with partial degree of composite action. The authors recommended that the findings of the study be used for design purposes.

1.4 FINITE ELEMENT MODELING OF STEEL AND STEEL-CONCRETE COMPOSITE BRIDGES

Finite element modeling of steel and steel-concrete composite bridges can provide a useful insight into the structural performance of the bridges and compensate the lack in full-scale tests on the bridges. Recent developments in computers and finite element general-purpose software make it possible to analyze structures having different nonlinear geometries, different material properties, different loading conditions, and different boundary conditions. This book presents the latest modeling techniques used to investigate the behavior of the bridge components and the whole bridge behavior. The presented finite element models in this book are intended to be efficient and accurate models, which are not too-detailed and are not too-simplified models. There are numerous finite element books published in the literature, with examples shown in [1.12–1.18]. These books are mainly devoted to the development of different finite elements and or the development of a numerical scheme to expedite the convergence of iterative procedures. These finite element books mostly focus on explaining the finite element method as a general technique to solve engineering problems. However, books involved in finite element modeling of the bridge superstructure are rarely found in the literature, leading to the writing of this book. However, in order to present how finite element modeling can be used efficiently to simulate the behavior of steel and steel-concrete composite bridges, knowledge of the different loads applied on bridges, material nonlinearity of the bridge components, and design rules specified in current codes of practice for different bridges is required.

Test data are used to verify and validate the accuracy of finite element models developed for steel and steel concrete composite bridges. In order to investigate the structural performance, stability, and failure modes of steel and steel-concrete composite bridges and their components, laboratory tests have to be conducted to observe the actual behavior or theoretical analyses have to be performed to obtain an exact closed-form solution. Getting an exact solution sometimes becomes very complicated and even impossible in some cases that involve highly nonlinear material and geometry analyses. Experimental investigations are also costly and time-consuming, which require specialized laboratories and expensive equipments as well as highly trained and skilled technicians. Without the afore-mentioned requirements, the test data and results will not be accurate and will be misleading to finite element development. Therefore, accurate finite element models should be validated and calibrated against accurate test results. Although extensive experimental investigations were presented in the literature on small-scale bridges, as well as small- and full-scale bridge components, the number of tests on some research topics related to steel and steel-concrete composite bridges is still limited. This is attributed to many factors comprising time, costs, labor, capacity of testing frame, capacity of loading jack, measurement equipment, and testing devices. Therefore, numerical investigations using finite element analysis are currently main research areas in the literature to compensate the lack of test data in the field of steel and steel-concrete composite bridges. However, detailed explanation on how successful finite element analysis can provide a good insight into the structural performance of the bridges was not fully addressed as a complete piece of work, which is credited to this book.

Following experimental investigations on steel and steel-concrete composite bridges and their components, finite element analyses can be performed and verified against available test results. Successful finite element models are those validated against sufficient number of tests, preferably from different sources. Finite element modeling can be extended, once validated, to conduct parametric studies investigating the effects of the different parameters affecting the behavior of steel and steel-concrete composite bridges. The analyses performed in the parametric studies must be well planned to predict the performance of the investigated bridges outside the ranges covered in the experimental program. The parametric studies will generate more data that fill in the gaps of the test results and will help designers to understand the performance of the bridges under different loading and boundary conditions and different geometries. Hence, one of the advantages of the finite element modeling is to extrapolate the test data. However, the more significant advantage of finite element modeling is to clarify and explain the test data, which is credited to successful finite element models only. Successful finite element models can critically analyze test results and explain reasons behind failure of steel and steel-concrete composite bridges and their components. The successful finite element models can go deeply in the test results to provide deformations, stresses, and strains at different locations in the test specimens, which is very difficult to determine by instrumentation.

1.5 CURRENT DESIGN CODES OF STEEL AND STEEL-CONCRETE COMPOSITE BRIDGES

Design rules and specifications are proposed in different countries to define standards and methods of analysis of steel and steel-concrete composite bridges. The design guides are commonly based on experimental investigations on small-scale bridges and small/full-scale bridge components. Many design formulas specified in current codes of practice are in the form of empirical equations proposed by experts in the field of bridges. However, the empirical equations only provide guidance for design of the bridges and their components in the ranges covered by the specifications. The ranges covered by the specification depend on the number of tests conducted on the bridges at the time of proposing the codes. Since there is continuing progress in research to discover new materials, sections, connections, and different loadings, the codes of practice need to be updated from time to time. Furthermore, test programs on steel and steel-concrete bridges and their components are dependent on the limits of the test specimens, loading, boundary conditions, and so on. Therefore, the design equations specified in current codes of practice always have limitations. Finite element analysis can provide a good insight into the behavior of steel and steel-concrete composite bridges outside the ranges covered by specifications. In addition, finite element analysis can check the validity of the empirical equations for sections affected by nonlinear material and geometry, which may be ignored in the specifications. Furthermore, design guides specified in current codes of practice contain some assumptions based on previous measurements, for example, assuming values for initial local and overall imperfections on the bridge beams and compression members. Also, finite element modeling can investigate the validity of these assumptions. As mentioned previously, an example of the shortcomings in current codes of practice for steel-concrete composite bridges is that, up-to-date, there are no design provisions to consider the actual load-slip characteristic curve of the shear connectors used in the bridges, which results in partial degree of composite action behavior. This book will detail, for the first time, how to consider the correct and actual slip occurring at the steel-concrete interface in composite bridges through finite element modeling. This book addresses the efficiency of finite element analyses and the numerical results are able to improve design equations in the current codes of practice more accurately. However, it should be noted that there are many specifications developed all over the world for steel and steel-concrete composite bridges. It is not the intention to include all these codes of practice in this book. Once again, this book focuses on finite element analysis of steel and steel-concrete composite bridges. Therefore, the book only highlights, as examples, the design rules specified in the American Specifications [1.23–1.25] and Eurocode [1.27, 1.28]. However, the finite element modeling presented in this book can be used to simulate any bridge designed using any current code of practice used in any country. Following the simulation of the investigated bridge, the design predictions can be compared and assessed against finite element results.

REFERENCES

[1.1] Department of Transport, Design of composite bridges, Use of BS 5400: part 5: 1979 for Department of Transport structures, Department of Transport, London, UK, 1987.
[1.2] D.C. Iles, Design Guide for Composite Box Girder Bridges, The Steel Construction Institute, Ascot, Berkshire, UK, 1994.
[1.3] A.R. Biddle, D.C. Iles, E. Yandzio, Integral Steel Bridges: Design Guidance, The Steel Construction Institute, Ascot, Berkshire, UK, 1997.
[1.4] S. Chatterjee, The Design of Modern Steel Bridges, second ed., Blackwell Science Ltd, UK, 2003.
[1.5] J.F. Unsworth, Design of Modern Steel Railway Bridges, first ed., CRC Press, USA, 2010.
[1.6] R.M. Barker, J.A. Puckett, Design of Highway Bridges: An LRFD Approach, third ed., Wiley, UK, 2013.
[1.7] W.F. Chen, L. Duan, Bridge Engineering: Seismic Design, Taylor & Francis, UK, 2003.
[1.8] W.F. Chen, L. Duan, Bridge Engineering Handbook, Taylor & Francis, UK, 2010.
[1.9] B. Davison, G.W. Owens, Steel Designer Manual, sixth ed., Blackwell Publishers, UK, 2003.
[1.10] R.L. Brockenbrough, F.S. Merritt, Structural Steel Designers' Handbook, third ed., McGraw-Hill Book Company, UK, 1999.
[1.11] U.S. Department of Transportation, Steel Bridge Design Handbook, Publication No. FHWA-IF-12-052. Federal Highway Administration, U.S.A, 2012.
[1.12] J.T. Oden, Finite Element of Nonlinear Continua, McGraw-Hill, New York, 1972.
[1.13] H. Kardestuncer, Elementary Matrix Analysis of Structures, McGraw-Hill, New York, 1974.
[1.14] J.R. Whiteman, A Bibliography for Finite Elements, Academic Press, London, 1975.
[1.15] D. Norrie, G. deVries, Finite Element Bibliography, IFI/Plenum, New York, 1976.
[1.16] O.C. Zienkiewicz, The Finite Element Method, third ed., McGraw Hill, London, 1977.
[1.17] R.D. Cook, Concepts and Applications of Finite Element Analysis, second ed., Wiley, New York, 1981.
[1.18] H. Logan, A First Course in the Finite Element Method, PWS Engineering, Boston, 1986.
[1.19] A. Yargicoglu, C.P. Johnson, Temperature induced stresses in highway bridges by finite element analysis and field tests, Research Project 3-5-74-23, Center for highway research, The University of Texas at Austin, Texas, USA, 1978.
[1.20] J. C. Hall, C. N. Kostem, Inelastic overload analysis of continuous steel multi-girder highway bridges by the finite element method. Fritz Engineering Laboratory Report No. 432.6, Department of Civil Engineering, Lehigh University, Bethlehem, Pennsylvania, USA, 1981.

[1.21] D.J. Hodson, Live load test and finite element analysis of a box girder bridge for the long term bridge performance program. A thesis submitted for the degree of M. Sc., Utah State University, Utah, USA, 2010.

[1.22] S. Tande, K. Parate, A. Bargir, Nonlinear Seismic Analysis of Bridges: Finite Element Analysis, LAP LAMBERT Academic Publishing, 2012.

[1.23] American Association of State Highway and Transportation Officials, AASHTO LRFD Bridge Design Specifications (SI Units), third ed., AASHTO, Washington, DC, 2005.

[1.24] American Association of State Highway and Transportation Officials, LRFDUS-6-E1, Errata to LRFD DESIGN, sixth ed., AASHTO, Washington, DC, 2012.

[1.25] AREMA, Manual for Railway Engineering, AREMA, Landover, MD, 2011.

[1.26] BS 5400-3, Code of Practice for Design of Steel Bridges, British Standards Institution, 2004.

[1.27] EC3. Eurocode 3—design of steel structures—part 2: steel bridges, Code of Practice for Design of Steel Bridges, BS EN 1993-2, British Standards Institution, 2006.

[1.28] EC4. Eurocode 4—design of composite steel and concrete structures—part 2: general rules and rules for bridges, Code of Practice for Design of Steel Bridges, BS EN 1994-2, British Standards Institution, 2005.

[1.29] ABAQUS Standard User's Manual, Hibbitt, Karlsson and Sorensen, Inc., vols. 1–3, Version 6.11-1, USA, 2011.

[1.30] A. Zureick, D. Linzell, R.T. Leon, J. Burrell, Curved steel I-girder bridges: experimental and analytical studies, Eng. Struct. 22 (2000) 180–190.

[1.31] P.F. McManus, G.A. Nasir, C.G. Culver, Horizontally curved girders-state of the art, J. Struct. Div. ASCE 95 (ST5) (1969) 853–870.

[1.32] ASCE-AASHTO, Task Committee on Curved Girders. Curved steel box-girder bridges: a survey, J. Struct. Div. ASCE 104 (ST11) (1978) 1719–1739.

[1.33] H. Nakai, C.H. Yoo, Analysis and Design of Curved Steel Bridges, McGraw-Hill, New York, 1988.

[1.34] A., Zureick, R. Naqib, J. Yadlosky, Curved steel bridge research project, interim report I; synthesis, Report No. FHWA-RD-93-129, Federal Highway Administration, December 1994.

[1.35] J.E. Padgett, R. DesRoches, Three-dimensional nonlinear seismic performance evaluation of retrofit measures for typical steel girder bridges, Eng. Struct. 30 (2008) 1869–1878.

[1.36] S.N. Shoukry, M.Y. Riad, G.W. William, Longterm sensor-based monitoring of an LRFD designed steel girder bridge, Eng. Struct. 31 (2009) 2954–2965.

[1.37] AASHTO, American Association of State Highway and Transportation Officials, Bridging the Gap, July 2008.

[1.38] J. Cheng, Q.S. Li, Reliability analysis of a long span steel arch bridge against wind-induced stability failure during construction, J. Constr. Steel Res. 65 (2009) 552–558.

[1.39] H. Yoo, D.H. Choi, Improved system buckling analysis of effective lengths of girder and tower members in steel cable-stayed bridges, Comput. Struct. 87 (2009) 847–860.

[1.40] J. Cheng, Optimum design of steel truss arch bridges using a hybrid genetic algorithm, J. Constr. Steel Res. 66 (2010) 1011–1017.

[1.41] S.A. Hamidi, F. Danshjoo, Determination of impact factor for steel railway bridges considering simultaneous effects of vehicle speed and axle distance to span length ratio, Eng. Struct. 32 (2010) 1369–1376.

[1.42] Y.H. Huang, R.H. Wang, J.H. Zou, Q. Gan, Finite element analysis and experimental study on high strength bolted friction grip connections in steel bridges, J. Constr. Steel Res. 66 (2010) 803–815.

[1.43] T. Guo, Y.W. Chen, Field stress/displacement monitoring and fatigue reliability assessment of retrofitted steel bridge details, Eng. Fail. Anal. 18 (2011) 354–363.

[1.44] S.H. Kim, J.H. Yoon, J.H. Kim, W.J. Choi, J.H. Ahn, Structural details of steel girder–abutment joints in integral bridges: an experimental study, J. Constr. Steel Res. 70 (2012) 190–212.

[1.45] K. Miyachi, S. Nakamura, A. Manda, Progressive collapse analysis of steel truss bridges and evaluation of ductility, J. Constr. Steel Res. 78 (2012) 192–200.

[1.46] X.W. Ye, Y.Q. Ni, J.M. Ko, Experimental evaluation of stress concentration factor of welded steel bridge T-joints, J. Constr. Steel Res. 70 (2012) 78–85.

[1.47] I.P. Zaforteza, M.E.M. Garlock, A numerical investigation on the fire response of a steel girder bridge, J. Constr. Steel Res. 75 (2012) 93–103.

[1.48] X.W. Ye, Y.Q. Ni, K.Y. Wong, J.M. Ko, Statistical analysis of stress spectra for fatigue life assessment of steel bridges with structural health monitoring data, Eng. Struct. 45 (2012) 166–176.

[1.49] L.T. Mo, X. Fang, D.P. Yan, M. Huurman, S.P. Wu, Investigation of mechanical properties of thin epoxy polymer overlay materials upon orthotropic steel bridge decks, Constr. Build. Mater. 33 (2012) 41–47.

[1.50] H. Yoo, H.S. Na, D.H. Choi, Approximate method for estimation of collapse loads of steel cable-stayed bridges, J. Constr. Steel Res. 72 (2012) 143–154.

[1.51] Y. Shifferaw, F.S. Fanous, Field testing and finite element analysis of steel bridge retrofits for distortion-induced fatigue, Eng. Struct. 49 (2013) 385–395.

[1.52] F. Cavadas, C. Rodrigues, C. Félix, J. Figueiras, Post-rehabilitation assessment of a centenary steel bridge through numerical and experimental analysis, J. Constr. Steel Res. 80 (2013) 264–277.

[1.53] G. Brunell, Y.J. Kim, Effect of local damage on the behavior of a laboratory-scale steel truss bridge, Eng. Struct. 48 (2013) 281–291.

[1.54] B. Cheng, Q. Qian, H. Sun, Steel truss bridges with welded box-section members and bowknot integral joints, Part I: linear and non-linear analysis, J. Constr. Steel Res. 80 (2013) 465–474.

[1.55] B. Cheng, Q. Qian, H. Sun, Steel truss bridges with welded box-section members and bowknot integral joints, Part II: minimum weight optimization, J. Constr. Steel Res. 80 (2013) 475–482.

[1.56] H.G. Kwak, Y.J. Seo, C.M. Jung, Effects of the slab casting sequences and the drying shrinkage of concrete slabs on the short-term and long-term behavior of composite steel box girder bridges. Part 2, Eng. Struct. 23 (2000) 1467–1480.

[1.57] I.S. Ahn, M. Chiewanichakorn, S.S. Chen, A.J. Aref, Effective flange width provisions for composite steel bridges, Eng. Struct. 26 (2004) 1843–1851.

[1.58] C.J. Chang, D.W. White, An assessment of modeling strategies for composite curved steel I-girder bridges, Eng. Struct. 30 (2008) 2991–3002.

[1.59] J.H. Ahn, C. Sim, Y.J. Jeong, S.H. Kim, Fatigue behavior and statistical evaluation of the stress category for a steel-concrete composite bridge deck, J. Constr. Steel Res. 65 (2009) 373–385.

[1.60] H.Y. Kim, Y.J. Jeong, Steel-concrete composite bridge deck slab with profiled sheeting, J. Constr. Steel Res. 65 (2009) 1751–1762.

[1.61] H.Y. Kim, Y.J. Jeong, Ultimate strength of a steel-concrete composite bridge deck slab with profiled sheeting, Eng. Struct. 32 (2010) 534–546.

[1.62] J.J.O. Pedro, A.J. Reis, Nonlinear analysis of composite steel-concrete cable-stayed bridges, Eng. Struct. 32 (2010) 2702–2716.

[1.63] S.T.D. Freitas, H. Kolstein, F. Bijlaard, Composite bonded systems for renovations of orthotropic steel bridge decks, Compos. Struct. 92 (2010) 853–862.

[1.64] S.H. Kim, J.H. Choi, Experimental study on shear connection in unfilled composite steel grid bridge deck, J. Constr. Steel Res. 66 (2010) 1339–1344.

[1.65] J. Machacek, M. Cudejko, Composite steel and concrete bridge trusses, Eng. Struct. 33 (2011) 3136–3142.

[1.66] F.N. Leitão, J.G.S. da Silva, P.C.G. da S. Vellasco, S.A.L. de Andrade, L.R.O. de Lima, Composite (steel-concrete) highway bridge fatigue assessment, J. Constr. Steel Res. 67 (2011) 14–24.

[1.67] Y. Okamoto, S. Nakamura, Static and seismic studies on steel/concrete hybrid towers for multi-span cable-stayed bridges, J. Constr. Steel Res. 67 (2011) 203–210.

[1.68] H.S. Ji, J.K. Byun, C.S. Lee, B.J. Son, Z.H. Ma, Structural performance of composite sandwich bridge decks with hybrid GFRP-steel core, Compos. Struct. 93 (2011) 430–442.

[1.69] A. Turer, B.M. Shahrooz, Load rating of concrete-deck-on-steel-stringer bridges using field-calibrated 2D-grid models, Eng. Struct. 33 (2011) 1267–1276.

[1.70] J.G. Nie, Y.H. Wang, X.J. Zhang, J.S. Fan, C.S. Cai, Mechanical behavior of composite joints for connecting existing concrete bridges and steel-concrete composite beams, J. Constr. Steel Res. 75 (2012) 11–20.

[1.71] J.F. Davalos, A. Chen, B. Zou, Performance of a scaled FRP deck-on-steel girder bridge model with partial degree of composite action, J. Eng. Struct. 40 (2012) 51–63.

CHAPTER 2

Nonlinear Material Behavior of the Bridge Components

2.1 GENERAL REMARKS

Chapter 1 has provided a brief background on steel and steel–concrete composite bridges and reviewed recent developments reported in the literature related to the design and finite element modeling of the bridges. This chapter highlights the nonlinear material behavior of the main components of steel and steel–concrete composite bridges, comprising structural steel, concrete, reinforcement bars, shear connectors, bolts, and welds. Overall, this chapter aims to provide a useful background regarding the stress–strain curves of the different materials used in the bridges. Also, this chapter aims to highlight the important parameters required for finite element modeling. The definitions of yield stresses, ultimate stresses, maximum strains at failure, initial stiffness, and proportional limit stresses are presented in this chapter. This chapter enables beginners to understand the fundamental behavior of the materials in order to correctly insert them in the finite element analyses. Covering the behavior of shear connectors in this chapter is also important to understand how the shear forces are transmitted at the steel–concrete slab interfaces in composite bridges. In addition, the material properties of the main components of joints used in steel and steel–concrete composite bridges such as bolts are highlighted in this chapter. Furthermore, this chapter presents how the different materials are treated in current codes of practice and the design values specified in current codes of practice. This chapter paves the way for Chapters 3 and 4, which address the design and stability issues related to steel and steel–concrete composite bridges. It should be noted that bridge components, such as structural steels, concrete, and reinforcement steels, are used in bridge and building constructions. However, when presenting the material behavior of a component in this chapter, it is presented as it is used in bridges. As an example, structural steels used in bridges generally have more rigid performance requirements compared with steels used in buildings. Bridge steels have to perform in an outdoor environment with relatively large temperature changes, are subjected to excessive cyclic live loading, and are often exposed to corrosive environments. In addition, steels

Finite Element Analysis and Design of Steel and Steel–Concrete Composite Bridges

47

are required to meet strength and ductility requirements for all structural applications. However, bridge steels have to provide adequate service with respect to the additional fatigue and fracture limit state. They also have to provide enhanced atmospheric corrosion resistance in many applications where they are used with normal protective coatings. For these reasons, structural steels for bridges are required to have fracture toughness and often corrosion resistance that exceed general structural requirements in building constructions. Overall, the author aims that this chapter acts as a basis for designing and finite element modeling of steel and steel–concrete composite bridges.

2.2 NONLINEAR MATERIAL PROPERTIES OF STRUCTURAL STEEL

2.2.1 General

The main component of steel and steel-concrete composite bridges investigated in this book is structural steel. Understanding the material behavior of the steel is quite important for designing and finite element modeling of the bridges. As a material composition, steel contains iron, a small percentage of carbon and manganese, impurities such as sulfur and phosphorus, and some alloying elements that are added to improve the properties of the finished steel such as copper, silicon, nickel, chromium, molybdenum, vanadium, columbium, and zirconium. The strength of the steel increases as the carbon content increases, but some other properties like ductility and weldability decrease. Steel used for bridges can be classified as carbon steels, which come with yield stresses up to 275 N/mm^2; high-strength steels, which cover steels having yield stresses up to 390 N/mm^2; heat-treated carbon steels, which cover steels having yield stresses greater than 390 N/mm^2; and weathering steels, which have improved resistance to corrosion. Steels used for bridges should have main properties including strength, ductility, fracture toughness, weldability, weather resistance, and residual stresses. These properties are briefly highlighted in the coming sections.

2.2.2 Steel Stresses

In the United States, the specifications for plate and rolled shape steels used for bridges are covered by the ASTM A709 [2.1] and AASHTO M270 [1.23, 1.24]. Table 2.1 shows the applicable AASHTO and ASTM standards for steel product categories, while Table 2.2 provides an overview of the various steel grades covered by the ASTM A709 [2.1]. The number in

Table 2.1 Examples of American Standards for Main Bridge Steel Products

Product	AASHTO	ASTM
Structural steel for bridges	M270/M 270M	A709/A709M
Pins, rollers, and rockers	M169	A108
	M102/M102M	A668/A668M
Bolts	M164	A307 grade A or B
	M253	A325
		A490
Anchor bolts	M314-90	A307 grade C
Nuts	M291	A563
Washers	M293	F436
		F959
Shear studs	M169	A108
Cast steel	M103/M103M	A27/A27M
	M163/M163M	A743/A743M
Cast iron	M105 class 30	A48 class 30
Cables		A510
Galvanized wire		A641
Bridge strand/bridge rope		A586
		A603
Wire rope	M277	
Seven-wire strand	M203/M203M	A416/A416M

Table 2.2 Examples of Bridge Steels Available in the ASTM A709 Specification

M270 A709 Grade	ASTM	Description	Weather Resistance	Product		
				Plates	Shapes	Bars
36	A36	Carbon steel	No	Yes	Yes	Yes
50	A572	HSLA[a] steel	No	Yes	Yes	Yes
50S	A992	Structural steel	No		Yes	
50W	A588	HSLA steel	Yes	Yes	Yes	Yes
HPS[b] 50W	A709	HSLA steel	Yes	Yes		
HPS 70W	A709	Heat-treated HSLA steel	Yes	Yes		
HPS 100W	A709	Q&T Cu-Ni steel[c]	Yes	Yes		

[a]HSLA high-strength low-alloy.
[b]High-performance steel (HPS) grades with enhanced weldability and toughness.
[c]Q&T Cu-Ni quenched and tempered copper-nickel steel.

the grade designation indicates the nominal yield strength in ksi (1 ksi is equal to 6.895 MPa). The A709M specification is the metric version of A709. According to ASTM, grade 36 and 50 steels have yield stresses of 36 and 50 ksi (248 and 344 MPa, respectively). Grade 50 steel is commonly used for primary bridge members, which can be painted or galvanized in service. Grade 50W steel is a weathering steel that has the same strength as grade 50 steel, but it has enhanced atmospheric corrosion resistance. The enhanced corrosion resistance was achieved by adding different combinations of copper, chromium, and nickel to the grade 50 chemistry. Grade 100 and 100W steels are high-strength steels having a yield stress of 100 ksi (689 MPa), if quenched and tempered. It is common that engineers specify the use of grade 100 and 100W steels for highly stressed parts of the bridge such as bearing components. High-performance steel (HPS) has enhanced weldability and toughness compared to grade 100 steel. The properties of HPS can be achieved by lowering the percentage of carbon in the steel chemistry. Since carbon is traditionally one of the primary strengthening elements in steel, the composition of other alloying elements must be more precisely controlled to meet the required strength and compensate for the reduced carbon content. Using HPS allows for increasing the span length of bridges. Grade 100W steels are the same as grade 100 steels but with enhanced weldability and toughness.

In the United States, structural bolts for members requiring slip critical connections in bridges are required to comply with either the ASTM A325 [2.2] or the A490 [2.3] specification. On the other hand, anchor bolts and nonslip critical connections are required to comply with the ASTM A307 [2.4] specification. Compatible nuts are required to be used with all bolts meeting provisions for the appropriate grade in the ASTM A563 [2.5] specification. Hardened steel washers meeting the ASTM F436M [2.6] specification are required underneath all parts of the bolt assembly that are turned during installation. The surface condition and presence of lubrication are important for proper installation of the bolt-nut assemblies. Table 2.3 shows the minimum specified tensile strength of structural bolts to be used for bridges. The A325 [2.2] and A490 [2.3] specifications have two different chemistry requirements for bolts: type 1 and type 3. Type 1 bolts are carbon-manganese steel with mainly silicon additions and are suitable for use with painted and galvanized coatings. On the other hand, type 3 bolts have additional requirements for copper, nickel, and chromium to be

Table 2.3 Nominal Tensile Strengths of American Structural Bolts Used in Bridges

Grade	Diameter (mm)	Tensile Strength (MPa)
A307 (Grade A or B)	All	414
A325	13–25	827
	28–38	724
Seven-wire strand	All	1034

compatible with the chemistry of weathering steel grades and are required for use in unpainted applications where both the bolts and the base metal can develop rust in service.

Cables and wires used in bridges in the United States are either strands, which are covered by ASTM A586 [2.7], or ropes, which are covered by ASTM A603 [2.8]. Cables and wires are constructed from individual cold-drawn wires that are spirally wound around a wire core. The commonly used nominal diameters are between 1/2 (12.7 mm) and 4 in. (101.6 mm) depending on the intended application. The capacities of the cables and wires are defined as the minimum breaking strength that depends on the nominal diameter of the cables or wires. Cables and wires are used as tension members in bridges. Because relative deformation between the individual wires will affect elongation, strands and ropes are preloaded to about 55% of the breaking strength after manufacturing to "seat" the wires and stabilize the deformation response. Following preloading, the axial deformation becomes linear and predictable based on an effective modulus for the wire bundles. A bridge rope has an elastic modulus of 20,000 ksi (138,000 MPa). The elastic modulus of a bridge strand is 24,000 ksi (165,000 MPa). Seven-wire steel strands (tendon) are commonly used for prestressed concrete bridge decks. They are also used as cable stays, hangers, and posttensioning members. They consist of seven individual cold-drawn round wires spirally wounded to form a strand with nominal diameters between 0.25. (6.4 mm) and 0.60 in. (15.2 mm). Two grades are available (250 and 270) where the grade indicates the tensile strength of the wires (f_{pu}). The net cross-sectional area of the seven-wire strand (area of the individual wires) should be used in all calculations, and prestress losses should be accounted for, either by measurements or based on specified values in current codes of practice. Mechanical properties of seven-wire strands are measured from tensile coupon tests. The tensile strength is calculated by dividing the breaking load by the net cross-sectional area of the seven-wire strand. Compared to structural steels, strands do not exhibit a yield plateau, and there is a gradual rounding of the stress–strain curve beyond the proportional limit. The yield stress in this case may be calculated as the stress at the 0.1% strain offset line ($f_{0.1}$). Strands are loaded provided that they do not reach the yield stress. AASHTO [1.23, 1.24] defines the yield strength as $f_{py} = 0.90 f_{pu}$.

The ASTM A370 [2.9] and the ASTM E8 [2.10] specifications cover tensile coupon testing procedures for determining the material properties of steel products. The main properties measured from a tensile coupon test are the yield strength (f_y), tensile strength (f_u), Young's modulus (E_s),

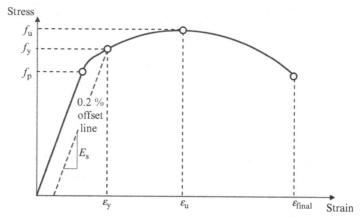

Figure 2.1 Engineering stress-strain curve for structural steels without a defined yield plateau.

ultimate strain at failure (ε_u), and full nonlinear stress–strain curve. The full nonlinear stress–strain curve is known as the engineering stress–strain curve, which can be measured by recording the load and elongation of an extensometer during the tensile coupon test. Young's modulus for steel can be determined by predicting the slope of the elastic initial portion of the stress–strain curve as shown in Figure 2.1. In the absence of the measured engineering stress–strain curve, Young's modulus for steel can be conservatively taken as $E_s = 29{,}000\text{-}30{,}000$ ksi ($200{,}000\text{-}207{,}000$ MPa) for structural calculations for all structural steels used in bridge construction. The yield strength of steel is determined by the 0.2% offset method. A line is plotted parallel to the elastic part of the stress–strain curve below the proportional limit with an x-axis offset of 0.2% (0.002) strain. The intersection of the offset line with the stress–strain curve defines the yield strength. Figures 2.1 and 2.2 show the 0.2% offset method applied to steels without a definite yield plateau and to steels that exhibit a yield plateau, respectively. For the steels that exhibit a yield plateau, there is an upper yield point that is greater than the yield strength. When yielding first occurs, there is typically a slight drop in load before the steel plastically deforms along the yield plateau (see Figure 2.2). Following the first yield, steels with $f_y \le 70$ ksi (483 MPa) undergo plastic deformation at a relatively constant load level defining the yield plateau. The length of this plateau varies for different steels, but approximately, ε_{st} is around $10\varepsilon_y$. Strain hardening begins at the end of the plateau and continues until the maximum load is achieved

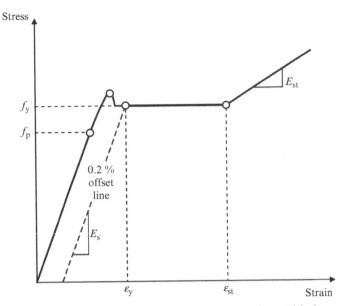

Figure 2.2 Initial part of the stress-strain curve for steels with a yield plateau.

corresponding to the tensile strength f_u. The slope of the stress–strain curve constantly varies during strain hardening. The tangent slope of the curve at the onset of strain hardening (E_{st}) is often used for analysis of steel behavior at high strain levels. Tensile coupon test results are usually presented by engineering stress–strain curves where stress is calculated based on the undeformed cross-sectional area of the specimen. As the specimen is loaded, the cross-sectional area is constantly being reduced, which is known as necking phenomena. The true stress at any given point can be calculated with respect to the contracted area at that point in time. In nonlinear structural analyses, true stress–strain curves should be used. Figure 2.3 shows typical stress–strain curves for steels in the A709 [2.1] specification. Steels with $f_y \leq 70$ ksi (483 MPa) show definite yield plateaus with similar ductility. The HPS 100W steel does not have a clearly defined yield plateau and shows slightly lower ductility compared to the lower-strength steels. The amount of strain hardening decreases with increasing yield strength. The minimum specified yield strength (f_y) and tensile strength (f_u) are shown in Table 2.4 for steel grades included in the A709 specification. Plates with thickness up to 4 in. (101.6 mm) are available in all grades, except for 50S. Rolled shapes are not available in the HPS grades. The shear yield stress (f_{yv}) can be determined

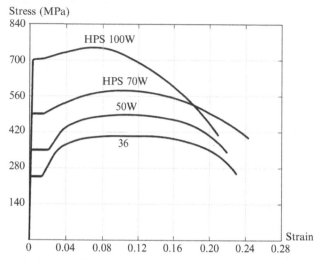

Figure 2.3 Typical example engineering stress-strain curves for American bridge structural steels.

Table 2.4 Nominal Strength of American ASTM A709 Steel Grades

Grade	36	50	50S	50W	HPS 50W	HPS 70W	HPS 100W	
Plate thickness (mm)	$t \leq 102$	$t \leq 102$	N/A	$t \leq 102$	$t \leq 102$	$t \leq 102$	$t \leq 64$	$64 < t \leq 102$
Shapes	All	All	All	All	N/A	N/A	N/A	N/A
f_u (MPa)	400	448	448	483	483	586	758	689
f_y (MPa)	248	345	345	345	345	483	689	620

using the von Mises yield criterion, which is commonly used to predict the onset of yielding in steel subject to multiaxial states of stress as follows:

$$f_y = \sqrt{\frac{(\sigma_x - \sigma_y)^2 + (\sigma_y - \sigma_z)^2 + (\sigma_z - \sigma_x)^2 + 6(\tau_{xy}^2 + \tau_{yz}^2 + \tau_{zx}^2)}{2}} \quad (2.1)$$

For the state of pure shear in one direction, the normal stresses are equal to zero; the shear yield stress (f_{yv}) can be determined as follows:

$$f_{yv} = \frac{1}{\sqrt{3}} f_y \approx 0.58 f_y \quad (2.2)$$

The shear modulus (G) based on Young's modulus (E) and Poisson's ratio (v) is given as

$$G = \frac{E_s}{2(1 + v)} = 11,200 \text{ ksi} (77,200 \text{ MPa}) \quad (2.3)$$

In Europe, EC3 (BS EN 1993-1-1) [2.11] specifies that the nominal values of the yield strength f_y and the ultimate strength f_u for structural steel used for buildings and bridges should be obtained either by adopting the values $f_y = R_{eh}$ and $f_u = R_m$ direct from the product standard or by using the simplification given in Table 2.5, where R_{eh} and R_m are yield and ultimate strengths to product standards. EC3 [2.11] also specifies that for structural steels, a minimum ductility is required, which should be expressed in terms of the limits for the ratio f_u/f_y (the specified minimum ultimate tensile strength f_u to the specified minimum yield strength f_y), the elongation at failure on a gauge length of $5.65\sqrt{A_o}$ (where A_o is the original cross-sectional area), and the ultimate strain ε_{us} (where ε_{us} corresponds to the ultimate strength f_u). It should be noted that EC3 [2.11] specifies that the limiting values of the ratio f_u/f_y, the elongation at failure, and the ultimate strain ε_u may be defined in the National Annex of the country of construction. Alternatively, EC3 recommends that $f_u/f_y \geq 1.10$, elongation at failure not less than 0.15, and $\varepsilon_u \geq 15\varepsilon_y$, where ε_y is the yield strain ($\varepsilon_y = f_y/E_s$). Generally, steels conforming with one of the steel grades listed in Table 2.5 should be accepted as satisfying these requirements. According to EC3, the modulus of elasticity of steel $E_s = 210\ 000$ N/mm^2, shear modulus $G = 81,000$ N/mm^2, Poisson's ratio in elastic stage $v = 0.3$, and the coefficient of linear thermal expansion $\alpha = 12 \times 10^{-6}$ per K (for $T \leq 100\ °C$). It should be noted that for calculating the structural effects of unequal temperatures in composite concrete-steel structures to EC4 [2.12], the coefficient of linear thermal expansion is taken as $\alpha = 10 \times 10^{-6}$ per K.

Bolts, nuts, and washers used for bridges, according to EC3 (BS EN 1993-2) [1.27], should conform to the reference standards given in EC3 (BS EN 1993-1-8) [2.13], 2.8: Group 4. The different bolt grades used in bridges are presented in Table 2.6. The nominal values of the yield strength f_{yb} and the ultimate tensile strength f_{ub} are shown in Table 2.6, and they should be adopted as characteristic values in calculations, while high-strength structural bolts of grades 8.8 and 10.9, which conform to the reference standards given in BS EN 1993-1-8 [2.13], 2.8: Group 4, may be used as preloaded bolts when controlled tightening is carried out in accordance with the reference standards given in BS EN 1993-1-8 [2.13], 2.8: Group 7. In addition, EC3 (BS EN 1993-2)[1.27] specifies that steel grades in accordance with the reference standards given in BS EN 1993-1-8 [2.13], 2.8: Group 1, steel grades in accordance with the reference standards given in BS EN 1993-1-8, 2.8: Group 4, and reinforcing bars conforming to EN 10080 [2.14] may be used for anchor bolts. The nominal yield strength for anchor bolts should not exceed 640 N/mm^2.

Table 2.5 Nominal Values of Yield Strength f_y and Ultimate Tensile Strength f_u

| | Nominal Thickness of the Element t (mm) | | | |
| | $t \leq 40$ mm | | $40 < t \leq 80$ mm | |
Standard and Steel Grade	f_y (MPa)	f_u (MPa)	f_y (MPa)	f_u (MPa)
(a) For European Hot-Rolled Structural Steel Specified in EC3 [2.11]				
EN 10025-2				
S235	235	460	215	360
S275	275	430	255	410
S355	355	510	335	470
S450	440	550	410	550
EN 10025-3				
S275 N/NL	275	390	255	370
S355 N/NL	355	490	335	470
S420 N/NL	420	520	390	520
S460 N/NL	460	540	430	540
EN 10025-4				
S275 M/ML	275	370	255	360
S355 M/ML	355	470	335	450
S420 M/ML	420	520	390	500
S460 M/ML	460	540	430	530
EN 10025-5				
S235 W	235	360	215	340
S355 W	355	510	335	490
EN 10025-6				
S460 Q/QL/QL1	460	570	440	550
(b) For European Structural Hollow Sections Specified in EC3 [2.11]				
EN 10210-1				
S235 H	235	360	215	340
S275 H	275	430	255	410
S355 H	355	510	335	490
S275 NH/NLH	275	390	255	370
S355 NH/NLH	355	490	335	470
S420 NH/NHL	420	540	390	520
S460 NH/NLH	460	560	430	550
EN 10219-1				
S235 H	235	360		
S275 H	275	430		
S355 H	355	510		

Table 2.5 Nominal Values of Yield Strength f_y and Ultimate Tensile Strength f_u—cont'd

	Nominal Thickness of the Element t (mm)			
	$t \leq 40$ mm		$40 < t \leq 80$ mm	
Standard and Steel Grade	f_y (MPa)	f_u (MPa)	f_y (MPa)	f_u (MPa)
S275 NH/NLH	275	370		
S355 NH/NLH	355	470		
S460 NH/NLH	460	550		
S275 MH/MLH	275	360		
S355 MH/MLH	355	470		
S420 MH/MLH	420	500		
S460 MH/MLH	460	530		

Table 2.6 Nominal Values of the Yield Strength f_{yb} and the Ultimate Tensile Strength f_{ub} for European Bolts Specified in EC3 [1.27]

Bolt grade	4.6	5.6	6.8	8.8	10.9
f_{yb} (N/mm^2)	240	300	480	640	900
f_{ub} (N/mm^2)	400	500	600	800	1000

2.2.3 Ductility

Steel ductility is the capacity of steel material to undergo large strains after the onset of yielding and before fracture, which provides an advance warning of possible failure. For steel products, relative ductility is measured as the percent elongation that occurs before rupture in a standard tensile coupon test. The percent elongation is dependent on the test specimen geometry and the gauge length used to measure elongation during tensile coupon test. In the United States, for the same material, tension specimens with a 2 in. (50.8 mm) gauge length will exhibit a lower percent elongation compared to those with an 8 in. (203.2 mm) gauge length. The ASTM A709 [2.1] specification specifies that structural steel for bridges has an adequate level of material ductility to perform well in structural applications. Steel material ductility is different from structural steel connections and overall structural ductility. For example, a steel member may be ductile on its own; however, if there are holes in the cross section, it may undergo brittle failure behavior. The yield-tensile stress ratio (YT ratio) defined as YT $= f_y/f_u$ can provide a reasonable measure to steel ductility. However, for steels specified in the A709 [2.1] specification, there is no need for special consideration of the YT ratio for most bridge structural applications.

2.2.4 Fracture Toughness

Steel members used in bridges must have sufficient fracture toughness to reduce the probability of brittle failure. The brittle failure may occur suddenly under a load, which may be below the load level to cause yielding. It may be initiated by the existence of a small crack or other forms of notch. Very high concentration of stress occurs at the root of a natural crack. Any sudden change that occurs in the cross section of a loaded member having a notch-like effect can disturb the stress pattern and cause a local stress concentration. If the local yielding at the tip of the crack or notch is insufficient to spread the load over a large area, a brittle fracture may be initiated. Once initiated, the fracture propagates at high speed driven by the release of the elastic strain energy in the structure. Linear-elastic fracture mechanics analysis is the basis for predicting brittle fracture in structural steels. The stress intensity factor (K_I) can characterize the crack tip singularity. For a given plate geometry, the stress intensity present at a crack tip is a function of the crack size and the applied stress. The material fracture resistance is characterized by the critical stress intensity factor (K_{Ic}) that can be sustained without fracture. When the applied stress intensity K_I equals or exceeds the material fracture resistance K_{Ic}, fracture is predicted. The Charpy V-notch (CVN) test can be used to measure the fracture toughness of structural steel [2.15]. A small 10×10 mm bending specimen with a machined notch is impacted by a hammer, and the energy required to initiate fracture is measured. CVN test data can be used to predict the K_{Ic} fracture toughness. The AASHTO [1.23, 1.24] specification classifies structural steel materials into two categories, which are fracture-critical material and non-fracture-critical material. Fracture-critical materials' fracture would cause collapse of the structure. The specification divides the United States into three temperature zones for specifying fracture toughness of bridge steels. The zones are defined by the lowest expected service temperature as shown in Table 2.7. It should be noted that the specified toughness requirements are higher with colder zones, thicker components, higher grades of steel, and fracture-critical components.

In Europe, EC3 (BS EN 1993-2) [1.27] requires that structural steels used for bridges should have the enough material toughness to prevent brittle fracture within the intended design to prolong the working life of the

Table 2.7 AASHTO Temperature Zones for Specifying CVN Toughness

Lowest Anticipated Service Temperature	Temperature Zone
0 °F and above	1
−1 to −30°F	2
−31 to −60 °F	3

Table 2.8 Example for Additional Requirement for Toughness of Base Material Specified in EC3 [1.27]

Example	Nominal Thickness	Additional Requirement
1	$t \leq 30$ mm	$T_{27J} = -20\ °C$ in accordance with EN 10025
	$30 < t \leq 80$ mm	Fine-grain steel in accordance with EN 10025, e.g., S355N/M
	$t > 80$ mm	Fine-grain steel in accordance with EN 10025, e.g., S355NL/ML

structure. The specification requires that no further checks against brittle fracture need to be made if the conditions given in EN 1993-1-10 [2.16] are met for the lowest service temperature. EC3 (BS EN 1993-2) [1.27] also recommends that additional requirements depending on the plate thickness, with an example given in Table 2.8, may be adopted.

2.2.5 Weldability

Steel weldability is defined as the ability of steel to be welded to serve its intended application. In the United States, the AASHTO/AWS D1.5 welding specifications [2.17] govern welding of bridge steels. Following the D1.5 provisions, all bridge steels in the A709 [2.1] specification can be considered weldable. It should be noted that increasing amounts of carbon and manganese, which are necessary for higher strengths, make the steel harder and consequently more difficult to weld. Also, the elements added to improve weathering resistance reduce weldability. In addition, the weldability of structural steels depends on the chemical composition. Graville [2.18] showed that the tendency of a heat-affected zone (HAZ) to crack depends on the carbon content and the carbon equivalent (CE) calculated using Equation (2.4) as recommended by the Bridge Welding Code D1.5, which considers other alloying elements in addition to carbon:

$$CE = C + \frac{Mn + Si}{6} + \frac{Ni + Cu}{15} + \frac{Cr + Mo + V}{5} \tag{2.4}$$

where C, Mn, etc., represent the percentage of the element concerned in the chemical composition of the steel. To obtain higher yield stresses, the percentage content of the various alloying elements is increased leading to the increase of the carbon equivalent value. Therefore, welding of higher-strength steels is more difficult compared with normal-strength steels. Specifications sometimes limit maximum values for carbon equivalent. Steels with carbon equivalent values higher than 0.53 should have special measures in welding.

2.2.6 Weather Resistance

As mentioned previously, steel grades with the "W" suffix used in the United States are called "weathering steels" since they are developed for increased weather resistance, such as corrosion resistance. Weathering steels form a thin iron oxide film on the surface when exposed to damp environment, which acts as a coating that resists any further rusting. Weathering steels can be used in bridge structures without special paints. Compared to normal steel grades, weathering steel grades provide around $3 \times$ corrosion resistance. However, this is greatly dependent on the severity of environment conditions. In the United States, the ASTM G101 [2.19] specification proposed a methodology for classification of steels as weathering. The specification proposed a corrosion index (I) based on the chemical composition of the steel. The ASTM A709 [2.1] specification indicates that steel grades with $I \geq 6$ can be classified as weathering steels indicated by the W suffix to the grade. It should be noted that although paint coatings for normal steels are commonly used for corrosion resistance, other options such as galvanizing steel may be effectively used. All of the A709 bridge steels are suitable for use with any of these coating options.

2.2.7 Residual Stresses

Residual stresses are initial internal stresses existing in cross sections without the application of an external load such as stresses resulting from manufacturing processes of metal structural members by cold forming. Residual stresses produce internal membrane forces and bending moments, which are in static equilibrium inside the cross sections. The force and the moment resulting from residual stresses in the cross sections must be zero. Residual stresses in structural cross sections are attributed to the uneven cooling of parts of cross sections after hot rolling. Uneven cooling of cross-sectional parts is subjected to internal stresses. The parts that cool quicker have residual compressive stresses, while parts that cool slower have residual tensile stresses. Residual stresses cannot be avoided and in most cases are not desirable. The measurement of residual stresses is therefore important for accurate finite element modeling of steel and steel-concrete composite bridges.

Extensive experimental investigations were conducted in the literature to determine the distribution and magnitude of residual stresses inside cross sections. The experimental investigations can be classified into two main categories: nondestructive and destructive methods. Examples of nondestructive methods are X-ray diffraction and neutron diffraction. Nondestructive methods are suitable for measuring stresses close to the outside surface of cross sections. On the other hand, destructive methods involve machining/cutting of the cross section to release internal stresses and measure resulting change of strains. Destructive methods are based on the destruction of the state of equilibrium of the residual stresses in the cross section. In this way, the residual stresses can be measured by relaxing these stresses. However, it is only possible to measure the consequences of the stress relaxation rather than the relaxation itself. One of the main destructive methods is to cut the cross section into slices and measure the change in strains before and after cutting. After measuring the strains, some simple analytic approaches can be used to evaluate resultant membrane forces and bending moments in the cross sections. Although the testing procedures to determine residual stresses are outside the scope of this book, it is important to detail how to incorporate residual stresses in finite element models. It should be noted that in some cases, incorporating residual stresses can have a small effect on the structural performance of metals. However, in some other cases, it may have a considerable effect. Structural steel cross sections used in bridges are subjected to more loading conditions than that commonly applied to buildings. Since the main objective of this book is to accurately model all parameters affecting the behavior and design of steel and steel–concrete composite bridges, the way to model residual stresses is highlighted.

Limited numerical methods were presented in the literature to simulate some typical and simple procedures introducing residual stresses. Dixit and Dixit [2.20] modeled cold rolling for steel and gave a simplified approach to find the longitudinal residual stress. The numerical simulation [2.20] has provided the scope to investigate the effects of different parameters on the magnitude and distribution of residual stresses such as material characteristics and boundary conditions. Kamamato et al. [2.21] have analyzed residual stresses and distortion of large steel shafts due to quenching. The results showed that residual stresses are strongly related to the transformational behavior. Toparli and Aksoy [2.22] analyzed residual stresses during water quenching of cylindrical solid steel bars of various diameters by using finite element technique. The authors have computed the transient temperature distribution for solid bars with general surface heat transfer. Jahanian [2.23] modeled heat treatment and calculated the residual stress in a long solid cylinder by using theoretical and numerical methods with different cooling speeds. Yuan and Wu [2.24] used a finite element program to analyze the transient temperature and residual stress fields for a metal specimen during quenching. They modified the elastic-plastic properties of specimen according to temperature fields. Yamada [2.25] presented a method of solving uncoupled quasistatic thermoplastic problems in perforated plates. In their analysis, a transient thermal stress problem was solved for an infinite plate containing two elliptic holes with prescribed temperature. An extensive survey of the aforementioned numerical investigations was presented by Ding [2.26]. However, to date, the effect of residual stresses on the structural behavior and strength of the components of steel and steel-concrete composite bridges was not fully understood, which is addressed in this book.

Residual stresses and their distribution are very important factors affecting the strength of axially loaded steel members. These stresses are of particular importance for slender columns, with slenderness ratio varying from approximately 40-120. As a column load is increased, some parts of the column will quickly reach the yield stress and go into the plastic range because of the presence of residual compression stresses. The stiffness will reduce and become a function of the part of the cross section that is still elastic. A column with residual stresses will behave as though it has a reduced cross section. This reduced cross section or elastic portion of the column will change as the applied load changes. The buckling analysis and postbuckling calculation can be carried out theoretically or numerically by using an effective moment of inertia of the elastic portion of the cross section or by using the tangent modulus. Figure 2.4 shows typical residual stress distributions in hot-rolled and built-up I-sections. It can be seen that although both cross sections are I-shaped sections, welding and cutting of plates of the built-up sections result in differences in the distributions of residual stresses in both sections.

2.3 NONLINEAR MATERIAL PROPERTIES OF CONCRETE

2.3.1 General

Understanding nonlinear material behavior of concrete is quite important in designing and finite element modeling of steel-concrete composite bridges investigated in this book. As a material composition, plain concrete is a composite material comprising a mixture of coarse and fine aggregates, cement, water, and additions. Numerous design approaches are available in the literature that can be effectively used to provide the mix proportions that produce plain concrete with a target strength, workability, permeability, etc. It should be noted that explaining these design approaches is outside the scope of this book. However, the nonlinear material properties of plain concrete required for designing and finite element modeling of steel-concrete composite bridges are highlighted in this book. Plain concrete behaves completely different when subjected to compressive and tensile stresses. Plain concrete is a brittle

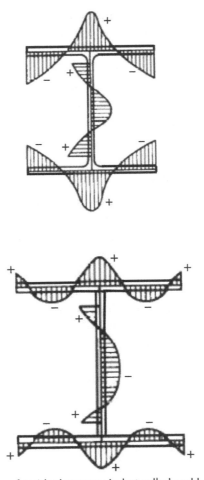

Figure 2.4 Distributions of residual stresses in hot-rolled and built-up I-sections.

material that has a considerably higher compressive strength compared with its tensile strength. Therefore, in steel–concrete composite bridges where massive concrete decks lies on the top of steel beams, it is more economical to use it to benefit from the composite action between steel and concrete in the regions where the concrete is subjected to compressive stresses. However, when plain concrete is subjected to tensile stresses, it fails prematurely if its tensile strength is exceeded. Therefore, reinforcement steel bars are commonly placed in the regions where plain concrete is subjected to tensile stresses to form reinforced concrete (RC). Nonlinear material properties of reinforcement bars are also briefly highlighted in this book. It should be also noted that detailing the nonlinear material properties of concrete in this book is based on specified properties provided in current codes of practice. This is attributed to that experimental and numerical investigations reported in the literature on non-linear material properties are unlimited and differ from a country to another. Since the main objective of this book is to provide a consistent and robust nonlinear approach for designing and finite element modeling of steel-concrete composite bridges, only specified values in current codes of practice are highlighted in this book.

2.3.2 Concrete Stresses

In Europe, EC2 (BS EN 1992-1-1 and BS EN 1992-2) [2.27, 2.28] specify that the compressive strength of concrete is denoted by concrete strength classes, which are based on the characteristic cylinder strength f_{ck} determined at 28 days with a recommended maximum value C_{max} ($f_{ck}/f_{ck,cube}$) of C90/105. The characteristic strengths for f_{ck} and the corresponding mechanical characteristics necessary for design according to EC2 are given in Table 2.9. In certain situations (e.g., prestressing), it may be appropriate to assess the compressive strength of concrete before or after 28 days with reference to test specimens stored under other conditions rather than those prescribed in EN 12390 [2.29]. If the concrete strength is determined at an age $t > 28$ days, the values α_{cc} and α_{ct}, which are coefficients taking account of long-term effects on the compressive and tensile strengths, respectively, should be reduced by a factor k_t. EC2 recommends the value of k_t as 0.85. According to EC2, when it is required to specify the concrete compressive strength, $f_{ck}(t)$, at time t for a number of stages (e.g., transfer of prestress), the following relationships can be used:

$$f_{ck}(t) = f_{cm}(t) - 8(\text{MPa}) \quad \text{for} \quad 3 < t < 28 \text{ days.} \tag{2.5}$$

$$f_{ck}(t) = f_{ck} \quad \text{for} \quad t \geq 28 \text{ days.} \tag{2.6}$$

The compressive strength of concrete at an age t depends on the type of cement, temperature, and curing conditions. For a mean temperature of 20 °C and curing in accordance with EN 12390 [2.29], the compressive strength of concrete at various ages $f_{cm}(t)$ may be estimated as follows:

$$f_{cm}(t) = \beta_{cc}(t) f_{cm} \tag{2.7}$$

with

$$\beta_{cc}(t) = \exp\left\{ s\left[1 - \left(\frac{28}{t}\right)^{1/2}\right] \right\} \tag{2.8}$$

Table 2.9 Strength and Deformation Characteristics for Concrete Specified in EC2 [2.27] Strength Classes for Concrete

f_{ck} (MPa)	12	16	20	25	30	35	40	45	50	55	60	70	80	90
$f_{ck,cube}$ (MPa)	15	20	25	30	37	45	50	55	60	67	75	85	95	105
f_{cm} (MPa)	20	24	28	33	38	43	48	53	58	63	68	78	88	98
f_{ctm} (MPa)	1.6	1.9	2.2	2.6	2.9	3.2	3.5	3.8	4.1	4.2	4.4	4.6	4.8	5.0
$f_{ctk,\,0.05}$ (MPa)	1.1	1.3	1.5	1.8	2.0	2.2	2.5	2.7	2.9	3.0	3.1	3.2	3.4	3.5
$f_{ctk,\,0.95}$ (MPa)	2.0	2.5	2.9	3.3	3.8	4.2	4.6	4.9	5.3	5.5	5.7	6.0	6.3	6.6
E_{cm} (GPa)	27	29	30	31	33	34	35	36	37	38	39	41	42	44
ε_{c1} (‰)	1.8	1.9	2.0	2.1	2.2	2.25	2.3	2.4	2.45	2.5	2.6	2.7	2.8	2.8
ε_{cu1} (‰)	3.5									3.2	3.0	2.8	2.8	2.8
ε_{c2} (‰)	2.0									2.2	2.3	2.4	2.5	2.6
ε_{cu2} (‰)	3.5									3.1	2.9	2.7	2.6	2.6
n	2.0									1.75	1.6	1.45	1.4	1.4
ε_{c3} (‰)	1.75									1.8	1.9	2.0	2.2	2.3
ε_{cu3} (‰)	3.5									3.1	2.9	2.7	2.6	2.6

where $f_{cm}(t)$ is the mean concrete compressive strength at an age of t days, f_{cm} is the mean compressive strength at 28 days according to Table 2.9, $\beta_{cc}(t)$ is a coefficient that depends on the age of the concrete t, t is the age of the concrete in days, and s is a coefficient that depends on the type of cement. The value of s is equal to 0.2 for cement of strength classes CEM 42,5 R, CEM 52,5N, and CEM 52,5 R (class R). The value of s is equal to 0.25 for cement of strength classes CEM 32,5 R and CEM 42,5N (class N). The value of s is equal to 0.38 for cement of strength class CEM 32.5N (class S). The tensile strength refers to the highest stress reached under concentric tensile loading. Where the tensile strength is determined as the splitting tensile strength, $f_{ct,sp}$, an approximate value of the axial tensile strength, f_{ct}, may be taken as

$$f_{ct} = 0.9 f_{ct,sp} \qquad (2.9)$$

The development of tensile strength with time is strongly influenced by curing and drying conditions and by the dimensions of the structural members. As a first approximation, it may be assumed that the tensile strength $f_{ctm}(t)$ can be calculated based on the values of f_{ctm}, given in Table 2.9, as follows:

$$f_{ctm}(t) = (\beta_{cc}(t))^{\alpha} \cdot f_{ctm} \qquad (2.10)$$

where

$$\alpha = 1 \quad \text{for} \quad t < 28$$
$$\alpha = 2/3 \quad \text{for} \quad t \geq 28 \qquad (2.11)$$

According to EC2 [2.27, 2.28], the elastic deformations of concrete largely depend on its composition (especially the aggregates). The values given in EC2 should be regarded as indicative for general applications. However, they should be specifically assessed if the structure is likely to be sensitive to deviations from these general values. The modulus of elasticity of a concrete is controlled by the moduli of elasticity of its components. Approximate values for the modulus of elasticity E_{cm}, secant value between $\sigma_c = 0$ and $0.4 f_{cm}$, for concretes with quartzite aggregates, are given in Table 2.9. For limestone and sandstone aggregates, the value should be reduced by 10% and 30%, respectively. For basalt aggregates, the value should be increased by 20%. According to EC2, variation of the modulus of elasticity with time can be estimated as follows:

$$E_{cm}(t) = (f_{cm}(t)/f_{cm})^{0.3} E_{cm} \qquad (2.12)$$

where $E_{cm}(t)$ and $f_{cm}(t)$ are the values at an age of t days and E_{cm} and f_{cm} are the values determined at an age of 28 days. Poisson's ratio (v_c) may be taken equal to 0.2 for uncracked concrete and 0.0 for cracked concrete. Unless more accurate information is available, the linear coefficient of thermal expansion may be taken equal to 10×10^{-6} per K.

2.3.3 Creep and Shrinkage

Creep and shrinkage of the concrete depend on the ambient humidity, the dimensions of the element, and the composition of the concrete. Creep is also influenced by the maturity of the concrete when the load is first applied and depends on the duration and magnitude of the loading. According to EC2 [2.27, 2.28], the creep coefficient, $\varphi(t, t_0)$, is related to E_c, the tangent modulus, which may be taken as $1.05 E_{cm}$. The code provides charts for determining the creep coefficient, provided that the concrete is not subjected to a compressive stress greater than $0.45 f_{ck}(t_0)$ at an age t_0, the age of concrete at the time of loading. The creep deformation of concrete ε_{cc} (∞, t_0) at a time $t = \infty$ for a constant compressive stress σ_c applied at the concrete age t_0 is given by

$$\varepsilon_{cc}(\infty,t_0) = \varphi(\infty,t_0)(\sigma_c/E_c) \tag{2.13}$$

When the compressive stress of concrete at an age to exceed the value $0.45 f_{ck}(t_0)$, EC2 [2.27, 2.28] requires that creep nonlinearity should be considered. The high stress can occur as a result of pretensioning, for example, in precast concrete members at tendon level. In such cases, the nonlinear notional creep coefficient should be obtained as follows:

$$\varphi_{nl}(\infty,t_0) = \varphi(\infty,t_0) \exp[1.5(k_\sigma - 0.45)] \tag{2.14}$$

where $\varphi_{nl}(\infty,t_0)$ is the nonlinear notional creep coefficient, which replaces $\varphi(\infty,t_0)$; k_σ is the stress–strength ratio $\sigma_c/f_{ck}(t_0)$, where σ_c is the compressive stress; and $f_{ck}(t_0)$ is the characteristic concrete compressive strength at the time of loading. The values of the final creep coefficient $\varphi(\infty,t_0)$ are given in EC2 and are valid for ambient temperatures between $-40\,°C$ and $+40\,°C$ and a mean relative humidity between RH=40% and RH=100%. In determining $\varphi(\infty,t_0)$, t_0 is the age of the concrete at time of loading in days and h_0 is the notional size $= 2A_c/u$, where A is the concrete cross-sectional area and u is the perimeter of that part that is exposed to drying.

The total shrinkage strain, according to EC2, is composed of two components, the drying shrinkage strain and the autogenous shrinkage strain. The drying shrinkage strain develops slowly, since it is a function of the migration of the water through the hardened concrete. The autogenous shrinkage strain develops during hardening of the concrete, with the major part therefore developing in the early days after casting. Autogenous shrinkage is a linear function of the concrete strength. It should be considered specifically when new concrete is cast against hardened concrete. Hence, the values of the total shrinkage strain ε_{cs} can be calculated as follows:

$$\varepsilon_{cs} = \varepsilon_{cd} + \varepsilon_{ca} \tag{2.15}$$

where ε_{cs} is the total shrinkage strain, ε_{cd} is the drying shrinkage strain, and ε_{ca} is the autogenous shrinkage strain. The final value of the drying shrinkage strain $\varepsilon_{cd,\infty}$ is equal to $k_h \cdot \varepsilon_{cd,0}$. $\varepsilon_{cd,0}$ may be taken from Table 2.10. The development of the drying shrinkage strain in time can be calculated as follows:

$$\varepsilon_{cd}(t) = \beta_{ds}(t,t_s) k \varepsilon_{cd,0} \tag{2.16}$$

where k_h is a coefficient depending on the notional size h_0 according to Table 2.11:

$$\varepsilon_{cd}(t) = \frac{(t - t_s)}{(t - t_s) + 0.04\sqrt{h_0^3}} \tag{2.17}$$

where t is the age of the concrete at the moment considered in days and t_s is the age of the concrete (days) at the beginning of drying shrinkage (or swelling). Normally, this is at the end of curing; h_0 is the notional size (mm) of the

Table 2.10 Nominal Unrestrained Drying Shrinkage Values ε_{cd0} (‰) for Concrete with Cement CEM Class N Specified in EC2 [2.27]

$f_{ck}/f_{ck,cube}$	Relative Humidity (%)					
	20	40	60	80	90	100
20/25	0.62	0.58	0.49	0.30	0.17	0.00
40/50	0.48	0.46	0.38	0.24	0.13	0.00
60/75	0.38	0.36	0.30	0.19	0.10	0.00
80/95	0.30	0.28	0.24	0.15	0.08	0.00
90/105	0.27	0.25	0.21	0.13	0.07	0.00

Table 2.11 Values for k_h Specified in EC2 [2.27]

h_0	K_h
100	1.00
200	0.85
300	0.75
≥ 500	0.70

cross section $=2A_c/u$, where A_c is the concrete cross-sectional area and u is the perimeter of that part of the cross section, which is exposed to drying. The autogenous shrinkage strain can be calculated as follows:

$$\varepsilon_{ca}(t) = \beta_{as}(t)\varepsilon_{ca}(\infty) \tag{2.18}$$

where

$$\varepsilon_{ca}(\infty) = 2.5(f_{ck} - 10)10^{-6} \tag{2.19}$$

and

$$\beta_{as}(t) = 1 - \exp\left(-0.2t^{0.5}\right) \tag{2.20}$$

2.3.4 Stress-Strain Relation of Concrete for Nonlinear Structural Analysis

In nonlinear structural analysis and in nonlinear finite element modeling, concrete material should be carefully treated. In the absence of experimental data, design rules specified in current codes of practice can be adopted. As an example, EC2 [2.27] provides the relation between σ_c and ε_c shown in Figure 2.5 (compressive stress and shortening strain shown as absolute values and the use of $0.4\,f_{cm}$ for the definition of E_{cm} are approximates) for short-term uniaxial loading, which is described by the following expression:

$$\frac{\sigma_c}{f_{cm}} = \frac{k\eta - \eta^2}{1 + (k-2)\eta} \tag{2.21}$$

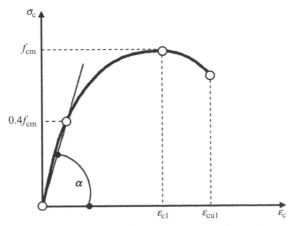

Figure 2.5 Schematic representation of the stress-strain relation for structural analysis specified in EC2 [2.27].

where $\eta = \varepsilon_c/\varepsilon_{c1}$, where ε_{c1} is the strain at peak stress according to Table 2.9, and $k = 1.05\, E_{cm} \times |\varepsilon_{c1}|/f_{cm}$, where f_{cm} is taken from Table 2.9. Expression (2.21) is valid for $0 < |\varepsilon_c| < \varepsilon_{cu1}$, where ε_{cu1} is the nominal ultimate strain.

According to EC2 [2.27, 2.28], the value of the design compressive strength is defined as follows:

$$f_{cd} = \alpha_{cc}\, f_{ck}/\gamma_C \qquad (2.22)$$

where γ_C is the partial safety factor for concrete and α_{cc} is the coefficient taking account of long-term effects on the compressive strength, which is recommended to be taken as 1.0. The value of the design tensile strength, f_{ctd}, is defined as follows:

$$f_{ctd} = \alpha_{ct}\, f_{ctk,0.05}/\gamma_C \qquad (2.23)$$

where α_{ct} is a coefficient taking account of long-term effects on the tensile strength, which is recommended to be taken as 1.0.

2.3.5 Stress-Strain Relations for the Design of Cross Sections

To design concrete cross section, simplified stress-strain curves can be adopted to ease hand calculations. As an example, for the design of cross sections using EC2, the following stress-strain relationship is recommended (see Figure 2.6) (compressive strain shown as positive):

$$\sigma_c = f_{cd}\left[1 - \left(1 - \frac{\varepsilon_c}{\varepsilon_{c2}}\right)^n\right] \quad \text{for } 0 \le \varepsilon_c \le \varepsilon_{c2} \qquad (2.24)$$

$$\sigma_c = f_{cd} \quad \text{for } \varepsilon_{c2} \le \varepsilon_c \le \varepsilon_{cu2} \qquad (2.25)$$

where n is the exponent according to Table 2.9, ε_{c2} is the strain at reaching the maximum strength according to Table 2.9, and ε_{cu2} is the ultimate strain according to Table 2.9. Other simplified stress-strain relationships may be used if equal to or more conservative than the nonlinear one, for instance, bilinear according to Figure 2.7 (compressive stress and shortening strain shown as absolute values) with values of ε_{c3} and ε_{cu3} according to Table 2.9.

A rectangular stress distribution (as given in Figure 2.8) may be assumed. The factor λ, defining the effective height of the compression zone, and the factor η, defining the effective strength, can be taken as follows:

$$\begin{aligned} \lambda &= 0.8 \quad \text{for } f_{ck} \le 50 \text{ Mpa} \\ \lambda &= 0.8 - (f_{ck} - 50)/400 \quad \text{for } 50 < f_{ck} \le 90 \text{ Mpa} \end{aligned} \qquad (2.26)$$

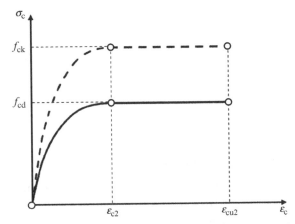

Figure 2.6 Parabola-rectangle diagram for concrete under compression specified in EC2 [2.27].

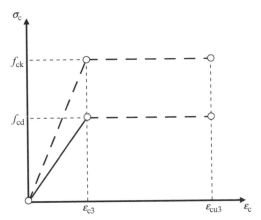

Figure 2.7 Bilinear stress-strain relation specified in EC2 [2.27].

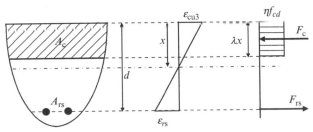

Figure 2.8 Rectangular stress distribution specified in EC2 [2.27].

and

$$\eta = 1.0 \quad \text{for } f_{ck} \leq 50 \text{ Mpa}$$
$$\eta = 1.0 - (f_{ck} - 50)/200 \quad \text{for } 50 < f_{ck} \leq 90 \text{ Mpa} \tag{2.27}$$

It should be noted that, according to EC2, if the width of the compression zone decreases in the direction of the extreme compression fiber, the value ηf_{cd} should be reduced by 10%.

2.3.6 Flexural Tensile Strength

The mean flexural tensile strength of reinforced concrete members depends on the mean axial tensile strength and the depth of the cross section. EC2 [2.27, 2.28] recommends the following relationship to be used in determining mean flexural tensile strength of reinforced concrete members:

$$f_{ctm,fl} = \max\{(1.6 - h/1000)f_{ctm}; f_{ctm}\} \tag{2.28}$$

where h is the total member depth in mm and f_{ctm} is the mean axial tensile strength following Table 2.9. The relation given earlier also applies for the characteristic tensile strength values.

2.3.7 Confined Concrete

In cases where concretes are surrounded by a stiffer material, such as concrete-filled steel tubular columns, the compressive strength and ductility of concrete are improved significantly. In this case, the concrete is called confined concrete, and depending on the yield stress and geometries of the surrounding stiffer material, the mechanical properties of this concrete improve considerably compared with unconfined concrete. To accurately model confined concrete, improved mechanical properties must be considered in nonlinear structural analysis and in nonlinear finite element modeling. Current codes of practice provide guidelines to account for concrete confinement. As an example, by adopting EC2 [2.27, 2.28], confinement of concrete results in a modification of the effective stress-strain relationship, achieving higher strength and higher critical strains. The other basic material characteristics may be considered as unaffected for design. In the absence of more precise data, the stress-strain relation shown in Figure 2.9 (compressive strain shown as positive) may be used, with increased characteristic strength and strains according to

$$f_{ck,c} = f_{ck}(1.0 + 5.0\sigma_2/f_{ck}) \quad \text{for } \sigma_2 \leq 0.05\, f_{ck} \tag{2.29}$$

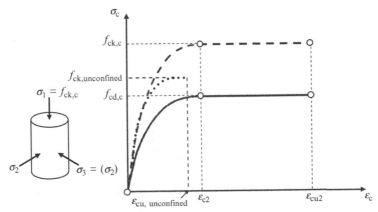

Figure 2.9 Stress-strain relationship for confined concrete specified in EC2 [2.27].

$$f_{ck,c} = f_{ck}(1.125 + 2.5\sigma_2/f_{ck}) \text{ for } \sigma_2 > 0.05\, f_{ck} \qquad (2.30)$$

$$\varepsilon_{c2,c} = \varepsilon_{c2}\left(f_{ck,c}/f_{ck}\right)^2 \qquad (2.31)$$

$$\varepsilon_{cu2,c} = \varepsilon_{cu2} + 0.2\sigma_2/f_{ck} \qquad (2.32)$$

where σ_2 ($=\sigma_3$) is the effective lateral compressive stress at the ultimate limit state due to confinement and ε_{c2} and ε_{cu2} follow Table 2.9. Confinement can be generated by adequately closed links or cross ties, which reach the plastic condition due to lateral extension of the concrete.

2.4 NONLINEAR MATERIAL PROPERTIES OF REINFORCEMENT BARS

2.4.1 General

The third main component of steel–concrete composite bridges is the reinforcement bars. Concrete slab decks used in steel–concrete composite bridges are strengthened with either reinforcement bars or prestressing tendons. In order to model the bridges accurately, it is quite important to understand the nonlinear material behavior of the reinforcement imbedded in the floor decks. Once again, when highlighting the nonlinear material behavior of the reinforcement bars, to use specified values recommended in current codes of practice to provide a consistent modeling approach is advisable. However, in the presence of detailed experimental investigations regarding the nonlinear material behavior of reinforcement bars, the experimental data can be also incorporated in designing and finite element modeling of the bridges.

In Europe, EC2 [2.27, 2.28] gives principles and rules for reinforcement, which is in the form of bars, decoiled rods, welded fabric, and lattice girders. They are not applicable to especially coated bars. One of the required properties of the reinforcement is that the material should be placed in the hardened concrete. The steels covered in EC2 are that in accordance with EN10080 [2.14]. The required properties of reinforcing steels shall be verified using the testing procedures in accordance with EN 10080. It should be noted that EN 10080 refers to a yield strength R_e, which relates to the characteristic, minimum and maximum values based on the long-term quality level of production. In contrast, f_{yk} is the characteristic yield stress based on only the reinforcement used in a particular structure. There is no direct relationship between f_{yk} and the characteristic R_e. However, the methods of evaluation and verification of yield strength given in EN 10080 provide a sufficient check for obtaining f_{yk}.

2.4.2 Properties

According to EC2 [2.27, 2.28], the behavior of reinforcing steel is specified by main properties comprising yield strength (f_{yk} or $f_{0.2k}$), maximum actual yield strength ($f_{y,max}$), tensile strength (f_t), ductility (ε_{uk} and f_t/f_{yk}), bendability, bond characteristics (f_R), section sizes and tolerances, fatigue strength, weldability, and shear and weld strength for welded fabric and lattice girders. EC2 [2.27, 2.28] applies to ribbed and weldable reinforcements, including fabric. The permitted welding methods are given in Table 2.12, while Table 2.13 gives the properties of reinforcement suitable for use with EC2. The properties are valid for temperatures between -40 and $100\,°C$ for the reinforcement in the finished structure. It should be noted that the values for the fatigue stress range with an upper limit of βf_{yk} are given in Table 2.14. The recommended value of β is 0.6. The rules for design and detailing specified in EC2 are valid for a specified yield strength range, $f_{yk} = 400\text{-}600$ MPa. The surface characteristics of ribbed bars shall be such to ensure adequate bond with the concrete. Adequate bond may be assumed by complying the specification of projected rib area, f_R, with minimum values of the relative rib area, f_R, given in Table 2.14. The yield strength f_{yk} (or the 0.2% proof stress, $f_{0.2k}$) and the tensile strength f_{tk} are defined, respectively, as the characteristic value of the yield load and the characteristic maximum load in direct axial tension, each divided by the nominal cross-sectional area. The reinforcement shall have adequate ductility as defined by the ratio of tensile strength

Table 2.12 Permitted Welding Processes and Examples of Application Specified in EC2 [2.27]

Loading Case	Welding Method	Bars in Tension[a]	Bars in Compression[b]
Predominantly static	Flash welding	Butt joint	
	Manual metal arc welding and metal arc welding with filling electrode	Butt joint with $\varphi \geq 20$ mm, splice, lap, cruciform joints[c]; joint with other steel members	
	Metal arc active welding[b]	Splice, lap, cruciform[c] joints; joint with other steel members	
		–	Butt joint with $\varphi \geq 20$ mm
	Friction welding	Butt joint and joint with other steels	
	Resistance spot welding	Lap joint[d] Cruciform joint[b,d]	
Not predominantly static	Flash welding	Butt joint	
	Manual metal arc welding	–	Butt joint with $\varphi \geq 14$ mm
	Metal arc active welding[b]	–	Butt joint with $\varphi \geq 14$ mm
	Resistance spot welding	Lap joint[d] Cruciform joint[b,d]	

[a]Only bars with approximately the same nominal diameter may be welded together.
[b]Permitted ratio of mixed diameter bars ≥ 0.57.
[c]For bearing joints $\varphi \geq 16$ mm.
[d]For bearing joints $\varphi \geq 28$ mm.

to the yield stress $(f_t/f_y)_k$ and the elongation at maximum force, ε_{uk}. Figure 2.10 shows stress-strain curves for typical hot-rolled and cold-worked steel. Values of $k = (f_t/f_y)_k$ and ε_{uk} for classes A, B, and C are shown in Table 2.13. Welding processes for reinforcing bars shall be in accordance with Table 2.12, and the weldability shall be in accordance with EN 10080 [2.14]. The strength of the welded joints along the anchorage length of welded fabric shall be sufficient to resist the design forces. The strength of the welded joints of welded fabric may be assumed to be adequate if each welded joint can withstand a shearing force not less than 25% of a force equivalent to the specified characteristic yield stress times the nominal cross-sectional area. This force should be based on

Table 2.13 Properties of Reinforcement Specified in EC2 [2.27]

Product Form	Bars and Decoiled Rods			Wire Fabrics			Requirement or Quantile Value (%)
Class	A	B	C	A	B	C	—
Characteristic yield strength f_{yk} or $f_{0.2k}$ (MPa)	400–600						5.0
Minimum value of $k=(f_t/f_y)_k$	≥1.05	≥1.08	≥1.15 ≥1.35	≥1.05	≥1.08	≥1.15 ≥1.35	10.0
Characteristic strain at maximum force, ε_{uk} (%)	≥2.50	≥5.0	≥7.5	≥2.5	≥5.0	≥7.5	10.0
Bendability	Bend/rebend test			—			
Shear strength	—			$0.3Af_{yk}$ (A is the area of wire)			Minimum
Maximum deviation from Nominal bar size (mm) Nominal mass (Individual bar or wire) (%)	≤8 ±6.0 >8 ±4.5						5.0

Table 2.14 Properties of reinforcement specified in EC2 [2.27]

Product Form Class	Bars and Decoiled Rods			Wire Fabrics			Requirement or Quantile Value (%)
	A	B	C	A	B	C	–
Fatigue stress range (MPa) (for $N \geq 2 \times 10^6$ cycles) with an upper limit of βf_{yk}	≥ 150			≥ 100			10.0
Bond	Nominal bar size (mm)						5.0
Minimum relative rib area, $f_{R,min}$	5–6			0.035			
	6.5–12			0.04			
	>12			0.056			

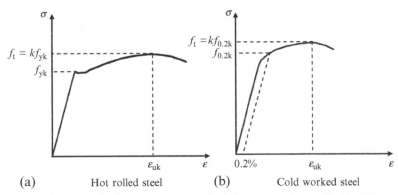

Figure 2.10 Stress-strain diagrams of typical reinforcing steel specified in EC2 [2.27].

(a) Hot rolled steel   (b) Cold worked steel

the area of the thicker wire if the two are different. EC2 specifies that where fatigue strength is required, it shall be verified in accordance with EN 10080 [2.14]. The design of concrete cross sections with reinforcement bars should be based on the nominal cross–sectional area of the reinforcement and the design values derived from the characteristic values given in EC2 [2.27, 2.28]. For normal design (see Figure 2.11), either an inclined top branch with a strain limit of ε_{ud} and a maximum stress of $k f_{yk}/\gamma_S$ at ε_{uk}, where $k = (f_t/f_y)_k$, or a horizontal top branch without the need to check the strain limit can be used. The recommended value specified in EC2 [2.27, 2.28] for ε_{ud} is $0.9\varepsilon_{uk}$. The mean value of density may be assumed to be 7850 kg/m^3. The design value of the modulus of elasticity, E_S, may be assumed to be 200 GPa.

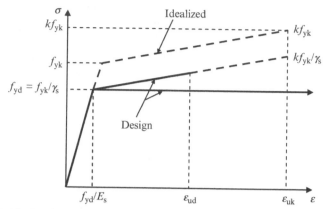

Figure 2.11 Idealized and design stress-strain diagrams for reinforcing steel (for tension and compression) specified in EC2 [2.27].

2.5 NONLINEAR MATERIAL PROPERTIES OF PRESTRESSING TENDONS

2.5.1 General

Prestressed concretes are commonly used in steel-concrete composite bridges. In this case, high-strength prestressing tendons are used to apply the initial stresses in concrete. To incorporate prestressed concrete decks in designing and finite element modeling of steel-concrete composite bridges, it is necessary to understand the nonlinear material behavior of the prestressing tendons imbedded in the floor decks. Once again, when highlighting the nonlinear material behavior of the reinforcement bars, it is decided to use specified values recommended in current codes of practice to provide a consistent modeling approach. However, in the presence of detailed experimental investigations regarding the nonlinear material behavior of prestressing tendons, the experimental data can be also incorporated into the designing and finite element modeling of the bridges. EC2 [2.27] also specifies rules that apply to wires, bars, and strands used as prestressing tendons in concrete structures. Prestressing tendons shall have an acceptably low level of susceptibility to stress corrosion. The level of susceptibility to stress corrosion may be assumed to be acceptably low if the prestressing tendons comply with the criteria specified in EN 10138 [2.30]. The requirements for the properties of the prestressing tendons are for the materials as placed in their final position in the structure, where the methods of production, testing, and attestation of conformity for prestressing tendons are in accordance with EN 10138 [2.30]. For steels complying with this Eurocode, tensile strength, 0.1% proof stress, and elongation at maximum load are specified in terms of characteristic values; these values are designated, respectively, f_{pk}, $f_{p0.1k}$, and ε_{uk}. It should be noted that EN 10138 [2.30] refers to the characteristic, minimum and maximum values based on the long-term quality level of production. In contrast, $f_{p0.1k}$ and f_{pk} are the characteristic proof stress and tensile strength based on only the prestressing steel required for the structure. There is no direct relationship between the two sets of values. However, the characteristic values for 0.1% proof force, $F_{p0.1k}$, divided by the cross-sectional area, S_n, given in EN 10138 together with the methods for evaluation and verification provide a sufficient check for obtaining the value of $f_{p0.1k}$. No welds in wires and bars are allowed. Individual wires of strands may contain staggered welds made only before cold drawing.

2.5.2 Properties

According to EC2 [2.27, 2.28], the properties of prestressing steel are given in EN 10138 [2.30]. The prestressing tendons (wires, strands, and bars) shall be classified according to strength, class, size, and surface characteristics. The strength denotes the value of the 0.1% proof stress ($f_{p0.1k}$) and the value of the ratio of tensile strength to proof strength ($f_{pk}/f_{p0.1k}$) and elongation at maximum load ε_{uk}. On the other hand, the class indicates the relaxation behavior. The actual mass of the prestressing tendons shall not differ from the nominal mass by more than the limits specified in EN 10138 [2.30]. EC2 specifies three classes of relaxation, which are class 1 (wire or strand with ordinary relaxation), class 2 (wire or strand with low relaxation), and class 3 (hot-rolled and processed bars). The design calculations for the losses due to relaxation of the prestressing steel should be based on the value of ρ_{1000}, the relaxation loss (in %) at 1000 h after tensioning and at a mean temperature of 20 °C. It should be noted that the value of ρ_{1000} is expressed as a percentage ratio of the initial stress and is obtained for an initial stress equal to $0.7f_p$, where f_p is the actual tensile strength of the prestressing steel samples. The values for ρ_{1000} can be either assumed equal to 8% for class 1, 2.5% for class 2, and 4% for class 3 or taken from the certificate. The relaxation loss may be obtained from the manufacturers' test certificates or defined as the percentage ratio of the variation of the prestressing stress over the initial prestressing stress, which should be determined by applying one of the expressions in the succeeding text. Expressions (2.33) and (2.34) apply to wires or strands for ordinary prestressing and low-relaxation tendons, respectively, whereas Expression (2.35) applies to hot-rolled and processed bars:

$$\text{Class 1} \frac{\Delta\sigma_{\text{pr}}}{\sigma_{\text{pi}}} = 5.39\rho_{1000}e^{6.7\mu}\left(\frac{t}{1000}\right)^{0.75(1-\mu)}10^{-5} \qquad (2.33)$$

$$\text{Class 2} \frac{\Delta\sigma_{\text{pr}}}{\sigma_{\text{pi}}} = 0.66\rho_{1000}e^{9.1\mu}\left(\frac{t}{1000}\right)^{0.75(1-\mu)}10^{-5} \qquad (2.34)$$

$$\text{Class 3} \frac{\Delta\sigma_{\text{pr}}}{\sigma_{\text{pi}}} = 1.98\rho_{1000}e^{8\mu}\left(\frac{t}{1000}\right)^{0.75(1-\mu)}10^{-5} \qquad (2.35)$$

where $\Delta\sigma_{\text{pr}}$ is the absolute value of the relaxation losses of the prestress, σ_{pi}; for posttensioning, σ_{pi} is the absolute value of the initial prestress $\sigma_{\text{pi}} = \sigma_{\text{pm0}}$ and, for pretensioning, is the maximum tensile stress applied to the tendon minus the immediate losses occurred during the stressing process; t is the time after tensioning (in hours); $\mu = \sigma_{\text{pi}}/f_{\text{pk}}$ where f_{pk} is the characteristic value of the tensile strength of the prestressing steel; and ρ_{1000} is the value of relaxation loss (in %), at 1000 h after tensioning and at a mean temperature of 20 °C. The long-term (final) values of the relaxation losses may be estimated for a time t equal to 500,000 h (i.e., around 57 years).

The 0.1% proof stress ($f_{\text{p0.1k}}$) and the specified value of the tensile strength (f_{pk}) are defined as the characteristic value of the 0.1% proof load and the characteristic maximum load in axial tension, respectively, divided by the nominal cross-sectional area as shown in Figure 2.12. According to EC2,

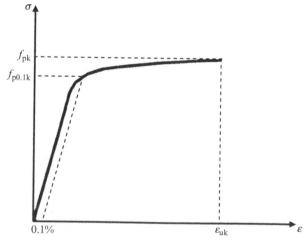

Figure 2.12 Stress-strain diagram for typical prestressing steel specified in EC2 [2.27].

the prestressing tendons shall have adequate ductility, as specified in EN 10138 [2.30]. Adequate ductility in elongation may be assumed if the prestressing tendons obtain the specified value of the elongation at maximum load given in EN 10138. Adequate ductility in tension may be assumed for the prestressing tendons if $f_{pk}/f_{p0.1k} \geq k$. The value of k specified in EC2 is 1.1. EC2 also specifies that prestressing tendons shall have adequate fatigue strength. The fatigue stress range for prestressing tendons shall be in accordance with EN 10138 [2.30]. Structural analysis is performed on the basis of the nominal cross-sectional area of the prestressing steel and the characteristic values $f_{p0.1k}$, f_{pk}, and ε_{uk}. The design value for the modulus of elasticity, E_P, may be assumed equal to 205 GPa for wires and bars. The actual value can range from 195 to 210 GPa, depending on the manufacturing process. The design value for the modulus of elasticity, E_P, may be assumed equal to 195 GPa for a strand. The actual value can range from 185 to 205 GPa, depending on the manufacturing process. The mean density of prestressing tendons for the purposes of design may normally be taken as 7850 kg/m^3. The values given earlier may be assumed to be valid within a temperature range between -40 and $+100\,°C$ for the prestressing steel in the finished structure. The design value for the steel stress, f_{pd}, is taken as $f_{p0.1k}/\gamma_s$ (see Figure 2.13). For cross-sectional design, either an inclined branch with a strain limit ε_{ud} or a horizontal top branch without strain limit can be utilized. The design may also be based on the actual stress-strain relationship, if this is known, with stress above the elastic limit reduced analogously (see Figure 2.11). It should be noted that, according to EC2, the recommended value for ε_{ud} is $0.9\varepsilon_{uk}$. If more accurate values are not known,

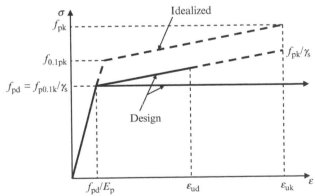

Figure 2.13 Idealized and design stress-strain diagrams for prestressing steel specified in EC2 [2.27].

the recommended values are $\varepsilon_{ud} = 0.02$ and $f_{p0.1k}/f_{pk} = 0.9$. Prestressing tendons in sheaths (e.g., bonded tendons in ducts and unbonded tendons) shall be adequately and permanently protected against corrosion. Prestressing tendons in sheaths shall be adequately protected against the effects of fire as specified in EC2 (BS EN 1992-1-2) [2.31].

2.6 NONLINEAR BEHAVIOR OF SHEAR CONNECTION

2.6.1 General

Steel-concrete composite construction is used extensively in highway bridges owing to its advantages in terms of saving in weight of steel, high strength, high stiffness, high resistance to seismic and cyclic loading, increasing load capacity, better fire resistance, and reduction in construction depth. In composite beam design, shear connectors are commonly used to transfer longitudinal shear forces across the steel-concrete interface. The shear strength of the connector and the resistance of the concrete slab against longitudinal cracking are the main factors affecting the shear stiffness and strength of the shear connection. Calculation of the structural behavior of composite beams depends on how much slip is assumed to occur at the interface between concrete and steel. Experimental push-off tests are the traditional source of knowledge about the load-slip behavior and the shear capacity of the shear stud in composite beams.

Up to the early 1950s, steel beams were designed to act as composite beams with solid concrete slabs of various thickness, connected to them using a variety of types of mechanical shear connectors. However, during this period, composite construction in buildings was generally uneconomical. This was due to the significant amount of formwork and propping required for the concrete slabs, along with the costly process in terms of time of having the shear connectors welded to the steel beams. In a composite steel-concrete beam, the floor slab tends to slide along the flange of the steel beam and the importance of the shear connectors arises from preventing this slippage. The structural behavior of a composite beam is affected by the slip at the steel-slab interface. Practically, the assumption that this slip may be completely eliminated cannot be ensured. Therefore, accurate calculation methods of the structural behavior of composite girders must take into consideration the effects of this slip. Push-off tests provide a convenient way to study the behavior of shear connector without carrying an expensive full bending test. Initially, the evaluation of the shear capacity of connectors was the main output of these tests. After that, researchers realized that the load-slip behavior of the connector was also of equal importance. The slip at the steel-concrete interface depends on many factors such as type of connector, size of shear connector, spacing between connectors, type of floor slab, and concrete strength of slab. To evaluate the load-slip behavior of the connector taking into account all parameters that affect the shear connection, an unlimited number of expensive push-off tests would need to be carried out. Numerical modeling of push-off tests can be used in carrying out extensive parametric studies to evaluate the load-slip behavior of the shear connector.

The development of the electric drawn arc stud welding apparatus in 1954 allowed a type of shear connector known as the headed stud connector (see Figure 2.14) to be rapidly fastened to the top flange of the steel beams *in situ*. Due to its advantages over other forms of shear connection, such as rapid installation and the fact that they were equally strong and stiff in shear in all directions normal to the axis of the stud, the stud connector became one of the most popular types of connector used in composite construction. Studies of stud connectors did not begin until 1956. Push-off tests on stud connectors were first carried out by Viest [2.32]. The study used straight studs with an upset head of diameter ranging from 0.5 (13 mm) to 1.25 in. (32 mm). Fatigue and static tests were also performed by Thurlimann [2.33]. These push-off studies used 0.5 in. (13 mm) diameter bent studs and, to a lesser extent, 0.75 in. (19 mm) diameter straight studs with an upset head. All the push-off tests showed that steel studs are suitable for use as shear connectors and that the behavior of a stud connector is similar to that of a flexible channel connector. The shear capacity was found to be a function of the diameter and height of the stud and of the strength of the concrete.

During the latter half of the 1950s, profiled steel sheeting (decking) (see Figure 2.15) was introduced in the North American steel construction

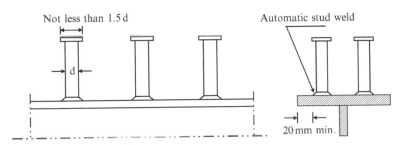

Figure 2.14 Headed stud shear connector.

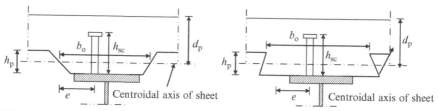

Figure 2.15 Composite beam with profiled steel decking spanning the same direction.

market that eliminated the use of traditional timber temporary forms. Initially, the new decking system served only as a replacement for the timber formwork due to its advantages of serving as a working platform to support the construction loads, as well as a permanent formwork for the concrete. Once the sheets had their surfaces suitably embossed with small indentations to ensure reliable bond with the concrete, it became an integral structural element of the slab by providing all or a part of the main tension reinforcement, and it was eventually incorporated into the overall composite floor and framing system. Since the decking created a barrier between the concrete slab and the steel beam, holes were initially cut or punched into the deck for the welding of the stud shear connectors, but it soon became possible to weld these connectors through the decking. The disadvantages of this form of construction were the operation and cost of welding the connectors through the decking on site, the limitations to maximum spans to about 3.5 m without propping, and the addition of framing, and a "wet trade" is involved in pouring the concrete floor that prevents a dry construction environment.

The use of composite construction can be seen now between steel beams and different concrete slabs. As an example, the use of prestressed hollow core concrete slabs in conjunction with steel beams to provide composite action is a new form of construction. In this construction, the prestressed hollow core concrete units are placed on the top of the steel beam as shown in Figure 2.16. Tie steel is placed on site into the slots made at the top of the hollow cores, which are filled with grade C25 (minimum) *in situ* concrete. The slab rests directly on the top of the flange of the steel beam as shown in Figure 2.16, and shear connectors are used to ensure the composite action between the prestressed hollow core concrete floor and the steel beam. Figure 2.17 shows the details of the precast *in situ* joint in a composite beam with hollow core concrete units. The longitudinal and transverse joints between the hollow core concrete units are filled with *in situ* concrete so that horizontal compressive membrane forces can be transferred through

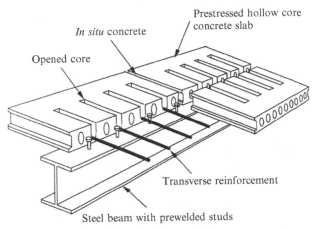

Figure 2.16 Prestressed hollow core concrete-steel beam construction.

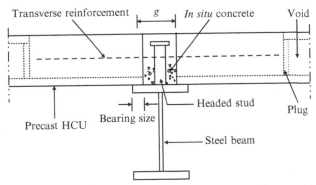

Figure 2.17 Details of the prestressed *in situ* joint of composite beam with hollow core concrete units.

the slab. Composite construction incorporating prestressed hollow core concrete units is intended to complement the traditional composite construction with steel decking and to offer advantages where, for reasons of design or environmental considerations, a steel decking system may be precluded. The main advantages of this form of construction are that the precast concrete slabs can span up to 15 m without propping, the erection of 1.2 m wide precast concrete units is simple, and quick and shear studs are prewelded on beams before delivery to site, thereby offering additional savings associated with shorter construction times.

2.6.2 Shear Connectors

In steel-to-concrete composite construction, longitudinal shear forces are transferred across the steel-concrete interface by the mechanical action of the shear connectors. The problem associated with this connection is that it is a region of severe and complex stress. The methods of connection have been developed empirically and verified by tests. These tests show that at low loads, most of the longitudinal shear is transferred by bond at the interface. This bond breaks down at higher loads, and once broken, it cannot be restored. So, in design calculations, bond strength is taken as zero. Also, greasing the steel flange before the concrete is cast destroys the bond between it and the concrete slab. The design of the connectors to ensure an adequate degree of interaction was first specified in CP 117 [2.34, 2.35]. There are numerous types of shear connectors available in the market such as channels, bent bars, and T-sections. However, the most widely used type of connector is the headed stud shown in Figure 2.14. The British Standards [2.36] require the steel from which the studs are manufactured to have an ultimate tensile strength of at least 450 N/mm^2 and an elongation of at least 15%. The advantages of stud connectors are that the welding process is rapid, they provide little obstruction to reinforcement in the concrete slab, and they are equally strong and stiff in shear in all directions normal to the axis of the stud. The property of a shear connector most relevant to design is the relationship between the longitudinal shear force transmitted, P, and the slip at the interface, δ. This load-slip curve should ideally be found from tests on composite beams. However, most of the data on connectors have been obtained from various types of "push-off" test. The flanges of a short length of steel I-section are connected to two small concrete slabs. The details of the "standard push-off test" given in Eurocode 4 (EC4) [2.37] are shown in Figure 2.18. The slabs are bedded onto the lower platen of a compression-testing machine or frame, and the load is applied to the upper end of the steel section. The slip between the steel member and the two slabs is measured at several points, and the average slip is plotted against the load per connector. The push-off test must be specified in detail for the load-slip relationship, which is influenced by many variables. The variables include the number of connectors in the test specimen; mean longitudinal stress in the concrete slab surrounding the connectors; size, arrangement, and strength of slab reinforcement; thickness of concrete surrounding the connectors: freedom of the base of each slab to move laterally; bond at the steel-concrete interface; strength of the concrete slab; and degree of compaction of the concrete surrounding the base of each connector. The amount of

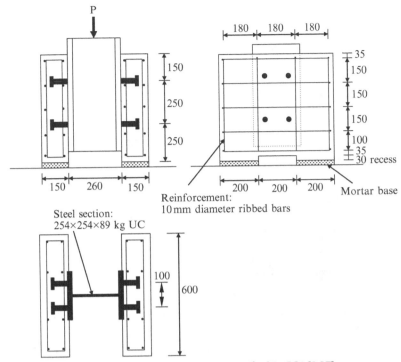

Figure 2.18 Standard push-off test specimen specified in EC4 [2.37].

reinforcement specified and the size of the slabs are greater than that of the earlier British Standard push–off test [2.38] shown in Figure 2.19. EC 4 [2.37] gives results that are less influenced by splitting of the slabs and so give better predictions of the behavior of connectors in beams as mentioned by Johnson [2.39, 2.40].

2.6.3 Complete and Partial Shear Concoction

The connection between the steel beam and the concrete slab is called "complete" in the sense that the slip and uplift at the interface of the two elements are negligible. Shear connection in composite beams is identified [2.41, 2.42] as complete when the beam has a bending strength that would not be increased by the addition of further connectors. On the other hand, the connection between steel beam and concrete is called partial when fewer connectors are used than are required for the complete shear connection. The term "partial connection" should not be considered to imply

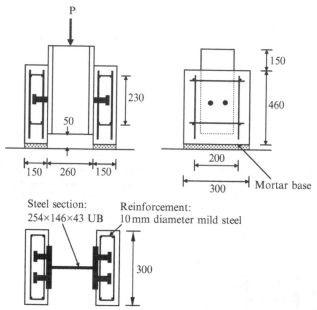

Figure 2.19 Standard push-off test specimen according to BS 5400 [2.38].

unsatisfactory shear connectors but rather a connection resulting in nonne-gligible slip at the steel beam-concrete slab interface. This slip has a great influence on both the strength and the deformations of the composite beam. Significant contributions have been made in this scientific area initiated by [2.43] where the influence of the slip on the ultimate plastic strength of the composite beam has been studied. The accurate analysis of the behavior of composite beams with partial connection is very important since the slip between steel and concrete may be big enough to cause fracture of some connectors at a serviceability state. Appropriate ductility of the shear con-nectors is the only way to sustain the likely big slip deformations without fracture. Since the main objective of this book is to accurately model the behavior of steel-concrete composite bridges, the correct shear connection behavior will be incorporated in the finite element models.

2.6.4 Main Investigations on Shear Connection in Composite Beams with Solid Slabs

Davies [2.44] showed that the ultimate capacity of a stud connector in a push-off test depends on a large extent upon the pattern and spacing of the connectors. It was observed that if the studs were oriented parallel to the direction of the load, its ultimate capacity was reduced. Also, a decrease in the longitudinal stud spacing resulted in a decrease in ultimate strength. A further study by Davies [2.45] showed that when transverse reinforcement is provided in a solid concrete slab, the cracking resistance of the slab is improved. The longitudinal cracks only develop when the yield stress of the reinforcement is reached. Therefore, a certain minimum amount of transverse reinforcement has to be used, to achieve the maximum load-carrying capacity of a composite beam. Johnson [2.39] found that the concrete strength influences the mode of failure of shear connection between steel and concrete, as well as the failure load. Menzies [2.46] compared the strengths of shear connectors given in CP 117 part 1 [2.34] and part 2 [2.35] with his results of push-off tests. It was found that CP 117 part 1 assumes linear relationships between the static strength of shear connectors and the concrete strength and CP 117 part 2 assumes that the variation of fatigue strength of stud connectors with concrete strength was inside limited range of concrete strengths. Therefore, the author conducted an experimental investigation comprising 34 push-off specimens to investigate the effect of concrete strength and density on the static and fatigue strength of connectors. The investigation was carried out over a wide range of concrete strengths. Different types of connectors, studs, channels, and bars were used, and the maximum load per connector, the mode of failure, and the slab in which the failure occurred were given. The maximum static loads per connector were plotted against the compressive strength of both water-stored and air-stored concrete cubes. The slip in the static tests, that is, the vertical movement of the slab relative to the steel beam, was plotted for each specimen against the load. The maximum and minimum values of the cyclic load per connector, the fatigue life, and the mode of failure were given. It was concluded that a modification of specified strengths of shear connector given in CP 117 is desirable for a larger range of concrete strengths; a distinction should be made in design between connectors embedded in normal-density concrete and lightweight concrete; the static strengths of studs in normal-density concrete are overestimated in CP 117; when the density of the lightweight concrete is below 1400 kg/m^3, there may be difficulty in ensuring adequate connection strength and an adequate degree of interaction in a composite beam; and the specification in CP 117 of the fatigue strength of stud connectors based on percentages of static strength is confirmed when embedded in normal-density concrete.

Jayas and Hosain [2.47] conducted tests on 18 full-size push-off specimens and four pull-out specimens. The objective of the project was mainly to study the behavior of headed studs in composite beams with ribbed metal decks perpendicular and parallel to the steel beam, but five of the push-off specimens had solid concrete slabs. These five push-off specimens were similar to those tested by Ollgaard *et al.* [2.48]. The used stud has a diameter of 16 mm and a height of 76 mm. They found that when the studs were spaced sufficiently far apart, the mode of failure is likely to be because of the shearing off of studs. On the other hand, concrete failure was observed in specimens when the studs were closely spaced (longitudinal spacing less than 6 × the stud diameter), and this led to a reduction in the stud strength by 7%. Oehlers [2.49] investigated the longitudinal shear flow in composite steel-concrete beams across the steel flange/concrete slab interface by the action of individual connectors. The authors have shown that shear connectors in steel-concrete composite beams act as steel dowels embedded in a concrete medium. These shear connectors are generally assumed to fail when the steel component fractures, which may be a consequence of the gradual reduction in strength and stiffness of the concrete in the bearing zone of high triaxial compressive stress (Oehlers and Johnson 1987 [2.50]). Oehlers [2.49] showed that the concentrated load *P* that a connector applies to a slab can induce three distinct modes of cracking of the slab shown in Figure 2.20. The modes of cracking comprise lateral, shear, and splitting cracks. The lateral cracks are the cracks extending from the sides of the connector and caused by the ripping action of the concentrated load on the slab. These cracks are assumed to have little effect on the connector strength since they occur away from the high triaxial compression bearing zone. The shear cracks occur near the compressive zone and hence may affect the triaxial restraint. Finally, the splitting cracks occur in front of the triaxial

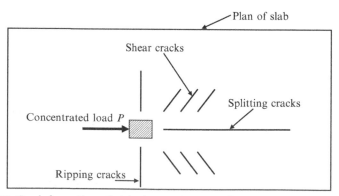

Figure 2.20 Crack formation of slabs in composite girders.

compression zone due to the concentrated load induced by the shear connector. These large lateral tensile stresses propagate and induce splitting behind the shear connector and also relieve the triaxial restraint to the bearing zone leading to connector failure through compressive failure of concrete. Oehlers [2.49] found that splitting cracks reduce the strength of the shear connection to less than 20% of its theoretical shear connector strength. Also, fully anchored transverse reinforcement placed in front of a heavily loaded single connector did not increase the splitting strength of the slab nor increase the strength after splitting. However, the transverse reinforcement was found to limit the strength of the split and allow a general gradual reduction in the shear load after splitting.

Oehlers and Park [2.51] found that composite steel–concrete connections that incorporate a haunch are prone to splitting failure. This is because the shear connectors have to transfer high concentrations of load into the concrete slab in the region of the haunch where the side cover to the connectors is limited. The experimental tests were on stud shear connectors encased in haunches with sloping sides. A similar study, reported by Johnson [2.52], determined the splitting resistance for haunches with vertical sides and obtained the load–slip curves of the connectors for different haunch slopes. It was concluded that these results can be used to design composite slabs made with steel decking when the ribs of the steel decking are parallel to the steel section of composite beam. Push-off tests on studs in high-strength and normal-strength concrete have been carried out by Li and Krister [2.53]. Eight push-off specimens were divided into four pairs, according to the concrete strength and the amount of reinforcement in concrete slabs. The authors found that the concrete compressive strength significantly affects the strength of the shear connections. The increase in the maximum shear load was about 34% when the cylinder compressive strength of the concrete increased from 30 to 81 MPa. The tests showed that the amount of slip at the maximum load was of the same level for both normal-strength and high-strength concrete. However, ductile behavior of the studs was observed for the normal-strength specimen after the maximum load. The descending branch of the load–slip curves for the high-strength concrete was short and steep. The reinforcement in the concrete slabs is a negligible influence on the capacity of normal-strength concrete (the increase was about 6%) but confined the concrete surrounding the studs. A negligible effect of reinforcement on the capacity of the shear connection was observed in high-strength concrete specimen.

2.6.5 Main Investigations on Shear Connection in Composite Beams with Profiled Steel Decking

Where profiled sheeting is used, stud connectors are located within concrete ribs that have a haunch shape. The sheeting normally run either transverse or parallel with the span of the beam. There are different parameters [2.39, 2.54] that affect the behavior of stud in composite beams with profiled sheeting in addition to geometrical data shown in Figure 2.15. The parameters comprise the compressive strength and density of concrete; ultimate tensile strength of stud; location of stud within the concrete rib, in relation to the direction of sheeting; shape of the steel profile, whether the studs are welded through it or through holes in it; and size, spacing, and level of any reinforcement in the slab. Hawkins and Mitchell [2.55] conducted 13 push-off tests to study the behavior of headed stud shear connectors in composite beams with profiled steel sheeting perpendicular to the beam. The diameter of the stud was 19 mm, the profiled sheeting depths were 38 and 76 mm, and the profiled sheeting widths ranged from 38 to 127 mm. Four different failure modes were observed by Hawkins and Mitchell [2.55]. The failure modes were stud shearing, concrete pull-out, rib shearing, and rib punching. Jayas and Hosain [2.47] conducted 18 push-off tests on full-size specimens having ribbed metal decks perpendicular and parallel to the steel beam. The 16 mm diameter × 75 mm height headed studs that were welded through profiled steel sheeting having a depth of 38 mm and a width ranged from 53.8 to 165.9 mm were used in most of the tests. The main parameters studied were longitudinal spacing of the headed studs and rib geometry. Jayas and Hosain [2.47] provided two separate empirical equations to calculate the shear capacity of headed shear stud in composite beams with profiled steel sheeting having depths of 38 and 76 mm. Jayas and Hosain [2.56] conducted two full-size push-off tests on composite beams. The profiled steel sheeting was placed perpendicular to the steel beam. The 19 × 127 mm headed stud was used in the tests, and the profiled steel sheeting had a depth of 76 mm and widths of 144 and 225 mm. The failure modes observed were concrete pull-out and mixed concrete pull-out and stud shearing. Lloyd and Wright [2.57] carried out 42 push-off tests on headed studs welded through-deck. The investigation focused on the amount and position of reinforcement and dimensions of the composite slab. The authors concluded that the capacity of shear connection in composite beams with profiled steel sheeting depends upon the geometry of the sheeting and stud height. It is also concluded that the capacity of shear connection is considerably less than that in solid slabs. Kim et al. [2.58, 2.59] conducted three push-off tests to study the behavior of through–deck–welded shear connectors. The headed stud used in the tests was 13×65 mm and the profiled steel sheeting had a depth of 38 mm. Kim *et al.* [2.58, 2.59] discussed concrete pull–out failure surface area and strength. The major failure modes found in the tests were concrete pull–out failure and local concrete crushing around the foot of the stud.

2.6.6 Main Investigations on Shear Connection in Composite Beams with Prestressed Hollow Core Concrete Slabs

Moy and Tayler [2.60] carried out 27 push–off tests to evaluate the shear strength of headed studs in solid prestressed concrete planks. The prestressed planks were used as permanent shutters for the *in situ* concrete. The planks had a depth of 65 mm and a bearing width of 50 mm on the steel beam flange. A 533 mm depth by 210 mm width by 92 kg steel beam was used with two studs welded on each flange. The diameter of the headed stud used was 19 mm, and different lengths ranging from 95 to 120 mm were used. The rest of the 150 mm depth of the slab was made with *in situ* concrete. Figure 2.21 shows the details of Moy and Tayler [2.60] push–off test specimen. A typical load–slip curve of the 19 mm stud was obtained, and the

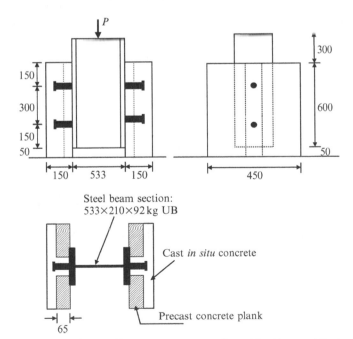

Figure 2.21 Details of push-off test specimen conducted by Moy and Tayler [2.60].

results showed a reduction in strength of connection as the volume of *in situ* concrete decreases. It was recommended that the width of the *in situ* concrete on the flange be a minimum of 100 mm. It was also recommended that two layers of reinforcement must be used in the slab to avoid concrete splitting.

Push-off tests on headed studs in precast HC slabs were reported by Lam *et al.* [2.61]. The authors carried out 12 full-scale push-off tests (10 tests on headed studs used with tapered-end precast HCU slabs and 2 on headed studs used with reinforced solid concrete slabs). The tests were carried out horizontally as shown in Figure 2.22 with the same cross section shown in Figure 2.17. The tests were carried out for different gap sizes "*g*" (40, 65, and 120 mm) between the ends of the precast slabs. Also, different transverse reinforcement sizes (8, 16, and 25 mm) were used. Two of the 10 tests consisted of two 1200 mm wide × 150 mm deep HCUs, whereas others consisted of four 600 mm wide × 150 mm deep hollow core units. The units were connected to grade 43 steel 356 × 171 UB with prewelded headed studs at 150 mm centers. Milled slots approximately 500 mm long were made in the second cores from the edges of the units. The characteristic cube strength for the precast concrete was taken as 50 N/mm². All studs were 19 mm diameter × 125 mm height (TRW–Nelson headed studs). The authors found that the capacity of the stud is reduced compared with that in a solid reinforced concrete slab. A reduction formula for the precast effect

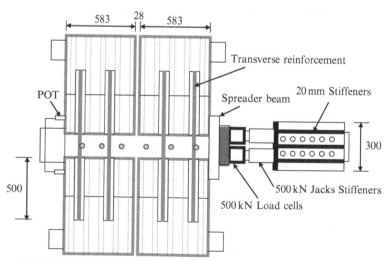

Figure 2.22 Horizontal push-off testing as carried by Lam *et al.* [2.61].

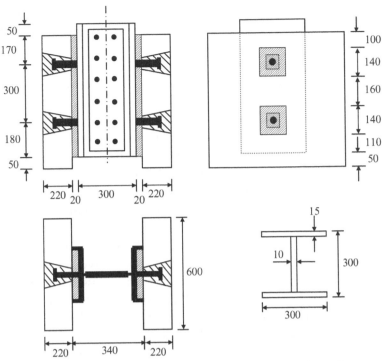

Figure 2.23 Details of push-off test specimen used by Shim *et al.* [2.62].

was derived, and the load–slip curves of the studs were plotted for the 12 push-off tests.

The behavior of shear connections in a composite beam with a full-depth prestressed slab was investigated by Shim *et al.* [2.62]. Eighteen push-off tests were performed with variations of the stud shank diameter and the compressive strength of the mortar. Figure 2.23 shows the push-off test specimen used by the authors to evaluate the shear stud properties in a composite beam with a full-depth precast slab. Different stud diameters (13, 16, 19, and 22 mm) were used with a stud height of 150 mm. The push-off tests were similar to BS 5400 [2.38], but it had shear pockets for stud shear connectors and a bedding mortar layer 20 mm thick between the precast slab and the steel beam. The load–slip curves of the stud were obtained, and the relationship between the shear stud capacity and the stud diameters was plotted. From the experimental work, the authors observed that the deformations of the stud in a full-depth precast slab were greater than its deformations in a cast in place slab. The static strengths of the shear connections agree approximately with those evaluated from the tensile strength of the stud shear connectors. Also, an empirical equation for the initial shear stiffness of a shear connection was proposed.

Nip and Lam [2.63] investigated the effect of end conditions of hollow core slabs on longitudinal shear capacity of composite beams. The published work was an extension for Lam *et al.* [2.61] and mainly concerned about push–off tests with precast hollow core concrete slab of square ends (see Figures 2.16 and 2.17). Eighteen push–off tests (12 push–off tests with precast hollow core slab of square ends, 2 push–off tests with precast hollow core slab of tapered ends, and 4 push–off tests with solid slab) were carried out by the authors. The same horizontal push–off testing approach used by Lam *et al.* [2.61] was used. The headed studs used had 19 mm diameter and 100 mm height. The precast floor specimens consisted of four 600 mm wide hollow core units connected to a $254 \times 254 \times 73$ UC. Each beam had six prewelded studs at 150 mm centers. The effects of transverse reinforcement size, gap width, and *in situ* concrete strength were discussed by the authors. The authors concluded that 100 mm high headed studs with square-end hollow core slabs performed as well as the 125 mm high headed studs with tapered-end hollow core slabs. It is also concluded that the optimum *in situ* gap width that should be used for square-end hollow core slab is 80 mm and 16 mm diameter high-tensile bars are recommended to be used as transverse reinforcement to ensure a slip ductility of 6 mm at the maximum load.

2.6.7 Main Investigations on Numerical Modeling of Shear Connection

Finite element modeling could provide a good insight into the behavior of shear connection and compensate the lack in the experimental data. Nethercot [2.64] highlighted the importance of combining experimental and numerical study in advancing structural engineering understanding. The author mentioned that there is a lack in the detailed numerical studies dealing with the behavior of the individual connector. It is also mentioned that the absence of experimental/numerical approach means that real understanding is lacking and design expressions are very limited. Limited numerical models have been found in the literature for push–off tests with different slabs. Initially, Johnson and Oehlers [2.65] used a simplified purpose-written program, developed originally by Oehlers [2.66], in their parametric study to predict the shank failure loads of headed stud in steel-solid slab push–off test and the influence of weld collar on forces acting on the stud. The program performed a step-by-step plane stress elastic analysis using triangular finite elements. Initially, one stud and one slab of a push-off test specimen were modeled. The stud shank and the weld collar were assumed to be of square cross section. The program took into account two types of local failure, cracking due to tensile stress and cracking due to tensile strain caused by normal compressive stress. The program was used in two modes: linear elastic analyses of isotropic materials with and without the provision for concrete to fail in tension. In both modes, the shear connection was assumed to have reached its maximum strength when the maximum stress in the shank of the stud reached the measured ultimate tensile strength of the shank. The failure of concrete in compression was not modeled in this study. The authors found that a weld collar less than 5 mm high attracts 70% of the total shear and reduces the bending moment at the base of the stud to one-third of the value found for a stud without a collar.

The inelastic behavior of shear connections was investigated by Kalfas et al. [2.67]. The authors used the finite element method to model the behavior of shear connectors in a steel-solid slab push-off test. The results were compared with a series of push-off tests performed in the steel-structures laboratory of Democritus University of Thrace. The model simulated the linear and nonlinear behavior of the materials (bilinear stress-strain curves were used for concrete and headed stud shear connector). The three components of a push-off test were simulated with different types of standard finite elements. The concrete slab was modeled by nonlinear volume elements, the steel beam by a rigid bar element, and the shear connectors by nonlinear beam elements as shown in Figure 2.24. The finite element package COSMOS was used in the analysis. The load-slip curve obtained from the finite element solution was compared with experimental results, and the maximum deviation between the results was about 14%. Although the concrete slab was modeled with 1920 elements, the results obtained were inaccurate and this may be attributed to the incompatibility of the FE elements used. The predicted shear stud capacity is considerably higher than that tabulated in BS 5950 and Eurocode 4, and the mode of failure was not investigated. The authors suggested to improve the material models by including the strain hardening of steel, contribution of concrete rebars, and small tensile branch of the stress-strain diagram of concrete to reduce the deviation between experimental and FE solution. The behavior of headed shear stud connector in a steel-full-depth precast slab push-off test was modeled numerically by Shim et al. [2.62] using the finite element method. The push-off test specimen shown in Figure 2.23, previously discussed, was constructed using the finite element method. The purpose of the model was to

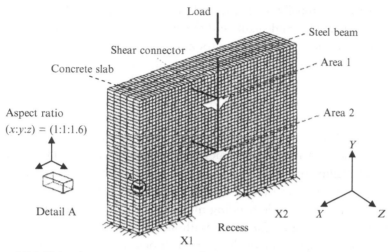

Figure 2.24 Finite element representation of push-off test specimen modeled by Kalfas *et al.* [2.67].

investigate the initial stress distribution of the shear stud connector in the push-off test under consideration. The distributions of flexural and shear stresses along the stud shank were given. The stresses were concentrated around the root of the stud shank below a height of 20 mm. The authors found that the flexural deformation of the stud shear connector was greater than that in the case of cast in place slabs, which can resist the splitting force better through adequate reinforcement. The study was based on linear elastic material properties to investigate initial stresses only. The load-slip curves of the stud, shear stud capacity, and modes of failure were not obtained from this finite element study.

Ellobody [2.68] and Lam and Ellobody [2.69] developed an accurate nonlinear finite element model to investigate the behavior of headed shear stud connector in solid slabs. The results obtained from the finite element analysis compared well with the experimental results conducted by Ellobody [2.68] and Lam and Ellobody [2.69]. The capacity of the shear connection, the load-slip behavior of the headed studs, and the failure modes were accurately predicted by the finite element model. A parametric study was conducted to investigate the effects of the change in headed stud diameter and height and concrete slab strength. The results of the finite element model were compared with the American, British, and European specifications for steel-concrete composite structures. It was concluded that the European code provides good agreement with experimental and finite element results, while the American and British specifications overestimated the resistance of headed stud shear connectors in composite beams with solid slabs. Ellobody [2.68] and Ellobody and Lam [2.70] extended the research to investigate the behavior of headed shear stud connector in composite beams with precast hollow core slabs. An accurate finite element model was developed, and the results obtained from the finite element analysis compared well with experimental results conducted by Ellobody [2.68] and Ellobody and Lam [2.70].

The performance of headed stud shear connectors in composite beams with profiled steel sheeting was investigated by Ellobody and Young [2.71]. The authors developed an efficient nonlinear three-dimensional finite element model to investigate the behavior of headed stud shear connectors in composite beams with profiled steel sheeting perpendicular to the steel beam. The finite element program ABAQUS [1.29] was used in the analysis. The results obtained from the finite element analysis were verified against the test results carried out by Lloyd and Wright [2.57] and Kim *et al.* [2.58, 2.59]. Parametric studies were performed to investigate the effects of the changes in profiled steel sheeting geometries, diameter and height of headed shear stud, concrete slab dimensions, and strength of concrete on the strength and behavior of shear connection in composite beams with profiled steel sheeting. The results obtained from the finite element analysis were compared with design strengths calculated using current codes of practice for headed stud shear connectors in composite slabs with profiled steel sheeting perpendicular to the steel beam.

2.6.8 Main Investigations on Numerical Modeling of Composite Girders

Many experimental and numerical investigations have been carried out to investigate the structural behavior of steel-concrete composite girders. Most of these investigations were concentrated on steel-solid concrete slab girders composite beams and limited investigations focused on composite girders with prestressed hollow core slabs and composite slabs with profiled steel sheeting. There is no intention to survey these investigations in general in this book. However, this chapter is concentrated on how researchers numerically simulated different components of composite girders including the shear connection between these components. Also, investigations dealing with the evaluation of the effective width, ultimate load capacity, and load–deflection curves of composite girders are also highlighted in this book.

Earlier investigations by Ansourian [2.72] used the finite element method in analyzing composite steel–solid slab floor systems. The author studied the full composite action between concrete slabs and steel joists in the elastic range. The principal variable of the study was the ratio of the flexural stiffness of the joist and slab. Two different methods of finite element models were investigated. In the first model, the slab was represented by a combination of 16-node and 8-node solid 3D prismatic elements and the joist was represented by plane stress elements. At a section of the structure, each steel beam flange was represented by one element and the web was represented by two elements. The authors found satisfactory convergence with this method, but the preparation of data and analysis of output were time-consuming. In the second model, the slab was represented by thin plate elements built up from the superposition of linear curvature triangles for the flexural stresses and linear strain triangles for the membrane stresses, and the joist was represented by beam elements. The author found that the second model gives applicable output of bending moments and plane stresses. Mofatt and Dowling [2.73] introduced a finite element study of the elastic longitudinal bending behavior of composite box girder bridges in which the use of flexible shear connectors results in incomplete interaction between the slab and girder components. The buckling and the inelastic behavior were not considered. Some preliminary investigations were carried out to determine suitable finite element meshes for use in analyzing the girders. The box girders, the slab, and the reinforcement were represented by shell elements. The shear connectors were represented by linkage elements that allowed slip in the plane of the concrete–steel interface. It was assumed that the shear connection would be provided by 19 mm × 100 mm headed studs. On the basis of information given in reference [2.74], the authors used constant slip modulus of the studs as 0.4×10^6 N/mm. The authors highlighted the need for codes of practice to include design information on the stiffness and distribution requirements of the connectors in composite girders.

Mistakidis et al. [2.75] introduced a numerical method taking into consideration the nonlinearities introduced in the analysis of long-span composite girders. The authors showed that the experimental data show a nonlinear behavior for the load–deformation curve of the shear connectors joining the steel beam with the concrete slab. The influence of the behavior of the connectors has been demonstrated through the analysis of a long-span composite girder spanning 30.0 m. The finite element method was used in modeling the composite girder. Bending finite elements with axial deformation possibility representing the concrete slab and the steel beam were used. Spring elements of zero length, which can bear only shear force and obey the load-deformation law of the shear connectors, were used to connect steel and concrete. The authors have concluded that consideration of the actual load-deformation diagram of studs in the design of composite girders is needed and increases the safety of the beam at the serviceability limit state.

Studies carried out by Oehlers et al. [2.76] and by Oven et al. [2.77] investigated numerically the behavior of composite steel–solid slab beams. Oehlers et al. [2.76] found that in the maximum flexural capacity of composite beams, where the axial strength of the concrete section is usually much larger than that of the steel section, partial interaction has virtually no effect on the strength. Conversely, partial interaction can reduce the strength of composite beams with very strong steel sections, where the axial strength of the steel section is much greater than that of the concrete section. Also, it has been found that the greatest effect of partial interaction is to reduce the strain in the steel element and hence limit the beneficial effects of strain hardening. The work of the authors was part of an ongoing study, and a computer model has been developed to carry out the parametric study, while Oven et al. [2.77] developed a 2D nonlinear inelastic finite element model for the structural analysis of steel–solid slab composite beams with flexible shear connection. The effects of slip between the steel beam and the concrete slab and the nonlinear nature of force–slip characteristics of the shear connectors were included. The model was based on a 2D nonlinear FE analysis program INSTAF, developed originally for steel frames by El-Zanaty [2.78]. The program used a line element with 4 degrees of freedom at each node to represent the steel I-section and the concrete slab. The material nonlinearities of the composite girder components have been incorporated. The author concluded that the model can be used to predict the load–deflection behavior and the slip distribution along the length of the beam and the model has been validated by comparing the results with published test data. Cai et al. [2.79] investigated the behavior of cable-stayed composite bridges. The author developed a finite element model to represent the shear connection between a steel beam and a concrete slab. A 2D finite element was used in their analysis to model the steel beam and the concrete slab. The two elements are then connected via rigid links to model the shear connectors, and hence, they eliminated slip at the steel–concrete interface. The composite connection between steel beam and solid concrete slab was modeled by Youn and Chang [2.80] using 3D finite elements. The model consisted of two layers of solid elements and 3D beam elements. Rigid links were used between the concrete slab and the girders.

Orthotropic reinforcing bars in the concrete slab were modeled by four layered elements as a smeared layer with constant thickness. The thickness of the smeared layer was equal to the area of each bar divided by the bar spacing. The finite element program ABAQUS was used to analyze this model. The use of rigid links shows that the interaction between the steel beam and concrete slab is complete and there is no slip between the shear connectors, which is not true and has been rejected by many researchers. Gattesco [2.81] studied numerically the nonlinear behavior of composite steel–solid slab beams with deformable shear connection. The numerical procedure accounted for the nonlinear behavior of concrete, steel, and shear connector. The finite element package COBENA was used in the analysis. The steel beam and the concrete slab were modeled by using beam-type elements that have four nodal points with 3 degrees of freedom per node (horizontal and vertical displacements and rotation in the x-y plane). The interface between the steel beam and the concrete slab was modeled by two horizontal springs. The uplift of the concrete slab with respect to the steel beam and the buckling effects of the steel beam were neglected. The model was verified by comparing the finite element solutions with the experimental work of Chapman and Balakrishnan [2.82]. The author found good agreement between numerical and experimental results and concluded that this model can be used for extensive parametric studies on composite beams with complete or partial shear connection.

Kwak and Seo [2.83] modeled the behavior of composite girder using the finite element method. The aim of the study was to predict the long-term behavior of composite steel–solid slab girders in bridges. A 2D beam element that has 3 degrees of freedom (two translations and one rotation) was used in the analysis to represent the steel beam and the concrete slab. Material nonlinearities have been taken into consideration. The elements were divided into imaginary layers to describe the material properties with the assumption that plane sections remain plane to represent the linearity in the strain distribution on any section at any time. The load–slip characteristic of the stud has been neglected in the analysis, and perfect interaction has been assumed between the steel beam and the concrete slab. The finite element method was used in the modeling of steel–solid composite beams curved in plane by Thevendran et al. [2.84]. The numerical model was used to verify the experimental testing aimed by the authors to study the ultimate load behavior of these composite girders. The finite element software, ABAQUS, has been used in the analysis. Full composite action between steel beam and concrete slab was assumed. Figure 2.25 shows a typical finite

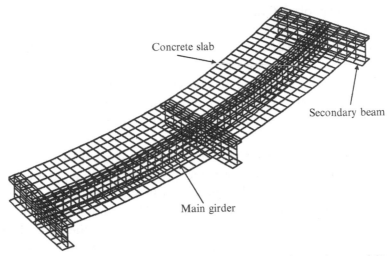

Concrete slab

Secondary beam

Main girder

Figure 2.25 FE model of the composite beam modeled by Thevendran *et al.* [2.84].

element mesh with 1257 elements used by the authors in the analysis. The-vendran *et al.* [2.84] used 3D finite elements to develop the finite element model. In the model, the concrete slab was simulated by four-node isopara-metric thick shell elements with the coupling of bending stiffness, while the steel flanges and web were modeled by four-node isoparametric thin ele-ments with the coupling of bending and stiffness. The shear connectors between concrete slab and steel flange were modeled by rigid beam ele-ments. Rigid connection beam elements were used to model the shear studs based on the assumption that no slip occurs between the concrete slab and the steel girder. The material nonlinearities of the steel beam and the con-crete slab were accounted in the analysis. The authors found good agreement between experimental and numerical results in most of the cases. The observed discrepancies in some of the results between the values predicted numerically and that predicted experimentally are attributed to neglecting the slip at the steel-concrete interface by using rigid elements to represent the studs. Amadio and Fragiacomo [2.85] used the finite element method to model steel-solid slab composite girders. The model was used in studying the evaluation of effective width for serviceability and ultimate analysis. In the model, the shell elements were used in modeling both the steel beam and the concrete slab. A nonlinear elastic law represented the behavior of the shear connection. The effects of steel and concrete material nonlinearities were taken into consideration. Although this research was applied on can-tilever beams, the authors concluded that the numerical study demonstrated that the connection deformability affected the evaluation of the effective width of steel–solid slab composite beams. A displacement-based finite element model for the analysis of steel-solid slab composite girders with flexible shear connection was developed by Faella *et al.* [2.86]. The model was originally developed by improving the two–node 1D displacement-based finite element. The research is still in press and the authors suggested that the model might be used in the accurate simulation of the behavior of composite girders.

Lam [2.87, 2.88] developed a 2D finite element model to verify the experimental testing of composite steel–prestressed hollow core concrete slab girders. In the model, the concrete slab was modeled by using 2D eight-node plane stress elements, while the steel beam was modeled by using 2D four-node plane stress elements. Figure 2.26 shows the finite element mesh developed in [2.87, 2.88]. The transverse reinforcement was not modeled as a separate element due to the limitation of the 2D element used, but its effect was taken indirectly into account in simulating the shear connection behavior. Both the precast hollow core concrete slab and *in situ* concrete were modeled as a single concrete element that had a breadth equal to the thickness of the prestressed hollow core concrete units and with combined material properties. The shear connectors were modeled by using spring elements that obeyed the load–slip characteristic of the shear stud connector used. Lam [2.87, 2.88] concluded that, although the composite precast hollow core beam was modeled in a simplified way, the results obtained from the model showed good agreement with the experimental results. The model was used in carrying out parametric studies that took into account the different parameters affecting the behavior of steel-prestressed hollow core

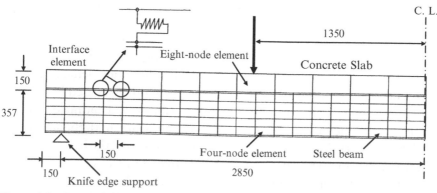

Figure 2.26 Finite element mesh of the composite beam used by Lam [2.87, 2.88].

concrete slab composite girders. The parametric study on composite girders with precast hollow core slabs was published by Lam *et al.* [2.89]. Also, from the results of the parametric study, design charts have been developed for initial sizing of composite girders by the authors [2.90]. Shim *et al.* [2.91] developed a finite element model to simulate shear connection in composite girders with full-depth prestressed decks. The numerical model was developed as a part of a study carried out to investigate design considerations of the shear connection in steel-concrete composite bridges with full-depth prestressed decks. The authors assumed that the shear connection was continuous and uniform along the beam and no separation took place at the interface. The finite element method has been used in the analysis, and a composite beam element that has 12 degrees of freedom was developed. The shear stiffness of the shear connection was evaluated from linear elastic analysis, and this, in addition to the assumption of full interaction between steel beam and concrete slab, limits the use of this model in an accurate finite element analysis. Ellobody and Lam [2.92] evaluated the effective width of composite steel beams with precast hollow core slabs numerically using the finite element method. A parametric study, carried out on 27 beams with different steel cross sections, hollow core unit depths, and spans, is presented. The effective width of the slab is predicted for both the elastic and the plastic ranges. Eight-node 3D solid elements are used to model the composite beam components. The material nonlinearity of all the components is taken into consideration. The nonlinear load-slip characteristics of the headed shear stud connectors are included in the analysis. The moment-deflection behavior of the composite beams, the ultimate moment capacity, and the modes of failure are also presented. In addition, the ultimate moment capacity of the beams evaluated using the present FE analysis was compared with the results calculated using the rigid-plastic method.

Ranzi and Bradford [2.93] presented a numerical model for the analysis of composite steel-concrete beams at elevated temperatures accounting for both longitudinal and transverse interaction. The model was derived by means of the principle of virtual work. A finite element was developed based on the formulation of partial interaction. The authors performed parametric studies investigating the effects of different thermal distributions on the structural response of a composite beam. Elastic material properties were assumed for all materials while still accounting for their degradation with temperature. A bilinear constitutive model was adopted for the transverse interface connection. Based on the proposed numerical model, it was concluded that it is important to account for the combined actions, that is, combined tension and shear force, to better identify the stress state at the interface connection. In addition, it was concluded that ignoring this coupling might lead to a significant underestimation of the connectors' available capacity. Furthermore, a prescriptive failure criterion based on the von Mises yield condition was proposed for shear connectors. Valipour and Bradford [2.94] presented the formulation of a force-based one-dimensional steel-concrete composite element that captured material nonlinearities and partial shear interaction between the steel profile and the reinforced concrete slab. A total secant solution strategy based on a direct iterative scheme was introduced by the authors. The slip forces along the element axis were calculated analytically. The accuracy and efficiency of the formulation are verified by some numerical examples reported by other researchers in the literature. It was shown that the formulation could lead to virtually closed form of analytic results as long as the integrals in the formulation were calculated accurately.

Recently, Erkmen and Saleh [2.95] have shown that when modeling composite or built-up beams using finite element software, analysts connected two standard Euler-Bernoulli beam elements at the nodes by using a rigid bar or master-slave-type kinematic constraints to express the degrees of freedoms of one of the members in terms of the other. The authors have shown that this type of modeling can lead to eccentricity-related numerical errors and special solutions that avoid eccentricity-related issues may not be available for a design engineer due to the limitations of the software. Therefore, a simple correction technique was introduced in the application of master-slave-type constraints. It was shown that the eccentricity-related numerical errors in the stiffness matrix can be completely corrected by using extra fictitious elements and springs. The correction terms were obtained by using the exact homogenous solution of the composite beam problem as the interpolation functions, which impose the zero-slip constraint between the two components in the point-wise sense. Yu-hang et al. [2.96] developed a steel-concrete composite fiber beam-column model. The model consisted of a preprocessor program that was used to divide a composite section into fibers. Uniaxial hysteretic material constitutive models were incorporated in the model. The authors showed that the steel-concrete composite fiber beam-column model can be used for global elastoplastic analysis on composite frames with rigid connections subjected to the combined action of gravity and cyclic lateral loads. The model was verified against a number of experiments, and the results showed that the developed composite fiber model behaved better compared with traditional finite element models. In addition, it was shown that although the fiber beam-column model neglected the slip between the steel beam and concrete slab, there were no effects on the global calculation results of steel-concrete composite frames. It was concluded that the proposed model can be used to analyze composite frames subjected to cyclic loading due to earthquake.

REFERENCES

[2.1] ASTM, Standard Specification for Structural Steel for Bridges. West Conshohocken: American Society for Testing and Materials, A 709/A709M, 2010.

[2.2] ASTM A325, Standard Specification for Structural Bolts, Steel, Heat Treated, 120/105 ksi Minimum Tensile Strength, American Society for Testing and Materials, 2010.

[2.3] ASTM A490, Standard Specification for Structural Bolts, Alloy Steel, Heat Treated, 150 ksi Minimum Tensile Strength. American Society for Testing and Materials, 2012.

[2.4] ASTM A307, Standard Specification for Carbon Steel Bolts, Studs, and Threaded Rod 60 000 PSI Tensile Strength. American Society for Testing and Materials, 2012.

[2.5] ASTM A563, Standard Specification for Carbon and Alloy Steel Nuts. American Society for Testing and Materials, 2007.

[2.6] ASTM F436M, Standard Specification for Hardened Steel Washers (Metric). American Society for Testing and Materials, 2011.

[2.7] ASTM A586, Standard Specification for Zinc Coated Parallel and Helical Steel Wire Structural Strand. American Society for Testing and Materials, 2009.

[2.8] ASTM A603, Standard Specification for Zinc Coated Steel Structural Wire Rope. American Society for Testing and Materials, 2009.

[2.9] ASTM A370, Standard Test Methods and Definitions for Mechanical Testing of Steel Products. American Society for Testing and Materials, 2012.

[2.10] ASTM E8/E8M, Standard Test Methods for Tension Testing of Metallic Materials. American Society for Testing and Materials, 2011.

[2.11] EC3, Eurocode 3—design of steel structures—part 1-1: general rules and rules for buildings. BS EN 1993-1-1, British Standards Institution, 2005.

[2.12] EC4, Eurocode 4: design of composite steel and concrete structures—part 1-2: general rules—structural fire design. British Standards Institution, BS EN 1994-1-2, London, UK, 2005.

[2.13] EC3, Eurocode 3—design of steel structures—part 1-8: design of joints. BS EN 1993-1-8, British Standards Institution, 2005.

[2.14] EN 10080, Steel for the reinforcement of concrete—weldable reinforcing steel—general. BS EN 10080, British Standards Institution, 2005.

[2.15] ASTM, Standard Test Methods for Notched Bar Impact testing of Metallic Materials, American Society for Testing and Materials, West Conshohocken, 2011.

[2.16] EC3, Eurocode 3—design of steel structures—part 1-10: material toughness and through-thickness properties. BS EN 1993-1-10, British Standards Institution, 2005.

[2.17] AASHTO/AWS, Bridge Welding Code. Washington, DC: American Association of State Highway and Transportation Officials, ANSI/AASHTO/AWS D1.5-10, 2010.

[2.18] B.A. Graville, Cold Cracking in Welds in HSLA Steels. s.l.: American Society for Metals, 1976. Welding of HSLA (Microalloyed) Structural Steels, 2012.

[2.19] ASTM, Standard Guide for Estimating Atmospheric Corrosion Resistance of Low Alloy Steels, American Society for Testing and Materials, West Conshohocken, 2004, G101.

[2.20] U.S. Dixit, P.M. Dixit, A study on residual stresses in rolling, Int. J. Mach. Tools Manufact. 37 (6) (1997) 837–853.

[2.21] S. Kamamato, T. Nihimori, S. Kinoshita, Analysis of residual stress and distortion resulting from quenching in large low-alloy steel shafts, J. Mater. Sci. Technol. 1 (1985) 798–804.

[2.22] M. Toparli, T. Aksoy, Calculation of residual stresses in cylindrical steel bars quenched in water from 600 °C, in: Proceedings of AMSE Conference, 4, New Delhi, India, 1991, pp. 93–104.

[2.23] S. Jahanian, Residual and thermo-elasto-plastic stress distributions in a heat treated solid cylinder, Mater. High Temp. 13 (2) (1995) 103–110.

[2.24] F.R. Yuan, S.L. Wu, Transient-temperature and residual-stress fields in axisymmetric metal components after hardening, J. Mater. Sci. Technol. 1 (1985) 851–856.

[2.25] K. Yamada, Transient thermal stresses in an infinite plate with two elliptic holes, J. Therm. Stresses 11 (1988) 367–379.

[2.26] Y. Ding, Residual stresses in hot-rolled solid round steel bars and their effect on the compressive resistance of members. Master Thesis, University of Windsor, Windsor, Ontario, Canada, 2000.

[2.27] EC2, Eurocode 2—design of concrete structures—part 1-1: General rules and rules for buildings. BS EN 1992-1-1, British Standards Institution, 2004.

[2.28] EC2, Eurocode 2—design of concrete structures—part 2: concrete bridges—design and detailing rules. BS EN 1992-2, British Standards Institution, 2005.

[2.29] EN 12390, Testing hardened concrete—part 1: part 1: shape, dimensions and other requirements for specimens and moulds. BS EN 12390, British Standards Institution, 2000.

[2.30] EN 10138, Prestressing steels—part 1: general requirements. BS EN 10138, British Standards Institution, 2006.

[2.31] EC2, Eurocode 2: design of concrete structures—part 1-2: general rules. Structural Fire Design. British Standards Institution, BS EN 1992-1-2, London, UK, 2004.

[2.32] I.M. Viest, Investigation of stud shear connectors for composite concrete and steel T-beams, ACI J. 27 (1956) 875–891, Title No.52-56.

[2.33] B. Thurlimann, Fatigue and static strength of stud shear connectors, ACI J. 30 (12) (1959) 1287–1302.

[2.34] CP 117, Part 1. Composite Construction in Structural Steel and Concrete: Simply Supported Beams in Building, British Standards Institution, London, 1965.

[2.35] CP 117, Part 2. Composite Construction in Structural Steel and Concrete: Beams for Bridges, British Standards Institution, London, 1967.

[2.36] BS 5950, Part 3. Code of Practice for Design of Simple and Continuous Composite Beams, British Standards Institution, London, 1990.

[2.37] EC4, Eurocode 4—design of composite steel and concrete structures—part 1-1: general rules and rules for buildings. BS EN 1994-1-1, British Standards Institution, 2004.

[2.38] BS 5400, Part 5. Design of Composite Bridges, British Standards Institution, London, 1979.

[2.39] R.P. Johnson, Composite structures of steel and concrete, In: Beams, Slabs, Columns and Frames for Building, vol. 1, Blackwell Scientific Publications, Oxford, 1971.

[2.40] R.P. Johnson, D. Anderson, Designers 'Handbook' to Eurocode 4: Part 1.1, 'Design of Steel and Composite Structures', Thomas Telford, London, 1993.

[2.41] ECCS, Composite Structures, The construction Press, London, 1981.

[2.42] C. Johnson, Deflection of steel-concrete composite engineering, Struct. Eng. 124 (10) (1998) 1159–1165.

[2.43] L.C.P. Yam, J.C. Chapman, The inelastic behavior of simply supported beams of steel and concrete, Proc. Inst. Civ. Eng. 41 (1968) 651–683.

[2.44] C. Davies, Small-scale push-out tests on welded stud shear connectors, Concrete J. 1 (1967) 311–316.

[2.45] C. Davies, Tests on half-scale steel-concrete composite beams with welded shear connectors, Struct. Eng. 47 (1) (1969) 20–40.

[2.46] J.B. Menzies, CP 117 and shear connectors in steel-concrete composite beams made with normal-density or lightweight concrete, Struct. Eng. 49 (3) (1971) 137–153.

[2.47] B.S. Jayas, M.U. Hosain, Behaviour of headed studs in composite beams: push-out tests, Civ. Eng. 15 (1987) 240–253.

[2.48] J.G. Ollgaard, R.G. Slutter, J.W. Fisher, Shear strength of stud connectors in lightweight and normal-weight concrete, Am. Inst. Steel Constr. Eng. J. 8 (2) (1968) 651–683.

[2.49] D.J. Oehlers, Splitting induced by shear connectors in composite beams, J. Struct. Eng. 115 (2) (1989) 341–360.

[2.50] D.J. Oehlers, R.P. Johnson, The Strength of stud shear connections in composite beams, Struct. Eng. 65B (2) (1987) 44–48.

[2.51] D.J. Oehlers, S.M. Park, Shear connection in haunched composite beams with sloping sides', J. Struct. Eng. 18 (3) (1994) 191–207.

[2.52] R.P. Johnson, Design of composite beams with deep haunches, Proc. Inst. Civ. Eng. 2 (51) (1972) 83–90.

[2.53] A. Li, C. Krister, Push-out tests on studs in high strength and normal strength concrete, J. Constr. Steel Res. 36 (1) (1996) 15–29.

[2.54] J.T. Mottram, R.P. Johnson, Push tests on studs welded through profiled steel sheeting, Struct. Eng. 68 (10) (1990) 187–193.

[2.55] N.M. Hawkins, D. Mitchell, Seismic response of composite shear connections, J. Struct. Eng. ASCE 110 (9) (1984) 2120–2136.

[2.56] B.S. Jayas, M.U. Hosain, Behaviour of headed studs in composite beams: full-size tests, Can. J. Civil Eng. 16 (1989) 712–724.

[2.57] R.M. Lloyd, H.D. Wright, Shear connection between composite slabs and steel beams, J. Constr. Steel Res. 15 (1990) 255–285.

[2.58] B. Kim, H.D. Wright, R. Cairns, The behaviour of through-deck welded shear connectors: an experimental and numerical study, J. Constr. Steel Res. 57 (2001) 1359–1380.

[2.59] B. Kim, H.D. Wright, R. Cairns, The behaviour of through-deck welded shear connectors: a numerical study, in: The first International Conference on Steel and Composite Structures, Pusan, Korea, 2001, pp. 1327–1334.

[2.60] S.S.J. Moy, C. Tayler, The effect of precast concrete planks on shear connector strength, J. Constr. Steel Res. 36 (3) (1996) 201–213.

[2.61] D. Lam, K.S. Elliott, D.A. Nethercot, Push-off tests on shear studs with hollow-cored floor slabs, Struct. Eng. 76 (9) (1998) 167–174.

[2.62] C.S. Shim, J.H. Kim, S.P. Chang, C.H. Chung, The behavior of shear connections in a composite beam with a full-depth precast slab', Proc. Inst. Civ. Eng. Struct. Build. 140 (1) (2000) 101–110.

[2.63] T.F. Nip, D. Lam, Effect of end condition of hollow core slabs on longitudinal shear capacity of composite beams', in: First International Conference on Steel and Composite Structures, Pusan, Korea, 2001, pp. 1229–1236.

[2.64] D.A. Nethercot, The importance of combining experimental and numerical study in advancing structural engineering understanding, Struct. Eng. Mech. Comput. 1 (2001) 15–26.

[2.65] R.P. Johnson, D.J. Oehlers, Analysis and design for longitudinal shear in composite T-beams', Proc. Inst. Civ. Eng. Part 2 71 (1981) 989–1021.

[2.66] D.J. Oehlers, Finite element simulation of the behavior of concrete subjected to con-
        centrated loads'. University of Warwick, Technical report CE10, 1981.
[2.67] C. Kalfas, P. Pavlidis, E. Galoussis, Inelastic behavior of shear connection by a method
        based on FEM, J. Constr. Steel Res. 44 (1–2) (1997) 107–114.
[2.68] E. Ellobody, Finite element modeling of shear connection for steel-concrete compos-
        ite girders. PhD thesis, School of Civil Engineering, The University of Leeds, Leeds,
        2002.
[2.69] D. Lam, E. Ellobody, Behavior of headed stud shear connectors in composite beam, J.
        Struct. Eng. ASCE 131 (1) (2005) 96–107.
[2.70] E. Ellobody, D. Lam, Modeling of headed stud in steel-precast composite beams, Steel
        Comp. Struct. 2 (5) (2002) 355–378.
[2.71] E. Ellobody, B. Young, Performance of shear connection in composite beams with
        profiled steel sheeting, J. Constr. Steel Res. 62 (7) (2006) 682–694.
[2.72] P. Ansourian, An Application of the method of finite element to the analysis of com-
        posite floor systems', Proc. Inst. Civ. Eng. Part 2 59 (1975) 699–726.
[2.73] U.R. Moffat, P.J. Dowling, The longitudinal bending behaviour of composite box
        girder bridges having incomplete interaction, Struct. Eng. 56B (3) (1978) 53–60.
[2.74] S. Balakrishnan, The behaviour of Composite Steel and Concrete Beams with
        Welded Stud Shear Connectors. PhD thesis, University of London, 1963.
[2.75] E.S. Mistakidis, K. Thomopoulos, A. Avdelas, P.D. Panagiotopoulos, Shear connec-
        tors in composite beams: a new accurate algorithm, Thin Wall. Struct. 18 (3) (1994)
        191–207.
[2.76] D.J. Oehlers, N.T. Tguyen, M. Ahmed, M.A. Bradford, Partial interaction in com-
        posite steel and concrete beams with full shear connection, J. Constr. Steel Res. 41 (2/
        3) (1997) 235–248.
[2.77] V.A. Oven, I.W. Burgess, R.J. Plank, A.A. Abdul Wali, An analytical model for the
        analysis of composite beams with partial interaction', Comp. Struct. 62 (3) (1997)
        493–504.
[2.78] M. H. El-Zanati, D.W. Murray, R. Bjorhovde, Inelastic behavior of multi-storey
        steel frames. Structural Engineering Report, No. 83, Department of Civil Engineer-
        ing, University of Alberta, Canada, 1980.
[2.79] C.S. Cai, Y. Zhang, J. Nie, Composite girder design of cable-stayed bridges, Pract.
        Period. Struct. Des. Constr. Proc. Eng. Found. Conf. 3 (1998) 158–163.
[2.80] S.G. Youn, S.P. Chang, Behaviour of composite bridge decks subjected to static and
        fatigue loading', ACI Struct. J. 95 (3) (1998) 249–258.
[2.81] N. Gattesco, Analytical modelling of nonlinear behaviour of composite beams with
        deformable connection, J. Constr. Steel Res. 52 (1999) 195–218.
[2.82] J.C. Chapman, S. Balakrishnan, Experiments on composite beams, Struct. Eng.
        42 (1964) 369–383.
[2.83] H.G. Kwak, Y.J. Seo, Long-term behaviour of composite girder bridges, Comp.
        Struct. 74 (2000) 583–599.
[2.84] V. Thevendran, N.E. Shanmugam, S. Chen, J.Y.R. Liew, Experimental study on
        steel-concrete composite beams curved in plane, Eng. Struct. 22 (2000) 877–889.
[2.85] C. Amando, M. Fragiacomo, Effective width evaluation for steel-concrete composite
        beams, J. Constr. Steel Res. 58 (2002) 373–388.
[2.86] C. Faella, E. Maeinelli, E. Nigro, Steel and concrete composite beams with flexible
        shear connection: 'Exact' analytical expression of the stiffness matrix and applications,
        Comput. Struct. 80 (11) (2002) 1001–1009.
[2.87] D. Lam, Composite steel beams using precast concrete hollow core floor slabs. PhD
        thesis, University of Nottingham, 1998.

[2.88] D. Lam, K.S. Elliott, D.A. Nethercot, Experiments on composite steel beams with precast concrete hollow core floor slabs, Proc. Inst. Civ. Eng. Struct. Build. 140 (2000) 127–138.

[2.89] D. Lam, K.S. Elliott, D.A. Nethercot, Parametric study on composite steel beams with precast concrete hollow core floor slabs, J. Constr. Steel Res. 54 (2000) 283–304.

[2.90] D. Lam, K.S. Elliott, D.A. Nethercot, Designing composite steel beams with precast concrete-hollow core slabs, Proc. Inst. Civ. Eng. Struct. Build. 140 (2000) 139–149.

[2.91] C.S. Shim, J.H. Kim, P.G. Lee, S.P. Chang, Design of shear connection in composite steel and concrete bridges with precast decks, J. Constr. Steel Res. 57 (2001) 203–219.

[2.92] E. Ellobody, D. Lam, Determining the effective width for composite beams with pre-cast hollow core slabs, Struct. Eng. Mech.Int. J. 21 (3) (2005) 295–313.

[2.93] G. Ranzi, M.A. Bradford, Composite beams with both longitudinal and transverse partial interaction subjected to elevated temperatures, Eng. Struct. 29 (2007) 2737–2750.

[2.94] H.R. Valipour, M.A. Bradford, A steel-concrete composite beam element with material nonlinearities and partial shear interaction, Finite Elem. Anal. Des. 45 (2009) 966–972.

[2.95] R.E. Erkmen, A. Saleh, Eccentricity effects in the finite element modelling of composite beams, Adv. Eng. Softw. 52 (2012) 55–59.

[2.96] W. Yu-hang, N. Jian-guo, C.S. Cai, Numerical modeling on concrete structures and steel–concrete composite frame structures, Compos. Part B 51 (2013) 58–67.

CHAPTER 3

Applied Loads and Stability of Steel and Steel-Concrete Composite Bridges

3.1 GENERAL REMARKS

The brief introduction of steel and steel-concrete composite bridges presented in Chapter 1 and the revision of the nonlinear material behavior of the main bridge components presented in Chapter 2 provide a useful background on bridges and the material behavior of the components of the bridges. It is now possible to detail applied loads acting on steel and steel-concrete composite bridges, which is highlighted in this chapter. This chapter presents different loads acting on railway and highway bridges and the stability of the bridges when subjected to these loads. The chapter starts by showing the dead loads of steel and steel-concrete composite bridges that are initially estimated for the design of bridges. Then, the chapter presents the live loads from traffic as specified in the American and European codes. After that, the chapter presents the calculation of horizontal loads and other loads acting on the bridges such as centrifugal forces, seismic loading, collision forces, and temperature effects. In addition, the chapter presents the load combinations specified in the current codes of practice to predict the worst case of loading for the calculation of different straining actions in the bridge components. Furthermore, different design approaches specified in the current codes of practice are highlighted in this chapter. Finally, the chapter addresses the main issues related to the stability of steel and steel-concrete composite plate girder and truss bridges such as buckling behavior of compression members, stability of thin-walled steel plate girders, lateral torsional buckling, and composite plate girder behavior. Once again, when highlighting the main issues related to the stability of the bridge components, it intends to review and present the issues based on the design rules specified in the current codes of practice, with particular focus on the Eurocode as an example. Overall, the author hopes that this chapter paves the way to the design examples of different bridge components presented in Chapter 4.

Finite Element Analysis and Design of Steel and Steel–Concrete Composite Bridges

3.2 DEAD LOADS OF STEEL AND STEEL-CONCRETE COMPOSITE BRIDGES

The dead loads acting on railway and highway bridges consist of the weight of all its structural parts, fittings, finishing, curbs, lighting and signing devices, gas and water mains, electricity and telephone cables, etc. These loads are permanent and remain constant in position and magnitude. To calculate the straining actions on the bridge components, the weight of the structural parts has to be initially assumed. The assumed weights have to be assessed after designing and predicting cross sections of all structural parts. When there is a considerable difference between the assumed and predicted weights, the calculation of the loads and design has to be repeated until close agreement is achieved between assumed and predicted weights. It should be noted that most current codes of practice provide guidance for the unit weights of commonly used materials in steel and steel-concrete composite bridges, which can also be used to estimate the dead loads acting on the bridges. Furthermore, the dead loads of previously designed existing bridges can be used to provide guidance to dead loads expected on similar bridges under construction.

3.2.1 Dead Loads of Railway Steel Bridges

As an example, let us estimate the dead loads acting on different components of the traditional double-track open-timber floor plate girder railway steel bridge shown in Figure 1.20. Starting with the dead loads acting on a stringer, these loads are half the weight of the track loads (train, sleepers, and rails), own weight of the stringer, and weight of stringer bracing. The track load varies from country to country and can be found in the national code of practice of the country of construction. A commonly assumed track load is 6 kN/m acting along the stringer length, which is the spacing between two adjacent cross girders. The own weight of a stringer depends on its length and the type of steel used. The weight of a stringer can be reasonably assumed to be from 1 to 1.5 kN/m all over the stringer length. Finally, the weight of stringer bracing can be reasonably assumed to be 0.2–0.3 kN/m acting along the stringer length. By knowing the assumed total dead load acting on the stringer, the straining actions resulting from dead loads comprising bending moment and shear force can be calculated.

The dead loads acting on an intermediate cross girder (see Figure 1.20) are the concentrated dead loads coming from the stringers, which are supported by the cross girders and own weight of cross girder. Once again, the own weight of cross girder depends on its length and the type of steel used. The own weight of cross girder can be reasonably assumed to be 2-3 kN/m of the length of cross girder, which is the pacing between the main girders. By knowing the assumed concentrated dead loads acting on the cross girder and its assumed own weight, the straining actions resulting from dead loads comprising bending moment and shear force can be calculated. Finally, the dead loads acting on the main girders (see Figure 1.20) are the weight of steel structural parts plus the weight of tracks. The weight of steel structural parts can be estimated from similar existing bridges or from some empirical formulas, which are obtained from the data available from previously designed railway steel bridges. The assumed weight of steel structural parts depends on many factors including the length of main girder, type of steel used, and type of bridge. A reasonable assumption of the weight of structural parts (w_s) for single-track open-timber floor bridges can be given by the following empirical equations:

$$w_s = 4 + 0.5L \, (\text{kN/m})$$
(for deck bridges without stringers and cross girders)
$$(3.1)$$

$$w_s = 9 + 0.5L \, (\text{kN/m})$$
(for deck bridges with stringers and cross girders)
$$(3.2)$$

$$w_s = 11 + 0.5L \, (\text{kN/m}) \quad \text{(for pony bridges)} \tag{3.3}$$

where L is the length of main girder in meters and the weight of structural parts (w_s) is divided into two main girders. For double-track bridges, the previously mentioned loads can be increased by 80-90%. By knowing the assumed total dead load acting on the main girder, the straining actions resulting from dead loads comprising bending moment and shear force can be calculated. For ballasted floor railway steel bridges, the weight of structural parts can be increased by 20-40%.

3.2.2 Dead Loads of Highway Steel and Steel-Concrete Composite Bridges

As an example, let us estimate the dead loads acting on different components of the traditional through-truss highway steel bridge as shown in Figure 1.21. Starting with the dead loads acting on a stringer, the dead loads include the weight of flooring (1.5-2 kN/m²), the weight of reinforced concrete slabs having a thickness of around 200 mm (5 kN/m²), the weight of reinforced concrete haunch (0.3 kN/m²), and the own weight of stringer

(1–1.5 kN/m). By knowing the assumed total dead load acting on the stringer, the straining actions resulting from dead loads comprising bending moment and shear force can be calculated.

The dead loads acting on an intermediate cross girder (see Figure 1.21) are the concentrated dead loads coming from the stringers, which are supported by the cross girder, and own weight of cross girder. Once again, the own weight of cross girder depends on its length and the type of steel used. The own weight of cross girder can be reasonably assumed to be 2.5–3 kN/m of the length of cross girder. By knowing the assumed concentrated dead loads acting on the cross girder and its assumed own weight, the straining actions resulting from dead loads comprising bending moment and shear force can be calculated. Finally, the dead loads acting on the main trusses are the weight of steel structural parts plus weight of finishing, reinforced concrete slabs, and haunches. Once again, the weight of steel structural parts can be estimated from similar existing bridges or from some empirical formulas, which are obtained from the data available from previously designed railway steel bridges. A reasonable assumption of the weight of structural parts (w_s) for highway bridges can be given by the following empirical equations:

$$w_{s1} = 1.75 + 0.04L + 0.0003L^2 \leq 3.5 \left(\text{kN/m}^2\right)$$
$$\text{(for part of bridge between main trusses)} \tag{3.4}$$

$$w_{s2} = 1 + 0.03L \left(\text{kN/m}^2\right)$$
$$\text{(for part of bridge outside main trusses)} \tag{3.5}$$

where L is the length of the main girder in meters. By knowing the assumed total dead load acting on the truss, the straining actions resulting from dead loads comprising axial tension and compression forces in the truss can be calculated.

3.3 LIVE LOADS ON STEEL AND STEEL-CONCRETE COMPOSITE BRIDGES

Live loads acting on steel and steel-concrete composite bridges differ from country to country. National codes of practice in any country specify design live loads that should be considered in the calculation of different straining actions on the bridge components. The design loads represent the worst cases of traffic loading permitted and expected to pass over a specific bridge. Since the main objective of this book is to provide a consistent and general approach for finite element analysis and design of steel and steel-concrete composite bridges, the design live loads specified in the current codes of practice are highlighted in this section. The design live loads for railway bridges depend on the types of trains passing on the bridges, while the design loads for highway bridges depend on the types of vehicles passing on the bridge. The design live loads for railway and highway bridges are either concentrated loads acting on the axles of the specified trains and vehicles, respectively, or equivalent uniformly distributed loads simulating the case of several closely spaced vehicles in a jam situation. The specified trains and vehicles have specified dimensions and axles spaced at definite locations in the trains or vehicles. Live loads resulting from moving trains or vehicles are magnified to account for the effect of impact and dynamic application of the loads on the bridges. Wide steel and steel-concrete composite bridges are designed on the worst cases of live loads acting on several adjacent lanes. Long steel and steel-concrete composite bridges are designed to carry multiple trains or vehicles that should be positioned to provide the maximum straining actions at a specific section.

3.3.1 Live Loads for Railway Steel Bridges

Let us start by highlighting the live loads specified in the Eurocode (EC1) [3.1] for railway bridges. EC1 is applicable to railway traffic on the standard track gauge and wide track gauge European mainline network. According to EC1, the design load models adopted do not describe actual loads. However, they have been selected so that their effects, with dynamic enhancements taken into account separately, represent the effects of service traffic. Where traffic outside the scope of the load models specified in EC1 needs to be considered, then alternative load models, with associated combination rules, should be specified. The live loads specified in EC1 are not applicable for actions due to narrow-gauge railway, tramways and other light railway, preservation railway, rack and pinion railways, and funicular railways. EC1 provides three standard mixes of railway traffic, which are recommended for calculating the fatigue life of bridges as detailed in Annex D of the code. EC1 specifies the general rules for the calculation of characteristic vertical load values (static effects) and eccentricity and distribution of loading and specifies the associated dynamic effects, centrifugal forces, nosing force, traction and braking forces, and aerodynamic actions due to passing railway traffic.

According to EC1, railway traffic actions are defined by means of five load models of railway loading that are as follows: "Load Model 71" and "Load Model SW/0" for continuous bridges to represent normal railway traffic on mainline railways, "Load Model SW/2" to represent heavy loads, "high-speed load model (HSLM)" to represent the loading from passenger trains at speeds exceeding 200 km/h, and finally Load Model "unloaded train" to represent the effect of an unloaded train. Load Model 71 represents the static effect of vertical loading due to normal railway traffic. The load arrangement and the characteristic values for vertical loads shall be taken into account as shown in Figure 3.1. The characteristic values given in Figure 3.1 shall be multiplied by factor α on lines carrying railway traffic that is heavier or lighter than normal railway traffic. When multiplied by the factor α, the loads are called "classified vertical loads." This factor α varies from 0.75 to 1.46. Also, the actions comprising equivalent vertical loading for earthworks and earth pressure effects, centrifugal forces, nosing force, traction and braking forces, combined response of structure and track to variable actions, derailment actions for accidental design situations, and Load Model SW/0 for continuous span bridges shall be multiplied by the same factor α. For checking the limits of deflection, classified vertical loads and other actions enhanced by α shall be used (except for passenger comfort where α shall be taken as unity). Load Model SW/0 represents the static effect of vertical loading due to normal railway traffic on continuous beams. Load Model SW/2 represents the static effect of vertical loading due to heavy railway traffic. The load arrangements SW/0 and SW/2 shall be taken as shown in Figure 3.2, with the characteristic values of the vertical loads according to Table 3.1. For some specific verifications, a particular load model is used, called "unloaded train." The Load Model "unloaded train" consists of a vertical uniformly distributed load, with a characteristic value of 10.0 kN/m.

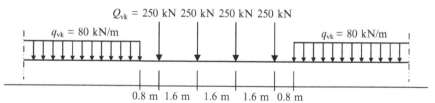

Figure 3.1 Load Model 71 and characteristic values for vertical loads of standard trains specified in EC1 [3.1].

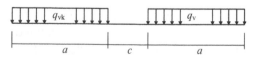

Figure 3.2 Load Models SW/0 and SW/2 of standard trains specified in EC1 [3.1].

Table 3.1 Characteristic Values for Vertical Load Models SW/0 and SW/2 of Standard Trains Specified in EC1 [3.1]

Load Model	q_{vk} (kN)	a (m)	c (m)
SW0	133	15.0	5.3
SW2	150	25.0	7.0

In the United States, design live train loads recommended by the American Railway Engineering and Maintenance-of-Way Association (AREMA) [1.25] are based on the Cooper E80 loading as shown in Figure 3.3. The 80 in E80 refers to the 80 kip (1 kip is equal to 4.448 kN) weight of the locomotive drive axles. An E60 load has the same axle locations, but all loads are factored by 60/80. New bridges may be designed to carry E90 or E100 loads. The designated steel bridge design live load also includes an "Alternate E80" load, consisting of four 100 kip axles, which is shown in Figure 3.4. This load controls over the regular Cooper load on shorter spans. AREMA [1.25] also presents formulas for the calculation of the impact, which is the dynamic amplification of the live load effects on the bridge caused by the movement of the train across the span. The design impact values are based on an assumed train speed of 60 mph. It should be noted that the steel design procedure allows the reduction of the calculated impact for ballast deck structures. Different values for impact from

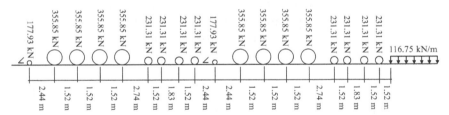

Figure 3.3 Cooper E80 live loading of standard trains specified in AREMA [1.25].

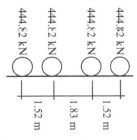

Figure 3.4 Alternate E80 live load of standard trains specified in AREMA [1.25].

steam and diesel locomotives are used. The steam impact values are significantly higher than diesel impact over most span lengths.

3.3.2 Live Loads for Highway Steel and Steel-Concrete Composite Bridges

The live loads on highway (roadway) steel and steel-concrete composite bridges are also specified in the Eurocode (EC1) [3.1]. Load models covered by the code should be used for the design of highway bridges with loaded lengths less than 200 m. The 200 m corresponds to the maximum length taken into account for the calibration of Load Model 1 (see Figure 3.5). In general, the use of Load Model 1 is safe-sided for loaded lengths over 200 m. The models and associated rules are intended to cover all normally foreseeable traffic situations (i.e., traffic conditions in either direction on any lane due to the road traffic), which should be taken into account for design. Specific models may be used for bridges equipped with appropriate means including road signs intended to strictly limit the weight of any vehicle (e.g., for local, agricultural, or private roads). Load models for abutments and walls adjacent to bridges are defined separately. The load models specified in EC1 derive from the road traffic models without any correction for dynamic effects. For frame bridges, loads on road embankments may also give rise to action effects in the bridge structure. The effects of loads on road construction sites (e.g., due to scrapers and motortrucks carrying earth) or of loads specifically for inspection and tests are not intended to be covered by the load models and should be separately specified, where relevant. According to EC1, loads due to the road traffic, consisting of cars, motortrucks, and special vehicles (e.g., for industrial transport), give rise to vertical and horizontal and static and dynamic forces. It should be noted that the load models defined in EC1 do not describe actual loads. They have been selected and calibrated so that their effects (with dynamic amplification included

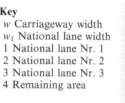
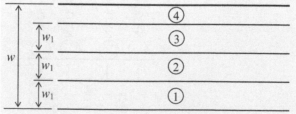

Key
w Carriageway width
w_1 National lane width
1 National lane Nr. 1
2 National lane Nr. 2
3 National lane Nr. 3
4 Remaining area

Figure 3.5 Example of lane numbering in the most general case according to EC1 [3.1].

where indicated) represent that of the actual traffic in the year 2000 in European countries. Where vehicles that do not comply with the regulations concerning the limits of weights and, possibly, dimensions of vehicles not requiring special permits, or military loads, have to be taken into account for the design of a bridge, they should be defined.

The carriageway width (w) (see Figure 3.5 specified in EC1 (3.1)) should be measured between curbs or between the inner limits of vehicle restraint systems and should not include the distance between fixed vehicle restraint systems or curbs of a central reservation nor the widths of these vehicle restraint systems. The recommended minimum value of the height of the curbs is 100 mm. The width w_1 of notional lanes on a carriageway and the greatest possible whole (integer) number n_1 of such lanes on this carriageway are defined in Table 3.2, which are specified in the code. For variable carriageway widths, the number of notional lanes should be defined in accordance with the principles used for Table 3.2. Where the carriageway on a bridge deck is physically divided into two parts separated by a central reservation, then each part, including all hard shoulders or strips, should be separately divided into notional lanes if the parts are separated by a permanent road restraint system. Where the whole carriageway, central reservation included, should be divided into notional lanes if the parts are separated by a temporary road restraint system, the locations of notional lanes should not be necessarily related to their numbering. For each individual verification (e.g., for a verification of the ultimate limit state of resistance of a cross section to bending), the number of lanes to be taken into account as loaded, their location on the carriageway, and their numbering should be so chosen so that the effects from the load models are the most adverse. For fatigue representative values and models, the location and the numbering of the lanes should be selected depending on the traffic to be expected in normal conditions. The lane giving the most unfavorable effect is Lane Number 1, the lane giving the second most unfavorable effect is Lane Number 2, etc.

Table 3.2 Load Model 1: Characteristic Values Specified in EC1 [3.1]

Location	Tandem System, TS Axle loads Q_{ik} (kN)	UDL System q_{ik} or (q_{ik}) (kN/m^2)
Lane number 1	300	9
Lane number 2	200	2.5
Lane number 3	100	2.5
Other lanes	0	2.5
Remaining area (q_{rk})	0	2.5

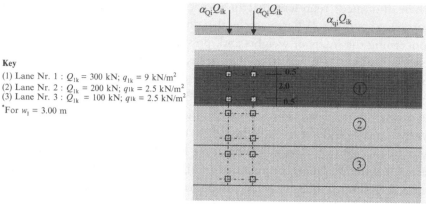

Key
(1) Lane Nr. 1 : $Q_{1k} = 300$ kN; $q_{1k} = 9$ kN/m^2
(2) Lane Nr. 2 : $Q_{1k} = 200$ kN; $q_{1k} = 2.5$ kN/m^2
(3) Lane Nr. 3 : $Q_{1k} = 100$ kN; $q_{1k} = 2.5$ kN/m^2
*For $w_1 = 3.00$ m

Figure 3.6 Application of Load Model 1 according to EC1 [3.1].

(see Figure 3.6). Where the carriageway consists of two separate parts on the same deck, only one numbering should be used for the whole carriageway. Hence, even if the carriageway is divided into two separate parts, there is only one Lane Number 1, which can be considered alternatively on the two parts. Where the carriageway consists of two separate parts on two independent decks, each part should be considered as a carriageway. Separate numbering should then be used for the design of each deck. If the two decks are supported by the same piers and/or abutments, there should be a single numbering for the two parts together for the design of the piers and/or the abutments. According to EC1, for each individual verification, the load models, on each notional lane, should be applied on such a length and so longitudinally located that the most adverse effect is obtained. On the remaining area, the associated load model should be applied on such lengths and widths in order to obtain the most adverse effect.

It should be noted that characteristic loads specified in EC1 (3.1) are intended for the determination of road traffic effects associated with ultimate limit state verifications and with particular serviceability verifications. The load models for vertical loads represent the following four traffic effects: Load Model 1 (LM1), Load Model 2 (LM2), Load Model 3 (LM3), and Load Model 4 (LM4). LM1 contains concentrated and uniformly distributed loads, which cover most of the effects of the traffic of motortrucks and cars. This model should be used for general and local verifications. LM2 contains a single axle load applied on specific tire contact areas, which covers the dynamic effects of the normal traffic on short structural members. As an order of magnitude, LM2 can be predominant in the range of loaded lengths up to 3-7 m. LM3 contains a set of assemblies of axle loads representing special vehicles (e.g., for industrial transport), which can travel on routes permitted for abnormal loads. It is intended for general and local verifications. Finally, LM4 contains a crowd loading, intended only for general verifications. This crowd loading is particularly relevant for bridges located in or near towns if its effects are not covered by Load Model 1. Load Models 1, 2, and 3, where relevant, should be taken into account for any type of design situation (e.g., for transient situations during repair works). Load Model 4 should be used only for some transient design situations. Load Model 1 consists of the following two partial systems: double-axle concentrated loads (tandem system, TS) and uniformly distributed loads (UDL system). For double-axle concentrated loads, each axle has a weight of $\alpha_Q Q_k$ where α_Q are adjustment factors. No more than one tandem system should be taken into account per notional lane and only complete tandem systems should be taken into account. For the assessment of general effects, each tandem system should be assumed to travel centrally along the axes of notional lanes. Each axle of the tandem system should be taken into account with two identical wheels, the load per wheel being therefore equal to $0.5\alpha_Q Q_k$. The contact surface of each wheel should be taken into account as a square of side 0.40 m (see Figure 3.7). On the other hand, uniformly distributed loads have a weight of $\alpha_q q_k$ per square meter of notional lane, where α_q are adjustment

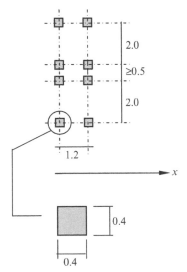

Figure 3.7 Application of tandem systems for local verifications according to EC1 [3.1].

factors. The uniformly distributed loads should be applied only in the unfavorable parts of the influence surface, longitudinally and transversally. LM1 is intended to cover flowing, congested, or traffic jam situations with a high percentage of heavy motortrucks. Load Model 1 should be applied on each notional lane and on the remaining areas. On notional Lane Number 1, the load magnitudes are referred to as $\alpha_{Qi}Q_{ik}$ and $\alpha_{qi}q_{ik}$ (see Table 3.1). On the remaining areas, the load magnitude is referred to as $\alpha_{qr}q_{rk}$. The values of adjustment factors α_{Qi}, α_{qi}, and α_{qr} should be selected depending on the expected traffic and possibly on different classes of routes. In the absence of specification, these factors should be taken equal to unity. The characteristic values of Q_{ik} and q_{ik}, dynamic amplification included, should be taken from Table 3.2. For local verifications, a tandem system should be applied at the most unfavorable location. Where two tandem systems on adjacent notional lanes are taken into account, they may be brought closer, with a distance between wheel axles not below 0.50 m (see Figure 3.7).

Load Model 2 consists of a single axle load $\beta_Q Q_{ak}$ with Q_{ak} equal to 400 kN, dynamic amplification included, which should be applied at any location on the carriageway. However, when relevant, only one wheel of $200\beta_Q$ (kN) may be taken into account. The value of β_Q should be specified. The contact surface of each wheel should be taken into account as a rectangle with sides 0.35 and 0.60 m (see Figure 3.8). The contact areas of Load Models 1 and 2 are different and correspond to different tire models, arrangements, and pressure distributions. The contact areas of Load Model 2, corresponding to twin tires, are normally relevant for orthotropic decks. For simplicity, the National Annex may adopt the same square contact

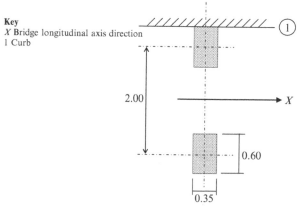

Figure 3.8 Load Model 2 according to EC1 [3.1].

surface for the wheels of Load Models 1 and 2. For Load Model 3 (special vehicles), where relevant, the models of special vehicles should be defined and taken into account. The National Annex may define Load Model 3 and its conditions of use. Finally, Load Model 4 (crowd loading), if relevant, should be represented by a load model consisting of a uniformly distributed load (which includes dynamic amplification) equal to 5 kN/m^2. The application of LM4 may be defined for the individual project. Load Model 4 should be applied on the relevant parts of the length and width of the road bridge deck, the central reservation being included, where relevant. This loading system, intended for general verifications, should be associated only with a transient design situation.

In the United States, the AASHTO [1.23] specifies vehicular live loading on the roadways of bridges based on designated HL-93 load model consisting of a combination of the design truck or design tandem and design lane load. Each design lane under consideration shall be occupied by either the design truck or tandem, coincident with the lane load. The weights and spacings of axles and wheels for the design truck specified in AASHTO are shown in Figure 3.9. A dynamic load allowance is considered for the design truck load. The spacing between the two 145,000 N axles varies between 4300 and 9000 mm to produce extreme force effects. It should be noted that the total design force effect is also a function of load factor, load modifier, load distribution, and dynamic load allowance. The design tandem shall consist of a pair of 110,000 N axles spaced 1200 mm apart. The transverse spacing of wheels shall be taken as 1800 mm. A dynamic load allowance is also considered for the design tandem load. The design lane load consists of a load of 9.3 N/mm uniformly distributed in the longitudinal direction. Transversely, the design lane load is assumed to be distributed over a 3000 mm width. The force effects from the design lane load are not subject to a dynamic load allowance. The tire contact area of a wheel consisting of one or two tires is assumed to be a single rectangle having a width of 510 mm and a length of 250 mm. The tire pressure is assumed to be uniformly distributed over the contact area.

The extreme force effect according to AASHTO [1.23] is taken as the larger of the effect of the design tandem combined with the effect of the design lane load or the effect of one design truck with the variable axle spacing combined with the effect of the design lane load and for both negative moment between points of contraflexure under a uniform load on all spans and reaction at interior piers only; 90% of the effect of two design trucks spaced a minimum P equal to the design wheel load (N). For the design

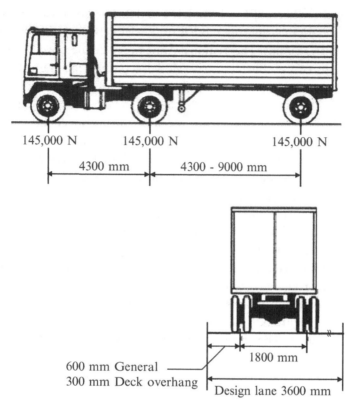

Figure 3.9 Characteristics of the design truck specified in AASHTO [1.23].

of deck overhangs with a cantilever, not exceeding 1800 mm from the centerline of the exterior girder to the face of a structurally concrete railing, the outside row of wheel loads may be replaced with a uniformly distributed line load of 14.6 N/mm intensity, located 300 mm from the face of the railing. To allow for dynamic effects, as specified in AASHTO, the static effects of the design truck or tandem, other than centrifugal and braking forces, shall be increased by the percentage specified in Table 3.3. For the dynamic load

Table 3.3 Dynamic Load Allowance (*IM*) Recommended by AASHTO [1.23]

Component	*IM* (%)
Deck joints—all limit states	75
All other components	
Fatigue and fracture limit state	15
All other limit states	33

allowance, the factor to be applied to the static load shall be taken as $(1 + IM/100)$. The dynamic load allowance is not applied to pedestrian loads or the design lane load.

3.4 HORIZONTAL FORCES ON STEEL AND STEEL-CONCRETE COMPOSITE BRIDGES

3.4.1 General

Steel and steel-concrete composite bridges are subjected to horizontal forces resulting from the moving trains or trucks and resulting from the environment. The horizontal forces may be transverse forces, acting transversely to the bridge direction, such as wind forces, lateral shock forces resulting from nosing of the trains, and centrifugal forces, or may be longitudinal forces, acting in the longitudinal direction of the bridge, such as traction and braking forces. In the next sections, the horizontal forces acting on the bridges will be highlighted. Once again, the forces presented are specified values in the current codes of practice.

3.4.2 Horizontal Forces on Railway Steel Bridges

3.4.2.1 Centrifugal Forces

Where the track on a bridge is curved over the whole or part of the length of the bridge, centrifugal force and track cant should be taken into account. According to EC1 [3.1], centrifugal forces should be taken to act outward in a horizontal direction at a height of 1.80 m above the running surface as shown in Figure 3.10. The centrifugal force shall always be combined with the vertical traffic load. The centrifugal force should not be multiplied by

Key
(1) Running surface
(2) Longitudinal forces acting along the centerline of the track
F_w^{**} Nosing force
h_t Height of centrifugal force over the running surface
h_w Height of wind force over the running surface
Q_{1a} Traction force
Q_{1b} Traction force
Q_s Nosing force
Q_t Centrifugal force
Q_v Vertical axle load
s Gauge
u ant

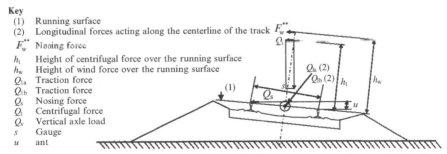

Figure 3.10 Notation and dimensions especially for railways according to EC1 [3.1].

any dynamic factors. The characteristic value of the centrifugal force shall be determined according to the following equations specified in EC1:

$$Q_{tk} = \frac{v^2}{g \times r}(f \times Q_{vk}) = \frac{V^2}{127r}(f \times Q_{vk}) \tag{3.6}$$

$$q_{tk} = \frac{v^2}{g \times r}(f \times q_{vk}) = \frac{V^2}{127r}(f \times q_{vk}) \tag{3.7}$$

where Q_{tk} and q_{tk} are characteristic values of the centrifugal forces in kN and kN/m, respectively. Q_{vk} and q_{vk} are characteristic values of the vertical loads (excluding any enhancement for dynamic effects) for Load Models 71, SW/ 0, SW/2, and "unloaded train." For HSLM, the characteristic value of the centrifugal force should be determined using Load Model 71. f is a reduction factor. v is the maximum speed in m/s. V is the maximum speed in km/h. g is the acceleration due to gravity 9.81 m/s^2. Finally, r is the radius of curvature in m. In the case of a curve of varying radii, suitable mean values may be taken for the value r.

The calculations shall be based on the specified maximum line speed at the site. In the case of Load Model SW/2, an alternative maximum speed may be assumed. For SW/2, a maximum speed of 80 km/h may be used. It is recommended that the individual project specify an increased maximum line speed at the site to take into account potential modifications to the infrastructure and future rolling stock. For Load Model 71 (and where required Load Model SW/0) and a maximum line speed at the site higher than 120 km/h, two cases should be considered. In the first case, Load Model 71 (and where required Load Model SW/0) with its dynamic factor and the centrifugal force for $V = 120$ km/h according to Equations (3.6) and (3.7) with $f = 1$ should be considered. While in the second case, Load Model 71 (and where required Load Model SW/0) with its dynamic factor and the centrifugal force according to Equations (3.6) and (3.7) for the maximum speed V specified should also be considered, with a value for the reduction factor f given by Equation (3.8). For Load Model 71 (and where required Load Model SW/0), the reduction factor f is subject to a minimum value of 0.35 and is given by

$$f = \left[1 - \frac{V - 120}{1000}\left(\frac{814}{V} + 1.75\right)\left(1 - \sqrt{\frac{2.88}{L_f}}\right)\right] \tag{3.8}$$

where L_f is the influence length of the loaded part of the curved track on the bridge, which is most unfavorable for the design of the structural element under consideration in meters as detailed in EC1 [3.1].

3.4.2.2 Nosing Force

The nosing (lateral shock) force shall be taken as a concentrated force acting horizontally, at the top of the rails, perpendicular to the centerline of the track. It shall be applied on both straight track and curved track. According to EC1 [3.1], the characteristic value of the nosing force shall be taken as $Q_{sk} = 100$ kN. It shall not be multiplied by any other factor. The nosing force shall always be combined with a vertical traffic load.

3.4.2.3 Traction and Braking Forces

Traction and braking forces act at the top of the rails in the longitudinal direction of the track. According to EC1, the forces shall be considered as uniformly distributed over the corresponding influence length $L_{a,b}$ for traction and braking effects for the structural element considered. The direction of the traction and braking forces shall take account of the permitted direction(s) of travel on each track. The characteristic values of traction and braking forces given in EC1 [3.1] are as follows:

$$\text{Traction force}: \quad Q_{lak} = 33(kN/m)L_{a,b}(m) \leq 1000(kN)$$
$$\text{For Load Models 71,SW/0,SW/2, and HSLM} \tag{3.9}$$

$$\text{Braking force}: \quad Q_{lbk} = 20(kN/m)L_{a,b}(m) \leq 6000(kN)$$
$$\text{For Load Models 71,SW/0,SW/2, and HSLM} \tag{3.10}$$

$$Q_{lbk} = 35(kN/m)L_{a,b}(m)$$
$$\text{For Load Model SW/2} \tag{3.11}$$

The characteristic values of traction and braking forces shall not be multiplied by any other factor.

EC1 [3.1] specifies that for Load Models SW/0 and SW/2, traction and braking forces need only to be applied to those parts of the bridge that are loaded according to Figure 3.2 and Table 3.1. Traction and braking may be neglected for the Load Model "unloaded train." These characteristic values are applicable to all types of track construction, e.g., continuous welded rails or jointed rails, with or without expansion devices. The earlier-mentioned traction and braking forces for Load Models 71 and SW/0 should be multiplied by the factor α in accordance with the requirements of Section 3.3.1, as specified in EC1 [3.1]. For loaded lengths greater than 300 m, additional requirements for taking into account the effects of braking should be specified. For lines carrying special traffic (e.g., restricted to high-speed passenger traffic), the traction and braking forces may be taken as equal to 25% of the sum of the axle loads (real train) acting on the influence length of the action effect of the structural element considered, with a maximum value of 1000 kN for Q_{lak} and 6000 kN for Q_{lbk}. The lines carrying special traffic and associated loading details may be specified. Traction and braking forces shall be combined with the corresponding vertical loads. In the case of a bridge carrying two or more tracks, the braking forces on one track shall be considered with the traction forces on the other track. Where two or more tracks have the same permitted direction of travel, either traction on two tracks or braking on two tracks shall be taken into account. It should be noted that braking and traction forces may be resisted using special systems of braking bracing added to the upper or lower wind bracing systems. In this case, their effect on the bridge components above the bearings can be neglected since the forces can be transmitted directly to the bearings.

3.4.2.4 Wind Forces

Wind actions on railway bridges fluctuate with time and act directly as pressures on the external surfaces of the main carrying systems of the bridge and on moving trains. Pressures act on areas of the surface resulting in forces normal to the surface of the main carrying systems of the bridge. The wind action is represented by a simplified set of pressures or forces whose effects are equivalent to the extreme effects of the turbulent wind. The wind actions calculated using the rules specified in EC1 (BS EN 1991-1-4) [3.2] are characteristic values determined from the basic values of wind velocity or the velocity pressure. The response of the bridge to wind actions depends on the size, shape, and dynamic properties of the bridge. EC1 [3.2] covers dynamic response due to along-wind turbulence in resonance with the along-wind vibrations of a fundamental flexural mode shape with constant sign. The response of the bridge should be calculated from the peak velocity pressure, q_p, at the reference height in the undisturbed wind field, the force and pressure coefficients, and the structural factor $c_s c_d$. q_p depends on the wind climate, the terrain roughness and topography, and the reference height. q_p is equal to the mean velocity pressure plus a contribution from short-term pressure fluctuations.

According to EC1 [3.2], wind forces are calculated for bridges of constant depth and with cross sections as shown in Figure 3.11 consisting of a single deck with one or more spans. Wind actions for other types of bridges (e.g., arch bridges, bridges with suspension cables or cable-stayed, roofed bridges, moving bridges, and bridges with multiple or significantly curved decks) may be defined in the National Annex. Wind actions on bridges produce forces in the x-, y-, and z-directions as shown in Figure 3.12, where

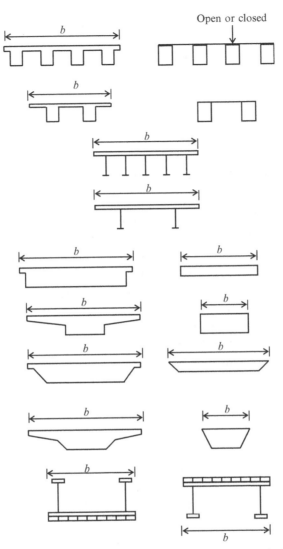

Figure 3.11 Limitations of cross sections of normal construction decks according to EC1 [3.2].

x-direction is the direction parallel to the deck width, perpendicular to the span; y-direction is the direction along the span; and z-direction is the direction perpendicular to the deck. The forces produced in the x- and y-directions are due to wind blowing in different directions and normally are not simultaneous. The forces produced in the z-direction can result from the wind blowing in a wide range of directions; if they are unfavorable and

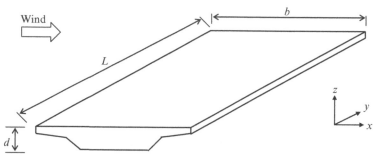

Figure 3.12 Directions of wind actions on bridges according to EC1 [3.1].

significant, they should be taken into account as simultaneous with the forces produced in any other direction. The notations used for bridges are L, length in y-direction; b, width in x-direction; and d, depth in z-direction.

It should be noted, according to EC1 [3.2], whether a dynamic response procedure for bridges is needed. If a dynamic response procedure is not needed, the factor $c_s c_d$ may be taken equal to 1.0. For normal road and railway bridge decks of less than 40 m span, a dynamic response procedure is generally not needed. For the purpose of this categorization, normal bridges may be considered to include bridges constructed of steel, concrete, aluminum, or timber, including composite construction, and whose shape of cross sections is generally covered by Figure 3.11. Force coefficients for wind actions on bridge decks in the x-direction are given by

$$c_{f,x} = c_{fx,0} \tag{3.12}$$

where $c_{fx,0}$ is the force coefficient without free-end flow specified in EC1 [3.2]. A bridge has usually no free-end flow because the flow is deviated only along two sides (over and under the bridge deck). For normal bridges, $c_{fx,0}$ may be taken equal to 1.3. Alternatively, $c_{fx,0}$ may be taken from Figure 3.13. Reference areas $A_{ref,x}$ are given (see Figure 3.13) and the following should be taken into account: for road bridges, a height of 2 m from the level of the carriageway, on the most unfavorable length, independently of the location of the vertical traffic loads for roadway bridges, and for railway bridges, a height of 4 m from the top of the rails, on the total length of the bridge. The reference height, z_e, may be taken as the distance from the lowest ground level to the center of the bridge deck structure, disregarding other parts (e.g., parapets) of the reference areas. Where it has been assessed that a dynamic response procedure is not necessary, the wind force in the x-direction may be obtained as follows:

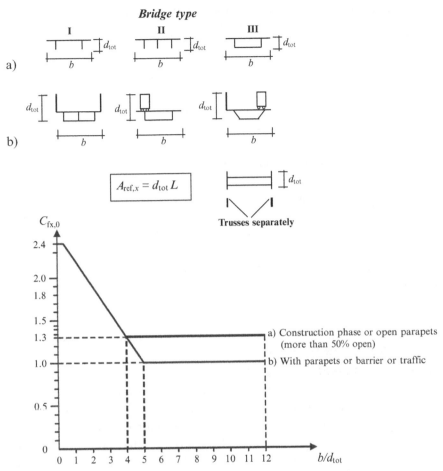

Figure 3.13 Force coefficient for bridges, $c_{fx,0}$, specified in EC1 [3.1].

$$F_w = \frac{1}{2}\rho v_b^2 C A_{ref,x} \qquad (3.13)$$

where v_b is the basic wind speed, C is the wind load factor, $A_{ref,x}$ is the reference area, ρ is the density of air, which is reasonably assumed as 1.25 kg/m^3.

$$v_b = c_{dir} \times c_{season} \times v_{b,0} \qquad (3.14)$$

The fundamental value of the basic wind velocity, $v_{b,0}$, is the characteristic 10 min mean wind velocity, irrespective of wind direction and time of year, at 10 m above ground level in open country terrain with low vegetation

Table 3.4 Values of the Force Factor C for Bridges Recommended by EC1 [3.1]

b/d_{tot}	$z_e \leq 20$ m	$z_e = 50$ m
≤ 0.5	5.7	7.1
≥ 4.0	3.1	3.8

such as grass and isolated obstacles with separations of at least 20 obstacle heights. The value of the directional factor, c_{dir}, for various wind directions may be found in the National Annex wherein the recommended value is 1.0, whereas the value of the season factor, c_{season}, may be given in the National Annex with a recommended value of 1.0. The wind load factor C can be calculated as follows based on c_e, which is the exposure factor given in EC1 [3.2]:

$$C = c_e \times c_{f,x} \tag{3.15}$$

The recommended values for C are shown in Table 3.4 as given by EC1 [3.2].

3.4.3 Horizontal Forces on Highway Steel and Steel-Concrete Composite Bridges

3.4.3.1 Braking and Acceleration Forces

Similar to railway bridges, highway steel and steel-concrete composite bridges are subjected to braking and acceleration forces. According to EC1 [3.1], braking force, Q_{lk}, shall be taken as a longitudinal force acting at the surfacing level of the carriageway. The characteristic value of Q_{lk} limited to 900 kN for the total width of the bridge and should be calculated as a fraction of the total maximum vertical loads corresponding to the Load Model 1 likely to be applied on Lane Number 1, as follows:

$$Q_{lk} = 0.6 \, \alpha_{Q1}(2Q_{1k}) + 0.1\alpha_{q1}q_{1k}w_1 L$$
$$180\alpha_{Q1}(kN) \leq Q_{1k} \leq 900(kN) \tag{3.16}$$

where L is the length of the deck or part of it under consideration. For example, $Q_{lk} = 360 + 2.7L$ (≤ 900 kN) for a 3 m wide lane and for a loaded length $L > 1.2$ m, if α factors are equal to unity. The upper limit (900 kN) may be adjusted in the National Annex. The value 900 kN is normally intended to cover the maximum braking force expected to pass over the bridge. Horizontal forces associated with Load Model 3 should be defined where appropriate. This force should be taken into account as located along the axis of any lane. However, if the eccentricity effects are not significant, the force may be considered to be applied only along the carriageway axis and uniformly distributed over the loaded length. Acceleration forces should be taken into account with the same magnitude as braking forces, but in the opposite direction.

In the United States, AASHTO [1.23,1.24] specifies that the braking force on highway bridges shall be taken as the greater than 25% of the axle weights of the design truck or design tandem or 5% of the design truck plus lane load or 5% of the design tandem plus lane load. The braking force shall be placed in all design lanes carrying traffic heading in the same direction. The forces shall act horizontally at a distance of 1.8 m above the roadway surface in either longitudinal direction to cause extreme force effects.

3.4.3.2 Centrifugal Forces

The centrifugal force Q_{tk}, specified in EC1, acting on highway bridges should be taken as a transverse force acting at the finished carriageway level and radially to the axis of the carriageway. The characteristic value of Q_{tk}, in which dynamic effects are included, should be taken from Table 3.5, where r is the horizontal radius of the carriageway centerline in meters and Q_v is the total maximum weight of vertical concentrated loads of the tandem systems of LM1, that is, $\sum_i \alpha_{Qi}(2Q_{ik})$ (see Table 3.2). Q_{tk} should be assumed to act as a point load at any deck cross section. Where relevant, lateral forces from skew braking or skidding should be taken into account. A transverse braking force, Q_{trk}, equal to 25% of the longitudinal braking or acceleration force, Q_{1k}, should be considered to act simultaneously with Q_{1k} at the finished carriageway level.

In the United States, AASHTO [1.23,1.24] recommends that for the purpose of computing the radial force or overturning effect on wheel loads, the centrifugal effect on live load shall be taken as the product of the axle weights of the design truck or tandem and the factor C, taken as

$$C = f\frac{v^2}{gR} \tag{3.17}$$

where v is the highway design speed in m/s, f is equal to 4/3 for load combinations other than fatigue and 1.0 for fatigue, g is gravitational acceleration

Table 3.5 Characteristic Values of Centrifugal Forces Recommended by EC1 [3.1]

$Q_{tk} = 0.2Q_v$ (kN)	If $r < 200$ m
$Q_{tk} = 40Q_v/r$ (kN)	If $200 \leq r \leq 1500$ m
$Q_{tk} = 0$	If $r > 1500$ m

(9.807 m/s^2), and R is radius of curvature of traffic lane in meters. Centrifugal forces shall be applied horizontally at a distance of 1.8 m above the roadway surface.

3.5 OTHER LOADS ON STEEL AND STEEL-CONCRETE COMPOSITE BRIDGES

3.5.1 Fatigue Loads

Steel and steel-concrete composite bridges are subjected to a stress spectrum and consequently fatigue owing to running traffic on the bridges. The stress spectrum depends on the geometry of the trains or trucks, the axle loads, the axle spacing, the composition of the traffic, and its dynamic effects. The current codes of practice specify fatigue load models as guidance for the assessment of fatigue load effects on highway and railway steel bridges.

3.5.1.1 Fatigue Loads on Highway Bridges

Specified in EC1 [3.1] are five fatigue load models of vertical forces on highway bridges. Fatigue Load Models 1, 2, and 3 are intended to be used to determine the maximum and minimum stresses resulting from the possible load arrangements on the bridge of any of these models. Fatigue Load Models 4 and 5 are intended to be used to determine stress range spectra resulting from the passage of motortrucks on the bridge. Fatigue Load Models 1 and 2 are intended to be used to check whether the fatigue life may be considered as unlimited when a constant stress amplitude fatigue limit is given. Therefore, they are appropriate for steel constructions and may be inappropriate for other materials. Fatigue Load Model 1 is generally conservative and covers multilane effects automatically. Fatigue Load Model 2 is more accurate than Fatigue Load Model 1 when the simultaneous presence of several motortrucks on the bridge can be neglected for fatigue verifications. Fatigue Load Models 3, 4, and 5 are not numerically comparable to Fatigue Load Models 1 and 2. Fatigue Load Model 3 may also be used for the direct verification of designs by simplified methods. Fatigue Load Model 4 is more accurate than Fatigue Load Model 3 for a variety of bridges and of the traffic when the simultaneous presence of several motortrucks on the bridge can be neglected. Fatigue Load Model 5 is the most general model, using actual traffic data. The load values given for Fatigue Load Models 1-3 are appropriate for typical heavy traffic on European main roads or motorways.

Table 3.6 Indicative Number of Heavy Vehicles Expected Per Year and Per Slow Lane Specified in EC1 [3.1]

Traffic Categories		N_{obs} Per year and Per Slow Lane
1	Roads and motorways with two or more lanes per direction with high flow rates of motortrucks	2.0×10^6
2	Roads and motorways with medium flow rates of motortrucks	0.5×10^6
3	Main roads and motorways with low flow rates of motortrucks	0.125×10^6
4	Local roads and motorways with low flow rates of motortrucks	0.05×10^6

A traffic category on a bridge should be defined according to EC1, for fatigue verifications, at least, by the number of slow lane and the number N_{obs} of heavy vehicles (maximum gross vehicle weight more than 100 kN), observed or estimated, per year and per slow lane (i.e., a traffic lane used predominantly by motortrucks). The traffic categories and values may be defined in the National Annex. Indicative values for N_{obs} are given in Table 3.6 for a slow lane when using Fatigue Load Models 3 and 4. On each fast lane (i.e., a traffic lane used predominantly by cars), additionally, 10% of N_{obs} may be taken into account. For the assessment of general action effects (e.g., in main girders), all fatigue load models should be placed centrally on the previously defined notional lanes. For the assessment of local action effects (e.g., in slabs), the models should be centered on notional lanes assumed to be located anywhere on the carriageway. According to EC1, fatigue Load Model 1 (similar to LM1) has the configuration of the characteristic Load Model 1, with the values of the axle loads equal to $0.7Q_{ik}$ and the values of the uniformly distributed loads equal to $0.3q_{ik}$ and (unless otherwise specified) $0.3q_{rk}$. The load values for Fatigue Load Model 1 are similar to those defined for the Frequent Load Model. However, adopting the Frequent Load Model without adjustment would have been excessively conservative in comparison with the other models, especially for large loaded areas. The maximum and minimum stresses should be determined from the possible load arrangements of the model on the bridge. Fatigue Load Model 2 consists of a set of idealized motortrucks, called "frequent" motortrucks, to be used as defined in the succeeding text. Each "frequent" motortruck" is defined by the number of axles and the axle spacing, the frequent load of each axle, the wheel contact areas, and the transverse distance between wheels (see Tables 3.7–3.9). The maximum and minimum stresses

Table 3.7 Set of Frequent Motortrucks Specified in EC1 [3.1]

1 Motortruck silhouette	2 Axle spacing (m)	3 Frequent axle loads (kN)	4 Wheel type (see table 3.9)
	4.5	90	A
		190	B
	4.2	80	A
	1.3	140	B
		140	B
	3.2	90	A
	5.3	180	B
	1.3	120	C
	1.3	120	C
		120	C
	3.4	90	A
	6.0	190	B
	1.8	140	B
		140	B
	4.8	90	A
	3.6	180	B
	4.4	120	C
	1.3	110	C
		110	C

should be determined from the most severe effects of different motortrucks, separately considered, traveling alone along the appropriate lane. Fatigue Load Model 3 (single-vehicle model) consists of four axles, each of them having two identical wheels. The geometry is shown in Figure 3.14. The weight of each axle is equal to 120 kN, and the contact surface of each wheel is a square of side 0.4 m. The maximum and minimum stresses and the stress ranges for each cycle of stress fluctuation, that is, their algebraic difference, resulting from the transit of the model along the bridge should be calculated. Fatigue Load Model 4 (set of "standard" motortrucks) consists of sets of standard motortrucks, which together produce effects equivalent to those of typical traffic on European roads. A set of motortrucks appropriate to the traffic mixes predicted for the route as defined in Tables 3.7 and 3.8 should be taken into account. Each standard motortruck is defined by the number of axles and the axle spacing, the equivalent load of each axle, the wheel contact areas, and the transverse distances between wheels. The calculations should be based on the following procedure: the percentage of each standard

Table 3.8 Set of Equivalent Motortrucks Specified in EC1 [3.1]

1 Vehicle Type Motortrucks	2 Axle spacing (m)	3 Equivalent axle loads (kN)	Traffic Type			7 Wheel type
			4 Long Distance Motortruck percentage	5 Medium Distance Motortruck percentage	6 Local Traffic Motortruck percentage	
	4.5	70	20.0	40.0	80.0	A
		130				B
	4.2	70	5.0	10.0	5.0	A
	1.3	120				B
		120				B
	3.2	70	50.0	30.0	5.0	A
	5.3	150				B
	1.3	90				C
	1.3	90				C
		90				C
	3.4	70	15.0	15.0	5.0	A
	6.0	140				B
	1.8	90				B
		90				B
	4.8	70	10.0	5.0	5.0	A
	3.6	130				B
	4.4	90				C
	1.3	80				C
		80				C

Table 3.9 Definition of Wheels and Axles According to EC1 [3.1]

Wheel/Axle Type	Geometric Definition

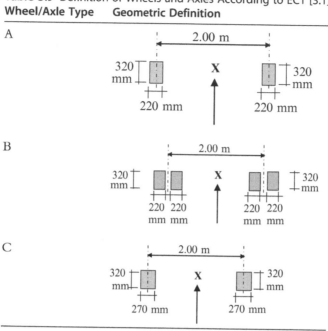

A

B

C

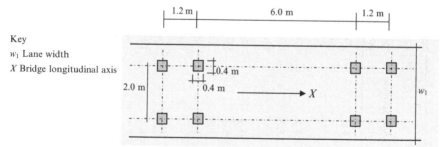

Key

w_l Lane width

X Bridge longitudinal axis

Figure 3.14 Fatigue Load Model 3 according to EC1 [3.1].

motortruck in the traffic flow, the total number of vehicles per year to be considered for the whole carriageway N_{obs} should be defined, and each standard motortruck is considered to cross the bridge in the absence of any other vehicle. Fatigue Load Model 5 (based on recorded road traffic data) consists of the direct application of recorded traffic data, supplemented, if relevant, by appropriate statistical and projected extrapolations.

In the United States, AASHTO [1.23,1.24] specifies that the fatigue load shall be one design truck or axles but with a constant spacing of 9000 mm between the 145,000 N axles, with the dynamic load allowance applied to the fatigue load. The frequency of the fatigue load shall be taken as the single-lane average daily truck traffic ($ADTT_{SL}$). This frequency shall be applied to all components of the bridge, even to those located under lanes that carry a lesser number of trucks. AASHTO specifies that in the absence of better information, the single-lane average daily truck traffic shall be taken as

$$ADTT_{SL} = p \times ADTT \qquad (3.18)$$

where ADTT is the number of trucks per day in one direction averaged over the design life and p is the fraction of truck traffic in a single lane, which is equal to 1.00, 0.85, and 0.8 for the number of lanes available to trucks equal to 1, 2, and 3 or more, respectively.

3.5.1.2 Fatigue Loads on Railway Bridges

Fatigue loads on railway steel bridges are also covered by EC1 [3.1] that recommends that a fatigue damage assessment shall be carried out for all structural elements of railway bridges subjected to fluctuations of stress. For normal traffic based on the characteristic values of Load Model 71, including the dynamic factor Φ, the fatigue assessment should be carried out on the basis of the traffic mixes, "standard traffic," "traffic with 250 kN axles," or "light traffic mix" depending on whether the structure carries mixed traffic, predominantly heavy freight traffic, or lightweight passenger traffic in accordance with the requirements specified. Details of the service trains and traffic mixes considered and the dynamic enhancement to be applied are given in Annex D of EC1. Each of the mixes is based on an annual traffic tonnage of 25×10^{6} t passing over the bridge on each track. For bridges carrying multiple tracks, the fatigue loading shall be applied to a maximum of two tracks in the most unfavorable positions. The fatigue damage should be assessed over the design working life. The design working life may be specified in the National Annex and 100 years is recommended. Vertical railway traffic actions including dynamic effects and centrifugal forces should be taken into account in the fatigue assessment. Generally, nosing and longitudinal traffic actions may be neglected in the fatigue assessment.

3.5.2 Dynamic Loads

3.5.2.1 General

The static stresses and deformations (and associated bridge deck acceleration) induced in a bridge are increased and decreased under the effects of moving traffic by the following three main parameters specified in EC1 [3.1]: The first parameter is the rapid rate of loading due to the speed of traffic crossing the structure and the inertial response (impact) of the structure. The second parameter is the passage of successive loads with approximately uniform spacing that can excite the structure and under certain circumstances create resonance (where the frequency of excitation (or a multiple thereof) matches a natural frequency of the structure (or a multiple thereof), there is a possibility that the vibrations caused by successive axles running onto the structure will be excessive). Finally, the variations in wheel loads result from track or vehicle imperfections, including wheel irregularities. For determining the effects (stresses, deflections, bridge deck acceleration, etc.) of railway traffic actions, the aforementioned parameters shall be taken into account. According to EC1, the main factors that influence dynamic behavior are the speed of traffic across the bridge; the span L of the structural element and the influence line length for deflection of the element being considered; the mass of the structure; the natural frequencies of the whole structure and relevant elements of the structure and the associated mode shapes (eigenforms) along the line of the track; the number of axles, axle loads, and the spacing of axles; the damping of the structure; vertical irregularities in the track; the unsprung/sprung mass and suspension characteristics of the vehicle; the presence of regularly spaced supports of the deck slab and/or track (cross girders, sleepers, etc.); vehicle imperfections (wheel flats, out of round wheels, suspension defects, etc.); and the dynamic characteristics of the track (ballast, sleepers, track components, etc.).

3.5.2.2 Dynamic Loads on Railway Bridges

Looking at dynamic loads acting on railway bridges, as an example, EC1 [3.1] provides some requirements for determining whether a static or a dynamic analysis is required based on V, which is the maximum line speed at the site in km/h; L, which is the span length in meters; n_0, which is the first natural bending frequency of the bridge loaded by permanent actions in Hz; n_T, which is the first natural torsional frequency of the bridge loaded by permanent actions in Hz; v, which is the maximum nominal speed in m/s; and $(v/n_0)_{lim}$, which is given in Annex F of EC1. The requirements are valid for simply supported bridges with only longitudinal line beam or simple plate behavior with negligible skew effects on rigid supports. A dynamic analysis is required where the frequent operating speed of a real train equals a resonant speed of the bridge. For bridges with a first natural frequency n_0 within the limits given by Figure 3.15 and a maximum line speed at the site not exceeding 200 km/h, a dynamic analysis is not required. For bridges with a first natural frequency n_0 exceeding the upper limit (1) in Figure 3.15, a dynamic analysis is required. For a simply supported bridge subjected to bending only, the natural frequency may be estimated using the following formula as specified in EC1:

$$n_0 = \frac{17.75}{\sqrt{\delta_0}} \tag{3.19}$$

where n_0 is in Hz and δ_0 is the deflection at midspan due to permanent actions in millimeters and is calculated, using a short-term modulus for

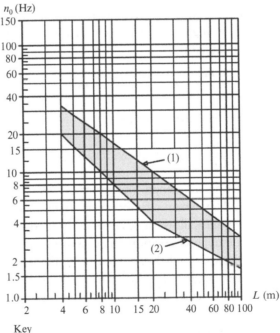

Key
(1) Upper limit of natural frequency
(2) Lower limit of natural frequency

Figure 3.15 Limits of bridge natural frequency n_0 (Hz) as a function of L (m) recommended by EC1 [3.1].

steel–concrete composite bridges, in accordance with a loading period appropriate to the natural frequency of the bridge.

According to EC1, the dynamic factor Φ takes account of the dynamic magnification of stresses and vibration effects in the structure but does not take account of resonance effects. Where a dynamic analysis is required, there is a risk that resonance or excessive vibration of the bridge may occur (with a possibility of excessive deck accelerations leading to ballast instability and excessive deflections and stresses). For such cases, a dynamic analysis shall be carried out to calculate impact and resonance effects. Quasi-static methods that use static load effects multiplied by the dynamic factor Φ are unable to predict resonance effects from high-speed trains. Dynamic analysis techniques, which take into account the time-dependent nature of the loading from the high-speed load model (HSLM) and real trains (e.g., by solving equations of motion), are required for predicting dynamic effects at resonance. Bridges carrying more than one track should be considered without any reduction of dynamic factor Φ. The dynamic factor Φ that enhances the static load effects under Load Models 71, SW/0, and SW/2 shall be taken as either Φ_2 or Φ_3. Generally, the dynamic factor Φ is taken as either Φ_2 or Φ_3 according to the quality of track maintenance as follows:
(a) For carefully maintained track,

$$\Phi_2 = \frac{1.44}{\sqrt{L_\Phi} - 0.2} + 0.82 \quad \text{with}: 1.00 \le \Phi_2 \le 1.67 \qquad (3.20)$$

(b) For track with standard maintenance,

$$\Phi_3 = \frac{2.16}{\sqrt{L_\Phi} - 0.2} + 0.73 \quad \text{with}: 1.0 \le \Phi_3 \le 2.0 \qquad (3.21)$$

where L_Φ is the "determinant" length (length associated with Φ) defined in Table 3.10 in meters. The dynamic factors were established for simply supported girders. The length L_Φ allows these factors to be used for other structural members with different support conditions. If no dynamic factor is specified, Φ_3 shall be used. The dynamic factor Φ shall not be used with the loading due to real trains, the loading due to fatigue trains, HSLM, and the load model "unloaded train." The determinant lengths L_Φ to be used are given in Table 3.10.

3.5.3 Accidental Forces

3.5.3.1 General

Steel and steel–concrete composite bridges may be subjected to forces resulting from accidental situations. The situations comprise vehicle collision with

Table 3.10 Determinant Lengths L_Φ According to EC1 [3.1]

Case	Structural Element	Determinant Length L_Φ
Steel deck plate: closed deck with ballast bed (orthotropic deck plate) (for local and transverse stresses)		
Deck with cross girders and continuous longitudinal ribs		
1.1	Deck plate (for both directions)	Three times the cross girder spacing
1.2	Continuous longitudinal ribs (including small cantilevers up to 0.50 m)	Three times the cross girder spacing
1.3	Cross girders	Twice the length of the cross girder
1.4	End cross girders	3.6 m
Deck plate with cross girders only		
2.1	Deck plate (for both directions)	Three times the cross girder spacing
2.2	Cross girders	Cross girder spacing + 3 m
2.3	End cross girders	3.6 m
Steel grillage: _open deck without ballast bed (for local and transverse stresses)_		
3.1	Rail bearers: – As an element of a continuous grillage – Simply supported	Three times the cross girder spacing Cross girder spacing + 3 m
3.2	Cantilever of rail bearer	3.6 m
3.3	Cross girders (as part of cross girder/continuous rail bearer grillage)	Twice the length of the cross girder
3.4	End cross girders	3.6 m

bridge piers, soffit of bridge, or decks; the presence of heavy wheels or vehicle on footways; and vehicle collision with curbs, vehicle parapets, and structural components. Since the main objective of this book is finite element analysis and design of steel and steel–concrete composite bridges, accidental forces can be easily applied as a load case and their effects on the bridges can be assessed.

3.5.3.2 Collision Forces from Vehicles Under the Bridge

Forces due to the collision of abnormal height or aberrant road vehicles with piers or with the supporting members of a bridge should be taken into account. The National Annex may define rules to protect the bridge from vehicular collision forces. When vehicular collision forces are to be taken into account (e.g., with reference to a safety distance between piers and the edge of the carriageway), the magnitude and location of vehicular collision forces and also the limit states should be considered. For stiff piers, the minimum values recommended in EC1 [3.1] are an impact force of 1000 kN in the direction of vehicle travel or 500 kN perpendicular to that direction with height above the level of adjacent ground surface equal to 1.25 m.

3.5.3.3 Collision Forces on Decks

Vehicle collision forces on bridge decks should also be specified as recommended in EC1. The National Annex may define the collision force on decks, possibly in relation to vertical clearance and other forms of protection. Collision loads on bridge decks and other structural components over roads may vary widely depending on structural and nonstructural parameters and their conditions of applicability. The possibility of collision by vehicles having an abnormal or illegal height may have to be envisaged and a crane swinging up while a vehicle is moving. Preventive or protective measures may be introduced as an alternative to designing for collision forces.

3.5.3.4 Actions from Vehicles on the Bridge

Collision forces from vehicles on footways and cycle tracks on road bridges are also covered by EC1 [3.1]. If a safety barrier of an appropriate containment level is provided, wheel or vehicle loading beyond this protection need not be taken into account. Where the protection mentioned is provided, one accidental axle load corresponding to $\alpha_{Q2} Q_{2k}$ should be so placed and oriented on the unprotected parts of the deck so as to give the most adverse effect adjacent to the safety barrier as shown in Figure 3.16. This axle load should not be taken into account simultaneously with any other variable load on the deck. A single wheel alone should be taken into account if geometric constraints make a two-wheel arrangement impossible.

3.5.3.5 Collision Forces on Curbs

The action from vehicle collision with curbs or pavement upstands should be taken as a lateral force equal to 100 kN acting at a depth of 0.05 m below the top of the curb, following the guidelines of EC1. This force should be considered as acting on a line 0.5 m long and is transmitted by the curbs to the structural members supporting them. In rigid structural members, the load should be assumed to have an angle of dispersal of 45°. When unfavorable, a vertical traffic load acting simultaneously with the collision force equal to $0.75\alpha_{Q1} Q_{1k}$ (see Figure 3.17) should be taken into account.

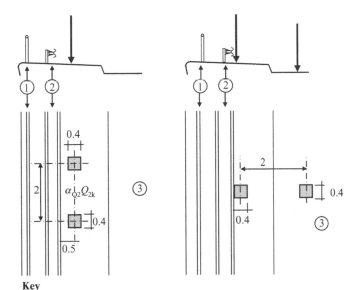

Key
(1) Pedestrian parapet (or vehicle parapet if a safety barrier
     is not provided)
(2) Safety barrier
(3) Carriageway

Figure 3.16 Examples showing locations of loads from vehicles on footways and cycle tracks of road bridges according to EC1 [3.1].

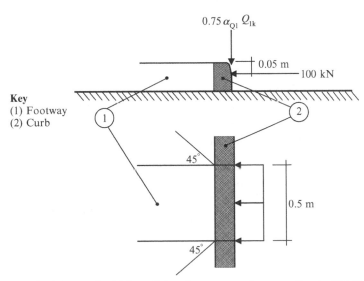

Figure 3.17 Definition of vehicle collision forces on curbs according to EC1 [3.1].

3.5.3.6 Collision Forces on Vehicle Restraint Systems

EC1 [3.1] recommends that horizontal and vertical forces transferred to the bridge deck by vehicle restraint systems should be taken into account for structural design. EC1 recommends four classes of values for the transferred horizontal force as given in Table 3.11. The horizontal force, acting transversely, may be applied 100 mm below the top of the selected vehicle restraint system or 1.0 m above the level of the carriageway or footway, whichever is the lower, and on a line 0.5 m long. The values of the horizontal forces given for the classes A-D derive from measurements during collision tests on real vehicle restraint systems used for bridges. There is no direct correlation between these values and performance classes of vehicle restraint systems. The proposed values depend rather on the stiffness of the connection between the vehicle restraint system and the curb or the part of the bridge to which it is connected. A very strong connection leads to the horizontal force given for class D. A very weak connection may lead to the horizontal force given for class A. The vertical force acting simultaneously with the horizontal collision force may be defined in the National Annex. The recommended values may be taken equal to $0.75\alpha_{Q1}Q_{1k}$. The calculations taking account of horizontal and vertical forces may be replaced, when possible, by detailing measures (for example, design of reinforcement). The structure supporting the vehicle parapet should be designed to sustain locally an accidental load effect corresponding to at least 1.25 times the characteristic local resistance of vehicle parapet (e.g., resistance of the connection of the parapet to the structure) and need not be combined with any other variable load.

3.5.3.7 Collision Forces on Structural Members

Vehicle collision forces on unprotected structural members above or beside the carriageway levels should be taken into account as recommended by EC1. The code recommends that the forces may act 1.25 m above the

Table 3.11 Four Classes for the Horizontal Force Transferred by Vehicle Restraint Systems Recommended by EC1 [3.1]

Recommended Class	Horizontal Force (kN)
A	100
B	200
C	400
D	600

carriageway level. The forces should not be considered to act simultaneously with any variable load. For some intermediate members where damage to one of which would not cause collapse (e.g., hangers or stays), smaller forces may be defined for the studied bridge.

3.5.3.8 Actions on Pedestrian Parapets

Forces that are transferred to the bridge deck by pedestrian parapets should be taken into account in structural design as variable loads and defined, depending on the selected loading class of the parapet as stated in EC1. For loading classes of pedestrian parapets, class C is the recommended minimum class. A line force of 1.0 kN/m acting, as a variable load, horizontally or vertically on the top of the parapet is a recommended minimum value for footways or footbridges. For service side paths, the recommended minimum value is 0.8 kN/m. For the design of the supporting structure, if pedestrian parapets are adequately protected against vehicle collision, the horizontal actions should be considered as simultaneous with the uniformly distributed vertical loads. Where pedestrian parapets cannot be considered as adequately protected against vehicle collisions, the supporting structure should be designed to sustain an accidental load effect corresponding to 1.25 times the characteristic resistance of the parapet, exclusive of any variable load.

3.5.4 Actions on Footways, Cycle Tracks, and Footbridges

Load models applicable to footways, cycle tracks, and footbridges are also covered by EC1. The models comprise a uniformly distributed load q_{fk} and a concentrated load Q_{fwk} that should be used for road and railway bridges and for footbridges, where relevant. All other variable actions and actions for accidental design situations defined in this section are intended only for footbridges. EC1 also specifies that for large footbridges (for example, more than 6 m width), load models defined in this section may not be appropriate and then complementary load models, with associated combination rules, may have to be defined for the individual project. Indeed, various human activities may take place on wide footbridges. Models and representative values given in this section should be used for serviceability and ultimate limit state calculations excluding fatigue limit states. EC1 specifies that the imposed loads defined in this section result from pedestrian and cycle traffic, minor common construction and maintenance loads (e.g., service vehicles), and accidental situations. These loads give rise to vertical and horizontal and static and dynamic forces. Loads due to cycle traffic are generally much lower than those due to pedestrian traffic, and the values given in this section are based on the frequent or occasional presence of pedestrians on cycle lanes. Special consideration may need to be given to loads due to horses or cattle for individual projects. The load models defined in this section do not describe actual loads. They have been selected so that their effects (with dynamic amplification included where mentioned) represent the effects of actual traffic. Actions for accidental design situations due to collision should be represented by static equivalent loads. Loads on footbridges may differ depending on their location and on the possible traffic flow of some vehicles. According to EC1, the same models should be used for pedestrian and cycle traffic on footbridges, on the areas of the deck of road bridges limited by pedestrian parapets and not included in the carriageway, and on the footpaths of railway bridges. Other appropriate models should be defined for inspection gangways within the bridges and for platforms on railway bridges. The recommended models, to be used separately in order to get the most unfavorable effects, are a uniformly distributed load of 2 kN/m^2 and a concentrated load of 3 kN applicable to a square surface of 0.20×0.20 m^2. Characteristic loads are intended for the determination of pedestrian or cycle track static load effects associated with ultimate limit state verifications and particular serviceability verifications.

According to EC1, three models should be taken into account as relevant on footways, cycle tracks, and footbridges. They consist of a uniformly distributed load q_{fk}, a concentrated load Q_{fwk}, and loads representing service vehicles Q_{serv}. For road bridges supporting footways or cycle tracks, a uniformly distributed load q_{fk} should be defined as shown in Figure 3.18. The recommended value is $q_{fk} = 5$ kN/m^2. For the design of footbridges, a uniformly distributed load q_{fk} should be defined and applied only in the

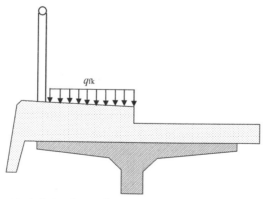

Figure 3.18 Characteristic load on a footway (or cycle track) according to EC1 [3.1].

unfavorable parts of the influence surface, longitudinally and transversally. The characteristic value of the concentrated load Q_{fwk} should be taken equal to 10 kN acting on a square surface of side 0.10 m. Finally, when service vehicles are to be carried on a footbridge or footway, one service vehicle Q_{serv} shall be taken into account. This vehicle may be a vehicle for maintenance, emergencies (e.g., ambulance and fire), or other services. The characteristics of this vehicle (axle weight and spacing and contact area of wheels), the dynamic amplification, and all other appropriate loading rules may be defined for the individual bridge.

The horizontal forces for footbridges are also specified in EC1. A horizontal force Q_{flk} should be taken into account, acting along the bridge deck axis at the pavement level, as defined in EC1. The characteristic value of the horizontal force should be taken equal to the greater than 10 percent of the total load corresponding to the uniformly distributed load and 60 percent of the total weight of the service vehicle. The horizontal force is considered as acting simultaneously with the corresponding vertical load and in no case with the concentrated load Q_{fwk}. EC1 states that this force is normally sufficient to ensure the horizontal longitudinal stability of footbridges. It does not ensure horizontal transverse stability, which should be ensured by considering other actions or by appropriate design measures. Accidental design situations for footbridges are also covered by EC1. Such situations are due to road traffic under the bridge (i.e., collision) or the accidental presence of a heavy vehicle on the bridge. Dynamic models of pedestrian loads are also specified in EC1.

3.5.5 Thermally Induced Loads

Temperature changes in bridges and their accompanied forces induced on the bridges are covered by EC1 (BS EN 1991-5) [3.3]. For the purposes of this part, steel and steel–concrete composite bridge decks are grouped into three types. Type 1 comprises steel bridges with steel decks supported by steel box girder or steel truss or plate girder main structural supporting systems (Figure 3.19). Type 2 comprises bridges with composite decks. Type 3 comprises bridges with concrete decks. Representative values of thermal actions should be assessed by the uniform temperature component and the temperature difference component. Where a horizontal temperature difference needs to be considered, a linear temperature difference component may be assumed in the absence of other information. The uniform temperature component depends on the minimum and maximum temperature that

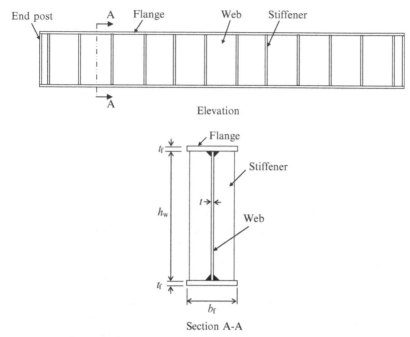

Figure 3.19 Definition of symbols for steel plate girders.

a bridge will achieve. This results in a range of uniform temperature changes, which, in an unrestrained structure, would result in a change in element length. The following effects should be taken into account: the restraint of associated expansion or contraction due to the type of construction (e.g., portal frame, arch, and elastomeric bearings); the friction at roller or sliding bearings; nonlinear geometric effects (second-order effects); and, for railway bridges, the interaction effects between the track and the bridge due to the variation of the temperature of the deck and of the rails that may induce supplementary horizontal forces in the bearings (and supplementary forces in the rails). Minimum shade air temperature (T_{min}) and maximum shade air temperature (T_{max}) for the site shall be derived from isotherms. The minimum and maximum uniform bridge temperature components $T_{e.min}$ and $T_{e.max}$ should be determined. EC1 specifies the recommended values for $T_{e.min}$ and $T_{e.max}$.

EC1 [3.3] states that characteristic values of minimum and maximum shade air temperatures for the site location shall be obtained, for example, from national maps of isotherms. Information on minimum and maximum shade air temperatures to be used in a country may be found in its National Annex.

These characteristic values should represent shade air temperatures for mean sea level in an open country with an annual probability exceeding 0.02. Where an annual probability exceeding 0.02 is deemed inappropriate, the minimum shade air temperatures and the maximum shade air temperatures should be modified in accordance with Annex A of EC1 [3.3]. The values of minimum and maximum uniform bridge temperature components for restraining forces shall be derived from the minimum (T_{min}) and maximum (T_{max}) shade air temperatures. The initial bridge temperature T_o at the time that the structure is restrained may be taken from Annex A of EC1 [3.3] for calculating contraction down to the minimum uniform bridge temperature component and expansion up to the maximum uniform bridge temperature component. Thus, the characteristic value of the maximum contraction range of the uniform bridge temperature component $\Delta T_{N,con}$ should be taken as

$$\Delta T_{N,con} = T_o - T_{e}.min \qquad (3.22)$$

and the characteristic value of the maximum expansion range of the uniform bridge temperature component $\Delta T_{N,exp}$ should be taken as

$$\Delta T_{N}, exp = T_{e}.max - T_o \qquad (3.23)$$

The overall range of the uniform bridge temperature component is expressed as follows:

$$\Delta T_{N} = T_{e}.max - T_{e}.min \qquad (3.24)$$

For bearings and expansion joints, the National Annex may specify the maximum expansion range of the uniform bridge temperature component and the maximum contraction range of the uniform bridge temperature component, if no other provisions are required. The recommended values are $(\Delta T_{N,exp} + 20)\,°C$ and $(\Delta T_{N,con} + 20)\,°C$, respectively. If the temperature at which the bearings and expansion joints are set is specified, then the recommended values are $(\Delta T_{N,exp} + 10)\,°C$ and $(\Delta T_{N,con} + 10)\,°C$, respectively. EC1 [3.3] states that for the design of bearings and expansion joints, the values of the coefficient of expansion given in Annex C (Table C.1 of EC1 [3.3]) may be modified if alternative values have been verified by tests or more detailed studies.

According to EC1 [3.3], over a prescribed time period, heating and cooling of a bridge deck's upper surface will result in a maximum heating (top surface warmer) and a maximum cooling (bottom surface warmer) temperature variation. The vertical temperature difference may produce effects within a structure due to restraint of free curvature due to the form of the structure (e.g., portal frame and continuous beams), friction at rotational bearings, and nonlinear geometric effects (second-order effects). EC1 [3.3] specifies two approaches for vertical linear component (Approaches 1 and 2). In Approach 1, the effect of vertical temperature differences should be considered by using an equivalent linear temperature difference component with $\Delta T_{M,heat}$ and $\Delta T_{M,cool}$. These values should be applied between the top and the bottom of the bridge deck. Values of $\Delta T_{M,heat}$ and $\Delta T_{M,cool}$ to be used in a country may be found in its National Annex. Recommended values for $\Delta T_{M,heat}$ and $\Delta T_{M,cool}$ are given in Table 3.12.

In Approach 2, vertical temperature components with nonlinear effects should be considered by including a nonlinear temperature difference component. Values of vertical temperature differences for bridge decks to be used in a country may be found in its National Annex. Recommended values are given in EC1 [3.3] and are valid for 40 mm surfacing depths for deck type 1 and 100 mm for deck types 2 and 3. For other depths of surfacing, see Annex B of EC1 [3.3]. Vertical temperature differences for bridge decks depends on "heating," which refers to conditions such that solar radiation and other effects cause a gain in heat through the top surface of the bridge deck, and "cooling," which refers to conditions such that heat is lost from the top surface of the bridge deck as a result of reradiation and other effects. In general, the temperature difference component need only be considered in the vertical direction. In particular cases, however (for example, when the orientation or configuration of the bridge results in one side being more highly exposed to sunlight than the other side), a horizontal temperature difference component should be considered. The National Annex may specify numerical values for the temperature difference. If no other information is available and no indications of higher values exist, 5 °C may be recommended as a linear temperature difference between the outer edges

Table 3.12 Values of Linear Temperature Difference Component for Different Types of Bridge Decks for Road, Foot, and Railway Bridges Recommended by EC1 [3.3]

Type of Deck	Top Warmer than Bottom $\Delta T_{M,heat}$ (°C)	Bottom Warmer than Top $\Delta T_{M,cool}$ (°C)
Type 1: steel deck	18	13
Type 1: composite deck	15	18
Type 1: concrete deck		
Concrete box girder	10	5
Concrete beam	15	8
Concrete slab	15	8

of the bridge independent of the width of the bridge. EC1 [3.3] recommends that care should be exercised in the design of large concrete box girder bridges where significant temperature differences can occur between the inner and outer web walls of such structures. The National Annex may specify numerical values for the temperature difference. The recommended value for a linear temperature difference is 15 °C.

3.6 LOAD COMBINATIONS

3.6.1 General

The different loads acting on steel and steel-concrete composite bridges, previously highlighted, should be grouped and superimposed to determine the worst case of loading that induce highest straining actions and consequently stresses at critical sections of the bridges. Grouping of the different loads acting on the bridges is commonly known as load combinations. The load combinations are dependent on the approaches adopted to design the bridge components, for example, allowable (permissible) stress design, limit state design, plastic design, and load and resistance factored design. The methods of design will be highlighted in the succeeding sections; however, in general, the concept of grouping different loads acting on the bridge is based on multiplying nominal or characteristic values of loads by partial safety factors to obtain the design value of the load. When several loads are to be grouped or combined, the partial safety factors should be reduced from their values for individual application of the loads in order to attain the same probability of occurrence of the combination as that of the individual loads.

3.6.2 Groups of Traffic Loads for Highway Bridges

Let us look at the grouping of traffic loads for highway bridges adopted in Europe, which is addressed in EC1 [3.1]. The code recommends that the simultaneity of the loading systems (Load Model 1, Load Model 2, Load Model 3, Load Model 4, and horizontal forces) and the loads of footways should be taken into account by considering the groups of loads defined in Table 3.13 specified the code. Each of these groups of loads, which are mutually exclusive, should be considered as defining a characteristic action for combination with nontraffic loads. The frequent action should consist only of either the frequent values of LM1 or the frequent value of LM2

Table 3.13 Assessment of Groups of Traffic Loads (Characteristic Values of the Multicomponent Action) Specified in EC1 [3.1]

Load Type	Carriageway						Footways and Cycle Tracks
	Vertical Forces				Horizontal Forces		Vertical Forces Only
Load system	LM1 (TS and UDL systems)	LM2 (single axle)	LM3 (special vehicles)	LM4 (crowd loading)	Braking and acceleration forces	Centrifugal and transverse forces	Uniformly distributed load
Groups of loads gr1a	Characteristic values						Combination value
gr1b		Characteristic value					
gr2	Frequent values				Characteristic value	Characteristic value	
gr3							Characteristic value
gr4				Characteristic value			Characteristic value
gr5	See Annex A		Characteristic value				

Table 3.14 Assessment of Groups of Traffic Loads (Frequent Values of the Multicomponent Action) Specified in EC1 [3.1]

Load Type		Carriageway		Footways and Cycle Tracks
		Vertical Forces		
Load system		LM1 (TS and UDL systems)	LM2 (Single axle)	Uniformly distributed load
Groups of loads	gr1a	Frequent values		
	gr1b		Frequent value	
	gr3			Frequent value

or the frequent values of loads on footways or cycle tracks, without any accompanying component, as defined in Table 3.14 specified in the code.

In the United States, AASHTO [1.24] adopts the load and resistance factor design (LRFD) methodology, where the total factored force effect shall be taken as

$$Q = \sum \eta_i \gamma_i Q_i \qquad (3.25)$$

where η_i is the load modifier, which is a factor relating to ductility, redundancy, and operational importance; γ_i are load factors specified in Tables 3.15 and 3.16 proposed by the specification; and Q_i are force effects from loads. The loads considered are classified as permanent and transient loads and forces. The permanent loads comprise downdrag (DD), dead loads of structural components and nonstructural attachments (DC), dead loads of wearing surfaces and utilities (DW), horizontal earth pressure load (EH), and accumulated locked-in force effects resulting from the construction process, including secondary forces from posttensioning (EL), earth surcharge load (ES), and vertical pressure from dead load of earth fill (EV). On the other hand, transient loads comprise vehicular braking force (BR), vehicular centrifugal force (CE), creep (CR), vehicular collision force (CT), earthquake (EQ), friction (FR), ice load (IC), vehicular dynamic load allowance (IM), vehicular live load (LL), live load surcharge, pedestrian live load (PL), settlement (SE), shrinkage (SH), temperature gradient (TG), uniform temperature (TU), water load and stream pressure (WA), wind on live load (WL), and wind on structure (WS). According to AASHTO [1.24], in the application of permanent loads, force effects of each of the six load types should be computed separately. Tables 3.15 and 3.16 present the load

Table 3.15 Load Combinations and Load Factors Specified in AASHTO [1.24]

Load combination Limit State	DC DD DW EH EV ES EL	LL IM CE BR PL LS	WA	WS	WL	FR	TU CR SH	TG	SE	Use One of These at a Time			
										EQ	IC	CT	CV
STRENGTH I (unless noted)	γ_P	1.75	1.00	—	—	1.00	0.5/1.2	γ_{TG}	γ_{SE}	—	—	—	—
STRENGTH II	γ_P	1.35	1.00	—	—	1.00	0.5/1.2	γ_{TG}	γ_{SE}	—	—	—	—
STRENGTH III	γ_P	—	1.00	1.40	—	1.00	0.5/1.2	γ_{TG}	γ_{SE}	—	—	—	—
STRENGTH IV EH, EV, ES, DW DC ONLY	γ_P 1.5	—	1.00	—	—	1.00	0.5/1.2	—	—	—	—	—	—
STRENGTH V	γ_P	1.35	1.00	0.40	1.00	1.00	0.5/1.2	γ_{TG}	γ_{SE}	—	—	—	—
EXTREME EVENT I	γ_P	γ_{EQ}	1.00	—	—	1.00	—	—	—	1.0	—	—	—
EXTREME EVENT II	γ_P	0.50	1.00	—	—	1.00	—	—	—	—	1.0	1.0	1.0
SERVICE I	1.00	1.00	1.00	0.30	1.00	1.00	0.5/1.2	γ_{TG}	γ_{SE}	—	—	—	—
SERVICE II	1.00	1.30	1.00	—	—	1.00	0.5/1.2	—	—	—	—	—	—
SERVICE III	1.00	0.80	1.00	—	—	1.00	0.5/1.2	γ_{TG}	γ_{SE}	—	—	—	—
SERVICE IV	1.00	—	1.00	0.70	—	1.00	0.5/1.2	—	1.0	—	—	—	—
FATIGUE I-LL, IM & CE ONLY	—	1.50	—	—	—	—	—	—	—	—	—	—	—
FATIGUE II-LL, IM & CE ONLY	—	0.75	—	—	—	—	—	—	—	—	—	—	—

Table 3.16 Load Factors for Permanent Loads, γ_p, Specified in AASHTO [1.24]

Type of Load	Load Factor	
	Maximum	Minimum
DC: component and attachments	1.25	0.90
DC: STRENGTH IV only	1.50	0.90
DD: downdrag, piles, α Tomlinson method	1.40	0.25
DD: downdrag, λ method	1.05	0.30
DD: downdrag, drilled shafts, O'Neil and Reese (1999) method	1.25	0.35
DW: wearing surfaces and utilities	1.50	0.65
EH: horizontal earth pressure		
• Active	1.50	0.90
• At rest	1.35	0.90
• *AEP* for anchored walls	1.35	N/A
EL: locked-in erection stresses	1.00	1.00
EV: vertical earth pressure		
• Overall stability	1.00	N/A
• Retaining walls and abutments	1.35	1.00
• Rigid buried structure	1.30	0.90
• Rigid frames	1.35	0.90
• Flexible buried structures	1.50	0.90
o Metal box culverts and structural plate culverts with deep corrugations	1.30	0.90
o Thermoplastic culverts	1.95	0.90
o All others		
ES: earth surcharge	1.50	0.75

combinations and load factors adopted by AASHTO for different loads and permanent loads, respectively.

3.6.3 Groups of Traffic Loads for Railway Bridges

The load combinations on railway bridges are also specified in EC1 [3.1] such that the simultaneity of the loading may be taken into account by considering the groups of loads defined in Table 3.17 specified in the code. EC1 recommends that each of these groups of loads, which are mutually exclusive, should be considered and applied as defining a single-variable characteristic action for combination with nontraffic loads. In some cases, it is necessary to consider other appropriate combinations of unfavorable individual traffic actions as specified by EC0 (BS EN 1990) [3.4]. The factors given in the table should be applied to the characteristic values of the

Table 3.17 Assessment of Groups of Loads for Railway Traffic (Characteristic Values of the Multicomponent Actions) Specified in EC1 [3.1]

Number of Tracks on Structure			Groups of Loads			Vertical Forces			Horizontal Forces		
1	2	≥3	Number of tracks	Load group	Loaded track	LM71 SW/0 HSLM	SW/2	Unloaded train	Traction braking	Centrifugal force	Nosing force
			1	gr11	T_1	1			1	0.5	0.5
			1	gr12	T_1	1			0.5	1	1
			1	gr13	T_1	1			1	0.5	0.5
			1	gr14	T_1	1			0.5	1	1
			1	gr15	T_1			1		1	1
			1	gr16	T_1		1		1	0.5	0.5
			1	gr17	T_1		1		0.5	1	1
			2	gr21	T_1	1			1	0.5	0.5
					T_2	1			1	0.5	0.5
			2	gr22	T_1	1			0.5	1	1
					T_2	1			0.5	1	1
			2	gr23	T_1	1			1	0.5	0.5
					T_2	1			1	0.5	0.5
			2	gr24	T_1	1			0.5	1	1
					T_2	1			0.5	1	1
			2	gr26	T_1		1		1	0.5	0.5
					T_2				1	0.5	0.5
			2	gr26	T_1	1	1		0.5	1	1
					T_2				0.5	1	1
			≥3	gr31	T_1	0.75			0.75	0.75	0.75

Dominant component action as appropriate
To be considered in designing a structure supporting one track
To be considered in designing structure supporting two tracks
To be considered in designing structure supporting three or more tracks

different actions considered in each group. It should be noted that where groups of loads are not taken into account, railway traffic actions shall be combined in accordance with EC0 [3.4].

3.7 DESIGN APPROACHES

3.7.1 General

The design of steel and steel–concrete composite bridges should fulfill the basic requirements of design, wherein, over its intended life, the bridges should sustain the applied loads and remain fit for use. Therefore, the current codes of practice proposed design rules and guides to ensure that the bridges, like other structures, must have a specified strength, perform in an acceptable manner, and be durable over the intended life. The design rules and guides were based on two main design approaches, which are briefly highlighted in this section. The design approaches are commonly known as *allowable (permissible) stress design approach* and *limit state design approach*.

3.7.2 Allowable Stress Design Approach

Earlier design rules specified in the current codes of practice were based on allowable (permissible) stress design approach. In this design approach, a factor of safety was adopted to account for the uncertainty in the loading, in material properties, in empirical design equations, and in the construction process. The allowable stresses were predicted by dividing the failure stress by the factor of safety. The failure stress may be taken as the yield stress or the proportional limit stress of the material of construction. In this design approach, a structural analysis could be performed to evaluate the stresses at the specified combination of loads, which were then checked against the specified allowable stresses. The allowable stress design approach was commonly used in the past owing to its simplicity and safety. Because stresses, and hence deformations/deflections, were kept at low levels, nonlinearity of material and/or structural behavior could be neglected and working stresses were calculated from linear elastic theories. In performing the structural analyses, stresses from various loads could be added together. However, this design approach has some disadvantages, which are mainly due to the use of a single factor of safety with different applied loads, and the analysis of the structure under working loads may not provide a realistic assessment of the behavior of the structure at failure. It should be noted that structures designed adopting the allowable stress design approach have moderate stresses in service conditions, and thus, the serviceability requirements such as deflections, cracking, slip, and deformations were not generally critical.

3.7.3 Limit State Design Approach

By gaining better understanding of the behavior of different loads acting on structures and material properties, which is accompanied by improved calculation and construction techniques, limit state design approach replaced the traditional allowable stress design approach in most current codes of practice. Limit state design approach considers that the structure should sustain all loads and deformations liable to occur during its construction, perform adequately in normal use, and have adequate durability. For most structures, the limit states can be classified into two main states that are the ultimate and serviceability limit states. The ultimate limit states are related to a collapse of the whole or a substantial part of the structure. On the other hand, the serviceability limit states are related to the disruption of the normal use of the structure. Ultimate limit states should have a very low probability of occurrence since they are considered failure situations. Examples of ultimate limit states include loss of static equilibrium of a part or the whole of the structure, loss of load-bearing capacity of a member due to its material strength being exceeded or due to buckling, or a combination of these two phenomena, or fatigue, and finally overall instability. While the serviceability limit states depend on the function of the structures and for bridges, they correspond to excessive deformation of the structure, or any of its parts, affecting the appearance and functional use or drainage or causing damage to nonstructural components like deck joints and surfacing; excessive local damage like cracking, splitting, spalling, yielding, or slipping, affecting appearance, use, or durability of the structure; and finally excessive vibration causing discomfort to pedestrians or drivers.

3.7.4 Limit State Design Codes

To provide design methods in the current codes of practice achieving the basic design requirements of structures, a reliability approach was commonly adopted. Design values are determined such that they have a known statistical probability of being achieved. The values of actions (loads) have a known (low) probability of not being exceeded and the values for strength have a known (high) probability level of being achieved. The design procedure is then to model and evaluate the behavior of a structural model in order to verify that calculated effects due to the actions do not exceed the design strength/deformation limits. The reliability approach is achieved through the use of limit state design principles adopted nowadays in most current codes of practice.

The design rules specified in the Eurocode are based on the limit state design approach. According to EC0 [3.4], the design verification of the ultimate limit states is governed by the following equation:

$$E_d \leq R_d \qquad (3.26)$$

where E_d is the design value of the effects of actions (internal moment, axial force, etc.) and R_d is the design value of the corresponding resistance. At ultimate limit states, actions (i.e., the internal bending moments and axial forces due to the applied loadings and displacements) are expressed in terms of combinations of actions that can occur simultaneously. The basic expression is expressed as follows:

$$E\left(\sum \gamma_{G,j} G_{k,j} + \gamma_p P + \gamma_{Q,1} Q_{k,1} + \sum_{i>1} \gamma_{Q,i} \psi_{0,i} Q_{k,i}\right) \qquad (3.27)$$

where $G_{k,j}$ is the characteristic value of the jth permanent action, P is the permanent action caused by controlled forces or deformation (prestressing), $Q_{k,1}$ is the characteristic value of the "leading" variable action, and $Q_{k,i}$ are the accompanying variable actions. The $E()$ denotes "the effect of" and the "+" signs denote the combination of effects due to the separate actions. Permanent actions are self-weight (typically the weight of steel, concrete, and superimposed load such as surfacing and parapets); the partial factors γ_G applied to each type of permanent action may be different, hence the summation term and the j index subscript. The γ_p factor is related to prestressing actions and may be ignored. The variable actions are either direct (the weight of traffic, the wind pressure, etc.) or indirect (expansion/contraction due to temperature). The partial factors γ_Q depend on the type of action and its predictability. It is unlikely that the most adverse loading from one action will occur simultaneously with that due to a different action. In recognition of this, EC refers to one action as a "leading action" and the other actions as "accompanying"; a reduction factor ψ is applied to accompanying actions. In principle, each different action should in turn be considered as the leading action, to determine which combination of leading and accompanying actions is the most onerous, but for simplified highway bridge design, it may be assumed that the traffic loading is the leading action. There are similar expressions for combinations of actions in accidental and seismic situations, each with a different set of partial factors, but these are not of concern for simplified design. The design value of resistance is given by Eurocode and its value is determined from characteristic resistance values divided by partial factors on material strength γ_M.

According to EC0 [3.4], the design verification of the serviceability limit states is governed by the following equation:

$$E_d \leq C_d \tag{3.28}$$

where E_d is the design value of the effects of actions in the serviceability limit state criterion and C_d is the limiting design value of the relevant criterion. At serviceability limit state, there are in principle three combinations of actions to consider: characteristic, frequent, and quasi-permanent. For bridges, the characteristic combination is used for checking that no inelastic response occurs; the frequent combination is used if deflection needs to be checked (this includes evaluation of dynamic response). The quasi-permanent combination relates to long-term effects; for bridges, provided that the appropriate modulus of elasticity is used for long-term actions, this combination only needs to be considered when determining crack widths in concrete. Only the characteristic combination is relevant to simplified design. For the characteristic loading combination, the same characteristic values of actions are used at ultimate limit states but all the γ factors are taken as unity. Thus, the expression becomes

$$E\left(\sum G_{k,j} + P + Q_{k,1} + \sum_{i>1} \psi_{0,i} Q_{k,i}\right) \tag{3.29}$$

The serviceability limit state criterion that might need to be considered in simplified design is the limitation that stresses in steel should not exceed the yield stress. This limitation would need to be considered if the ultimate limit state design resistance were based on plastic bending resistance. It must then be verified that the stress calculated elastically at serviceability limit states does not exceed yield. No partial factor is applied to yield stress (strictly, $\gamma_M = 1$).

In the United States, AASHTO [1.23] also specifies that bridges shall be designed for specialized limit states to achieve the objectives of constructability, safety, and serviceability. AASHTO adopts the load and resistance factor design, which is based on limit state design approach. Each component and connection in the bridge shall satisfy the following condition, which assumes that all limit states shall be considered of equal importance:

$$\sum \eta_i \gamma_i Q_i \leq \varphi R_n = R_r \qquad (3.30)$$

in which for loads for which a maximum value of γ_i is appropriate

$$\eta_i = \eta_D \eta_R \eta_I \geq 0.95 \qquad (3.31)$$

and for loads for which a minimum value of γ_i is appropriate

$$\eta_i = \frac{1}{\eta_D \eta_R \eta_I} \leq 1.0 \qquad (3.32)$$

where η_i is the load modifier, which is a factor relating to ductility, redundancy, and operational importance; η_D is a factor relating to ductility; η_R is a factor relating to redundancy; η_I is a factor relating to operational importance; γ_i are load factors specified in Tables 3.15 and 3.16 given by the specification; Q_i are force effects from loads; ϕ is resistance factor; R_n is the nominal resistance; and R_r is the factored resistance (φR_n). For service and extreme event limit states, resistance factors shall be taken as 1.0, except for bolts. According to AASHTO [1.23], the limit states are intended to provide a buildable, serviceable bridge, capable of safely carrying design loads for a specified lifetime. The resistances of components and connections can be determined on the basis of inelastic behavior, although the force effects are determined using elastic analysis. Equation (3.30) is the basis of LRFD methodology. Assign resistance factor $\phi = 1.0$ to all nonstrength limit states. Components and connections of a bridge shall satisfy Equation (3.30) for the applicable combinations of factored extreme force effects as specified at each of the limit states denoted as STRENGTH I, which is basic load combinations related to the normal vehicular use of the bridge without wind; STRENGTH II, which is load combinations related to the use of the bridge by owner-specified special design vehicles, evaluation permit vehicles, or both without wind; STRENGTH III, which is load combinations related to the bridge exposed to wind velocity exceeding 90 km/h; STRENGTH IV, which is load combination related to very high dead load to live load force effect ratios; STRENGTH V, which is load combination related to normal vehicular use of the bridge with wind of 90 km/h velocity; EXTREME EVENT V, which is load combination including earthquake; EXTREME EVENT II, which is load combination related to ice load, collision by vessels and vehicles, and certain hydraulic events with a reduced live load; SERVICE I, which is load combination related to the normal operational use of the bridge with a 90 km/h, deflection control in buried metal structures, tunnel liner plate, and slope satiability; SERVICE II, which is load combination related to control yielding of steel structures and slip of slip–critical connections due to vehicular live load; SERVICE III, which is load combination for longitudinal analysis related to tension in prestressed concrete superstructures with the objective of crack control; SERVICE IV, which is load combination only to tension in prestressed concrete structures with the objective of crack control; and finally, FATIGUE–fatigue and fracture load combination relating to repetitive gravitational vehicular live load and dynamic responses under a single design truck having specific axle spacing. As stated in AASHTO [1.23], the load factors of various loads comprising a design load combination shall be taken as specified in Table 3.15. For permanent force effects, the load factor that produces the more critical combination shall be selected from Table 3.16.

3.8 STABILITY OF STEEL AND STEEL-CONCRETE COMPOSITE PLATE GIRDER BRIDGES

3.8.1 General

In order to design the components of steel and steel-concrete composite bridges, it would helpful to review the design rules specified in the current codes of practice. As an example, a review of the rules specified in EC3 [1.27,2.11] is presented in this section. However, it should be noted that the main objective of this book is to highlight the finite element analysis and design of steel and steel-concrete composite bridges. Therefore, the finite element analysis results can be compared with the design results obtained using any current code of practice. According to EC3 [1.27,2.11], the internal forces and moments may be determined using either elastic global analysis or plastic global analysis, with elastic global analysis being used in all cases. Plastic global analysis may be used only where the structure has sufficient rotation capacity at the actual locations of the plastic hinges, whether this is in the members or in the joints. Where a plastic hinge occurs in a member, the member cross sections should be double symmetrical or single symmetrical with a plane of symmetry in the same plane as the rotation of the plastic hinge. Where a plastic hinge occurs in a joint, the joint should either have sufficient strength to ensure the hinge remains in the member or should be able to sustain the plastic resistance for a sufficient rotation.

According to EC3 [1.27,2.11], elastic global analysis should be based on the assumption that the stress–strain behavior of the material is linear, whatever the stress level is. Internal forces and moments may be calculated according to elastic global analysis even if the resistance of a cross section is based on its plastic resistance. Elastic global analysis may also be used for cross sectioning the resistances of which are limited by local buckling. On the other hand, plastic global analysis allows for the effects of material nonlinearity in calculating the action effects of a structural system. The behavior should be modeled by elastic-plastic analysis with plastified sections and/or joints as plastic hinges, by nonlinear plastic analysis considering the partial plastification of members in plastic zones, or by rigid plastic analysis neglecting the elastic behavior between hinges. Plastic global analysis may be used where the members are capable of sufficient rotation capacity to enable the required redistributions of bending moments to develop. Also, plastic global analysis should only be used where the stability of members at plastic hinges can be assured. A bilinear stress-strain relationship may be used for different structural steel grades. Rigid plastic analysis may be applied if no effects of the deformed geometry (e.g., second-order effects) have to be considered.

The role of cross-sectional classification is to identify the extent to which the resistance and rotation capacity of cross sections is limited by its local buckling resistance. Four classes of cross sections are defined in EC3 [2.11]: class 1 cross sections, which are sections that can form a plastic hinge with the rotation capacity required from plastic analysis without reduction of the resistance; class 2 cross sections, which are sections that can develop their plastic moment resistance, but have limited rotation capacity because of local buckling; class 3 cross sections, which are sections in which the stress in the extreme compression fiber of the steel member assuming an elastic distribution of stresses can reach the yield strength, but local buckling is liable to prevent the development of the plastic moment resistance; and, finally, class 4 cross sections, which are sections in which local buckling will occur before the attainment of yield stress in one or more parts of the cross section. In class 4 cross sections, effective widths may be used to make the necessary allowances for reductions in resistance due to the effects of local buckling. The classification of a cross section depends on the width to thickness ratio of the parts subject to compression. Compression parts include every part of a cross section that is either totally or partially in compression under the load combination considered. The various compression parts in a cross section (such as a web or flange) can, in general, be in different classes. A cross section is classified according to the highest (least favorable) class of its compression parts. The limiting proportions for class 1, 2, and 3 compression parts are specified in EC3. A part that fails to satisfy the limits for class 3 should be regarded as class 4. Where the web is considered to resist shear forces only and is assumed not to contribute to the bending and normal force resistance of the cross section, the cross section may be designed as class 2, 3, or 4, depending only on the flange class. EC3 also specifies cross-sectional requirements for plastic global analysis.

3.8.2 Bending Moment Resistance of Steel Plate Girders

Steel plate girders are used to carry larger loads over longer spans that are possible with traditional universal rolled I-sections. Plate girders are used in bridges mainly as main girders (see Figure 1.20) and as cross girders in bridges having wide cross sections and bridges carrying multiple traffic tracks. Steel plate girders are constructed by welding steel plates together to form I-sections as shown in Figure 3.17. The web of a plate girder is relatively thin, and stiffeners are required mainly to prevent buckling due to compression from bending and shear. Stiffeners are also required at load points, supports, and within panels. The depth of a plate girder (h) may be taken as one-tenth to one-twelfth of the span (L/10 to L/12) for main girders and as one-seventh to one-ninth of the span (L/7 to L/9) for cross girders. The breadth of flange plate is made about one-fifth of the depth (h/5). The deeper the girder is made, the smaller the flange plates required. However, the web plate must then be made thicker or additional stiffeners are provided to meet particular design requirements. Flange cover plates can be kept constant throughout, curtailed as shown in Figure 3.20, or single flange plates can be reduced in thickness when reduction in bending moment permits as shown in Figure 3.21. Long plate girders are commonly divided into parts to facilitate transportation process, connected at the field of construction using splices, as shown in Figure 3.22. The figure shows typical welded and bolted splices of plate girders.

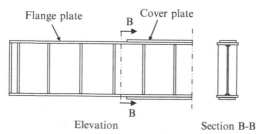

Figure 3.20 Curtailed cover plates.

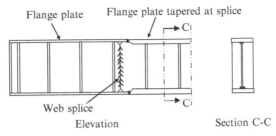

Figure 3.21 Constant depth plate girder.

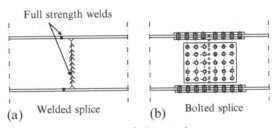

Figure 3.22 Welded and bolted splices of plate girders.

According to EC3 [1.27,2.11], the design value of the bending moment M_{Ed} for rolled beams and plate girders at each cross section shall satisfy

$$\frac{M_{Ed}}{M_{c,Rd}} \leq 1.0 \tag{3.33}$$

where $M_{c,Rd}$ is the design resistance for bending about one principal axis of a cross section considering fastener holes calculated as follows:

$$M_{c,Rd} = M_{pl,Rd} = \frac{W_{pl}f_y}{\gamma_{M0}} \quad \text{for class 1 or 2 cross sections} \tag{3.34}$$

$$M_{c,Rd} = M_{el,Rd} = \frac{W_{el,min}f_y}{\gamma_{M0}} \quad \text{for class 3 cross sections} \tag{3.35}$$

$$M_{c,Rd} = M_{el,Rd} = \frac{W_{eff,min}f_y}{\gamma_{M0}} \quad \text{for class 4 cross sections} \tag{3.36}$$

where W_{pl} is the plastic section modulus, $W_{el,min}$ is the minimum elastic section modulus, and $W_{eff,min}$ is the minimum effective section modulus. $W_{el,min}$ and $W_{eff,min}$ corresponds to the fiber with the maximum elastic stress. Fastener holes in the tension flange may be ignored provided that for the tension flange,

$$\frac{A_{f,net}0.9f_u}{\gamma_{M2}} \geq \frac{A_f f_y}{\gamma_{M0}} \tag{3.37}$$

where A_f is the area of the tension flange; γ_{M0} is a partial factor related to the resistance of cross section whatever the class is, which is taken equal to 1.0; and γ_{M2} is a partial factor related to the resistance of cross section in tension to fracture, which is equal to 1.25. Fastener holes in tension zone of the web need not be allowed for. Fastener holes except for oversize and slotted holes in the compression zone of the cross section need not be allowed for, provided that they are filled by fasteners.

3.8.3 Lateral Torsional Buckling of Plate Girders in Bending

When beams and plate girders are subjected to bending moment, the compression flange will be subjected to lateral torsional buckling. The lateral torsional buckling of the compression flange depends on the loading conditions, lateral restraint conditions, and geometries of the compression flange. EC3 [1.27,2.11] recommends that a laterally unrestrained member subject to major axis bending should be verified against lateral torsional buckling as follows:

$$\frac{M_{Ed}}{M_{b,Rd}} \leq 1.0 \tag{3.38}$$

where M_{Ed} is the design value of the moment and $M_{b,Rd}$ is the design buckling resistance moment. Beams with sufficient restraint to the compression flange are not susceptible to lateral torsional buckling. In addition, beams with certain types of cross sections, such as square or circular hollow sections, fabricated circular tubes, or square box sections, are not susceptible to lateral torsional buckling. The design buckling resistance moment of a laterally unrestrained beam should be taken as

$$M_{b,Rd} = \chi_{LT} W_y \frac{f_y}{\gamma_{M1}} \tag{3.39}$$

where W_y is the appropriate section modulus, which is taken as $W_{pl,y}$ for class 1 or 2 cross sections or $W_{el,y}$ for class 3 cross sections or $W_{eff,y}$ for class 4 cross sections, and χ_{LT} is the reduction factor for lateral torsional buckling. It should be noted that, according to EC3 [1.27,2.11], determining W_y holes for fasteners at the beam end need not be taken into account. Also, for bending members of constant cross section, the value of χ_{LT} for the appropriate nondimensional slenderness $\bar{\lambda}_{LT}$ should be determined from

$$\chi_{LT} = \frac{1}{\Phi_{LT} + \sqrt{\Phi_{LT}^2 - \bar{\lambda}_{LT}^2}} \quad \text{but} \quad \chi_{LT} \leq 1.0 \qquad (3.40)$$

$$\text{with } \Phi_{LT} = 0.5\left[1 + \alpha_{LT}(\bar{\lambda}_{LT} - 0.2) + \bar{\lambda}_{LT}^2\right] \qquad (3.41)$$

$$\bar{\lambda}_{LT} = \sqrt{\frac{W_y f_y}{M_{cr}}} \qquad (3.42)$$

where α_{LT} is an imperfection factor and M_{cr} is the elastic critical moment for lateral torsional buckling. M_{cr} is based on gross cross-sectional properties and takes into account the loading conditions, the real moment distribution, and the lateral restraints. The imperfection factor α_{LT} corresponding to the appropriate buckling curve can be taken from Table 3.18 as specified in EC3 [2.11]. The recommendations for buckling curves are given in Table 3.19 as specified in EC3 [2.11].

3.8.4 Shear Resistance of Steel Plate Girders

Checking the safety of steel plate girders against shear stresses is quite important in steel and steel–concrete composite bridges. Maximum shear stresses are normally located near the supports. Main girders made of steel plate girders are deep and thin, which make it vulnerable to fail owing to shear stresses, which is normally concentrate in panels near the supports.

Table 3.18 Recommended Values for Imperfection Factors for Lateral Torsional Buckling Curves as Given by EC3 [3.5]

Buckling curve	a	b	c	d
Imperfection factor α_{LT}	0.21	0.34	0.49	0.76

Table 3.19 Values for Lateral Torsional Buckling Curves for Different Cross Sections Recommended by EC3 [3.5]

Cross Section	Limits	Buckling Curve
Rolled I-sections	$h/b \leq 2$	a
	$h/b > 2$	b
Welded I-sections	$h/b \leq 2$	c
	$h/b > 2$	d
Other cross sections	–	d

According to EC3 [1.27,2.11], the design value of the shear force V_{Ed} at each cross section shall satisfy

$$\frac{V_{Ed}}{V_{c,Rd}} \leq 1.0 \tag{3.43}$$

where $V_{c,Rd}$ is the design shear resistance. For plastic design, $V_{c,Rd}$ is the design plastic shear resistance $V_{pl,Rd}$ calculated using Equation (3.44):

$$V_{pl,Rd} = \frac{A_v \left(f_y / \sqrt{3} \right)}{\gamma_{M0}} \tag{3.44}$$

where A_v is the shear area, which can be taken conservatively equal to $h_w t$, and h_w and t are the height and plate thickness of the web, respectively.

EC3 (3.6) provides the rules for shear resistance of plates considering shear buckling at the ultimate limit state where the panels are rectangular; stiffeners, if any, are provided in the longitudinal or transverse direction or both; all holes and cutouts are small; and members are of uniform cross section. Plates with h_w/t greater than $\frac{72}{\eta}\varepsilon$ for an unstiffened web, or $\frac{31}{\eta}\varepsilon\sqrt{k_\tau}$ for a stiffened web, should be checked for resistance to shear buckling and should be provided with transverse stiffeners at the supports, where $\varepsilon = \sqrt{\frac{235}{f_y}}$, where f_y is the yield stress in MPa. h_w is shown in Figure 3.19 and k_τ is the minimum shear buckling coefficient for the web panel as given in Annex A of EC3 [3.5]. The National Annex will define η. The value $\eta = 1.2$ is recommended for steel grades up to and including S460. For higher steel grades, $\eta = 1.0$ is recommended. For unstiffened or stiffened webs, the design resistance for shear should be taken as

$$V_{b,Rd} = V_{bw,Rd} + V_{bf,Rd} \leq \frac{\eta f_{yw} h_w t}{\sqrt{3}\gamma_{M1}} \tag{3.45}$$

in which the contribution from the web is given by

$$V_{bw,Rd} = \frac{\chi_w f_{yw} h_w t}{\sqrt{3}\gamma_{M1}} \tag{3.46}$$

For webs with transverse stiffeners at supports only and for webs with either intermediate transverse stiffeners or longitudinal stiffeners or both, the factor χ_w for the contribution of the web to the shear buckling resistance should be obtained from Table 3.20 according to EC3 based on the type of end support as shown in Figure 3.23. The slenderness parameter $\bar{\lambda}_w$ in Table 3.20 should be taken as

Table 3.20 Contribution from the Web χ_w to Shear Buckling Resistance According to EC3 [3.6]

	Rigid End Post	Nonrigid End Post
$\bar{\lambda}_w < 0.83/\eta$	η	η
$0.83/\eta \leq \bar{\lambda}_w < 1.08$	$0.83/\bar{\lambda}_w$	$0.83/\bar{\lambda}_w$
$\bar{\lambda}_w \geq 1.08$	$1.37/(0.7 + \bar{\lambda}_w)$	$0.83/\bar{\lambda}_w$

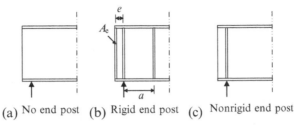

(a) No end post   (b) Rigid end post   (c) Nonrigid end post

Figure 3.23 Types of end supports.

$$\lambda^- = 0.76\sqrt{\frac{f_{yw}}{\tau_{cr}}} \tag{3.47}$$

where $\tau_{cr} = k_\tau \sigma_E$ and the values for k_τ and σ_E may be taken from Annex A of EC3 [3.5].

When the flange resistance is not completely utilized in resisting the bending moment ($M_{Ed} < M_{f,Rd}$), the contribution from the flanges $V_{bf,Rd}$ should be obtained as follows:

$$V_{bf,Rd} = \frac{b_f t_f^2 f_{yf}}{c\gamma_{M1}}\left(1 - \left(\frac{M_{Ed}}{M_{f,Rd}}\right)^2\right) \tag{3.48}$$

where b_f and t_f are taken for the flange that provides the least axial resistance, b_f being taken as not larger than $15\varepsilon t_f$ on each side of the web, and $M_{f,Rd}$ is the moment of resistance of the cross section consisting of the effective area of the flanges only calculated as follows:

$$M_{f,Rd} = \frac{M_{f,k}}{\gamma_{M0}} \tag{3.49}$$

$$c = a\left(0.25 + \frac{1.6 b_f t_f^2 f_{yf}}{t h_w^2 f_{yw}}\right) \tag{3.50}$$

The verification should be performed as follows:

$$\frac{V_{Ed}}{V_{b,Rd}} \leq 1.0 \tag{3.51}$$

where V_{Ed} is the design shear force including shear from torque.

3.8.5 Plate Buckling Effects Due to Direct Stresses
3.8.5.1 General
Steel plate girders are vulnerable to buckling owing to direct stresses. As an example, plate girder panels at midspan where maximum bending moments are expected are likely to fail due to mainly pure bending stresses. Design rules accounting for plate buckling effects from direct stresses at the ultimate limit state are covered by EC3 [3.5]. Rectangular panels should be provided by stiffeners in the longitudinal or transverse direction or both to avoid plate buckling. The resistance of plated members may be determined using the effective areas of plate elements in compression for class 4 sections using cross-sectional data (A_{eff}, I_{eff}, and W_{eff}) for cross-sectional verifications and member verifications for column buckling and lateral torsional buckling according to EC3 [3.5]. Effective[P] areas should be determined on the basis of the linear strain distributions with the attainment of yield strain in the mid-plane of the compression plate. In calculating longitudinal stresses, the combined effect of shear lag and plate buckling should be taken into account. The effective area A_{eff} should be determined assuming that the cross section is subject only to stresses due to uniform axial compression. The effective section modulus W_{eff} should be determined assuming the cross section is subject only to bending stresses. The effective[P] areas of flat compression elements should be obtained using Table 4.1, given by EC3, for internal elements and Table 4.2, recommended by EC3, for outstand elements. The effective[P] area of the compression zone of a plate with the gross cross-sectional area A_c should be obtained from the following:

For internal compression elements,

$$\rho = 1.0 \quad \text{for} \quad \bar{\lambda}_p \leq 0.673 \tag{3.52}$$

$$\rho = \frac{\bar{\lambda}_p - 0.055(3 + \psi)}{\bar{\lambda}_p^2} \leq 1.0 \quad \text{for} \quad \bar{\lambda}_p > 0.673, \text{ where } (3 + \psi) \geq 0$$

For outstand compression elements,

$$\rho = 1.0 \quad \text{for} \quad \bar{\lambda}_p \leq 0.748 \tag{3.53}$$

$$\rho = \frac{\bar{\lambda}_p - 0.188}{\bar{\lambda}_p^2} \leq 1.0 \text{ for } \bar{\lambda}_p > 0.748$$

where

$$\bar{\lambda}_p = \sqrt{\frac{f_y}{\sigma_{cr}}} = \frac{\bar{b}/t}{28.4\varepsilon\sqrt{k_\sigma}} \tag{3.54}$$

where ψ is the stress ratio used in Tables 4.1 and 4.2, $\bar{b}$ is the appropriate width, and k_σ is the buckling factor corresponding to the stress ratio ψ and boundary conditions. For long plates, k_σ is given in Table 4.1 or Table 4.2 given in EC3 [3.5] as appropriate, t is the thickness, σ_{cr} is the elastic critical plate buckling stress, and $\varepsilon = \sqrt{\frac{235}{f_y}}$, where f_y is the yield stress in MPa. According to EC3 [3.5], for aspect ratios $a/b < 1$, a column type of buckling may occur (see Figures 3.24 and 3.25), and the check should be performed

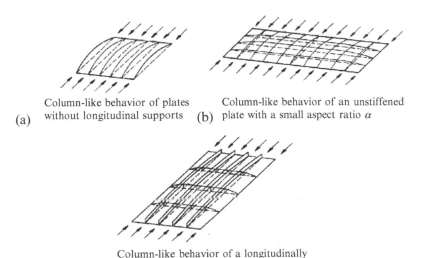

Figure 3.24 Definition of aspect ratio $\alpha = a/b$ of rectangular plates.

(a) Column-like behavior of plates without longitudinal supports

(b) Column-like behavior of an unstiffened plate with a small aspect ratio α

(c) Column-like behavior of a longitudinally stiffened plate with a large aspect ratio α

Figure 3.25 Column-like behavior according to EC4 [3.6].

considering the interaction between plate and column buckling using the reduction factor ρ_c.

3.8.5.2 Stiffened Plate Elements with Longitudinal Stiffeners

According to EC3 [3.5], for plates with longitudinal stiffeners, the effective[P] areas from local buckling of the various subpanels between the stiffeners and the effective[P] areas from the global buckling of the stiffened panel should be accounted for. The effective[P] section area of each subpanel should be determined by a reduction factor to account for local plate buckling. The stiffened plate with effective[P] section areas for the stiffeners should be checked for global plate buckling (by modeling it as an equivalent orthotropic plate) and a reduction factor ρ should be determined for overall plate buckling. The effective[P] area of the compression zone of the stiffened plate should be taken as

$$A_{c,\text{eff}} = \rho_c A_{c,\text{eff,loc}} + \sum b_{\text{edge,eff}} t \tag{3.55}$$

where $A_{c,\text{eff,loc}}$ is the effective[P] section areas of all the stiffeners and subpanels that are fully or partially in the compression zone except the effective parts supported by an adjacent plate element with the width $b_{\text{edge,eff}}$ (see example in Figure 3.26). The area $A_{c,\text{eff,loc}}$ should be obtained from

$$A_{c,\text{eff,loc}} = A_{\text{sl,eff}} + \sum_c \rho_{\text{loc}} b_{c,\text{loc}} t \tag{3.56}$$

where $\sum_c$ applies to the part of the stiffened panel width that is in compression except the parts $b_{\text{edge,eff}}$ (see Figure 3.26), $A_{\text{sl,eff}}$ is the sum of the effective[P] sections of all longitudinal stiffeners with gross area A_{sl} located in the compression zone, $b_{c,\text{loc}}$ is the width of the compressed part of each subpanel, and ρ_{loc} is the reduction factor for each subpanel. According to EC3

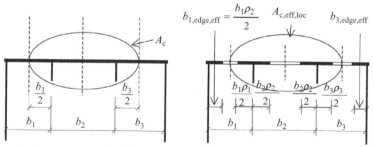

Figure 3.26 An example of stiffened plate under uniform compression according to EC3 [3.5].

[3.5], in determining the reduction factor ρ_c for overall buckling, the reduction factor for column-type buckling, which is more severe than the reduction factor than for plate buckling, should be considered. Interpolation should be carried out between the reduction factor ρ for plate buckling and the reduction factor χ_c for column buckling to determine ρ_c. The reduction of the compressed area $A_{c,eff,loc}$ through ρ_c may be taken as a uniform reduction across the whole cross section.

3.8.5.3 Plate-Type Behavior
The relative plate slenderness $\bar{\lambda}_p$ of the equivalent plate specified in EC3 [3.5] is defined as

$$\bar{\lambda}_p = \sqrt{\frac{\beta_{A,c}\, f_y}{\sigma_{cr,p}}} \tag{3.57}$$

$$\text{with } \beta_{A,c} = \frac{A_{c,eff,loc}}{A_c} \tag{3.58}$$

where A_c is the gross area of the compression zone of the stiffened plate and $A_{c,eff,loc}$ is the effective area of the same part of the plate with due allowance made for possible plate buckling of subpanels and/or stiffeners.

3.8.5.4 Column-Type Buckling Behavior
On the other hand, the elastic critical column buckling stress $\sigma_{cr,c}$ of an unstiffened or stiffened plate should be taken as the buckling stress with the supports along the longitudinal edges removed, as specified in EC3 [3.5]. For an unstiffened plate, the elastic critical column buckling stress $\sigma_{cr,c}$ may be obtained from

$$\sigma_{cr,c} = \frac{\pi^2 E t^2}{12(1-v^2)a^2} \tag{3.59}$$

For a stiffened plate, $\sigma_{cr,c}$ may be determined from the elastic critical column buckling stress $\sigma_{cr,sl}$ of the stiffener closest to the panel edge with the highest compressive stress as follows:

$$\sigma_{cr,sl} = \frac{\pi^2 E I_{sl,1}}{A_{sl,1} a^2} \tag{3.60}$$

where $I_{sl,1}$ is the second moment of area of the gross cross section of the stiffener and the adjacent parts of the plate, relative to the out-of-plane bending of the plate, and $A_{sl,1}$ is the gross cross-sectional area of the stiffener and the adjacent parts of the plate according to Annex A of EC3 [3.5]:

$$\bar{\lambda}_c = \sqrt{\frac{f_y}{\sigma_{cr,c}}} \quad \text{for unstiffened plates} \tag{3.61}$$

$$\bar{\lambda}_c = \sqrt{\frac{\beta_{A,c} f_y}{\sigma_{cr,c}}} \quad \text{for stiffened plates} \tag{3.62}$$

$$\text{with} \, \beta_{A,c} = \frac{A_{sl,1,\text{eff}}}{A_{sl,1}} \tag{3.63}$$

where $A_{sl,1,\text{eff}}$ is the effective cross-sectional area of the stiffener and the adjacent parts of the plate with due allowance for plate buckling detailed in Annex A of EC3 [3.5].

The reduction factor χ_c should be obtained from EC3 [2.11]. For unstiffened plates, $\alpha = 0.21$ corresponding to buckling curve a should be used. For stiffened plates, its value should be increased to

$$\alpha_e = \alpha + \frac{0.09}{i/e} \tag{3.64}$$

$$\text{with} \, i = \sqrt{\frac{I_{sl,1}}{A_{sl,1}}} \tag{3.65}$$

where e is the max of (e_1, e_2), which is the largest distance from the respective centroids of the plating and the one-sided stiffener (or of the centroids of either set of stiffeners when present on both sides) to the neutral axis of the effective column (see Annex A of EC3 [3.5]); $\alpha = 0.34$ for (curve b) for closed section stiffeners; and $\alpha = 0.49$ for (curve c) for open section stiffeners.

3.8.5.5 Interaction Between Plate and Column Buckling

The final reduction factor ρ_c should be obtained by interpolation between χ_c and ρ:

$$\rho_c = (\rho - \chi_c)\xi(2 - \xi) + \chi_c \tag{3.66}$$

$$\text{where} \, \xi = \frac{\sigma_{cr,p}}{\sigma_{cr,c}} - 1 \quad \text{but} \quad 0 \le \xi \le 1 \tag{3.67}$$

where $\sigma_{cr,p}$ is the elastic critical plate buckling stress detailed in Annex A of EC3 [3.5], $\sigma_{cr,c}$ is the elastic critical column buckling stress, χ_c is the reduction factor due to column buckling, and ρ is the reduction factor due to plate buckling.

3.8.5.6 Verification

Member verification for uniaxial bending should be performed, according to EC3 [3.5], as follows:

$$\eta_1 = \frac{N_{Ed}}{\frac{f_y A_{eff}}{\gamma_{M0}}} + \frac{M_{Ed} + N_{Ed} e_N}{\frac{f_y W_{eff}}{\gamma_{M0}}} \leq 1.0 \qquad (3.68)$$

where A_{eff} is the effective cross-sectional area, e_N is the shift in the position of neutral axis, M_{Ed} is the design bending moment, N_{Ed} is the design axial force, W_{eff} is the effective elastic section modulus, and γ_{M0} is the partial factor.

3.8.6 Behavior of Steel-Concrete Composite Plate Girders

Steel-concrete composite constructions offer many advantages including high strength, full usage of materials, high stiffness and ductility, toughness against seismic loads, and significant savings in construction time. In addition to the aforementioned advantages, steel-concrete composite constructions are gaining popularity due to the higher fire resistance compared to the conventional steel constructions that require additional protection against fire. Mainly in highway bridges, it is very common to benefit from the thick concrete deck on top of the floor beams (see Figure 1.21) and join them together using shear connectors to ensure that the two components act together in resisting traffic loads. Steel is known for its higher tensile resistance, while concrete is known for its higher compressive resistance. Therefore, joining the two components leads to the aforementioned benefits. In addition, steel parts are thin-walled structures, which make it vulnerable to local and overall buckling failure modes. The presence of a concrete deck on top of the steel beams eliminates lateral torsional buckling and local buckling of the top flange of the steel girders.

3.8.6.1 Effective Width of Flanges for Shear Lag

EC4 [3.6] covers steel-concrete composite bridges. The code recommends that allowance shall be made for the flexibility of steel or concrete flanges affected by shear in their plane (shear lag) either by means of rigorous analysis or by using an effective width of flange. The effective width of concrete flanges should be determined such that when elastic global analysis is used,

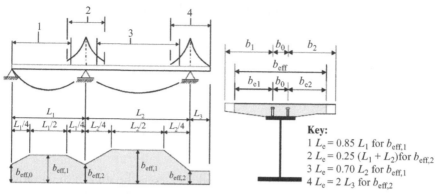

Figure 3.27 Equivalent spans for effective width of concrete flange according to EC4 [3.6].

a constant effective width may be assumed over the whole of each span. This value may be taken as the value $b_{eff,1}$ at midspan for a span supported at both ends or the value $b_{eff,2}$ at the support for a cantilever. According to EC4 [3.6], at midspan or an internal support, the total effective width b_{eff} (see Figure 3.27) may be determined as

$$b_{eff} = b_o + \sum b_{ei} \qquad (3.69)$$

where b_0 is the distance between the centers of the outstand shear connectors and b_{ei} is the value of the effective width of the concrete flange on each side of the web and taken as $L_e/8$ (but not greater than the geometric width b_i). The value b_i should be taken as the distance from the outstand shear connector to a point midway between adjacent webs, measured at middepth of the concrete flange, except that at a free edge b_i is the distance to the free edge. The length L_e should be taken as the approximate distance between points of zero bending moment. For typical continuous composite beams, where a moment envelope from various load arrangements governs the design, and for cantilevers, L_e may be assumed to be as shown in Figure 3.27. The effective width at an end support may be determined as

$$b_{eff} = b_o + \sum \beta_i b_{ei} \qquad (3.70)$$

$$\text{with } \beta_i = (0.55 + 0.025 L_e / b_{ei}) \le 1.0 \qquad (3.71)$$

where b_{ei} is the effective width of the end span at midspan and L_e is the equivalent span of the end span according to Figure 3.27. The distribution of the effective width between supports and midspan regions may be assumed as shown in Figure 3.27. The transverse distribution of stresses due to shear lag may be taken in accordance with EC3 [3.6] for both concrete and steel flanges. For cross sections with bending moments resulting from the main girder system and from a local system (for example, in composite trusses with direct actions on the chord between nodes), the relevant effective widths for the main girder system and the local system should be used for the relevant bending moments.

3.8.6.2 Bending Resistance of Composite Plate Girders

Let us now calculate the bending resistance of composite plate girders, according to EC4 [3.6]. The design bending resistance shall be determined by rigid plastic theory only where the effective composite cross section is in class 1 or 2. On the other hand, elastic analysis and nonlinear theory for bending resistance may be applied to cross sections of any class. For elastic analysis and nonlinear theory, it may be assumed that the composite cross section remains plane if the shear connection and the transverse reinforcement are designed considering appropriate distributions of design longitudinal shear force. The tensile strength of concrete shall be neglected. According to EC4 [3.6], the calculation of plastic moment resistance $M_{pl, Rd}$ can be performed assuming there is full interaction between structural steel, reinforcement, and concrete; the effective area of the structural steel member is stressed to its design yield strength f_{yd} in tension or compression; and the effective areas of longitudinal reinforcement in tension and in compression are stressed to their design yield strength f_{sd} in tension or compression. Alternatively, reinforcement in compression in a concrete slab may be neglected; the effective area of concrete in compression resists a stress of 0.85 f_{cd}, constant over the whole depth between the plastic neutral axis and the most compressed fiber of the concrete, where f_{cd} is the design cylinder compressive strength of concrete. Typical plastic stress distributions are shown in Figure 3.28 as given in EC4 [3.6]. For composite cross sections with structural steel grade S420 or S460, where the distance x_{pl} between the plastic neutral axis and the extreme fiber of the concrete slab in compression exceeds 15% of the overall depth h of the member, the design resistance moment M_{Rd} should be taken as $\beta M_{pl,Rd}$ where β is the reduction factor as shown in Figure 3.29 given by EC4. For values of x_{pl}/h greater than 0.4, the resistance to bending should be determined from nonlinear or elastic resistance to bending.

Where the bending resistance of a composite cross section is determined by nonlinear theory, EC4 [3.6] recommends the stress-strain relationships of the materials shall be taken into account. It should be assumed that the

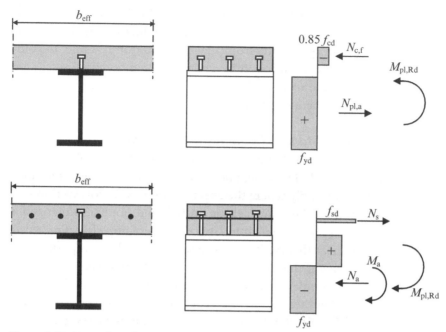

Figure 3.28 Examples of plastic stress distributions for a composite beam with a solid slab and full shear connection in sagging and hogging bending according to EC4 [3.6].

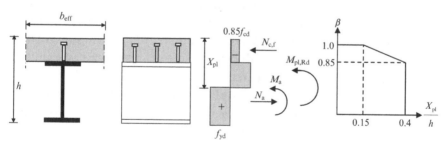

Figure 3.29 Reduction factor β for $M_{pl,Rd}$ recommended by EC4 [3.6].

composite cross section remains plane and that the strain in bonded reinforcement, whether in tension or compression, is the same as the mean strain in the surrounding concrete. The stresses in the concrete in compression should be derived from the stress-strain curves given in EC2 [2.27] and the stresses in the reinforcement should be derived from the bilinear diagrams given in the same specification. The stresses in structural steel in compression or tension should be derived from the bilinear diagram given in EC3 [3.5] and should take account of the effects of the method of construction (e.g., propped or unpropped). For class 1 and 2 composite cross sections with the concrete flange in compression, the nonlinear resistance to bending M_{Rd} may be determined as a function of the compressive force in the concrete N_c using the following simplified expressions, as shown in Figure 3.30:

$$M_{Rd} = M_{a,Ed} + \left(M_{el,Rd} - M_{a,Ed}\right)\frac{N_c}{N_{c,el}} \quad \text{for} \quad N_c \leq N_{c,el} \tag{3.72}$$

$$M_{Rd} = M_{el,Rd} + \left(M_{pl,Rd} - M_{el,Rd}\right)\frac{N_c - N_{c,el}}{N_{c,f} - N_{c,el}} \quad \text{for} \quad N_{c,el} \leq N_c \leq N_{c,f} \tag{3.73}$$

$$M_{el,Rd} = M_{a,Ed} + kM_{c,Ed} \tag{3.74}$$

where $M_{a,Ed}$ is the design bending moment applied to structural steel section before composite behavior, $M_{c,Ed}$ is the part of the design bending moment acting on the composite section, and k is the lowest factor such that a stress limit is reached, where unpropped construction is used, the sequence of construction should be taken into account, and $N_{c,el}$ is the compressive force in the concrete flange corresponding to moment $M_{el,Rd}$.

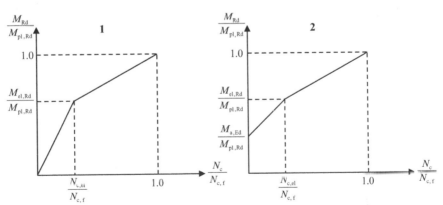

Key:
1 Propped construction
2 Unpropped construction

Figure 3.30 Simplified relationship between M_{Rd} and N_c for sections with the concrete slab in compression recommended by EC4 [3.6].

3.8.6.3 Resistance to Vertical Shear

To check the safety of composite plate girders against shear stresses, EC4 [3.6] proposes rules that apply to composite beams with a rolled or welded structural steel section with a solid web, which may be stiffened. According to EC4 [3.6], the plastic resistance to vertical shear $V_{pl,Rd}$ should be taken as the resistance of the structural steel section $V_{pl,a,Rd}$ unless the value for a contribution from the reinforced concrete part of the beam has been established. The design plastic shear resistance $V_{pl,a,Rd}$ of the structural steel section should be determined as previously detailed. The shear buckling resistance $V_{b,Rd}$ of an uncased steel web should be determined in accordance with EC3 [3.5] as previously detailed. No account should be taken of a contribution from the concrete slab in the calculation of the shear buckling resistance.

According to EC4 [3.6], where the vertical shear force V_{Ed} exceeds half the shear resistance V_{Rd} given by $V_{pl,Rd}$ or $V_{b,Rd}$, whichever is the smaller, allowance should be made for its effect on the resistance moment. For cross sections in class 1 or 2, the influence of the vertical shear on the resistance to bending may be taken into account by a reduced design steel strength $(1 - \rho)$ f_{yd} in the shear area as shown in Figure 3.31 where

$$\rho = (2V_{Ed}/V_{Rd} - 1)^2 \tag{3.75}$$

and V_{Rd} is the appropriate resistance to vertical shear. For cross sections in classes 3 and 4, EC3 [3.5] design rules are applicable using the calculated stresses of the composite section. No account should be taken of the change in the position of the plastic neutral axis of the cross section caused by the reduced yield strength when classifying the web.

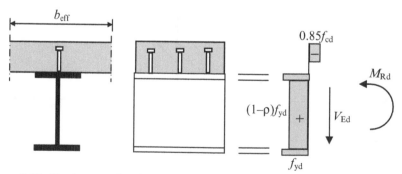

Figure 3.31 Plastic stress distribution modified by the effect of vertical shear according to EC4 [3.6].

3.8.6.4 Shear Connection

As mentioned previously, Section 2.6 of Chapter 2, the behavior of shear connection is of great importance in the design of steel-concrete composite bridges. The design rules governing shear connection is also covered by EC4 [3.6]. The code recommends that shear connection and transverse reinforcement shall be provided in composite beams to transmit the longitudinal shear force between the concrete and the structural steel element, ignoring the effect of natural bond between the two. Shear connectors shall have sufficient deformation capacity to justify any inelastic redistribution of shear assumed in design. Ductile connectors are those with sufficient deformation capacity to justify the assumption of ideal plastic behavior of the shear connection in the structure considered. A connector may be taken as ductile if the characteristic slip capacity δ_{uk} is at least 6 mm, with the evaluation of δ_{uk} given in Annex B of EC4 [2.37]. Where two or more different types of shear connection are used within the same span of a beam, account shall be taken of any significant difference in their load-slip properties. Shear connectors shall be capable of preventing separation of the concrete element from the steel element, except where separation is prevented by other means. To prevent separation of the slab, shear connectors should be designed to resist a nominal ultimate tensile force, perpendicular to the plane of the steel flange, of at least 0.1 times the design ultimate shear resistance of the connectors. If necessary, they should be supplemented by anchoring devices. Headed stud shear connectors may be assumed to provide sufficient resistance to uplift, unless the shear connection is subjected to direct tension. Longitudinal shear failure and splitting of the concrete slab due to concentrated forces applied by the connectors shall be prevented.

According to EC4 [3.6], for verifications for ultimate limit states, the size and spacing of shear connectors may be kept constant over any length where the design longitudinal shear per unit length does not exceed the longitudinal design shear resistance by more than 10%. Over every such length, the total design longitudinal shear force should not exceed the total design shear resistance. EC4 [3.6] specifies that for any load combination and arrangement of design actions, the longitudinal shear per unit length at the interface between steel and concrete in a composite member, $v_{L,Ed}$, should be determined from the rate of change of the longitudinal force in either the steel or the concrete element of the composite section. Where elastic theory is used for calculating resistances of sections, the envelope of transverse shear force in the relevant direction may be used. In general, the elastic properties of the uncracked section should be used for the determination of the longitudinal shear force, even where cracking of concrete is assumed in global analysis. The effects of cracking of concrete on the longitudinal shear force may be taken into account, if in global analysis, and the effects of tension stiffening and possible overstrength of concrete for the determination of the longitudinal shear force. Where concentrated longitudinal shear forces occur, the local effects of longitudinal slip should be taken into account. Otherwise, the effects of longitudinal slip may be neglected. According to EC4, in members with cross sections in class 1 or 2, if the total design bending moment $M_{\text{Ed,max}} = M_{\text{a,Ed}} + M_{\text{c,Ed}}$ exceeds the elastic bending resistance $M_{\text{el,Rd}}$, the nonlinear relationship between transverse shear and longitudinal shear within the inelastic lengths of the member should be taken into account. This applies in regions where the concrete slab is in compression, as shown in Figure 3.27. Shear connectors should be provided within the inelastic length $L_{\text{A-B}}$ to resist the longitudinal shear force $V_{\text{L,Ed}}$, resulting from the difference between the normal forces N_{cd} and $N_{\text{c,el}}$ in the concrete slab at the cross sections B and A, respectively. If the maximum bending moment $M_{\text{Ed,max}}$ at section B is smaller than the plastic bending resistance $M_{\text{pl,Rd}}$, the normal force N_{cd} at section B may be determined using the simplified linear relationship according to Figure 3.32.

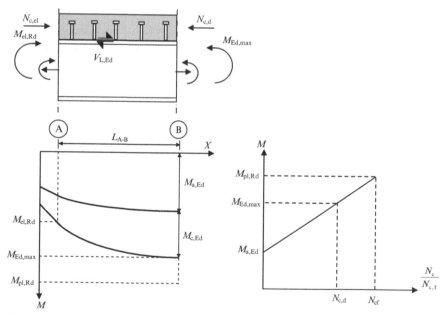

Figure 3.32 Determination of longitudinal shear in beams with inelastic behavior of cross sections according to EC4 [3.6].

3.8.6.5 Design Equations for the Evaluation of Headed Stud Capacities

As mentioned previously, Section 2.6 of Chapter 2, headed stud shear connectors are widely used in steel–concrete composite constructions owing to many advantages including rapid installation, equal strength, and stiffness in shear in all directions normal to the axis of the stud and high ductility. Therefore, the development of design equations used in evaluating headed stud capacities is discussed in the next paragraphs. The capacities depend on the type of the concrete slab used with the steel beam to form the composite interaction.

(a) *Composite beams with solid reinforced concrete slabs*

The strength of shear studs in solid reinforced concrete slab was first determined by Ollgaard et al. [2.48] and was presented in terms of an empirical formula after carrying out 48 push-off tests. The ultimate shear force resistance Q_u (in N units) of the headed studs was given as follows:

$$Q_u = 0.5A_s\sqrt{f_c E_c} \tag{3.76}$$

where A_s is the cross-sectional area of the stud diameter d (mm^2 units), f_c is the concrete cylinder compressive strength (N/mm^2), and E_c is the static modulus of elasticity of the concrete (N/mm^2). This equation, which was adopted in CP 117 [2.34,2.35], assumes concrete crushing failure rather than a shear failure of the headed stud. Later, in BS 5950 [2.36], data presented by Menzies [2.46] were used to develop the characteristic shear force resistance Q_K. There is no theoretical basis to these data and values given in BS 5950 [2.36] reflect only the size of the stud and strength of the concrete. Currently, the commonly used design equations for headed stud shear connectors are given in EC4 [3.6]. The resistance of headed stud (P_{Rd}) is defined using two equations. The equations represent concrete and stud failures. The lower of the following values should be used in design:

$$P_{Rd} = \frac{0.8f_u\pi d^2/4}{\gamma_v} \tag{3.77}$$

$$P_{Rd} = \frac{0.29\alpha d^2\sqrt{f_{ck}E_{cm}}}{\gamma_v} \tag{3.78}$$

whichever is smaller with

$$\alpha = 0.2\left(\frac{h_{sc}}{d}+1\right) \quad \text{for } 3 \le h_{sc}/d \le 4 \tag{3.79}$$

$$\alpha = 1 \quad \text{for } h_{sc}/d > 4 \tag{3.80}$$

where γ_v is the partial factor; d is the diameter of the shank of the stud, $16 \leq d \leq 25$ mm; f_u is the specified ultimate tensile strength of the material of the stud but not greater than 500 N/mm^2; f_{ck} is the characteristic cylinder compressive strength of the concrete at the age considered, of density not less than 1750 kg/m^3; and h_{sc} is the overall nominal height of the stud. The value for γ_v may be taken as 1.25.

(b) *Composite beams with profiled steel sheeting*

The experimental investigations, highlighted in Section 2.6 of Chapter 2, for headed stud shear connectors in composite beams with profiled steel sheeting show that the shear resistance of headed studs is sometimes lower than it is in a solid slab, for materials of the same strength, because of local failure of the concrete rib. For this reason, EC4 [2.37] specifies reduction factors, applied to the resistance P_{Rd} to determine the capacities of headed studs in composite beams with profiled steel sheeting as follows:

$$P_r = k P_{Rd} \qquad (3.81)$$

where (P_r) is resistance of a stud in a trough and k is the reduction factor that depends on the direction of sheeting.

For sheeting with ribs parallel to the beam, the factor (k_l) is calculated as follows:

$$k_l = 0.6 \frac{b_o}{h_p} \left(\frac{h_{sc}}{h_p} - 1 \right) \qquad (3.82)$$

where the dimensions b_o, h_p, and h are illustrated in Figure 2.15 and h_{sc} is taken as not greater than $h_p + 75$ mm. EC4 rules are discussed by Johnson and Anderson [2.40]. Systematic theoretical and finite element studies since 1981, mainly by Oehlers [2.51] and initially for solid and haunched slabs, have been extended to parallel sheeting.

For sheeting with ribs transverse to the beam, the factor (k_t) is calculated as follows:

$$k_t = \frac{0.7}{\sqrt{n_r}} \frac{b_o}{h_p} \left(\frac{h_{sc}}{h_p} - 1 \right) \qquad (3.83)$$

where n_r is the number of connectors in one rib where it crosses a beam, not to be taken greater than 2 in calculations. It is recommended that off-center studs should be placed on alternative sides of the trough, but no other account was taken of the important influence of dimension (e) in Figure 2.15. The reinforcement in a composite slab is usually a light welded mesh.

Tests [2.54] show that when placed below the heads of the studs, the mesh can increase the shear resistance of the studs. In practice, the control of its level is poor and its detailing is not related to that of the shear connection. Johnson and Yuan [3.7–3.9] considered the results of 269 push-off tests in a study of existing design rules for the static shear resistance of stud connectors in profiled steel sheeting. It was found [3.9] that test data were scarce for the influence of the thickness of the profiled sheeting (t) and of lightweight aggregate and for the influence of the position of the studs in each trough, values of (b_o) less than ($2h_p$) and parallel sheeting. Therefore, the authors reported the results for 34 push-off tests and identified seven distinct modes of failure. The design rules for the static shear resistance of stud connectors in profiled sheeting were studied, and it is found that they are limited in the case of studs placed off-center in the steel troughs. Developed equations based on theoretical models were obtained for the observed modes of failure. The modes are shown to give good performance when compared with reported test results.

(c) *Composite beams with prestressed hollow core concrete slabs*

Design equations developed for determining the capacity of the connectors in a composite beam consisting of prestressed hollow–cored concrete slabs were detailed in Lam *et al.* [2.61]. Twelve full-scale push-off tests were carried out to study the effects of the size of the gap between the ends of the precast slabs, the amount of tie steel placed transversely across the joint, and the strength of concrete in-fill on the capacity of the shear stud. The following design equation, modified from EC4 [2.37], "Equation (**3.78**)," was given and there was no modification in Equation (**3.77**) as the strength of the shear stud is thought not to be influenced by the precast construction:

$$P_{Rd} = \frac{0.29\alpha\beta\varepsilon d^2 \sqrt{\omega f_{cp} E_{cp}}}{\gamma_v} \tag{3.84}$$

where β is a factor that takes into account the gap width (g) in millimeters (see Figure 2.17) and is given as $0.5(g/70+1) \leq 1.0$ and $g > 30$ mm, ε is a factor that takes into account the diameter (ϕ) of transverse high tensile steel (grade 460) and is given by $0.5(\phi/20+1) \leq 1.0$ and $\psi \geq 8$ mm, ω is the transverse joint factor $= 0.5(\omega/600+1)$, ω is the width of hollow core concrete units, f_{cp} is the average concrete cylinder strength $= 0.8 \times$ average cube strength of the *in situ* and prestressed concrete, and E_{cp} is the average value of elastic modulus of the *in situ* and precast concrete. All other terms are as for Equation (3.78).

3.9 STABILITY OF STEEL AND STEEL-CONCRETE COMPOSITE TRUSS BRIDGES

3.9.1 General

Truss bridges are generally used for spans over 40 m. For spans between 40 and 70 m, parallel chord trusses are used, while for spans greater than 70 m, polygonal chord trusses are used. Trusses are, normally, designed to carry axial forces in its members, which are either tension or compression or reversible tension/compression depending on the worst cases of loading and load combinations. Truss members are connected at joints using welds or bolts. Joints are designed as pins and the forces in truss members are in full equilibrium at the joints. In practice, gusset plates are used at the joints to collect the forces in the members meeting at the joints, where equilibrium takes place. Therefore, the size of the gusset plates should be as small as possible to simulate the behavior of pins. If the maximum force in a truss is less than 3000 kN, single gusset plate trusses are used and truss members are designed as angles. On the other hand, if the maximum force in truss members is greater than 3000 kN, double gusset plate trusses are used and chord members are designed as box sections, while diagonals and verticals are designed as I-sections or box sections in case of long diagonals carrying compressive forces. Cross girders are located at the joints of trusses to eliminate bending moments on truss members. Figure 1.21 shows an example of a highway truss bridge.

3.9.2 Design of Tension Members

The design of tension members is covered by EC3 [1.27,2.11]. The code recommends that the design value of the tension force N_{Ed} at each cross section shall satisfy

$$\frac{N_{Ed}}{N_{t,Rd}} \leq 1.0 \tag{3.85}$$

For sections with holes, the design tension resistance $N_{t,Rd}$ should be taken as the smaller of the following:
(a) The design plastic resistance of the gross cross section:

$$N_{pl,Rd} = \frac{Af_y}{\gamma_{M0}} \tag{3.86}$$

(b) The design ultimate resistance of the net cross section at holes for fasteners:

$$N_{u,Rd} = \frac{0.9 A_{net} f_u}{\gamma_{M2}} \qquad (3.87)$$

where A_{net} is the net area of a cross section (its gross area A less appropriate deductions for all holes and other openings) and f_y and f_u are the yield and ultimate stresses of steel, respectively. Where capacity design is requested, the design plastic resistance $N_{pl,Rd}$ should be less than the design ultimate resistance of the net section at fasteners holes $N_{u,Rd}$. In category C connections, detailed in EC3 (BS EN 1993-1-8) [2.13], the design tension resistance $N_{t,Rd}$ of the net section at holes for fasteners should be taken as $N_{net,Rd}$, where

$$N_{net,Rd} = \frac{A_{net} f_y}{\gamma_{M0}} \qquad (3.88)$$

3.9.3 Design of Compression Members

Similar to tension members, the design of compression members is also covered by EC3 [1.27,2.11]. The code recommends that the design strength of the compression force N_{Ed} at each cross section shall satisfy

$$\frac{N_{Ed}}{N_{c,Rd}} \leq 1.0 \qquad (3.89)$$

The design resistance of the cross section for uniform compression $N_{c,Rd}$ should be determined as follows:

$$N_{c,Rd} = \frac{A f_y}{\gamma_{M0}} \quad \text{for class 1, 2, or 3 cross sections} \qquad (3.90)$$

$$N_{c,Rd} = \frac{A_{eff} f_y}{\gamma_{M0}} \quad \text{for class 4 cross sections} \qquad (3.91)$$

According to EC3 [2.11], a compression member should be verified against buckling as follows:

$$\frac{N_{Ed}}{N_{b,Rd}} \leq 1.0 \qquad (3.92)$$

where N_{Ed} is the design value of the compression force and $N_{b,Rd}$ is the design buckling resistance of the compression member. The design buckling resistance of a compression member should be taken as

$$N_{b,Rd} = \frac{\chi A f_y}{\gamma_{M1}} \quad \text{for class 1, 2, and 3 cross sections} \qquad (3.93)$$

$$N_{b,Rd} = \frac{\chi A_{eff} f_y}{\gamma_{M1}} \quad \text{for class 4 cross sections} \quad (3.94)$$

where χ is the reduction factor for the relevant buckling mode. In determining A and A_{eff}, holes for fasteners at the column ends need not be taken into account. For axial compression in members, the value of χ for the appropriate nondimensional slenderness $\bar{\lambda}$ should be determined from the relevant buckling curve according to

$$\chi = \frac{1}{\Phi + \sqrt{\Phi^2 - \bar{\lambda}^2}} \quad \text{but} \quad \chi \leq 1.0 \quad (3.95)$$

$$\text{where} \quad \Phi = 0.5\left[1 + \alpha(\bar{\lambda} - 0.2) + \bar{\lambda}^2\right] \quad (3.96)$$

$$\text{and} \quad \bar{\lambda} = \sqrt{\frac{A f_y}{N_{cr}}} \quad \text{for class 1, 2, and 3 cross sections} \quad (3.97)$$

$$\bar{\lambda} = \sqrt{\frac{A_{eff} f_y}{N_{cr}}} \quad \text{for class 4 cross sections} \quad (3.98)$$

where α is an imperfection factor and N_{cr} is the elastic critical force for the relevant buckling mode based on the gross cross-sectional properties. The imperfection factor α corresponding to the appropriate buckling curve should be obtained from Tables 6.1 and 6.2, given in EC3 [2.11]. The nondimensional slenderness $\bar{\lambda}$ is given by

$$\bar{\lambda} = \sqrt{\frac{A f_y}{N_{cr}}} = \frac{L_{cr}}{i} \frac{1}{\lambda_1} \quad \text{for class 1, 2, and 3 cross sections} \quad (3.99)$$

$$\bar{\lambda} = \sqrt{\frac{A_{eff} f_y}{N_{cr}}} = \frac{L_{cr}}{i} \frac{\sqrt{\frac{A_{eff}}{A}}}{\lambda_1} \quad \text{for class 4 cross sections} \quad (3.100)$$

where L_{cr} is the buckling length in the buckling plane considered and i is the radius of gyration about the relevant axis, determined using the properties of the gross cross section. The nondimensional slenderness λ_1 is given by

$$\lambda_1 = \pi \sqrt{\frac{E}{f_y}} = 93.9\varepsilon \quad (3.101)$$

with $\sqrt{\dfrac{235}{f_y}}$ and f_y in MPa.

3.10 DESIGN OF BOLTED AND WELDED JOINTS

3.10.1 General

The previously mentioned brief survey of the design rules specified in the current codes of practice provided a general background to the design of main components of steel and steel-concrete composite bridges. However, in order to get a complete background, it is also important to review the design rules on how these components are connected together, that is, the design of joints connecting the different components. Therefore, in the succeeding sections, it is decided to highlight the design rules specified, as an example, in the Eurocodes on bolted and welded joints. According to EC3 (BS EN 1993-1-8) [2.13], all joints shall have a design resistance such that the structure is capable of satisfying all the basic design requirements given in EC3 (BS EN 1993-1-1) [2.11]. The partial safety factors γ_M for joints are given in Table 3.21. The forces and moments applied to joints at the ultimate limit state shall be determined according to the principles in EC3 (BS EN 1993-1-1) [2.11]. EC3 (BS EN 1993-1-8) [2.13] specifies that the resistance of a joint should be determined on the basis of the resistances of its basic components. Linear-elastic or elastic-plastic analysis may be used in the design of joints. Joints shall be designed on the basis of a realistic assumption of the distribution of internal forces and moments. The main assumptions used to determine the distribution of forces are as follows: (a) the internal forces and moments assumed in the analysis are in equilibrium with the forces and moments applied to the joints, (b) each element

Table 3.21 Partial Safety Factors for Joints Specified in EC3 (BS EN 1993-1-8) [2.13]

Resistance of members and cross sections	γ_{M1}, γ_{M1}, and γ_{M2}
Resistance of bolts	γ_{M2}
Resistance of rivets	
Resistance of pins	
Resistance of welds	
Resistance of plates in bearing	
Slip resistance	
– At ultimate limit state (category C)	γ_{M3}
At serviceability limit state (category B)	$\gamma_{M3,ser}$
Bearing resistance of an injection bolt	γ_{M4}
Resistance of joints in hollow section lattice girder	γ_{M5}
Resistance of pins at serviceability limit state	$\gamma_{M6,ser}$
Preload of high-strength bolts	γ_{M7}
Resistance of concrete	γ_c

in the joint is capable of resisting the internal forces and moments, (c) the deformations implied by this distribution do not exceed the deformation capacity of the fasteners or welds and the connected parts, (d) the assumed distribution of internal forces shall be realistic with regard to relative stiffnesses within the joint, (e) the deformations assumed in any design model based on elastic-plastic analysis are based on rigid body rotations and/or in-plane deformations that are physically possible, and (f) any model used is in compliance with the evaluation of test results.

3.10.2 Connections Made with Bolts or Pins

Let us now review the design rules specified in Eurocode for connections made with bolts or pins. The rules specified in EC3 (BS EN 1993-1-8) [2.13] are valid for the bolt classes given in Table 2.6 of EC3 (BS EN 1993-2) [1.27]. The yield strength f_{yb} and the ultimate tensile strength f_{ub} for different bolt classes are given in Table 2.6. These values should be adopted as characteristic values in design calculations.

3.10.2.1 Bolted Connections

According to EC3 (BS EN 1993-1-8) [2.13], bolted connections loaded in shear should be designed as categories A, B, and C. In category A (bearing type), all bolts from class 4.6 up to and including class 10.9 can be used. No preloading and special provisions for contact surfaces are required. The design ultimate shear load should not exceed the design shear resistance nor the design bearing resistance. In category B (slip-resistant at serviceability limit state), preloaded bolts should be used. Slip should not occur at the serviceability limit state. The design serviceability shear load should not exceed the design slip resistance. The design ultimate shear load should not exceed the design shear resistance nor the design bearing resistance. Finally, in category C (slip-resistant at ultimate limit state), preloaded bolts should be used. Slip should not occur at the ultimate limit state. The design ultimate shear load should not exceed the design slip resistance nor the design bearing resistance. In addition, for a connection in tension, the design plastic resistance of the net cross section at bolt holes $N_{net,Rd}$ should be checked at the ultimate limit state. The design checks for these connections are summarized in Table 3.22 specified in EC3 (BS EN 1993-1-8) [2.13]. Bolted connection loaded in tension should be designed as categories D and E. In category D (non-preloaded), bolts from class 4.6 up to and including class 10.9 can be used. No preloading is required. This category should not be used where the connections are frequently subjected to variations of

Table 3.22 Categories of Bolted Connections Specified in EC3 (BS EN 1993-1-8) [2.13]

Category	Criteria	Remarks
Shear connections		
A	$F_{v,Ed} \geq F_{v,Rd}$	No preloading required
Bearing type	$F_{v,Ed} \geq F_{b,Rd}$	Bolt classes from 4.6 to 10.9 can be used
B	$F_{v,Ed,ser} \geq F_{s,Rd,ser}$	Preloaded 8.8 or 10.9 bolts should
Slip-resistant at	$F_{v,Ed} \geq F_{v,Rd}$	be used
serviceability	$F_{v,Ed} \geq F_{b,Rd}$	
C	$F_{v,Ed} \geq F_{s,Rd}$	Preloaded 8.8 or 10.9 bolts should
Slip-resistant at	$F_{v,Ed} \geq F_{b,Rd}$	be used
ultimate	$F_{v,Ed} \geq N_{net,Rd}$	
Tension connections		
D	$F_{t,Ed} \geq F_{t,Rd}$	No preloading required
Non-preloaded	$F_{t,Ed} \geq B_{p,Rd}$	Bolt classes from 4.6 to 10.9 can be used
E	$F_{t,Ed} \geq F_{t,Rd}$	Preloaded 8.8 or 10.9 bolts should
Preloaded	$F_{t,Ed} \geq B_{p,Rd}$	be used

tensile loading. However, they may be used in connections designed to resist normal wind loads. On the other hand, in category E (preloaded), preloaded 8.8 and 10.9 bolts with controlled tightening should be used. The design checks for these connections are also summarized in Table 3.22 specified in EC3 (BS EN 1993-1-8) [2.13]. Bolt holes should have limiting values for the spacing between two adjacent holes and for the distance between a hole and an adjacent edge to avoid local failures. EC3 (BS EN 1993-1-8) [2.13] specifies the minimum and maximum spacing and end and edge distances for bolts as given in Table 3.23. Minimum and maximum spacing and end and edge distances for structures subjected to fatigue are detailed in EC3 (BS EN 1993-1-9) [3.10].

The design shear resistance of bolts per shear plane can be calculated, adopting EC3 (BS EN 1993-1-8) [2.13], as follows:

$$F_{v,Rd} = \frac{\alpha_V f_{ub} A}{\gamma_{M2}} \tag{3.102}$$

wherein for the shear plane that passes through the threaded portion of the bolt (A is the tensile stress area of the bolt A_s), $\alpha_v = 0.6$ for classes 4.6, 5.6, and 8.8 and $\alpha_v = 0.5$ for classes 4.8, 5.8, 6.8, and 10.9, and for the shear plane that passes through the unthreaded portion of the bolt (A is the gross cross section of the bolt), $\alpha_v = 0.6$.

Table 3.23 Minimum and Maximum Spacing and End and Edge Distances Specified in EC3 (BS EN 1993-1-8) [2.13]

Distances and Spacings		Minimum	Maximum		
			Structures Made from Steels Conforming to EN 10025-5 Except Steels Conforming to EN 10025-5		Structures Made from Steels Conforming to EN 10025-5
			Steel exposed to the weather or other corrosive influences	Steel not exposed to the weather or other corrosive influences	Steel used unprotected
End distance e_1	$1.2d_o$		$4t + 40$ mm		The larger of $8t$ or 125 mm
Edge distance e_2	$1.2d_o$		$4t + 40$ mm		The larger of $8t$ or 125 mm
Distance e_3 in slotted holes	$1.5d_o$				
Distance e_4 in slotted holes	$1.5d_o$				
Spacing p_1	$2.2d_o$		The smaller of $14t$ or 200 mm	The smaller of $14t$ or 200 mm	The smaller of $14t_{min}$ or 175 mm
Spacing $p_{1,0}$			The smaller of $14t$ or 200 mm		
Spacing $p_{1,i}$			The smaller of $28t$ or 400 mm		
Spacing p_2	$2.4d_o$		The smaller of $14t$ or 200 mm	The smaller of $14t$ or 200 mm	The smaller of $14t_{min}$ or 175 mm

The design bearing resistance of bolts can be calculated according to EC3 (BS EN 1993-1-8) [2.13] as follows:

$$F_{b,Rd} = \frac{k_1 \alpha_b f_u d t}{\gamma_{M2}} \tag{3.103}$$

where α_b is the smallest of $\alpha_d; \dfrac{f_{ub}}{f_u}$ or 1.0

in the direction of load transfer : $\alpha_d = \dfrac{e_1}{3d_o}$ for end bolts

$$\alpha_d = \frac{p_1}{3d_o} - \frac{1}{4} \text{ for inner bolts} \tag{3.104}$$

perpendicular to the direction of load transfer

k_1 is the smallest of $2.8\dfrac{e_2}{d_o} - 1.7$ or 2.5 for edge bolts

k_1 is the smallest of $1.4\dfrac{p_2}{d_o} - 1.7$ or 2.5 for inner bolts

$$\tag{3.105}$$

where d_o is the hole diameter for a bolt (see Figure 3.33). It should be noted that according to EC3 (BS EN 1993-1-8) [2.13], the bearing resistance $F_{b,Rd}$ for bolts in oversized holes is 0.8 times that of bolts in normal holes and, in slotted holes, where the longitudinal axis of the slotted hole is perpendicular to the direction of the force transfer, and is 0.6 times that of bolts in round, normal holes. For countersunk bolt, the bearing resistance $F_{b,Rd}$ should be based on a plate thickness t equal to the thickness of the connected plate minus half the depth of the countersinking.

The design tension resistance of bolts specified in EC3 (BS EN 1993-1-8) [2.13] can be calculated as follows:

$$F_{t,Rd} = \frac{k_2 f_{ub} A_s}{\gamma_{M2}} \tag{3.106}$$

where $k_2 = 0.63$ for countersunk bolt; otherwise, $k_2 = 0.9$. The punching shear resistance can be calculated as follows:

$$B_{p,Rd} = \frac{0.6 \pi d_m t_p f_u}{\gamma_{M2}} \tag{3.107}$$

where d_m is the mean of the across points and across flats dimensions of the bolt head or the nut, whichever is smaller, and t_p is the thickness of the plate under the bolt or the nut. For bolts in combined shear and tension, the following interaction equation should be satisfied:

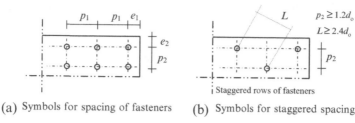

(a) Symbols for spacing of fasteners   (b) Symbols for staggered spacing

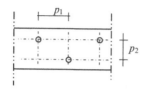

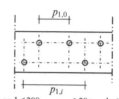

$p_1 \le 14t$ and ≤ 200 mm $p_2 \le 14t$ and ≤ 200 mm

(c) Staggered spacing in compression members

$p_{1,0} \le 14t$ and ≤ 200 $p_{1,i} \le 28t$ and ≤ 400 mm

(d) Staggered spacing in tension members

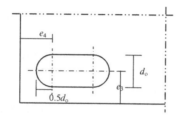

(e) End and edge distances for slotted holes

Figure 3.33 Symbols for end and edge distances and spacing of fasteners specified in EC3 (BS EN 1993-1-8) [2.13].

$$\frac{F_{v,Ed}}{F_{v,Rd}} + \frac{F_{t,Ed}}{1.4F_{t,Rd}} \le 1.0 \qquad (3.108)$$

where $F_{v,Ed}$ and $F_{t,Ed}$ are the design shear and tensile forces per bolt for the ultimate limit state, respectively.

EC3 (BS EN 1993-1-8) [2.13] also specifies design rules for preloaded bolts. The design preload, $F_{p,Cd}$, to be used in design calculations should be taken as

$$F_{p,Cd} = 0.7 f_{ub} A_s / \gamma_{M7} \qquad (3.109)$$

In single-lap joints with only one bolt row, the bolts should be provided with washers under both the head and the nut. The design bearing resistance $F_{b,Rd}$ for each bolt should be limited to

$$F_{b,Rd} \leq \frac{1.5 f_u d t}{\gamma_{M2}} \qquad (3.110)$$

where d is the nominal bolt diameter. In the case of class 8.8 or 10.9 bolts, hardened washers should be used for single-lap joints with only one bolt or one row of bolts. Where bolts transmitting load in shear and bearing pass through packing of total thickness t_p greater than one-third of the nominal diameter d, the design shear resistance $F_{v,Rd}$ should be multiplied by a reduction factor β_p given by

$$\beta_p = \frac{9d}{8d + 3t_p} \quad \text{with} \quad \beta_p \leq 1.0 \qquad (3.111)$$

For double-shear connections with packing on both sides of the splice, t_p should be taken as the thickness of the thicker packing.

For slip-resistant connections using class 8.8 or 10.9 bolts, EC3 (BS EN 1993-1-8) [2.13] specifies that the design slip resistance of a preloaded class 8.8 or 10.9 bolt should be taken as

$$F_{s,Rd} = \frac{k_s n \mu}{\gamma_{M3}} F_{p,C} \qquad (3.112)$$

where k_s is given in Table 3.24, n is the number of the friction surfaces, and μ is the slip factor obtained either by specific tests for the friction surface or when relevant as given in Table 3.25. For class 8.8 and 10.9 bolts with controlled tightening, the preloading force $F_{p,C}$ to be used in Equation (3.112) should be taken as

$$F_{p,C} = 0.7 f_{ub} A_s \qquad (3.113)$$

Table 3.24 Values of k_s Specified in EC3 (BS EN 1993-1-8) [2.13]

Description	k_s
Bolts in normal holes	1.00
Bolts in either oversized holes or short slotted holes with the axis of the slot perpendicular to the direction of load transfer	0.85
Bolts in long slotted holes with the axis of the slot perpendicular to the direction of load transfer	0.70
Bolts in short slotted holes with the axis of the slot parallel to the direction of load transfer	0.76
Bolts in long slotted holes with the axis of the slot parallel to the direction of load transfer	0.63

Table 3.25 Slip Factor μ for Preloaded Bolts Specified in EC3 (BS EN 1993-1-8) [2.13]

Class of Friction Surfaces	Slip Factor μ
A	0.5
B	0.4
C	0.3
D	0.2

If a slip-resistant connection is subjected to an applied tensile force, $F_{t,Ed}$ or $F_{t,Ed,ser}$, in addition to the shear force, $F_{v,Ed}$ or $F_{v,Ed,ser}$, tending to produce slip, the design slip resistance per bolt should be taken as follows:

$$\text{for a category B connection}: F_{s,Rd,ser} = \frac{k_s n \mu \left(F_{p,C} - 0.8 F_{t,Ed,ser}\right)}{\gamma_{M3,ser}} \quad (3.114)$$

$$\text{for a category B connection}: F_{s,Rd} = \frac{k_s n \mu \left(F_{p,C} - 0.8 F_{t,Ed}\right)}{\gamma_{M3}} \quad (3.115)$$

If, in a moment connection, a contact force on the compression side counterbalances the applied tensile force, no reduction in slip resistance is required.

3.10.2.2 Connections Made with Pins

Connections made with pins are also covered by EC3 (BS EN 1993-1-8) [2.13]. The code provides rules for designing connections with pins. According to the code, wherever there is a risk of pins becoming loose, they should be secured. Pin connections in which no rotation is required may be designed as single-bolted connections, provided that the length of the pin is less than three times the diameter of the pin. In pin-connected members, the geometry of the unstiffened element that contains a hole for the pin should satisfy the dimensional requirements given in Table 3.26. Pin-connected members should be arranged to avoid eccentricity and should be of sufficient size to distribute the load from the area of the member with the pin hole into the member away from the pin.

The design of solid circular pins specified in EC3 (BS EN 1993-1-8) [2.13] is dependent on the failure mode expected. The shear resistance of the pin can be calculated as follows:

$$F_{v,Rd} = 0.6 A f_{up} / \gamma_{M2} \geq F_{v,Ed} \quad (3.116)$$

The bearing resistance of the plate and the pin specified in EC3 (BS EN 1993-1-8) [2.13] can be calculated as follows:

Table 3.26 Geometric Requirements for Pin-Connected Members Specified in EC3 (BS EN 1993-1-8) [2.13]

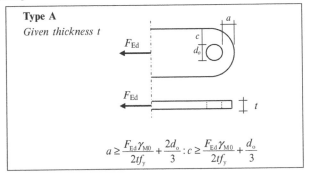

Type A

Given thickness t

$$a \geq \frac{F_{Ed}\gamma_{M0}}{2tf_y} + \frac{2d_o}{3} : c \geq \frac{F_{Ed}\gamma_{M0}}{2tf_y} + \frac{d_o}{3}$$

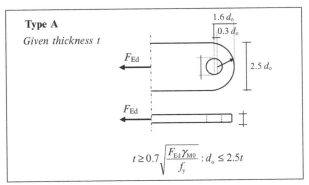

Type A

Given thickness t

$$t \geq 0.7\sqrt{\frac{F_{Ed}\gamma_{M0}}{f_y}} : d_o \leq 2.5t$$

$$F_{b,Rd} = 1.5tdf_y/\gamma_{M0} \geq F_{b,Ed} \tag{3.117}$$

It should be noted that if the pin is intended to be replaceable, EC3 (BS EN 1993-1-8) [2.13] specifies that the following requirement should also be satisfied:

$$F_{b,Rd,ser} = 0.6tdf_y/M_{M6,ser} \geq F_{b,Ed,ser} \tag{3.118}$$

The bending resistance of the pin should be calculated based on EC3 (BS EN 1993-1-8) [2.13] as follows:

$$M_{Rd} = 1.5W_{el}f_{yp}/\gamma_{M0} \geq M_{Ed} \tag{3.119}$$

If the pin is intended to be replaceable, the following requirement should also be satisfied:

$$M_{Rd,ser} = 0.8W_{el}f_{yp}/\gamma_{M6,ser} \geq M_{Ed,ser} \tag{3.120}$$

For pins subjected to combined shear and bending, the following interaction equation specified in EC3 (BS EN 1993-1-8) [2.13] should be satisfied:

$$\left[\frac{M_{Ed}}{M_{Rd}}\right]^2 + \left[\frac{F_{v,Ed}}{F_{v,Rd}}\right]^2 \leq 1 \tag{3.121}$$

where d is the diameter of the pin, f_y is the lower of the design strengths of the pin and the connected part, f_{up} is the ultimate tensile strength of the pin, f_{yp} is the yield strength of the pin, t is the thickness of the connected part, and A is the cross-sectional area of a pin. The moments in a pin should be calculated on the basis that the connected parts form simple supports. It should be generally assumed that the reactions between the pin and the connected parts are uniformly distributed along the length in contact on each part as indicated in Figure 3.34. If the pin is intended to be replaceable, the contact bearing stress should satisfy

$$\sigma_{h,Ed} \leq f_{h,Rd} \tag{3.122}$$

$$\text{with, } \sigma_{h,Ed} = 0.591\sqrt{\frac{EF_{Ed,ser}(d_0 - d)}{d^2 t}} \tag{3.123}$$

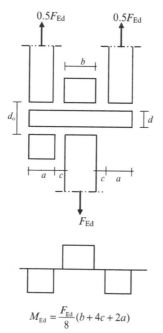

$$M_{Ed} = \frac{F_{Ed}}{8}(b + 4c + 2a)$$

Figure 3.34 Bending moment in a pin specified in EC3 (BS EN 1993-1-8) [2.13].

$$f_{h,Ed} = \frac{2.5 f_y}{\gamma_{M6,ser}} \qquad (3.124)$$

where d is the diameter of the pin, d_0 is the diameter of the pin hole, and $F_{Ed,ser}$ is the design value of the force to be transferred in the bearing, under the characteristic load combination for serviceability limit states.

3.10.3 Design of Welded Joints

The design of welded joints is also covered by EC3 (BS EN 1993-1-8) [2.13]. The code provides rules applicable to weldable structural steels conforming to EC3 (BS EN 1993-1-1) [2.11] and to material thicknesses of 4 mm and over. The rules also apply to joints in which the mechanical properties of the weld metal are compatible with those of the parent metal. For stud welding, reference should be made to EC4 (BS EN 1994-1-1) [2.37]. Welds subject to fatigue shall also satisfy the principles given in EC3 (BS EN 1993-1- 9) [3.10]. EC3 (BS EN 1993-1-8) requires that lamellar tearing should be avoided, with guidance on lamellar tearing given in EC3 (BS EN 1993-1-10) [2.16]. Also, the specified yield strength, ultimate tensile strength, elongation at failure, and minimum Charpy V-notch energy value of the filler metal should be equivalent to or better than that specified for the parent material. Generally, EC3 (BS EN 1993-1-8) [2.13] recommends to use electrodes that are overmatched with regard to the steel grades being used. This code covers the design of fillet welds, fillet welds all round, butt welds, plug welds, and flare groove welds. Butt welds may be either full penetration butt welds or partial penetration butt welds. Both fillet welds all round and plug welds may be either in circular holes or in elongated holes.

According to EC3 (BS EN 1993-1-8) [2.13], fillet welds may be used for connecting parts where the fusion faces form an angle between 60° and 120°. Angles smaller than 60° are also permitted. However, in such cases, the weld should be considered to be a partial penetration butt weld. For angles greater than 120°, the resistance of fillet welds should be determined by testing. Fillet welds finishing at the ends or sides of parts should be returned continuously, full size, around the corner for a distance of at least twice the leg length of the weld, unless access or the configuration of the joint renders this impracticable. End returns should be indicated on the drawings. Intermittent fillet welds should not be used in corrosive conditions. In an intermittent fillet weld, the gaps (L_1 or L_2) between the ends of each length of weld L_w should fulfill the requirement given in

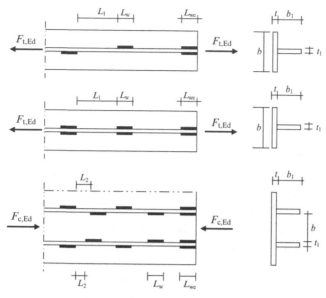

The smaller of $L_{we} \geq 0.75b$ and $0.75b$

For build-up members in tension:
The smallest of $L_1 \leq 16t$ and $16t_1$ and 200 mm

For build-up members in compression or shear:
The smallest of $L_2 \leq 12t$ and $12t_1$ and $0.25b$ and 200 mm

Figure 3.35 Intermittent fillet welds specified in EC3 (BS EN 1993-1-8) [2.13].

Figure 3.35. In an intermittent fillet weld, the gap (L_1 or L_2) should be taken as the smaller of the distances between the ends of the welds on opposite sides and the distance between the ends of the welds on the same side. In any run of intermittent fillet weld, there should always be a length of weld at each end of the part connected. In a built-up member in which plates are connected by means of intermittent fillet welds, a continuous fillet weld should be provided on each side of the plate for a length at each end equal to at least three-quarters of the width of the narrower plate concerned as shown in Figure 3.35. Fillet welds all round, comprising fillet welds in circular or elongated holes, may be used only to transmit shear or to prevent the buckling or separation of lapped parts. The diameter of a circular hole, or width of an elongated hole, for a fillet weld all round should not be less than four times the thickness of the part containing it. The ends of elongated holes should be semicircular, except for those ends that extend to the edge of the part concerned. The center-to-center spacing of fillet welds all round should not exceed the value necessary to prevent local buckling.

Also, according to EC3 (BS EN 1993-1-8) [2.13], a full penetration butt weld is defined as a weld that has complete penetration and fusion of weld and parent metal throughout the thickness of the joint. A partial penetration butt weld is defined as a weld that has joint penetration that is less than the full thickness of the parent material. Intermittent butt welds should not be used. Plug welds may be used to transmit shear, to prevent the buckling or separation of lapped parts, and to interconnect the components of built-up members but should not be used to resist externally applied tension. The diameter of a circular hole, or width of an elongated hole, for a plug weld should be at least 8 mm more than the thickness of the part containing it. The ends of elongated holes either should be semicircular or should have corners that are rounded to a radius of not less than the thickness of the part containing the slot, except for those ends that extend to the edge of the part concerned. The thickness of a plug weld in parent material up to 16 mm thick should be equal to the thickness of the parent material. The thickness of a plug weld in parent material over 16 mm thick should be at least half the thickness of the parent material and not less than 16 mm. The center-to-center spacing of plug welds should not exceed the value necessary to prevent local buckling. For solid bars, the design effective throat thickness of flare groove welds, when fitted flush to the surface of the solid section of the bars, is defined in Figure 3.36. In the case of welds with packing, the packing should be trimmed flush with the edge of the part that is to be welded. Where two parts connected by welding are separated by packing having a thickness less than the leg length of weld necessary to transmit the force, the required leg length should be increased by the thickness of the packing. Where two parts connected by welding are separated by packing having a thickness equal to, or greater than, the leg length of weld necessary

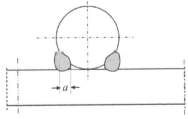

Figure 3.36 Effective throat thickness of flare groove welds in solid sections specified in EC3 (BS EN 1993-1-8) [2.13].

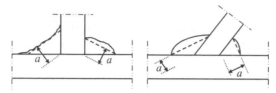

Figure 3.37 Throat thickness of a fillet weld specified in EC3 (BS EN 1993-1-8) [2.13].

to transmit the force, each of the parts should be connected to the packing by a weld capable of transmitting the design force.

The design of fillet welds requires the calculation of required lengths (l) of weld and the effective throat thickness (a). Following the design rules specified in EC3 (BS EN 1993-1-8) [2.13], the effective length of a fillet weld l should be taken as the length over which the fillet is full-sized. This may be taken as the overall length of the weld reduced by twice the effective throat thickness (a). Provided that the weld is full size throughout its length including starts and terminations, no reduction in effective length need be made for either the start or the termination of the weld. A fillet weld with an effective length less than 30 mm or less than six times its throat thickness, whichever is larger, should not be designed to carry load. On the other hand, the effective throat thickness (a) of a fillet weld should be taken as the height of the largest triangle (with equal or unequal legs) that can be inscribed within the fusion faces and the weld surface, measured perpendicular to the outer side of this triangle (see Figure 3.37). The effective throat thickness of a fillet weld should not be less than 3 mm. In determining the design resistance of a deep penetration fillet weld, its additional throat thickness may be taken of (see Figure 3.38), provided that preliminary tests show that the required penetration can consistently be achieved.

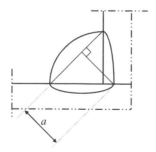

Figure 3.38 Throat thickness of a deep penetration fillet weld specified in EC3 (BS EN 1993-1-8) [2.13].

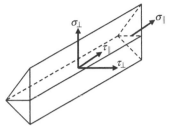

Figure 3.39 Stresses on the throat section of a fillet weld specified in EC3 (BS EN 1993-1-8) [2.13].

The design resistance of a fillet weld specified in EC3 (BS EN 1993-1-8) [2.13] should be determined using either the directional method given or the simplified method. In the directional method, the forces transmitted by a unit length of weld are resolved into components parallel and transverse to the longitudinal axis of the weld and normal and transverse to the plane of its throat. The design throat area A_w should be taken as $A_w = \sum al_{\text{eff}}$. The location of the design throat area should be assumed to be concentrated in the root. A uniform distribution of stress is assumed on the throat section of the weld, leading to the normal stresses and shear stresses shown in Figure 3.39. The normal stresses are denoted as $\sigma\perp$ (the normal stress perpendicular to the throat) and $\sigma\|$ (the normal stress parallel to the axis of the weld). The shear stresses are denoted as $\tau\perp$ (the shear stress, in the plane of the throat, perpendicular to the axis of the weld) and $\tau\|$ (the shear stress, in the plane of the throat, parallel to the axis of the weld). The normal stress $\sigma\|$ parallel to the axis is not considered when verifying the design resistance of the weld. The design resistance of the fillet weld will be sufficient if the following are both satisfied:

$$\left[\sigma\perp^2 + 3\left(\tau\perp^2 + \tau\|^2\right)\right]^{0.5} \leq f_u/(\beta_w \gamma_{M2}) \quad \text{and} \quad \sigma\perp \leq 0.9 f_u/\gamma_{M2} \quad (3.125)$$

where f_u is the nominal ultimate tensile strength of the weaker part joined and β_w is the appropriate correlation factor taken from Table 3.27 as recommended by EC3. Welds between parts with different material strength grades should be designed using the properties of the material with the lower strength grade.

EC3 (BS EN 1993-1-8) [2.13] also specifies a simplified method for design resistance of fillet weld. In the simplified method of weld design, the design resistance of a fillet weld may be assumed to be adequate if, at

Table 3.27 Correlation Factor β_w for Fillet Welds Recommended by EC3 (BS EN 1993-1-8) [2.13]

Standard and Steel Grade

EN 10025	EN 10210	EN 10219	Correlation Factor β_w
S 235	S 235H	S 235H	0.8
S 235 W			
S 275	S 275H	S 275H	0.85
S 275 N/NL	S 275 NH/NLH	S 275 NH/NLH	
S 275 M/ML		S 275 MH/MLH	
S 355	S 355H	S 355H	0.9
S 355 N/NL	S 355 NH/NLH	S 355 NH/NLH	
S 355 M/ML		S 355 MH/MLH	
S 355 W			
S 420 N/NL		S 420 MH/MLH	1.0
S 420 M/ML			
S 460 N/NL	S 460 NH/NLH	S 460 NH/NLH	1.0
S 460 M/ML		S 460 MH/MLH	
S 460 Q/QL/QL1			

every point along its length, the resultant of all the forces per unit length transmitted by the weld satisfies the following criterion:

$$F_{w,Ed} \leq F_{w,Rd} \tag{3.126}$$

where $F_{w,Ed}$ is the design value of the weld force per unit length and $F_{w,Rd}$ is the design weld resistance per unit length. Independent of the orientation of the weld throat plane to the applied force, the design resistance per unit length $F_{w,Rd}$ should be determined from

$$F_{w,Rd} = f_{vw.d}a \tag{3.127}$$

where $f_{vw.d}$ is the design shear strength of the weld. The design shear strength $f_{vw.d}$ of the weld should be determined from

$$f_{vw.d} = \frac{f_u/\sqrt{3}}{\beta_w \gamma_{M2}} \tag{3.128}$$

According to EC3 (BS EN 1993-1-8) [2.13], the distribution of forces in a welded connection may be calculated on the assumption of either elastic or plastic behavior. It is acceptable to assume a simplified load distribution within the welds. Residual stresses and stresses not subjected to the transfer of load need not be included when checking the resistance of a weld.

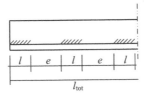

Figure 3.40 Calculation of weld forces for intermittent welds specified in EC3 (BS EN 1993-1-8) [2.13].

This applies specifically to the normal stress parallel to the axis of a weld. Welded joints should be designed to have adequate deformation capacity. However, ductility of the welds should not be relied upon. In joints where plastic hinges may form, the welds should be designed to provide at least the same design resistance as the weakest of the connected parts. In other joints where deformation capacity for joint rotation is required due to the possibility of excessive straining, the welds require sufficient strength not to rupture before general yielding in the adjacent parent material. If the design resistance of an intermittent weld is determined by using the total length l_{tot}, the weld shear force per unit length $F_{w,Ed}$ should be multiplied by the factor $(e+l)/l$ (see Figure 3.40).

3.11 DESIGN OF BRIDGE BEARINGS

3.11.1 General

This section provides a brief review for the types and design of bearings commonly used in steel bridges. The bearings used in steel bridges can be classified according to its main supply source to proprietary and steel-fabricated bearings. Proprietary bearings are commonly made of elastomeric material, which is either natural or synthetic rubber. Elastomeric materials are flexible when subjected to shearing forces; however, they are very stiff against volumetric change. On the other hand, steel-fabricated bearings are made of designed steel parts preventing or allowing applied translations and rotations. Proprietary bearings are efficiently used with most of steel bridges. However, steel-fabricated bearings can be economic in uplift situations or in situations where large rotations are expected to occur. Bearings used in bridges can be also classified according to their restraint performance to fixed, hinged, and expansion bearings. Fixed bearings prevent rotations and translations of the supported structure at their locations, while hinged bearings allow rotations and prevent translations of the supported structure

Table 3.28 Brief Comparisons of Commonly Used Bridge Bearings

Type	Capacity (kN)	Supply	Friction	Use	Limitations
Pot or disk	500–30,000	Proprietary	0.05	Span >20 m	Rotation 0.01 rad
Elastomeric laminated	100–1000	Proprietary	0.5-6 kN/mm	Short spans	Heavy loads
Cylindrical roller	1000–1500	Proprietary	0.01	Minimal friction	Nil lateral translation or rotation
Multiple roller	1000–10,000	Fabricated	0.25	Roller bearing/ railway bridges	High-friction, nil lateral rotation
Line rocker	1000–10,000	Fabricated	0.25	Hinged bearing/ railway bridges	High-friction, nil lateral rotation
Spherical sliding	1000–12,000	Proprietary	0.05	Span >20 m	More expensive than pot

at their locations. Finally, expansion bearings allow rotations of the supported structure at their locations and allow translations in particular directions. Expansion bearings can be sliding, roller, or rocker bearings. Table 3.28 shows the main types of bearings commonly used in steel bridges. The most frequently used type of bearing for highway bridges is the proprietary pot or disk type, which is able to accommodate rotation and, where required, lateral movement in either longitudinal or transverse directions or in both directions. Such bearings are particularly suitable for continuous and curved bridges. While for railway bridges or footbridges, fabricated line rocker bearings are often suitable at both ends of bridges. For rail bridges of span greater than 20 m, fabricated roller/rocker bearings can be used at the free end. For footbridges, elastomeric bearings are often used.

3.11.2 Examples of Proprietary Bearings

Proprietary pot or disk bearings are commonly used in practice all over the world. The bearings comprise a circular elastomeric disk confined by a metal housing (forming a cylinder and piston). The bearings can be combined with

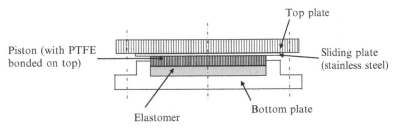

Figure 3.41 Pot bridge bearings.

Figure 3.41 Pot bridge bearings.

a sliding element to accommodate translational movements in one or any direction. This can be achieved by a PTFE (polytetrafluoroethylene)/stainless steel interface, usually arranged as shown in Figure 3.41. The coefficient of friction on the sliding surface depends on the PTFE interface pressure and is typically 5%. Pot bearings may be fixed, or guided, by providing suitable lateral restraints between the top and bottom plates. Proprietary elastomeric bearings may be of strip, rectangular pad, or laminated type. Laminated elastomeric bearings (see Figure 3.42) are economic for loads up to 1000 kN capacity. For loads greater than 1000 kN, the bearings may become uneconomically large. Therefore, elastomeric bearings are rarely used for steel highway or railway bridges. The design of elastomeric bearings is governed by serviceability limit state requirements, to control excessive distortion of the material. Movements and rotations are achieved by deformation of the elastomeric material (see Figure 3.43). Movement is restricted to about 40 mm from the mean position.

Proprietary cylindrical bearings consist of a backing plate with a convex cylindrical surface (rotational element) and a backing plate with a concave

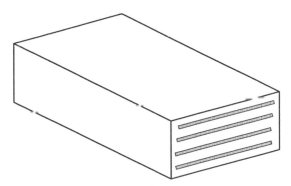

Figure 3.42 Laminated elastomeric bridge bearings.

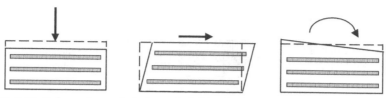

Figure 3.43 Deformations of elastomeric bridge bearings.

cylindrical surface between which a PTFE sheet and the mating material form a curved sliding surface. Cylindrical PTFE bearings are also used in combination with flat sliding elements and guides to form free or guided bearings. On the other hand, spherical PTFE bearing consist of a backing plate with a convex spherical surface (rotational element) and a backing plate with a concave spherical surface between which a PTFE sheet and the mating material form a curved sliding surface. Spherical bearings allow three-dimensional movements. They are designed for very high vertical, horizontal, and lateral loads and also for large rotational displacements. Like pot bearings, they can be fixed, free sliding, or guided sliding depending on the design. Spherical bearings have been structurally improved and designed for use as incremental launch bearing, which is applicable for bridge construction by launching system; force measuring bearing, which can be used for measuring and monitoring forces acting on the structure electronically; and uplift protection bearing, which can be used to accommodate high uplift loads encountered during construction or service life of a structure. Figure 3.44 shows an example of spherical bearings.

3.11.3 Examples of Steel-Fabricated Bearings

Steel-fabricated bearings are the oldest types of bearing. The most common types of steel bearings are the roller/rocker bearings, which may be hinged or expansion bearings. Roller/rocker bearings can support high loads and can be used where pot, spherical, and other high-capacity bearings cannot be used due to limited space. Roller/rocker bearings are applicable for conditions where only longitudinal movement is allowed and where transverse movement is to be prevented. They operate by the movement of a roller/rocker in between a sole plate and a lower bearing plate (see Figure 3.45). As an example of fabricated steel hinged bearing, Figure 3.46 shows fabricated line rocker bearings. The bearings provide a very economic solution in that they can be supplied by the steelwork fabricator and ensure a good match between hole positions in bearings, upper bearing plates, and girder flanges.

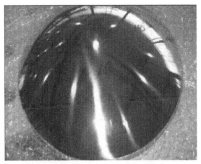

(a) Stainless steel convex surface plate

(b)   concave surface plate

Figure 3.44 Concave and concave components of a spherical bearing (www.mageba. ch). (For color version of this figure, the reader is referred to the online version of this chapter.)

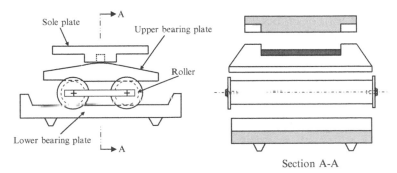

Figure 3.45 Detailing of twin roller-fabricated steel bridge bearings.

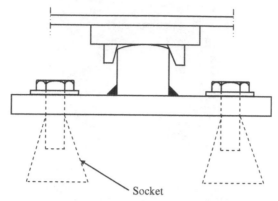

Socket

Figure 3.46 Detailing of hinged line rocker-fabricated steel bridge bearings.

When designing a line rocker, the maximum eccentricity of the reaction (due to the restraining torque that the bearing provides) needs to be considered carefully (there is no tensile restraint at the line of contact).

3.11.4 Design Rules for Bearings

The current technical specifications for designing different bearings are given in EC3 (BS EN 1993-2) [1.27]. The code gives guidance for designing bearings, which comply with BS EN 1337-1 [3.11]. Five bearings are not covered by the code, which are bearings that transmit moments as a primary function, bearings that resist uplift, bearings for moving bridges, bearings for concrete hinges, and bearings for seismic devices. According to the code [1.27], hinged (fixed) bearings prevent movements but other bearings such as guided bearings allow movements in one direction while free bearings allow movements in all directions. Detailed information on bearings is presented in 11 parts of BS EN 1337-1 [3.11]. Part 1 of the code provides general design rules for bearings. Parts 2-8 cover design rules for sliding elements, elastomeric bearings, roller bearings, pot bearings, rocker bearings, spherical and cylindrical PTFE bearings, and guided bearings and restraint bearings, respectively. Part 9 covers protection, part 10 covers inspection and maintenance, and finally, part 11 provides guides on transport, storage, and installation. According to EC3 (BS EN 1993-2) [1.27], the bearing layout should be designed to permit the specified movement of a bridge with the minimum possible resistance to such movements. The arrangement of bearings for a structure should be considered in conjunction with the design of the bridge as a whole. The forces and movements in bearings should be given to the bearing manufacturer to ensure that the bearings provided meet the requirements. A drawing showing the bearing layout should include the following: a simplified general arrangement of the bridge showing the bearings in plan, details at the bearing location, a clear indication of the type of bearing at each location, a table giving the detailed requirements for each bearing, and bedding and fixing details. Bearings should not normally be expected to resist moments due to rotational movement. Uplift may cause excessive wear in bearings if such conditions occur frequently. Where uplift is unavoidable, prestressing may be used to provide the necessary additional vertical force. Bearings and supports should be designed in such a way that they can be inspected, maintained, and replaced if necessary.

EC3 (BS EN 1993-2) [1.27] requires that for line rocker and single roller bearings, the full implications of uneven pressure along the length of the roller or rocker should be taken into account in the design of the structure and the bearing. Also, particular care should be taken in the design of bridges curved in plan, bridges with slender piers, bridges without transverse beams, bridges with transverse beams where the line rocker or single roller could effectively act as a built-in support for the transverse beam, and bridges with a transverse temperature gradient. Anchorages of bridge bearings shall be designed at the ultimate limit state. Where the position of a bearing or part of a bearing is retained either completely or partially by friction, its safety against sliding shall be checked in accordance with the following:

$$V_{Ed} \leq V_{Rd} \tag{3.129}$$

where V_{Ed} is the design value of the shear force acting at the bridge bearing

$$V_{Rd} = \frac{\mu_K}{\gamma_\mu} N_{Ed} + V_{pd} \tag{3.130}$$

where N_{Ed} is the minimum design force acting normal to the joint in conjunction with V_{Ed}, V_{pd} is the design value of shear resistance of any fixing device in accordance with the Eurocodes, μ_K is the characteristic value of the friction coefficient (see Table 3.29), and γ_μ is the partial factor for friction. The code [1.27] recommends the following values. $\gamma_\mu = 2.0$ for steel on steel and $\gamma_\mu = 1.2$ for steel on concrete. For dynamically loaded structures, the value of N_{Ed} should be determined, taking into account any dynamic variations in traffic loads. For railway bridges and structures subjected to seismic situations, friction should not be taken into account ($N_{Ed} = 0$). Where the bearings are designed to resist horizontal forces, some movements will

Table 3.29 Characteristic Values of the Friction Coefficient μ_K Recommended by [C] (BS EN 1993-2) [1.27]

Surface Treatment of Steel Components	Steel on Steel	Steel on Concrete
Uncoated and free from grease	0.4	0.6
Metal-sprayed		
Coated with fully hardened zinc silicate		
Other treatments	From tests	From tests

take place before clearances are taken up. The total clearance between the extremes of movements may be up to 2 mm unless otherwise specified or agreed with the manufacturer.

3.11.5 Design Rules for Fabricated Steel Bearings

The design rules for fabricated steel roller bearing are provided in EC1 (BS EN 1337-4) [3.12]. According to the code, only ferrous materials (see Table 3.30) shall be used in the manufacture of rollers and roller plates. Rollers and roller plates shall have a surface hardness less than that specified by the code. Carbon steel shall be in accordance with the requirements of EN 10025 [3.13] or EN 10083-1 [3.14] and EN 10083-2 [3.15], with a minimum yield strength of 240 N/mm^2. Stainless steel shall be in accordance with EN 10088-2 [3.16], with a minimum tensile strength of 490 N/mm^2 for any component. Cast steel shall be in accordance with ISO 3755 [3.17]. The design of roller bearings is based on the assumption that load passes through a Hertzian contact area between two surfaces with dissimilar radii. Design verification with respect to loading and rotation (movement) should be determined in accordance with BS EN 1337-1 [3.11]. The design values of the effects (forces, deformations, and movements) from the actions at the supports of the structure shall be calculated from the relevant

Table 3.30 Ferrous Material Classes According to BS EN 1337-4 [3.12]

Material Class	Tensile Strength (Minimum) (N/mm²)	Yield Strength (Minimum) (N/mm²)	Impact/At Temperature (Minimum) (J)	Surface Hardness (Maximum) (HV 10)	Elongation (Minimum) (%)	Friction Coefficient (Maximum)
A	340	240	27/0 °C	150	25	0.05
B	490	335	27/−20 °C	250	21	0.05
C	600	420	27/−20 °C	450	14	0.02
D	1350	1200	11/−20 °C	480	12	0.02

combination of actions according to BS EN 1990 [3.4]. Sliding elements should be designed and manufactured in accordance with EN 1337-2 [3.18]. The recommended material partial safety factor $\gamma_m = 1$. Roller bearings provide for translation in one direction only. Single rollers permit rotation about the line of contact but multiple rollers require additional elements to accommodate rotation. Roller bearings for use in curved parts of structures shall have additional sliding elements and/or rotation elements to ensure uniform distribution of load across the roller. The axis of rotation shall be perpendicular to the direction of movement. The curved surfaces shall be of cylindrical shape. Surfaces in contact shall have the same nominal strength and hardness. The length of a roller shall not be less than twice its diameter nor greater than six times its diameter. Guidance shall be provided to ensure that the axis of rolling is maintained correctly. Location shall be such that true rolling occurs during movement. Where gearing is used as security, the pitch circle diameter of the gear teeth shall be the same as the diameter of the rollers. The design axial force per unit length of roller contact N'_{Sd} specified in BS EN 1337-1 [3.11] shall meet the following condition under the fundamental combination of actions:

$$N'_{Sd} \leq N'_{Rd} \tag{3.131}$$

where N'_{Rd} is the design value of resistance per unit length of roller contact, which is calculated as

$$N'_{Rd} = \frac{N'_{Rk}}{\gamma_m^2} \tag{3.132}$$

where N'_{Rk} is the characteristic value of resistance of the contact surface per unit length calculated as

$$N'_{Rk} = 23 \times R \times \frac{f_u^2}{E_d} \tag{3.133}$$

where R is the radius of contact surface (mm), f_u is the ultimate strength of material (N/mm^2), and E_d is the design modulus of elasticity (N/mm^2). In determining the values of N'_{Sd}, the effects of asymmetrical loading due to transverse eccentricities and applied moments shall be considered. Roller plates shall be dimensioned in the direction of displacement to allow for movement calculated for the fundamental combination of actions plus an additional roller design movement of $2 \times t_p$, the thickness of the roller bearing plate, or 20 mm whichever is greater. The length of the plates parallel to the roller axis shall not be less than the length of the roller. In determining the thickness of the roller plates, the following shall be satisfied using the load distribution shown in Figure 3.48 under the fundamental combination of actions:

$$N_{Sd} \leq N_{Rd} \tag{3.134}$$

$$\text{where } N_{Rd} = \frac{N_{Rk}}{\gamma_m} \tag{3.135}$$

$$\text{and } N_{Rk} = f_y \left(2t_p + b\right) L \tag{3.136}$$

where b can be calculated according to Hertzian stress analysis principles or taken as equal to 0, L is the effective length of roller (mm), and $\gamma_m = 1.1$.

BS EN 1337-4 [3.12] specifies that for roller bearings, the stiffness of the supporting plates is of paramount importance; therefore, the roller plates shall be so proportioned that loads are adequately distributed to adjacent components (Figure 3.47). The maximum load dispersion through a component shall be taken as 45° unless a greater angle is justified by calculations that take into account the characteristics of the adjacent components and materials. In no case shall load dispersion be assumed beyond a line drawn at 60° to the vertical axis (see Figure 3.48). Where movement requirements permit, flat-sided rollers may be used. Such rollers shall be symmetrical about the vertical plane passing through the axis of the roller. The minimum width shall not be less than one-third of the diameter nor such that the bearing contact area falls outside the middle third of the rolling surface when the roller is

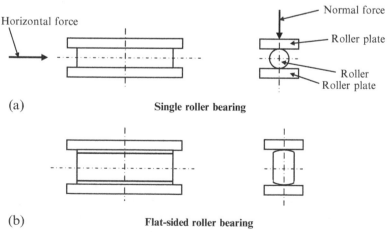

Figure 3.47 Cylindrical and flat-sided roller bearings.

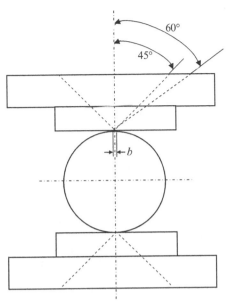

Figure 3.48 Load distribution to components according to BS EN 1337-4 [3.12].

at the extremes of movement determined in accordance with EN 1337-1 [3.11]. It should be noted that according to the code, flat-sided rollers can be mounted at closer centers than circular rollers of the same load capacity resulting in more compact bearings. Where a bearing has more than one roller, an additional bearing in accordance with other parts of EN 1337 shall be included to accommodate rotation (see Figure 3.45). The effects of any rotation moments from this element shall be included when calculating the roller forces by taking into account the corresponding eccentricities. The load per roller shall be calculated at the extreme of the expected movement. In addition, where a bearing has more than two rollers, the limiting values for design load effects shall be taken as two-thirds of N'_{Rd}. The design friction coefficient μ_d shall be taken as 0.02 for steel with a hardness ≥ 300 HV and 0.05 for all other steels.

Hinged line rocker bearings (see Figure 3.46) are capable of transferring applied vertical and horizontal forces between the superstructure and the substructure. Hinged line rockers permit rotation in one direction about the rocker axis. Hinged line rocker bearings resist horizontal forces by means of positive mechanical restraint such as shear dowels. The design of rocker bearings is covered by BS EN 1337-6 [3.19]. The rotation capability of the rocker bearing is an inherent characteristic of the system based on its geometry and shall be declared by the manufacturer. Its maximum value shall be 0.05 rad. The radius of the curved part of the liner rocker bearing is determined in the same way as roller bearing.

REFERENCES

[3.1] EC1, Eurocode 1—Actions on structures—Part 2: traffic loads on bridges. BS EN 1991-2, British Standards Institution, 2003.

[3.2] EC1, Eurocode 1: Actions on structures—General actions—Part 1-4: Wind actions. BS EN 1991-4, British Standards Institution, 2004.

[3.3] EC1, Eurocode 1: Actions on structures—Part 1-5: General actions—Thermal actions. BS EN 1991-5, British Standards Institution, 2003.

[3.4] EC0, Eurocode 0: UK National Annex for Eurocode—Basis of structural design. BS EN 1990, British Standards Institution, 2005.

[3.5] EC3, Eurocode 3—Design of steel structures—Part 1-5: Plated structural elements. BS EN 1993-1-5, British Standards Institution, 2006.

[3.6] EC4, Eurocode 4—Design of composite steel and concrete structures—Part 2: General rules and rules for bridges. BS EN 1994-2, British Standards Institution, 2005.

[3.7] R.P. Johnson, H. Yuan, Existing rules and new tests for stud shear connectors in troughs of profiled sheeting, Proc. Inst. Civil Eng. Struct. Build. 128 (11506) (1998) 244–251.

[3.8] R.P. Johnson, H. Yuan, Models and design rules for stud shear connectors in troughs of profiled sheeting', Proc. Inst. Civil Eng. Struct. Build. 128 (11507) (1998) 252–263.

[3.9] H. Yuan, The resistances of stud shear connectors with profiled sheeting. PhD Thesis, University of Warwick, 1996.

[3.10] EC3, Eurocode 3—Design of steel structures—Part 1-9: Fatigue. BS EN 1993-1-9, British Standards Institution, 2005.

[3.11] EN 1337-1, Structural bearings—Part 1: General design rules. BS EN 1337-1, British Standards Institution, 2000.

[3.12] EN 1337-4, Structural bearings—Part 4: Roller bearings. BS EN 1337-4, British Standards Institution, 2004.

[3.13] EN 10025, European structural steel standard EN 10025. BS EN 10025, British Standards Institution, 2004.

[3.14] EN 10083-1, Steels for quenching and tempering. General technical delivery conditions. BS EN 10083-1, British Standards Institution, 2006.

[3.15] EN 10083-2, Steels for quenching and tempering. BS EN 10083-2, British Standards Institution, 2006.

[3.16] EN 10088-2, Stainless steels. Technical delivery conditions for sheet/plate and strip of corrosion resisting steels for general purposes. BS EN 10088-2, British Standards Institution, 2005.

[3.17] ISO 3755, Cast carbon steels for general engineering purposes. ISO International Standard, Geneva, Switzerland, 1991.

[3.18] EN 1337-2, Structural bearings—Part 2: Sliding elements. BS EN 1337-2, British Standards Institution, 2001.

[3.19] EN 1337-6, Structural bearings—Part 2: Rocker bearings. BS EN 1337-6, British Standards Institution, 2004.

CHAPTER 4

Design Examples of Steel and Steel-Concrete Composite Bridges

4.1 GENERAL REMARKS

The previous Chapters 1–3 highlighted the main issues regarding the general background, layout, classification, literature review, nonlinear material behavior of the bridge components, shear connection behavior, applied loads, and stability and design of steel and steel-concrete composite bridges. Therefore, it is now possible in this chapter to present detailed design examples for the bridges. The design examples were carefully chosen to cover railway and highway bridges, plate girder steel bridges, truss steel bridges, and steel-concrete composite bridges. The presented examples cover the design of the bridge components comprising stringers (longitudinal floor beams), cross girders (lateral floor girders), main girders, connections, bracing members, stiffeners, splices, and bearings. The design examples are calculated, as an example, based on the design rules specified in EC3 [1.27, 2.11], which were previously highlighted in Chapter 3. The examples addressed in this chapter represent hand calculations performed by the author. Overall, the design examples detail how the cross sections are initially assumed, how the straining actions are calculated, and how the stresses are checked and assessed against the design rules. One of the designed bridges presented in this chapter will be modeled using the finite element method in Chapter 6, which is credited to this book. Once again, the main objective of this book is to introduce a complete piece of work regarding both the design and finite modeling of the bridges.

This chapter starts with a brief introduction of the presented design examples for steel and steel-concrete composite bridges. After that, the chapter details five detailed design examples for the bridges. The first design example presented is for a double-track open-timber floor plate girder deck railway steel bridge; the second, for a through truss highway steel bridge; the third, for a highway steel-concrete composite bridge; the fourth, for a double-track open-timber floor plate girder pony railway steel bridge; finally, the fifth, for a deck truss highway steel bridge. The author hopes that

Finite Element Analysis and Design of Steel and Steel–Concrete Composite Bridges

the chapter provides readers with sufficient background needed for future studies. It should be noted that the design examples are presented for specific bridges; however, the design procedures can be adopted for different steel and steel-concrete composite bridges. It should also be noted that the author purposely avoided complex bridge geometries, supports, and long spans to use hand calculations, which make it easy for readers to apply the design rules highlighted in Chapter 3. Finally, the author hopes that the presented design examples in this Chapter provide all the basic fundamentals for students interested in the structural analysis and design of steel and steel-concrete composite bridges.

4.2 DESIGN EXAMPLE OF A DOUBLE-TRACK PLATE GIRDER DECK RAILWAY STEEL BRIDGE

Let us start by presenting the first design example, which is for a double-track open–timber floor plate girder deck railway steel bridge. The general layout of the double-track railway bridge is shown in Figures 4.1 and 4.2, with a brief introduction of the bridge components previously highlighted in Figure 1.20. The bridge has simply supported ends, a length between supports of 30 m and an overall length of 31 m. The width of the bridge (spacing between main plate girders) is 7.2 m as shown in Figure 4.1. It is required to design the bridge components adopting the design rules specified in EC3 [1.27]. The steel material of construction of the double-track railway bridge conformed to standard steel grade EN 10025-2 (S 275) having a yield stress of 275 MPa and an ultimate strength of 430 MPa. The bridge has upper and lower wind bracings of K–shaped truss members as well as cross bracings of X–shaped truss members as shown in Figure 4.1. In addition, the bridge has lateral shock (nosing force) bracing for the stringers as well as braking force bracing at the level of upper wind bracing as shown in Figure 4.2. The lateral shock bracing eliminates bending moments around the vertical axis of the stringers, while the braking force bracing eliminates bending moments around the vertical axis of the cross girder. The plate girder web is stiffened by vertical stiffeners, to safeguard against shear stresses and web buckling, spaced at a constant distance of 1.667 m. The expected live loads on the bridge conform to Load Model 71, which represents the static effect of vertical loading due to normal rail traffic as specified in EC1 [3.1]. The bolts used in connections and field splices are M27 high-strength pretensioned bolts of grade 8.8.

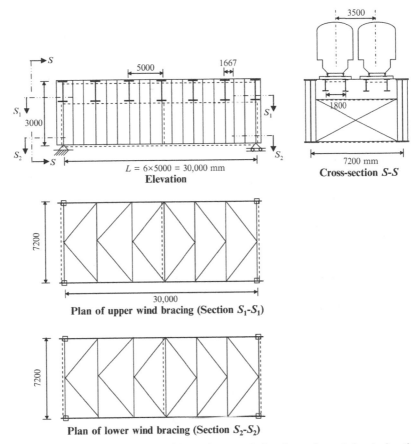

Figure 4.1 General layout of a double track open-timber floor plate girder deck railway steel bridge (the first design example).

4.2.1 Design of the Stringers (Longitudinal Floor Beams)

Let us start by designing the stringers, the longitudinal steel beams, supporting the track as shown in Figure 4.1.

Dead Loads

$$\text{Half weight of the track load} = 3 \, \text{kN/m}$$

$$\text{Weight of stringer bracing} = 0.3 \, \text{kN/m}$$

$$\text{Own weight of stringer} = 1.5 \, \text{kN/m}$$

$$\text{Total dead load} = g_{vk} = 4.8 \, \text{kN/m}$$

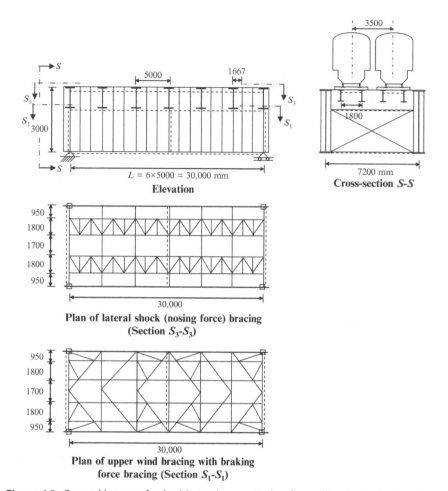

Figure 4.2 General layout of a double track open-timber floor plate girder deck railway steel bridge (the first design example).

Assuming the stringers are simply supported by the cross girders (lateral floor girders), we can calculate the maximum shear force and bending moment due to dead loads on a stringer (see Figure 4.3) as follows:

$$Q_{D.L.} = g_{vk} \times L/2 = 4.8 \times 5/2 = 12\,kN$$
$$M_{D.L.} = g_{vk} \times L^2/8 = 4.8 \times 5^2/8 = 15\,kN\,m$$

Live Loads
Considering the axle live loads on the bridge components according to Load Model 71, which represents the static effect of vertical loading due to normal

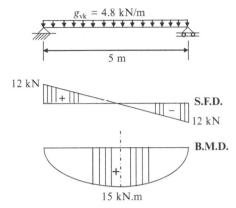

Figure 4.3 Straining actions from dead loads acting on a stringer.

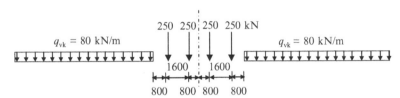

Figure 4.4 Axle live loads on the bridge conforming to Load Model 71 specified in EC1 [3.1].

rail traffic as specified in EC1 [3.1] (see Figure 4.4), three cases of loading for the evaluation of maximum bending moment due to the live loads on a stringer can be studied. The first case of loading is that the centerline at midspan of a stringer divides the spacing between the resultant of the concentrated live loads and the closest load, with maximum bending moment located at the closest load (point a in Figure 4.5), while the second case of loading is that the centerline of the stringer is located under one of the intermediate concentrated loads, with maximum bending moment located at the midspan, and finally, the third case of loading is that the stringer span is covered by the distributed live loads, with maximum bending moment located at the midspan. The three cases of loading are shown in Figure 4.5:

$$M_{L.L.}(\text{case of loading 1}) = 217.5 \times 2.1 - 125 \times 1.6 = 256.75 \, \text{kN m}$$

$$M_{L.L.}(\text{case of loading 2}) = 187.5 \times 2.5 - 125 \times 1.6 = 268.75 \, \text{kN m}$$

$$M_{L.L.}(\text{case of loading 3}) = 40 \times 5^2/8 = 125 \, \text{kN m}$$

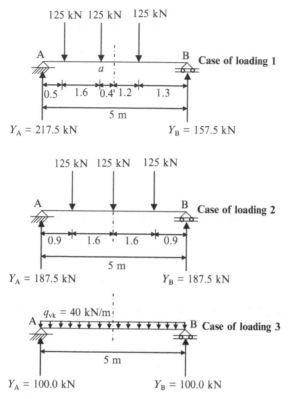

Figure 4.5 Cases of loading for the maximum bending moment acting on a stringer.

Dynamic Factor Φ

Assuming a track with standard maintenance, therefore,

$$L_\Phi = 5 + 3 = 8\,\text{m}$$

$$\Phi_3 = \frac{2.16}{\sqrt{8} - 0.2} + 0.73 = 1.552, \quad \Phi_3 \geq 1.0 \text{ and } \leq 2.0.$$

Bending Moment Due to Dead and Live Loads with Dynamic Effect Added $(M_{D+L+\Phi})$

$$M_{D+L+\Phi} = M_{D.L.} \times \gamma_g + \Phi \times M_{L.L.} \times \gamma_q$$
$$= 15 \times 1.2 + 1.552 \times 268.75 \times 1.45 = 622.8\,\text{kN m}$$

It should be noted that the load factors adopted in this study are that of the ultimate limit state. This is attributed to the fact that the finite element

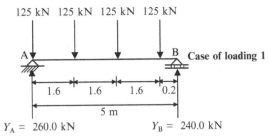

125 kN 125 kN 125 kN 125 kN

A   B   Case of loading 1

1.6   1.6   1.6   0.2

5 m

$Y_A = 260.0$ kN $Y_B = 240.0$ kN

Figure 4.6 Cases of loading for the maximum shear force acting on a stringer.

models presented in Chapters 6 and 7 can be used to analyze the bridges and provide more accurate predictions for the deflections and other serviceability limit state cases of loading.

Shearing Force Due to Dead and Live Loads with Dynamic Effect Added ($Q_{D+L+\Phi}$)

There is only a single case of loading for live loads to produce a maximum shear force at the supports of the stringer, which is shown in Figure 4.6:

$$Q_{L.L.} = 260\,\text{kN}$$
$$Q_{D+L+\Phi} = Q_{D.L.} \times \gamma_g + \Phi \times Q_{L.L.} \times \gamma_q$$
$$= 12 \times 1.2 + 1.552 \times 260 \times 1.45 = 599.5\,\text{kN}$$

Design Bending Moment (M_{Ed}) and Shear Force (Q_{Ed})

$$M_{Ed} = M_{D+L+\Phi} = 622.8\,\text{kN m}$$
$$Q_{Ed} = Q_{D+L+\Phi} = 599.5\,\text{kN}$$

Design of Stringer Cross Section

$$M_{c,Rd} = \frac{W_{pl} \times f_y}{\gamma_{M0}} \quad \text{for classes 1 and 2}$$

$$622.8 \times 10^6 - \frac{W_{pl} \times 275}{1.0}$$

$$W_{PL} = 2{,}264{,}727\,\text{mm}^3 = 2264.7\,\text{cm}^3$$

Choose UB 533 × 210 × 92 (equivalent to American W21 × 62), shown in Figure 4.7. W_{PL} around $x\text{-}x = 2360\,\text{cm}^3$. To classify the cross section chosen,

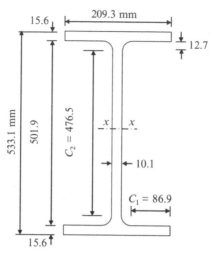

Figure 4.7 The cross-section of stringers (UB 533 × 210 × 92).

$$\varepsilon = \sqrt{\frac{235}{f_y}} = \sqrt{\frac{235}{275}} = 0.924$$

$C_1 = 86.9\,\text{mm}, t_{fl} = 15.6, C_1/t_{fl} = 86.9/15.6 = 5.6 \leq 9 \times 0.924$
$= 8.316\,(\text{Stringer flange is class 1})$

$C_2 = 476.5\,\text{mm}, t_w = 10.1, C_1/t_{fl} = 476.5/10.1 = 47.2 \leq 72 \times 0.924$
$= 66.5\,(\text{Stringer web is class 1})$

Check of Bending Resistance

$$M_{c,Rd} = \frac{W_{pl} \times f_y}{\gamma_{M0}} = \frac{2360 \times 10^3 \times 275}{1.0} = 649,000,000\,\text{N\,mm}$$
$$= 649.0\,\text{kN\,m} > M_{Ed} = 622.8\,\text{kN\,m}\,(\text{Then O.K.})$$

Check of Shear Resistance

$$V_{pl,Rd} = \frac{A_v\left(f_y/\sqrt{3}\right)}{\gamma_{M0}} = \frac{(501.9 \times 10.1) \times \left(275/\sqrt{3}\right)}{1.0} = 804,842\,\text{N}$$
$$= 804.8\,\text{kN} > Q_{Ed} = 599.5\,\text{kN}\,(\text{Then O.K.})$$

4.2.2 Design of the Cross Girders (Lateral Floor Girders)

The cross girders, the lateral floor beams, carry concentrated loads from the stringers as shown in Figure 4.1. Therefore, we can analyze an intermediate cross girder as follows:

Dead Loads

$$\text{Reaction from stringers due to dead loads} = 4.8 \times 5 = 24\,\text{kN}$$

$$\text{Own weight of cross girder} = 3.0\,\text{kN}\,\text{m}^{-1}$$

Assuming the cross girders are simply supported by the main plate girders, we can calculate the maximum shear force and bending moment due to dead loads on an intermediate cross girder (see Figure 4.8) as follows:

$$Q_{D.L.} = 3 \times 7.2/2 + 2 \times 24 = 58.8\,\text{kN}$$
$$M_{D.L.} = 3 \times 7.2^2/8 + 24 \times 0.95 + 24 \times 2.75 = 108.24\,\text{kN}\,\text{m}$$

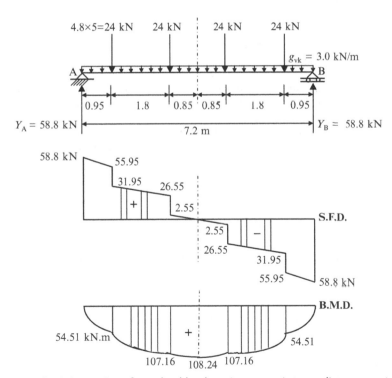

Figure 4.8 Straining actions from dead loads acting on an intermediate cross girder.

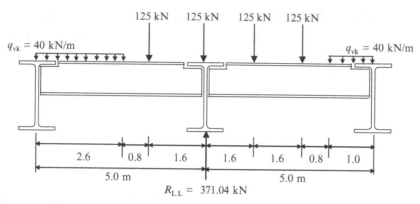

Figure 4.9 The case of loading producing maximum straining actions from live loads on an intermediate cross girder.

Live Loads

To determine the maximum reactions due to live loads transferred by the stringers to the cross girders, the case of loading shown in Figure 4.9 is studied. The maximum reaction $R_{L.L.}$ can be calculated as follows:

$$R_{L.L.} = 125 + 2 \times 125 \times (5 - 1.6)/5 + 125 \times (5 - 3.2)/5 + 40 \times 1$$
$$\times 0.5/5 + 40 \times 2.6 \times 1.3/5$$
$$= 371.04\,kN$$

The maximum straining actions due to live loads on an intermediate cross girder can be then calculated (see Figure 4.10) as follows:

$$Q_{L.L.} = 2 \times 371.04 = 742.08\,kN$$
$$M_{L.L.} = 371.04 \times 0.95 + 371.04 \times 2.75 = 1372.85\,kN\,m$$

Dynamic Factor Φ

$$L_\Phi = 2 \times 7.2 = 14.4\,m$$

$$\Phi_3 = \frac{2.16}{\sqrt{14.4 - 0.2}} + 0.73 = 1.331, \quad \Phi_3 \geq 1.0\,and \leq 2.0.$$

Bending Moment Due to Dead and Live Loads with Dynamic Effect Added ($M_{D+L+\Phi}$)

$$M_{D+L+\Phi} = M_{D.L.} \times \gamma_g + \Phi \times M_{L.L.} \times \gamma_q$$
$$= 108.24 \times 1.2 + 1.331 \times 1372.85 \times 1.45 = 2779.42\,kN\,m$$

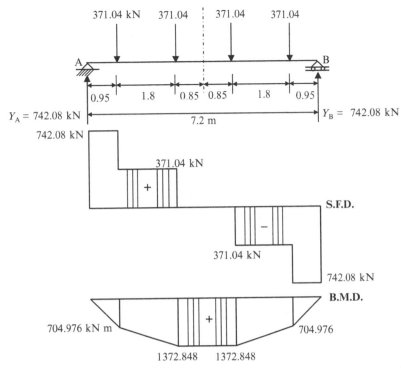

Figure 4.10 Straining actions from live loads acting on an intermediate cross girder.

Shearing Force Due to Dead and Live Loads with Dynamic Effect Added ($Q_{D+L+\Phi}$)

$$Q_{D+L+\Phi} = Q_{D.L.} \times \gamma_g + \Phi \times Q_{L.L.} \times \gamma_q$$
$$= 58.8 \times 1.2 + 1.331 \times 742.08 \times 1.45 = 1502.7\,\text{kN}$$

Design Bending Moment (M_{Ed}) and Shear Force (Q_{Ed})

$$M_{Ed} = M_{D+L+\Phi} = 2779.42\,\text{kN m}$$
$$Q_{Ed} = Q_{D+L+\Phi} = 1502.7\,\text{kN}$$

Design of the Cross Girder Cross Section

$$M_{c,Rd} = \frac{W_{pl} \times f_y}{\gamma_{M0}} \quad \text{for classes 1 and 2}$$

$$2779.42 \times 10^6 = \frac{W_{pl} \times 275}{1.0}$$

$$W_{PL} = 10,106,981.8\,\text{mm}^3 = 10,107\,\text{cm}^3$$

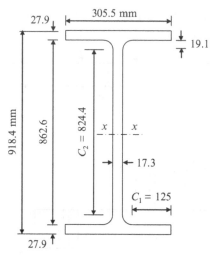

Figure 4.11 The cross-section of cross girders (UB 914 × 305 × 253).

Choose UB $914 \times 305 \times 253$ (equivalent to American $W36 \times 170$), shown in Figure 4.11. W_{PL} around $x\text{-}x = 10{,}940$ cm^3. To classify the cross section chosen,

$$\varepsilon = \sqrt{\frac{235}{f_y}} = \sqrt{\frac{235}{275}} = 0.924$$

$C_1 = 125$ mm, $t_{fl} = 27.9, C_1/t_{fl} = 125/27.9 = 4.48 \le 9 \times 0.924$
$\quad = 8.316$ (Cross girder flange is class 1)

$C_2 = 824.4$ mm, $t_w = 17.3, C_1/t_{fl} = 824.4/17.3 = 47.7 \le 72 \times 0.924$
$\quad = 66.5$ (Cross girder web is class 1)

Check of Bending Resistance

$$M_{c,Rd} = \frac{W_{pl} \times f_y}{\gamma_{M0}} = \frac{10940 \times 10^3 \times 275}{1.0} = 3{,}008{,}500{,}000\,\text{N mm}$$
$$= 3008.5\,\text{kN m} > M_{Ed} = 2779.42\,\text{kN m (Then O.K.)}$$

Check of Shear Resistance

$$V_{pl,Rd} = \frac{A_v\left(f_y/\sqrt{3}\right)}{\gamma_{M0}} = \frac{(862.6 \times 17.3) \times \left(275/\sqrt{3}\right)}{1.0} = 2{,}369{,}341\,\text{N}$$
$$= 2369.3\,\text{kN} > Q_{Ed} = 1502.7\,\text{kN (Then O.K.)}$$

4.2.3 Design of the Main Plate Girders

Let us now design the main plate girders supporting the cross girders as shown in Figure 4.1. We can estimate the dead and live loads acting on a main plate girder as follows:

Dead Loads

$$\text{Weight of steel structure} = 9 + 0.5 \times 30 = 24 \, \text{kN} \, \text{m}^{-1}$$

$$\text{Track load} = 6 \, \text{kN} \, \text{m}^{-1}$$

$$\text{Total dead load} = g_{vk} = 1.8 \times 24/2 + 6 = 27.6 \, \text{kN} \, \text{m}^{-1}$$

The main plate girders are simply supported; hence, we can calculate the maximum shear force and bending moment due to dead loads on a main plate girder (see Figure 4.12) as follows:

$$Q_{D.L.} = g_{vk} \times L/2 = 27.6 \times 30/2 = 414 \, \text{kN}$$
$$M_{D.L.} = g_{vk} \times L^2/8 = 27.6 \times 30^2/8 = 3105 \, \text{kN} \, \text{m}$$

Live Loads

Considering the axle loads on the bridge components according to Load Model 71 (see Figure 4.4), two cases of loading for the evaluation of maximum bending moment due to live loads on a main plate girder can be studied.

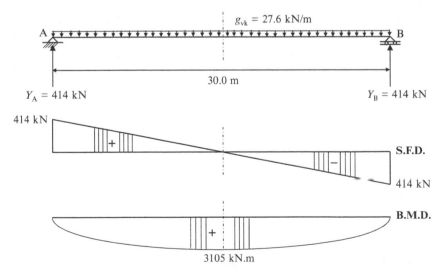

Figure 4.12 Straining actions from dead loads acting on one main plate girder.

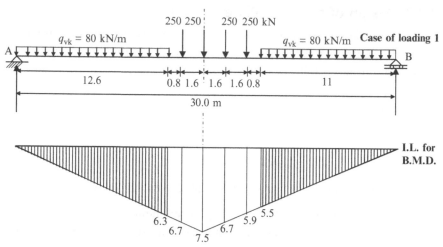

Figure 4.13 Determination of the maximum bending moment on one main plate girder due to live loads using the influence line method (case of loading 1).

The first case of loading is that the centerline of the main plate girder is located under one of the intermediate concentrated live loads, with maximum bending moment calculated at midspan (see Figure 4.13). On the other hand, the second case of loading is that the centerline of a main plate girder divides the spacing between the resultant of the concentrated live loads and the closest load, with maximum bending moment located at the closest load (point a in Figure 4.14). The maximum bending moment under the first case of loading is calculated using the influence line method (by multiplying the concentrated loads by the companion coordinates on the bending moment diagram and by multiplying the distributed loads by the companion areas on the bending moment diagram), while that under the second case of loading is calculated

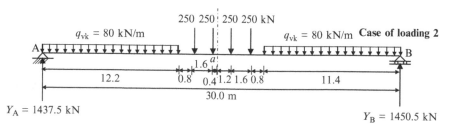

Figure 4.14 Determination of the maximum bending moment on one main plate girder due to live loads using the analytical method (case of loading 2).

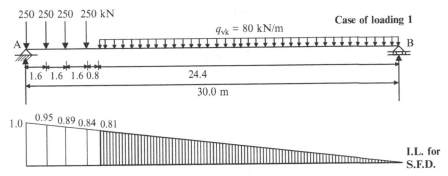

Figure 4.15 Determination of the maximum shear force on one main plate girder due to live loads using the influence line method (case of loading 1).

analytically using structural analysis. Hence, the bending moments due to live loads can be calculated as follows:

$$M_{L.L.}(\text{case of loading } 1) = 250 \times [2 \times 6.3 + 2 \times 7.1] + 2 \times 80 \times 0.5$$
$$\times 11.8 \times 5.9$$
$$= 12295.2\,\text{kN m}$$

$$M_{L.L.}(\text{case of loading } 2) = 1437.5 \times 14.6 - 80 \times 12.2 \times 8.5 - 250 \times 1.6$$
$$= 12291.5\,\text{kN m}$$

There is only a single case of loading for the live loads to produce a maximum shear force at the supports of a main plate girder, which is shown in Figure 4.15. Once again, we can use the influence line method to calculate the maximum shear force due to this case of loading or analytically by taking moment at support B and evaluate the reaction at A:

$$Q_{L.L.} = 1713.8\,\text{kN}$$

Dynamic Factor Φ

$$L_\Phi = 30\,\text{m}$$

$$\Phi_3 = \frac{2.16}{\sqrt{30 - 0.2}} + 0.73 = 1.139, \quad \Phi_3 \geq 1.0 \text{ and } \leq 2.0.$$

Bending Moment Due to Dead and Live Loads with Dynamic Effect Added ($M_{D+L+\Phi}$)

$$M_{D+L+\Phi} = M_{D.L.} \times \gamma_g + \Phi \times M_{L.L.} \times \gamma_q$$
$$= 3105 \times 1.2 + 1.139 \times 12{,}295.2 \times 1.45 = 24{,}032\,\text{kN m}$$

Shearing Force Due to Dead and Live Loads with Dynamic Effect Added ($Q_{D+L+\Phi}$)

$$Q_{D+L+\Phi} = Q_{D.L.} \times \gamma_g + \Phi \times Q_{L.L.} \times \gamma_q$$
$$= 414 \times 1.2 + 1.139 \times 1713.8 \times 1.45 = 3327.2\,kN$$

Design Bending Moment (M_{Ed}) and Shear Force (Q_{Ed})

$$M_{Ed} = M_{D+L+\Phi} = 24{,}032\,kN\,m$$
$$Q_{Ed} = Q_{D+L+\Phi} = 3327.2\,kN$$

Design of the Main Plate Girder Cross Section

Let us assume the main plate girder cross section shown in Figure 4.16. The cross section consists of two flange plates for the upper and lower flanges and a web plate. The web plate height is taken as equal to $L/10 = 30{,}000/10 = 3000$ mm, with a plate thickness of 16 mm. The width of the bottom

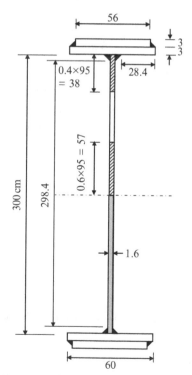

Figure 4.16 Reduced cross-section of plate girder.

plate of the upper and lower flanges of the cross section is taken as 0.2 the web height, which is equal to 600 mm, while the top plate width is taken as 560 mm, to allow for welding with the bottom flange plate. The flange plates have the same plate thickness of 30 mm. The choice of two flange plates for the upper and lower flanges is intended to curtail the top flange plate approximately at quarter-span as detailed in the coming sections. It should be noted that the web height value ($L/10$) is an acceptable recommended [1.9] value for railway steel bridges constructed in Great Britain and Europe. This value is an initial value for preliminary cross-sectional estimation. The cross section has to be checked, classified, designed, and assessed against deflection limits set by serviceability limit states. To classify the cross section chosen,

$$\varepsilon = \sqrt{\frac{235}{f_y}} = \sqrt{\frac{235}{275}} = 0.924$$

$C_1 = 284\,\text{mm}, t_{fl} = 60, C_1/t_{fl} = 284/60 = 4.73 \le 9 \times 0.924$
$= 8.316$ (Main plate girder flange is class 1).

$C_2 = 2984\,\text{mm}, t_w = 16, C_1/t_{fl} = 2984/16 = 186.5 > 124 \times 0.924$
$= 114.58$ (Main plate girder web is class 4).

To calculate the bending moment resistance, the effective area should be used. Considering web plate buckling, the effective area of the web part in compression (see Figure 4.16) can be calculated as follows:

$$k_\sigma = 23.9$$

$$\bar{\lambda}_p = \frac{300/1.6}{28.4 \times 0.924 \times \sqrt{23.9}} = 1.462 > 0.673$$

$$\rho = \frac{1.462 - 0.055(3-1)}{1.462^2} = 0.633$$

$$b_{eff} = 0.633 \times 300/2 = 95\,\text{cm},$$

Then, $b_{eff1} = 0.6 \times 95 = 57$ cm and $b_{eff2} = 0.4 \times 95 = 38$ cm as shown in Figure 4.17.

To calculate the elastic section modulus, the elastic centroid of the section has to be located by taking the first area moment, as an example, around axis y_0-y_0 shown in Figure 4.17, as follows:

$$A = 60 \times 3 \times 2 + 56 \times 3 \times 2 + 207 \times 1.6 + 38 \times 1.6 = 1088\,\text{cm}^2$$

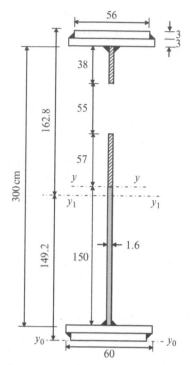

Figure 4.17 Calculation of properties of area for main plate girder.

$$y_c = \dfrac{\left[\begin{array}{c} 56 \times 3 \times 1.5 + 60 \times 3 \times 4.5 + 60 \times 3 \times 307.5 + 56 \times 3 \times 310.5 + 207 \\ \times 1.6 \times 109.5 + 38 \times 1.6 \times 287 \end{array}\right]}{1088}$$

$y_c = 149.2\,\text{cm}$

$$\begin{aligned}
\text{Inertia about } y_1\text{-}y_1 &= [56 \times 3^3/12 + 56 \times 3 \times 147.7^2] \\
&\quad + [60 \times 3^3/12 + 60 \times 3 \times 144.7^2] \\
&\quad + [1.6 \times 207^3/12 + 1.6 \times 207 \times 39.7^2] \\
&\quad + [1.6 \times 38^3/12 + 1.6 \times 38 \times 137.8^2] \\
&\quad + [60 \times 3^3/12 + 60 \times 3 \times 158.3^2] \\
&\quad + [56 \times 3^3/12 + 56 \times 3 \times 162.8^2] = 19,264,063\,\text{cm}^4
\end{aligned}$$

$$W_{\text{eff,min}} = 19,264,063/162.8 = 118,330\,\text{cm}^3$$

Check of Bending Resistance

$$M_{c,Rd} = \frac{W_{\text{eff,min}} \times f_y}{\gamma_{M0}} = \frac{118,330 \times 10^3 \times 275}{1.0} = 32,540,750,000\,\text{N mm}$$
$$= 32,541\,\text{kN m} > M_{Ed} = 24,032\,\text{kN m (Then O.K.)}$$

Check of Shear Resistance

$$V_{b,Rd} = V_{bw,Rd} + V_{bf,Rd} \leq \frac{\eta f_{yw} h_w t_w}{\sqrt{3}\gamma_{M1}}$$

By neglecting the flange contribution,

$$V_{b,Rd} = V_{bw,Rd} \leq \frac{1.2 \times 275 \times 3000 \times 16}{\sqrt{3} \times 1.1} = 8{,}313{,}843.9\,N$$

$$V_{bw,Rd} = \frac{\chi_w f_{yw} h_w t_w}{\sqrt{3}\gamma_{M1}}$$

$$\bar{\lambda}_w = 0.76\sqrt{\frac{f_{yw}}{\tau_{cr}}}, \quad \tau_{cr} = k_\tau \sigma_E$$

$$\sigma_E = 190{,}000\,(16/3000)^2 = 5.404\,N\,mm^{-2}$$

$$k_\tau = 4 + 5.34\,(3000/1666.7)^2 = 21.3$$

$$\bar{\lambda}_w = 0.76\sqrt{\frac{275}{21.3 \times 5.404}} = 1.175 > 1.08$$

$$\text{Then, } \chi_w = \frac{1.37}{0.7 + \bar{\lambda}_w} = \frac{1.37}{0.7 + 1.175} = 0.731$$

$$V_{bw,Rd} = \frac{0.731 \times 275 \times 3000 \times 16}{\sqrt{3} \times 1.1} = 5{,}064{,}516.6\,N = 5064.5\,kN$$
$$< 8313.8\,kN$$

$$\eta_3 = \frac{V_{Ed}}{V_{b,Rd}} = \frac{3327.2}{5064.5} = 0.657 < 1.0\,(\text{Then O.K.})$$

It should be noted that for this type of bridges, it is recommended that further checks regarding the assessment of fatigue loading have to be performed. However, this can be done using advanced finite element modeling of the bridge.

4.2.4 Curtailment (Transition) of the Flange Plates of the Main Plate Girder

The critical cross section of the main plate girder at midspan, which is subjected to the maximum bending moment, was designed previously with two flange plates. Since the main plate girder is simply supported, the bending moment is decreased towards the supports. Therefore, we can stop the top flange plate at a certain distance to get the most benefit from the material. This process is commonly called as curtailment (transition) of flange plates. It should be noted that, theoretically, curtailment (transition) of flange plates can be conducted by reducing the flange plate width, thickness, or both. However, in practice, fabricators prefer to keep the flange widths constant and vary the thickness because this option costs much less than reducing the flange width that might require a very heavy grinding work. To avoid lateral torsional buckling of the compression top flange at the reduction zone, it is recommended practically to reduce the width or thickness by 40% of the original with a smooth transition zone sloping at 1 (vertical) to 10 (horizontal). It is also recommended that bridges with lengths of 20-30 m are curtailed (transitioned) in one step. While for bridges with spans greater than 30 m, two steps of curtailment (transition) are recommended. For the investigated design example, we can conduct one-step curtailment (transition) by reducing the top flange plate of the upper and lower flanges, as shown in Figure 4.18. To classify the reduced cross section,

$$\varepsilon = \sqrt{\frac{235}{f_y}} = \sqrt{\frac{235}{275}} = 0.924$$

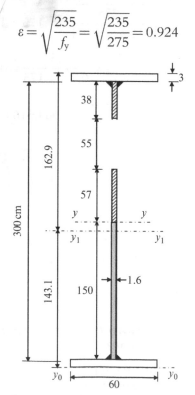

Figure 4.18 Calculation of properties of area for curtailed main plate girder.

$C_1 = 284 \, \text{mm}, \, t_{fl} = 30, C_1/t_{fl} = 284/30 = 9.47 \leq 14 \times 0.924$
$= 12.94 \, (\text{Class } 3).$

$C_2 = 2984 \, \text{mm}, \, t_w = 16, C_1/t_{fl} = 2984/16 = 186.5 > 124 \times 0.924$
$= 114.58 \, (\text{Class } 4).$

To calculate the bending moment resistance, the effective area should be used. Considering web plate buckling, the effective area of the part of web plate in compression (see Figure 4.18) can be calculated as follows:

$$k_\sigma = 23.9$$

$$\bar{\lambda}_p = \frac{300/1.6}{28.4 \times 0.924 \times \sqrt{23.9}} = 1.462 > 0.673$$

$$\rho = \frac{1.462 - 0.055(3 - 1)}{1.462^2} = 0.633$$

$$b_{\text{eff}} = 0.633 \times 300/2 = 95 \, \text{cm},$$

Then, $b_{\text{eff1}} = 0.6 \times 95 = 57$ cm and $b_{\text{eff2}} = 0.4 \times 95 = 38$ cm as shown in Figure 4.18.

To calculate the elastic section modulus, the elastic centroid of the section has to be located by taking the first area moment, as an example, around axis y_0-y_0 shown in Figure 4.18, as follows:

$$A = 60 \times 3 \times 2 + 245 \times 1.6 = 752 \, \text{cm}^2$$

$$y_c = \frac{[60 \times 3 \times 1.5 + 207 \times 1.6 \times 106.5 + 38 \times 1.6 \times 284 + 60 \times 3 \times 304.5]}{752}$$

$$y_c = 143.1 \, \text{cm}$$

$$\begin{aligned}
\text{Inertia about } y_1\text{-}y_1 &= [60 \times 3^3/12 + 60 \times 3 \times 141.6^2] \\
&+ [60 \times 3^3/12 + 60 \times 3 \times 161.4^2] \\
&+ [1.6 \times 207^3/12 + 1.6 \times 207 \times 36.6^2] \\
&+ [1.6 \times 38^3/12 + 1.6 \times 38 \times 140.9^2] \\
&= 11,139,025.4 \, \text{cm}^4
\end{aligned}$$

$$W_{\text{eff,min}} = 11,139,025.4/162.9 = 68,379.5 \, \text{cm}^3$$

Bending Moment Resistance

$$M_{c,Rd} = \frac{W_{eff,min} \times f_y}{\gamma_{M0}} = \frac{68,379.5 \times 10^3 \times 275}{1.0} = 18,804,362,500\,\text{N mm}$$

$$= 18,804.4\,\text{kN m}$$

Length of Flange Plates

Assuming the overall bending moment diagram of the main plate girder is a second-degree parabola (see Figure 4.19), we can determine the length of the curtailed top flange plate of the upper and lower flanges as follows:

$$\left(\frac{x}{L/2}\right)^2 = \frac{24,032 - 18,804.4}{24,032} = \frac{5221.6}{24,032}$$

$$\frac{x}{15} = 0.466, \quad \text{then } x = 6.99\,\text{m taken as 7 m.}$$

Hence, the length of the smaller top plate is 14 m.

4.2.5 Design of the Fillet Weld Between Flange Plates and Web

To determine the size of fillet weld connecting the bottom flange plates of the upper and lower flanges with the web plate for the investigated bridge, we can calculate the maximum shear flow at the support for the reduced cross section, shown in Figure 4.20, as follows:

Inertia about y-$y = 1.6 \times 300^3/12 + 2 \times [60 \times 3^3/12 + 60 \times 3 \times 151.5^2]$
$= 11,863,080$ cm^4.

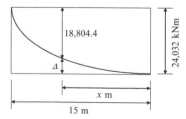

Figure 4.19 Calculation of curtailed flange plate lengths.

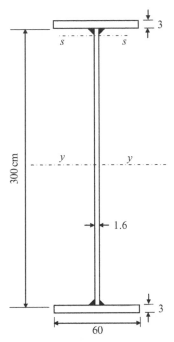

Figure 4.20 Calculation of flange fillet weld size at supports.

Shear flow at section *s-s*:

$$q = \frac{Q_{Ed} \times S_{ss}}{I_{yy}} = 2 \times a \times f_{vw,d}$$

$$f_{vw,d} = \frac{f_u/\sqrt{3}}{\beta_w \gamma_{M2}} = \frac{430/\sqrt{30}}{0.85 \times 1.25} = 233.7 \, N/mm^2$$

$$q = \frac{3327.2 \times 10^3 \times (60 \times 3 \times 151.5) \times 10^3}{11,863,080 \times 10^4} = 2 \times a \times 233.7$$

Then, $a = 1.64$ mm, taken as 8 mm, which is the minimum size.

4.2.6 Check of Lateral Torsional Buckling of the Plate Girder Compression Flange

To check the safety of the upper compression flange against lateral tor-sional buckling, we have to calculate the elastic critical moment for lateral

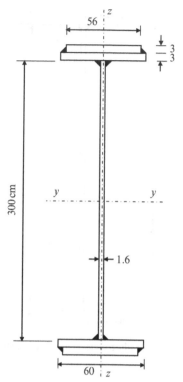

Figure 4.21 Check of lateral torsional buckling of plate girder.

torsional buckling (M_{cr}). Considering the cross section at midspan shown in Figure 4.21, we can calculate M_{cr} as follows:

$$M_{cr} = C_b \frac{\pi}{k l_b} \sqrt{\left(\frac{\pi E_y}{k l_b}\right)^2 C_w I_z + E_z I_z G J}$$

Given: $C_b = 1.13$, $E = 210$ GPa, $G = 81$ GPa, $l_b = 5000$ mm, and $k = 1$

Inertia about z-z $(I_z) = 2 \times 3 \times 60^3/12 + 2 \times 3 \times 56^2/12 + 300 \times 1.6^3/12$
$$= 195{,}910 \, \text{cm}^4$$

$$C_w = \frac{h^2 \times I_z}{4} = \frac{3120^2 \times 195{,}910 \times 10^4}{4} = 4.76766576 \times 10^{15} \, \text{mm}^6$$

$$j = \frac{1}{3}\left(2 \times 600 \times 30^3 + 2 \times 560 \times 30^3 + 3120 \times 16^3\right) = 25{,}139{,}840 \, \text{mm}^2$$

$$M_{cr} = 1.13 \frac{3.14}{5000} \sqrt{\left(\frac{3.14 \times 210}{5000}\right)^2 4.76766576 \times 10^{15} \times 195{,}910 \times 10^4}$$

$$\sqrt{+ 210 \times 195{,}910 \times 10^4 \times 81 \times 25{,}139{,}840}$$

$$M_{cr} = 0.00070964 \sqrt{1.624502122 \times 10^{23} + 8.377673439 \times 10^{20}}$$

$$M_{cr} = 286{,}757{,}770 \, \text{N mm} = 286{,}758 \, \text{kN m}$$

We can now check the safety against lateral torsional buckling following the rules specified in EC3 [1.27, 2.11] as follows:

$$\frac{M_{Ed}}{M_{b,Rd}} \leq 1.0$$

Given: $M_{Ed} = 24{,}026$ kN m and $W_y = 118{,}330$ cm^3

$$M_{b,Rd} = \chi_{LT} W_y \frac{f_y}{\gamma_{M1}}$$

$$\bar{\lambda}_{LT} = \sqrt{\frac{W_y f_y}{M_{cr}}} = \sqrt{\frac{32{,}541}{286{,}758}} = 0.337$$

$$\Phi_{LT} = 0.5\left[1 + \alpha_{LT}\left(\bar{\lambda}_{LT} - 0.2\right) + \bar{\lambda}_{LT}^2\right]$$
$$= 0.5\left[1 + 0.76(0.337 - 0.2) + 0.337^2\right] = 0.609$$

$$\chi_{LT} = \frac{1}{\Phi_{LT} + \sqrt{\Phi_{LT}^2 - \bar{\lambda}_{LT}^2}} \quad \text{but } \chi_{LT} \leq 1.0$$

$$\chi_{LT} = \frac{1}{0.609 + \sqrt{0.609^2 - 0.337^2}} \quad \text{but } \chi_{LT} \leq 1.0$$

$$\chi_{LT} = 0.896$$

$$M_{b,Rd} = \frac{0.896 \times 32{,}541}{1.0} = 29{,}156.7 \, \text{kN m} > 24{,}032 \, \text{kN m}$$

4.2.7 Design of Web Stiffeners

There are two types of stiffeners used to strengthen the thin web plate of the main plate girder against buckling due to shear stresses, bending stresses, or both. The stiffeners at the supports are commonly known as load bearing stiffeners, while intermediate stiffeners are commonly known as stability stiffeners (intermediate transverse stiffeners). The design of the stiffeners can be performed as follows:

4.2.7.1 Load Bearing Stiffeners

To design the load bearing stiffener at supports (see Figure 4.22), we can also follow the design rules specified in EC3 [1.27, 2.11] for concentrically loaded compression members. The axial force in the stiffener is the maximum reaction at supports ($N_{Ed} = R_{D+L+\Phi}$), which is equal to 3327.2 kN. The design procedures can be performed as follows:

$$\frac{N_{Ed}}{N_{b,Rd}} \leq 1.0$$

where, $N_{b,Rd} = \dfrac{\chi A f_y}{\gamma_{M1}}$

$$A = 2 \times 25 \times 2.4 + 46.4 \times 1.6 = 194.24 \, \text{cm}^2$$

$$\chi = \frac{1}{\Phi + \sqrt{\Phi^2 - \bar{\lambda}^2}} \quad \text{but } \chi \leq 1.0$$

$$\Phi = 0.5\left[1 + \alpha(\bar{\lambda} - 0.2) + \bar{\lambda}^2\right]$$

$$\bar{\lambda} = \sqrt{\frac{A f_y}{N_{cr}}}$$

$$N_{cr} = \frac{\pi^2 \times EI}{L^2} = \frac{3.14^2 \times 210,000 \times 27,492.6 \times 10^4}{3000^2} = 63,248,742 \, \text{N}$$

$$\bar{\lambda} = \sqrt{\frac{194.24 \times 100 \times 275}{63,248,742}} = 0.29$$

$$\Phi = 0.5\left[1 + 0.49(0.29 - 0.2) + 0.29^2\right] = 0.564$$

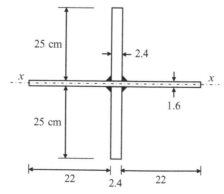

Figure 4.22 Load bearing web stiffeners at supports.

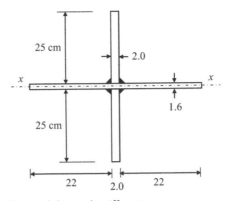

Figure 4.23 Intermediate stability web stiffeners.

$$\chi = \frac{1}{0.564 + \sqrt{0.564^2 - 0.29^2}} = 0.954 \text{ but } \chi \leq 1.0$$

$$\text{Then, } N_{b,Rd} = \frac{0.954 \times 194.26 \times 100 \times 275}{1.1} = 4,633,101 \text{ N}$$

$$N_{b,Rd} = 4633.1 \text{ kN} > N_{Ed} = 3327.2 \text{ kN (Then O.K.)}$$

4.2.7.2 Intermediate Stiffeners

Intermediate stiffeners (see Figure 4.23) can be designed by choosing its dimensions such that

$$\frac{a_1}{h_w} = \frac{1667}{3000} = 0.556 < \sqrt{2} = 1.414 \text{ (Then O.K.)}$$

$$\text{and } I_{st} \geq \frac{1.5 h_w^2 t_w^3}{a_1^2} = \frac{1.5 \times 300^3 \times 1.6^3}{166.7^2} = 5969.6 \text{ cm}^4$$

$$I_{st} = 46 \times 1.6^3/12 + 2 \times \left[2 \times 25^3/12 + 50 \times 13.3^2\right] = 22913 \text{ cm}^4$$
$$> 5969.6 \text{ cm}^4 \text{(Then O.K.)}$$

4.2.8 Design of Stringer Bracing (Lateral Shock or Nosing Force Bracings)

The stringer bracings are subjected to lateral moving reversible force of 100 kN. The bracing members carry either tensile or compressive forces according to the changing direction of the lateral shock force (transverse horizontal force) (see Figure 4.24). The cross section of the bracing member can be determined from designing the critical diagonal member for the compressive force as follows:

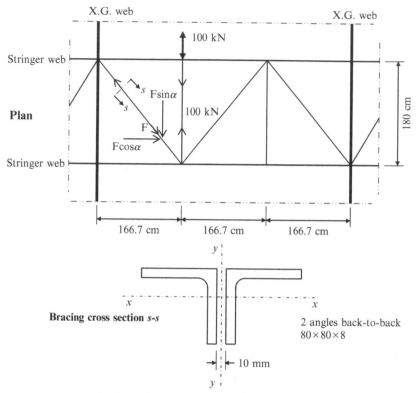

Figure 4.24 Lateral shock (nosing force) bracing for stringers.

Assume the cross section of the stringer bracing as two angles back-to-back $80 \times 80 \times 8$ (see Figure 4.24); then,

$$\alpha = \tan^{-1} \frac{1.8}{1.667} = 47.2^{\circ}$$

$$l_b = 2453 \, \text{mm}$$

$$\varepsilon = \sqrt{\frac{235}{275}} = 0.924$$

$$\bar{\lambda} = \frac{L_{cr}}{i} \frac{1}{\lambda_1}$$

$$\lambda_1 = 93.9 \times 0.924 = 86.7636$$

$$\bar{\lambda} = \frac{2435}{24.3} \frac{1}{86.7636} = 1.163$$

The axial compressive force in the diagonal bracing member ($N_{Ed} = 90.9$ kN):

$$\frac{N_{Ed}}{N_{b,Rd}} \leq 1.0$$

where $N_{b,Rd} = \dfrac{\chi A f_y}{\gamma_{M1}}$

$$A = 2 \times 12.3 = 24.6 \, \text{cm}^2$$

$$\chi = \frac{1}{\Phi + \sqrt{\Phi^2 - \bar{\lambda}^2}} \quad \text{but } \chi \leq 1.0$$

$$\Phi = 0.5\left[1 + \alpha\left(\bar{\lambda} - 0.2\right) + \bar{\lambda}^2\right]$$

$$\Phi = 0.5\left[1 + 0.34(1.163 - 0.2) + 1.163^2\right] = 1.34$$

$$\chi = \frac{1}{1.34 + \sqrt{1.34^2 - 1.163^2}} = 0.499 \quad \text{but } \chi \leq 1.0$$

Then, $N_{b,Rd} = \dfrac{0.499 \times 24.6 \times 100 \times 275}{1.1} = 306{,}885 \, \text{N}$

$$N_{b,Rd} = 306.9 \, \text{kN} > N_{Ed} = 90.9 \, \text{kN} \, (\text{Then O.K.})$$

4.2.9 Design of Wind Bracings

Wind forces acting on the double-track railway bridge (see Figure 4.25) as well as any other lateral forces directly applied to the bridge are transmitted to the bearings by systems of upper and lower wind bracings as well as cross bracings. The upper wind bracing carries wind forces on the moving train,

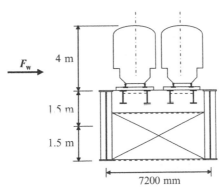

Figure 4.25 Design heights for upper and lower wind bracings.

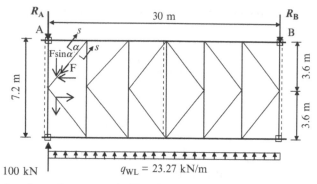

Figure 4.26 Loads on the upper wind bracing.

wind forces on upper half of the main plate girder, and lateral shock (nosing force) applied to the tracks (see Figure 4.26). On the other hand, wind forces acting on the lower half of the main plate girder are transmitted by the lower wind bracing (see Figure 4.27). Wind bracings are quite important to the lateral stability of the bridges, and therefore, it is recommended to use identical cross sections for the upper and lower wind bracings. Wind forces applied to this bridge can be sufficiently estimated using the design rules specified in EC1 [3.2] as follows:

$$F_w = \frac{1}{2}\rho v_b^2 C A_{\text{ref},x}$$

$$v_b = c_{\text{dir}} \times c_{\text{season}} \times v_{b,0} = 1.0 \times 1.0 \times 26 = 26\,\text{m/s}$$

$$A_{\text{ref},x} = 7 \times 31 = 217\,\text{m}^2$$

$$F_w = \frac{1}{2} \times 1.25 \times 26^2 \times 5.7 \times 217 = 522{,}590.3\,\text{N} = 522.6\,\text{kN}$$

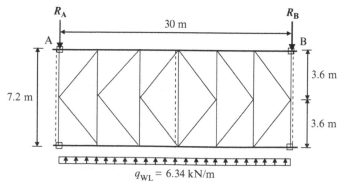

Figure 4.27 Loads on the lower wind bracing.

Considering the structural analysis for the upper wind bracing system shown in Figure 4.26, the critical design wind force in the diagonal bracing members can be calculated as follows:

$$\text{Distributed wind loads } (q_{wL}) = 522.6 \times (5.5/7)/30 = 13.69\,\text{kN/m}$$

$$\text{Factored distributed wind loads} = q_{wL} \times \gamma_q = 13.69 \times 1.7 = 23.27\,\text{kN/m}$$

$$R_A = 100 + 23.27 \times 15 = 449.05\,\text{kN}$$

$$\alpha = \tan^{-1}(3.6/5) = 35.75°$$

$$F_D = 349.05/(2 \times \sin 35.75) = 298.7\,\text{kN}$$

The cross section of the bracing member (see Figure 4.28) can be determined as follows:

$$l_{bx} = 6160\,\text{mm}, \quad l_{by} = 1.2 \times 6160 = 739.2\,\text{mm}$$

Choose two angles back-to-back $150 \times 150 \times 15$, with 10 mm gusset plate between them:

$$A = 2 \times 43.2 = 86.4\,\text{cm}^2, \quad i_x = 4.59\,\text{cm}, \quad e = 4.26\,\text{cm},$$

$$i_y = \sqrt{4.59^2 + (4.26 + 1/2)^2} = 6.61\,\text{cm}$$

$$\varepsilon = \sqrt{\frac{235}{275}} = 0.924$$

$$\bar{\lambda} = \frac{L_{cr}}{i}\frac{1}{\lambda_1}$$

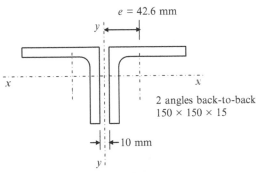

Figure 4.28 Upper wind bracing cross section s-s.

$$\lambda_1 = 93.9 \times 0.924 = 86.7636$$

$$\bar{\lambda} = \frac{6160}{45.9} \frac{1}{86.7636} = 1.547$$

The axial compressive force in the diagonal bracing member ($N_{Ed} = 298.7$ kN):

$$\frac{N_{Ed}}{N_{b,Rd}} \leq 1.0$$

where $N_{b,Rd} = \dfrac{\chi A f_y}{\gamma_{M1}}$

$$A = 2 \times 43.2 = 86.4 \, \text{cm}^2$$

$$\chi = \frac{1}{\Phi + \sqrt{\Phi^2 - \bar{\lambda}^2}} \quad \text{but } \chi \leq 1.0$$

$$\Phi = 0.5 \left[1 + \alpha(\bar{\lambda} - 0.2) + \bar{\lambda}^2 \right]$$

$$\Phi = 0.5 \left[1 + 0.34(1.547 - 0.2) + 1.547^2 \right] = 1.926$$

$$\chi = \frac{1}{1.926 + \sqrt{1.926^2 - 1.547^2}} = 0.325 \quad \text{but } \chi \leq 1.0$$

Then, $N_{b,Rd} = \dfrac{0.325 \times 86.4 \times 100 \times 275}{1.1} = 702,000 \, \text{N}$

$$N_{b,Rd} = 702 \, \text{kN} > N_{Ed} = 298.7 \, \text{kN (Then O.K.)}$$

4.2.10 Design of Stringer-Cross Girder Connection

The stringer is designed as a simply supported beam on cross girders; therefore, the connection is mainly transferring shear forces (maximum reaction from stringers of 599.5 kN) (see Figure 4.29). Using M27 high-strength pretensioned bolts of grade 8.8, having f_{ub} of 800 MPa, shear area A of 4.59 cm^2, and gross area A_g of 5.73 cm^2, we can determine the required number of bolts, following the rules specified in EC3 (BS EN 1993-1-8) [2.13], as follows:

$$F_{v,Rd} = \frac{\alpha_V f_{ub} A}{\gamma_{M2}}$$

$$F_{v,Rd} = \frac{0.6 \times 800 \times 459}{1.25} = 176,256 \, \text{N}$$

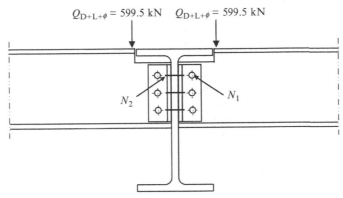

$Q_{D+L+\phi} = 599.5$ kN $Q_{D+L+\phi} = 599.5$ kN

Figure 4.29 The connection between stringer and cross girder.

Then, $F_{v,Rd}$ equals 176 kN (for bolts in single shear) and 353 kN (for bolts in double shear):

$$F_{s,Rd} = \frac{k_s n \mu}{\gamma_{M3}} F_{p,C}$$

$$F_{p,C} = 0.7 f_{ub} A_s = 0.7 \times 800 \times 573 = 320,880 \text{ N}$$

$$F_{s,Rd,ser} = \frac{1.0 \times 1.0 \times 0.4}{1.1} 320,880 = 116,683.6 \text{ N}.$$

Then, $F_{s,Rd} = 117$ kN (for bolts in single shear at serviceability limit states) and $F_{s,Rd} = 234$ kN (for bolts in double shear at serviceability limit states). At ultimate limit states, $F_{s,Rd,ult}$ can be calculated as follows:

$$F_{s,Rd,ult} = \frac{1.0 \times 1.0 \times 0.4}{1.25} 320,880 = 102,682 \text{ N}.$$

Then, $F_{s,Rd} = 103$ kN (for bolts in single shear at ultimate limit states) and $F_{s,Rd} = 206$ kN (for bolts in double shear at ultimate limit states):

$$N_1 = \frac{599.5}{206} = 2.9 \text{ taken as 3 bolts,}$$

$$N_2 = \frac{599.5}{103} = 5.8 \text{ taken as 6 bolts}$$

4.2.11 Design of Cross Girder-Main Plate Girder connection

The cross girder is designed as a simply supported beam on main plate girders; therefore, once again, the connection is mainly transferring shear forces (maximum reaction from cross girders of 1502.7 kN) (see Figure 4.30). We can determine the required number of bolts as follows:

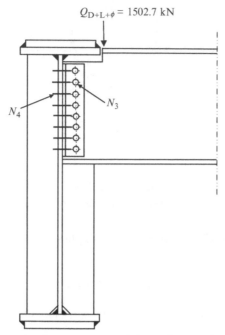

$Q_{D+L+\phi} = 1502.7$ kN

Figure 4.30 The connection between cross girder and main plate girder.

$$N_3 = \frac{1502.7}{206} = 7.3 \text{ taken as 8 bolts,}$$

$$N_2 = \frac{1502.7}{103} = 14.9 \text{ taken as 16 bolts}$$

4.2.12 Design of Field Splices

Figure 4.31 shows the locations of filed splices for the investigated bridge. Designing the splice requires determination of size of connecting plates as well as the number of bolts of the filed splice shown in Figure 4.32. The area of the flange plate equals to $60 \times 3 = 180$ cm^2; this can be compensated by three flange splice plates having a cross-sectional area of 60×1.6 and $2 \times 27 \times 1.6$ cm^2 with a total area of 182.4 cm^2, which is greater than the original area, while the area of web plate $= 300 \times 1.6 = 480$ cm^2 can be compensated by two web splice plates having cross-sectional area of $2 \times 290 \times 1.0$ cm^2 with a total area of 580 cm^2, which is governed by the minimum thickness (10 mm) of plates used in railway steel bridges.

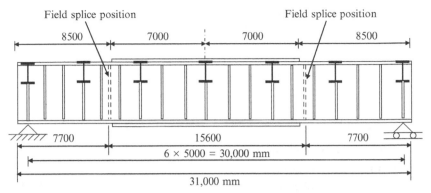

Figure 4.31 Positions of field splices in the main plate girder.

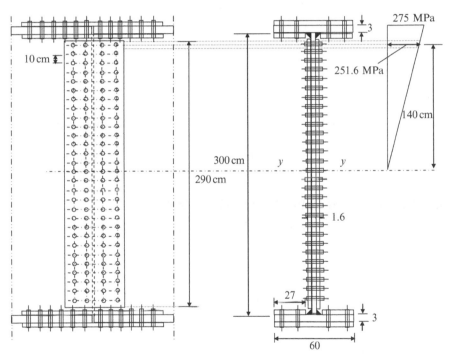

Figure 4.32 The field splice of the main plate girder.

The top row of bolts in the web (see Figure 4.32) is subjected to horizontal shear from the bending moment distribution, assuming the yield stress reached at the extreme and lower fibers of the flanges, and vertical shear from the applied loads. Using a spacing of 10 cm between two adjacent bolts, an edge spacing of 5 cm, and a hole of 3 cm (2.7 cm bolt diameter plus

0.3 cm clearance), we can determine the horizontal shear force (H) per bolt and the vertical shear per bolt (V) as follows:

$$H = \text{Area from centrelines between bolts}$$
$$\times \text{average stress at the bolt location } (f_{av})$$

$$f_{av} = 140 \times 275/153 = 251.6\,\text{MPa}$$

$$H = (100 - 30) \times 16 \times 251.6/2 = 140896\,\text{N} = 140.9\,\text{kN}$$

V = maximum shear resisted by web/total number of bolts.

Maximum shear resisted by web was previously calculated in the check of the safety of the plate girder against shear stresses and was 8313.8 kN. The total number of bolts in one side of the splice is 58 bolts:

$$V = 8313.8/58 = 143.3\,\text{kN}$$

The resultant of the forces per bolt (R) is equal to $\sqrt{140.9^2 + 143.3^2} = 201\,\text{kN}$, which is less than 206 kN (the resistance of the bolt in double shear). Then O.K.

Flange Splices

Maximum force in the upper flange $= 180 \times 275 \times 100/1000 = 4950\,\text{kN}$

$N(\text{flange}) = 4950/206 = 24\,\text{bolts}(6\text{ rows of four bolts in double shear})$

4.2.13 Design of Roller Steel Fabricated Bearings

Let us now design the roller steel fabricated bearings shown in Figure 4.1 and detailed in Figure 4.33. The maximum vertical reaction at the supports of the main plate girder was previously calculated under dead and live loads with dynamic effect ($R_{D+L+\Phi}$), which was 3327.2 kN. The material of construction for the bearings is cast iron steel (ISO 3755) 340-550 having a yield stress of 340 MPa and an ultimate stress of 550 MPa.

Design of the Sole Plate

The reaction ($R_{D+L+\Phi}$) can be assumed as two equal concentrated loads at two points, which are the centers of gravity of half of the load bearing stiffener section shown in Figure 4.33. To determine the centers of gravity (distance e), we can take the first area moment around the axis z-z, shown in Figure 4.33, as follows:

$$e = \frac{2 \times 25 \times 1.2 \times 0.6 + 23.2 \times 1.6 \times 11.6}{2 \times 25 \times 1.2 + 23.2 \times 1.6} = \frac{466.592}{97.12} = 4.8\,\text{cm}$$

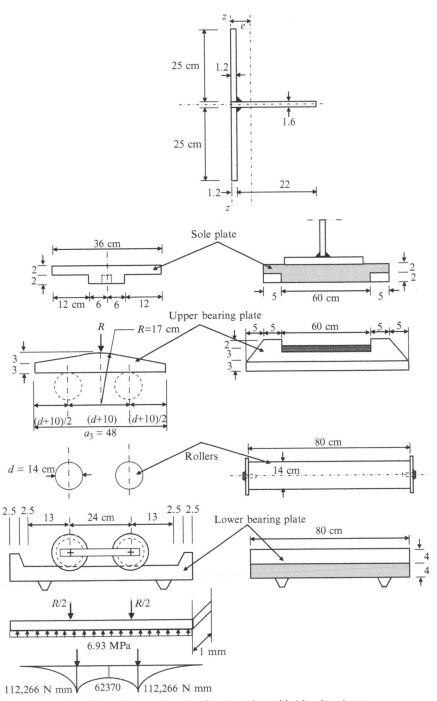

Figure 4.33 Detailing of the twin roller fabricated steel bridge bearings.

Assuming that the thickness of the sole plate is t_1, with detailed dimensions shown in Figure 4.33 based on the flange plate girder dimensions, we can determine the maximum moment applied to the sole plate (M) as follows:

$$M = R_{D+L+\Phi} \times e/2 = 3{,}327{,}200 \times 48/2 = 78{,}952{,}800 \, \text{N mm}.$$

Section plastic modulus $(W_{pl}) = b_1 t_1^2/4 = 700 \times t_1^2/4 = 175 \times t_1^2$

The plate thickness t_1 can be calculated now as follows:

$$\frac{M}{W_{pl}} = \frac{f_y}{\gamma_{M0}}$$

$$\frac{79{,}852{,}800}{175 \times t_1^2} = \frac{340}{1.0}$$

Then, $t_1 = 36.6$ mm, taken as 40 mm, as shown in Figure 4.33.

Design of the Rollers

The design of rollers requires determination of the diameter, length, and number of rollers that can resist the vertical load, as well as the arrangement, and allowed movement in the direction of rollers. The design axial force per unit length of roller contact N'_{Sd} specified in BS EN 1337-1 [3.11] shall satisfy

$$N'_{Sd} \leq N'_{Rd}$$

where N'_{Rd} is the design value of resistance per unit length of roller contact, which is calculated as

$$N'_{Rd} = 23 \times R \times \frac{f_u^2}{E_d} \times \frac{1}{\gamma_m^2} = 23 \times R \times \frac{550^2}{210{,}000} \times \frac{1}{1} = 33.131 \times R$$

Assume the number of rollers is 2 and their length is 800 mm as shown in Figure 4.33:

$$N'_{Sd} = \frac{R_{D+L+\Phi}}{2 \times 800} = \frac{3{,}327{,}200}{1600} = 2079.5 \, \text{N/mm}$$

Then, the radius of rollers can be determined by equalizing N'_{Sd} with N'_{Rd} as follows:

$$2079.5 = 33.131 \times R$$

Then, $R = 62.8$ mm, taken as 70 mm and the diameter D is 140 mm.

Design of Upper Bearing Plate

The upper bearing plate is shown in Figure 4.33. The width and length of the plate are dependent on the spacing between rollers and the length of rollers as well as the allowed movement in the direction of rollers. The thickness of the upper bearing plate can be determined as follows:

$$M = \frac{R_{D+L+\Phi}}{2} \times \frac{(D+100)}{2} = \frac{3327.2 \times 10^3}{2} \times \frac{240}{2} = 199,632,000 \, N \, mm.$$

$$W_{pl} = \frac{b_2 t_2^2}{4} = \frac{800 t_2^2}{4} = 200 \times t_2^2 \, mm^3$$

The plate thickness t_2 can be calculated now as follows:

$$\frac{M}{W_{pl}} = \frac{f_y}{\gamma_{M0}}$$

$$\frac{199,632,000}{200 \times t_1^2} = \frac{340}{1.0}$$

Then, $t_1 = 54.2$ mm, taken as 60 mm, as shown in Figure 4.33.

The radius of the curved part of the upper bearing plate, which has a length of 600 mm as shown in Figure 4.33, can be determined the same way as that adopted for the design of the rollers:

$$N'_{Rd} = 23 \times R \times \frac{f_u^2}{E_d} \times \frac{1}{\gamma_m^2} = 23 \times R \times \frac{550^2}{210,000} \times \frac{1}{1} = 33.131 \times R$$

$$N'_{Sd} = \frac{R_{D+L+\Phi}}{600} = \frac{3,327,200}{600} = 5545.33 \, N/mm$$

Then, the radius of rollers can be determined by equalizing N'_{Sd} with N'_{Rd} as follows:

$$5545.33 = 33.131 \times R$$

Then, $R = 167.4$ mm, taken as 170 mm.

Design of Lower Bearing Plate

The lower bearing plate is shown in Figure 4.33. The width and length of the plate are dependent on the strength of concrete and are dependent on the spacing between rollers and the length of rollers as well as the allowed movement in the direction of rollers. The thickness of the upper bearing plate can be determined as follows:

$$f_c = \frac{R_{D+L+\Phi}}{a_3 b_3} = \frac{3327.2 \times 10^3}{600 \times 800} = 6.93 \text{ MPa} < \frac{f_c}{\gamma_c} = \frac{40}{1.5}$$

$$= 26.7 \text{ MPa (for a typical concrete in bridges of C40/50 with } f_{ck})$$

The plate thickness t_3 can be calculated from the distribution of bending moment, caused by the pressure on the concrete foundation, as follows:

$$M = 112{,}266 \text{ N mm per unit width of the plate.}$$

$$W_{pl} = \frac{b_3 t_3^2}{4} = \frac{1 \times t_3^2}{4} = 0.25 \times t_3^2 \text{ mm}^3$$

$$\frac{M}{W_{pl}} = \frac{f_y}{\gamma_{M0}}$$

$$\frac{112{,}266}{0.25 \times t_3^2} = \frac{340}{1.0}$$

Then, $t_3 = 36.3$ mm, taken as 40 mm, as shown in Figure 4.33.

4.2.14 Design of Hinged Line Rocker Steel Fabricated Bearings

Finally, we can now design the hinged line rocker steel fabricated bearings shown in Figure 4.1 and detailed in Figure 4.34. The maximum vertical reaction at the support of the main plate girder was previously calculated under dead and live loads with dynamic effect ($R_{D+L+\Phi}$), which was 3327.2 kN. The bearing is also subjected to a lateral force from the braking and traction forces from tracks as well as subjected to a longitudinal force from the reactions of the upper and lower wind bracings, which cause moments around longitudinal and lateral directions of the bearing base, respectively. Similar to the roller bearing, the material of construction for the bearings is cast iron steel (ISO 3755) 340-550 having a yield stress of 340 MPa and an ultimate stress of 550 MPa. It should be noted that the overall height of the hinged bearing must be exactly the same as that of the roller bearing. The general layout and assumed dimensions of the hinged line rocker bearing are shown in Figure 4.34. It should be noted that the hinged line rocker bearings may not be the best hinged bearings for use nowadays. However, the main advantage of choosing this bearing is to illustrate for readers the applied loads on hinged bearings and review of the fundamentals of checking the stresses on hinged bearing. Steel fabricated bearings consist of designed parts, which are the best for teaching purposes. The traction Q_{lak} and braking Q_{lbk} forces can be calculated as follows:

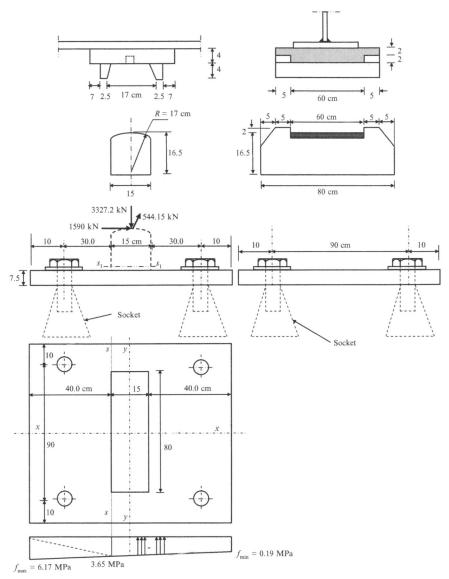

Figure 4.34 Detailing of the hinged line rocker fabricated steel bridge bearings

$$Q_{lak} = 33 \times L_{a,b} = 33 \times 30 = 990\,kN \leq 1000\,[kN], \quad \text{for Load Models 71}$$

$$Q_{lbk} = 20 \times L_{a,b} = 20 \times 30 = 600 \leq 6000\,[kN],$$

for Load Models 71,SW/0,SW/2 and HSLM

Total the braking and traction forces $(Q_{tot}) = 1590\,kN$ (see Figure 4.34 for the direction of the forces). Also, the reactions from upper and lower wind bracings (R_{tot}) (see Figure 4.34 for the direction of the forces) were previously calculated as follows:

$$R_{tot} = 449.05 + 95.1 = 544.15\,kN$$

We can now determine the normal stress distribution due to the applied loads, shown in Figure 4.34, on the concrete foundation as follows:

$$f = -\frac{N}{A} \pm \frac{M_x}{I_x}y \pm \frac{M_y}{I_y}x$$

$$\frac{N}{A} = \frac{3,327,200}{950 \times 1100} = 3.18\,MPa$$

$$\frac{M_x}{I_x}y = \frac{544.15 \times 10^3 \times 240}{950 \times 1100^3/12}550 = 0.68\,MPa$$

$$\frac{M_y}{I_y}x = \frac{1590 \times 10^3 \times 240}{1100 \times 950^3/12}475 = 2.31\,MPa$$

$$f_{max} = -3.18 - 0.68 - 2.31 = -6.17\,MPa$$

$$f_{min} = -3.18 + 0.68 + 2.31 = -0.19\,MPa$$

The critical bending moment on the base plate of the hinged bearing is at section s-s, shown in Figure 4.34:

$$M = (0.5 \times 400 \times 3.65) \times 1100 \times 400/3 + (0.5 \times 400 \times 6.17) \times 1100 \\ \times 400 \times 2/3 \\ = 469,040,000\,N\,mm$$

$$W_{pl} = 1100 \times t_4^2/4 = 275t_4^2$$

$$\frac{M}{W_{pl}} = \frac{f_y}{\gamma_{M0}}$$

$$\frac{469,040,000}{275t_4^2} = \frac{340}{1.0}$$

Then, $t_4 = 70.8$ mm, taken as 75 mm.

The normal stresses at section s_1-s_1, shown in Figure 4.34, of the line rocker bearing can be checked as follows:

$$M_x = 544.15 \times 10^3 \times 165 = 89,784,750\,N\,mm.$$

$$M_y = 1590 \times 10^3 \times 165 = 262,350,000\,N\,mm.$$

$$\frac{N}{A} = \frac{3,327,200}{150 \times 800} = 27.73\,\text{MPa}$$

$$\frac{M_x}{I_x}y = \frac{89,784,750}{150 \times 800^3/12}400 = 5.61\,\text{MPa}$$

$$\frac{M_y}{I_y}x = \frac{262,350,000}{800 \times 150^3/12}75 = 87.45\,\text{MPa}$$

$$f_{max} = -(27.73 + 5.61 + 87.45) = -120.79\,\text{MPa} < 340\,\text{MPa}\,(\text{Then O.K.})$$

4.3 DESIGN EXAMPLE OF A THROUGH TRUSS HIGHWAY STEEL BRIDGE

The second design example presented in this chapter is for a through truss highway steel bridge (Figure 4.35). The general layout of the through bridge is shown in Figures 4.36 and 4.37, with a brief introduction to the bridge components previously explained in Figure 1.21. This type of trusses is a Pratt truss bridge first designed by Thomas and Caleb Pratt in 1844. A Pratt truss has parallel top and bottom chords and is an efficient form of a truss arranged such that long diagonals are subjected to tension and verticals in compression. The truss bridge has simply supported ends with a length between supports of 60 m. The truss bridge has an N-shaped truss with 10 equal panels of 6 m. It is required to design the bridge adopting the design rules specified in EC3 [1.27]. The steel material of construction of the bridge conformed to standard steel grade EN 10025-2 (S 275) having a yield stress of 275 MPa and an ultimate strength of 430 MPa. The dimensions and general layout of the bridge are shown in Figures 4.36 and 4.37. The bridge has upper and lower wind bracings of K-shaped truss members. The expected live loads on the highway bridge conform to Load Model 1, which

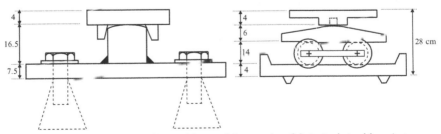

Figure 4.35 The designed roller and hinged line rocker fabricated steel bearings.

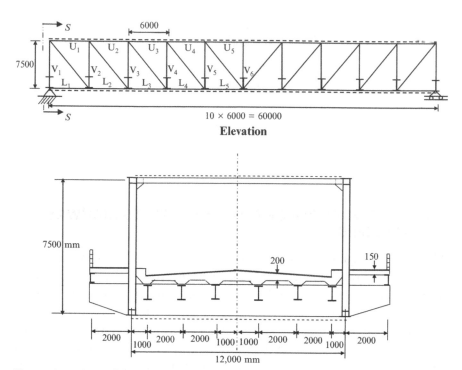

Elevation

Figure 4.36 General layout of a through truss highway steel bridge (the second design example).

represents the static and dynamic effects of vertical loading due to normal road traffic as specified in EC1 [3.1]. The bolts used in connections and field splices are M27 high-strength pretensioned bolts. The unit weight of reinforced concrete slab decks used is 25 kN/m³.

4.3.1 Design of the Stringers (Longitudinal Floor Beams)

Let us start by designing the stringers, the longitudinal steel beams, supporting the reinforced concrete slab deck as shown in Figure 4.36.

Dead Loads

The general layout of an intermediate stringer is shown in Figure 4.38. The dead loads acting on an intermediate stringer can be calculated as follows:

$$\text{Flooring} \left(1.75\,\text{kN/m}^2\right) = 1.75 \times 2 = 3.5\,\text{kN/m}$$

$$\text{Reinforced concrete slab deck} \left(0.2\,\text{m thickness}\right) = 5 \times 2 = 10\,\text{kN/m}$$

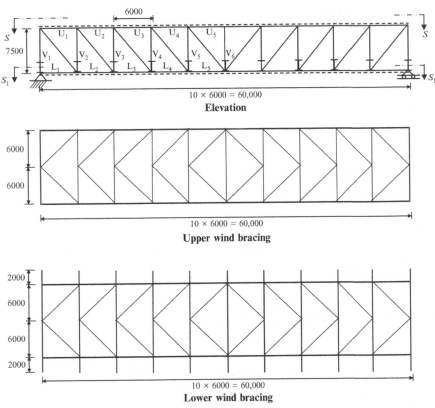

Figure 4.37 General layout of a through truss highway steel bridge (the second design example).

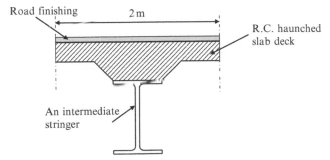

Figure 4.38 General layout of a an intermediate stringer.

Haunch (Equivalent to 1 cm slab thickness) $= 0.25 \times 2 = 0.5\,\text{kN/m}$

$$\text{Own weight of stringer} = 1.5\,\text{kN/m}$$

$$\text{Total dead load} = g_{vk} = 15.5\,\text{kN/m}$$

Assuming the stringers are simply supported by the cross girders, we can calculate the maximum shear force and bending moment due to dead loads on an intermediate stringer (see Figure 4.39) as follows:

$$Q_{\text{D.L.}} = g_{vk} \times L/2 = 15.5 \times 6/2 = 46.5\,\text{kN}$$
$$M_{\text{D.L.}} = g_{vk} \times L^2/8 = 15.5 \times 6^2/8 = 69.75\,\text{kN m}$$

Live Loads

The live loads acting on the highway bridge conform to Load Model 1, which represents the static and dynamic effects of vertical loading due to normal road traffic as specified in EC1 [3.1]. To determine the worst cases of loading on an intermediate stringer due to live loads, we can study a lateral section through vehicles and a lateral section through distributed loads of Load Model 1 acting on the bridge, as shown in Figure 4.40. From the section through vehicles, we find that the maximum concentrated load transferred to the stringer is 200 kN, while from the section through distributed loads, we find that the maximum distributed load transferred to the stringer is 14.34 kN/m. Therefore, the load distribution transferred to the stringer in the longitudinal direction is as shown in Figure 4.41. Two cases of loading for the evaluation of maximum bending moment due to live loads on a

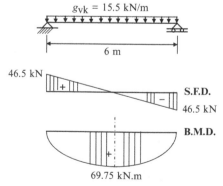

Figure 4.39 Straining actions from dead loads acting on an intermediate stringer.

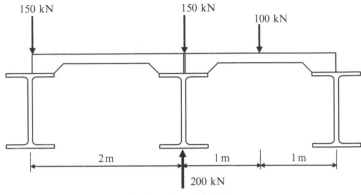

Section through vehicles

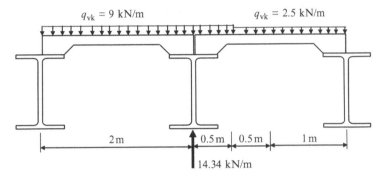

Section through distributed loads

Figure 4.40 Calculation of straining actions from live loads transferred on an intermediate stringer.

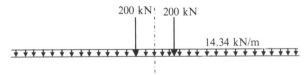

Figure 4.41 Transferred live loads on an intermediate stringer.

stringer can be studied. The first case of loading is that the centerline of the stringer divides the spacing between the resultant of the concentrated live loads and the closest load, with maximum bending moment calculated at the closest load (point a in Figure 4.42), while the second case of loading is that the centerline of the stringer is located in the middle between the

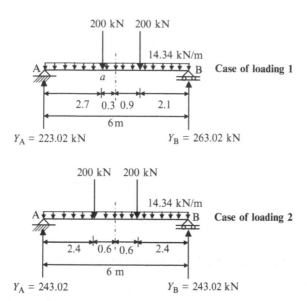

Figure 4.42 Cases of loading for the maximum bending moment acting on an intermediate stringer.

concentrated live loads, with maximum bending moment located at midspan as shown in Figure 4.42:

$$M_{L.L.}(\text{case of loading 1}) = 223.02 \times 2.7 - 14.34 \times 2.7^2/2 = 549.9\,\text{kN m}$$

$$M_{L.L.}(\text{case of loading 2}) = 200 \times 2.4 + 14.34 \times 6^2/8 = 544.5\,\text{kN m}$$

There is a single case of loading for live loads to produce a maximum shear force at the supports of the stringer, which is shown in Figure 4.43:

$$Q_{L.L.} = 403.02\,\text{kN}$$

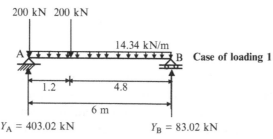

Figure 4.43 Cases of loading for the maximum shear force acting on a stringer.

Bending Moment Due to Dead and Live Loads with Dynamic Effect Added (M_{Ed})

$$M_{Ed} = M_{D.L.} \times \gamma_g + M_{L.L.} \times \gamma_q = 69.75 \times 1.3 + 549.9 \times 1.35$$
$$= 833.04 \, kN \, m$$

It should be noted that, according to EC0 (BS EN 1990) [3.4], the permanent actions of steel self-weight and superimposed load should be multiplied by 1.2 at the ultimate limit state, while the permanent actions of concrete weight should be multiplied by 1.35. Therefore, the total dead load is calibrated and multiplied by 1.3. On the other hand, variable actions comprising road traffic actions are multiplied by 1.35 at the ultimate limit state. Once again, it should be noted that the load factors adopted in this study are that of the ultimate limit state. This is attributed to the fact that the finite element models presented in Chapters 6 and 7 can be used to analyze the bridges and provide more accurate predictions for the deflections and other serviceability limit state cases of loading.

Shearing Force Due to Dead and Live Loads with Dynamic Effect Added (Q_{Ed})

$$Q_{Ed} = Q_{D.L.} \times \gamma_g + Q_{L.L.} \times \gamma_q = 46.5 \times 1.3 + 403.02 \times 1.35 = 604.5 \, kN$$

Design Bending Moment (M_{Ed}) and Shear Force (Q_{Ed})

$$\begin{aligned} M_{Ed} &= 833.04 \, kN \, m \\ Q_{Ed} &= 604.5 \, kN \end{aligned} \quad \text{for classes 1 and 2}$$

Design of Stringer Cross Section

$$M_{c,Rd} = \frac{W_{pl} \times f_y}{\gamma_{M0}}$$

$$833.04 \times 10^6 = \frac{W_{pl} \times 275}{1.0}$$

$$W_{PL} = 3,029,236 \, mm^3 = 3029 \, cm^3$$

Choose UB $610 \times 229 \times 113$ (equivalent to American W24 × 76), shown in Figure 4.44. W_{PL} around $x\text{-}x = 3281 \, cm^3$. To classify the cross section chosen,

$$\varepsilon = \sqrt{\frac{235}{f_y}} = \sqrt{\frac{235}{275}} = 0.924$$

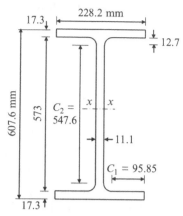

Figure 4.44 The cross-section of stringers (UB 610 × 229 × 113).

$C_1 = 95.85$ mm, $t_{fl} = 17.3$, $C_1/t_{fl} = 95.85/17.3 = 5.54 \leq 9 \times 0.924$
$= 8.316$ (Stringer flange is class 1)

$C_2 = 547.6$ mm, $t_w = 11.1$, $C_1/t_{fl} = 547.6/11.1 = 49.3 \leq 72 \times 0.924$
$= 66.5$ (Stringer web is class 1)

Check of Bending Resistance

$$M_{c,Rd} = \frac{W_{pl} \times f_y}{\gamma_{M0}} = \frac{3281 \times 10^3 \times 275}{1.0} = 902,275,000 \, \text{N mm}$$
$$= 902.3 \, \text{kN m} > M_{Ed} = 833.04 \, \text{kN m} \, (\text{Then O.K.})$$

Check of Shear Resistance

$$V_{pl,Rd} = \frac{A_v \left(f_y/\sqrt{3}\right)}{\gamma_{M0}} = \frac{(573 \times 11.1) \times \left(275/\sqrt{3}\right)}{1.0} = 1,009,833 \, \text{N}$$
$$= 1009.8 \, \text{kN} > Q_{Ed} = 604.5 \, \text{kN} \, (\text{Then O.K.})$$

4.3.2 Design of the Cross Girders (Lateral Floor Girders)

The cross girders (the lateral floor beams) carry concentrated loads from the stringers as shown in Figure 4.45. Therefore, the dead and live loads acting on an intermediate cross girder can be calculated as follows:

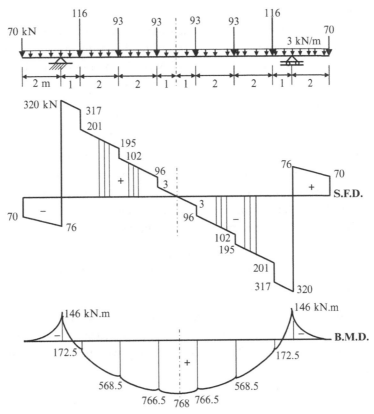

Figure 4.45 Straining actions from dead loads acting on an intermediate cross girder.

Dead Loads

Intermediate reactions from stringers due to dead loads $= 15.5 \times 6 = 93\,\text{kN}$

Reactions from stringers near supports due to dead loads $= 93 \times 2.5/2$
$= 116\,\text{kN}$

Reactions from stringers at edges due to dead loads $= 93 \times 1.5/2 = 70\,\text{kN}$

Own weight of cross girder $= 3.0\,\text{kN/m}$

Assuming the cross girders are simply supported by the main plate girders, we find that the maximum shear force and bending moment due to dead loads on an intermediate cross girder (see Figure 4.45) are as follows:

$$Q_{\text{D.L.}} = 320\,\text{kN}$$
$$M_{\text{D.L.}} = 768\,\text{kN}\,\text{m}$$

Live Loads

To determine the worst cases of loading on an intermediate cross girder due to live loads, we can study different longitudinal sections through vehicles, distributed loads, and sidewalks of Load Model 1 acting on the bridge, as shown in Figure 4.46. From the different sections, we can find that the maximum concentrated and distributed loads transferred to the intermediate cross girder are shown in Figure 4.46. The case of loading for the evaluation of maximum positive bending moment due to live loads on an intermediate cross girder can be studied, as shown in Figure 4.46. The case of loading is that the larger concentrated load from vehicles transferred is located at the centerline (midspan) of an intermediate cross girder, with maximum bending moment located at the midspan as shown in Figure 4.46. The maximum positive bending moment is calculated as follows:

$$M_{L.L.}(\text{maximum positive bending moment})$$
$$= 623.25 \times 6 - 270 \times 2 - 54 \times 2.5 \times 1.25 - 15 \times 2.5 \times 3.75 - 30 \times 1 \times 5.5 = 2725.1\,\text{kN m}$$

The case of loading for the evaluation of maximum negative bending moment due to live loads on an intermediate cross girder can be also studied, as shown in Figure 4.47. The maximum negative bending moment is calculated as follows:

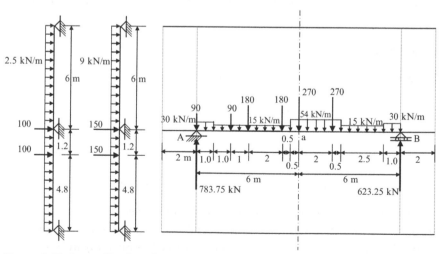

Figure 4.46 Case of loading for maximum positive bending moment from live loads acting on an intermediate cross girder.

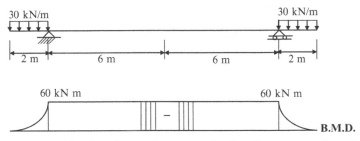

Figure 4.47 Case of loading for maximum negative bending moment from live loads acting on an intermediate cross girder.

$M_{L.L.}$(maximum negative bending moment) $= 30 \times 2 \times 1 = 60\,kN\,m$

The case of loading for live loads to produce a maximum shear force at the supports of an intermediate cross girder is shown in Figure 4.48. It should be noted that for this through bridge, cars are not allowed to go on top of the supports, which are the main trusses, by the presence of sidewalks to avoid direct collision forces with the main trusses, as shown in Figure 4.48:

$$Q_{L.L.} = 877.6\,kN$$

Bending Moment Due to Dead and Live Loads with Dynamic Effect Added (M_{Ed})

$$M_{Ed} = M_{D.L.} \times \gamma_g + M_{L.L.} \times \gamma_q = 768 \times 1.3 + 2725.1 \times 1.35 = 4677.3\,kN\,m$$

Shearing Force Due to Dead and Live Loads with Dynamic Effect Added (Q_{Ed})

$$Q_{Ed} = Q_{D.L.} \times \gamma_g + Q_{L.L.} \times \gamma_q = 320 \times 1.3 + 877.6 \times 1.35 = 1600.8\,kN$$

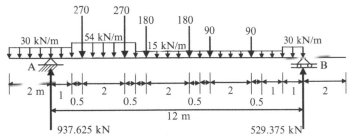

Figure 4.48 Case of loading for maximum shearing force from live loads acting on an intermediate cross girder.

Design Bending Moment (M_{Ed}) and Shear Force (Q_{Ed})

$$M_{Ed} = 4677.3 \, kN \, m$$
$$Q_{Ed} = 1600.8 \, kN$$

Design of the Cross Girder Cross Section

The cross girder is designed as a welded plate girder as shown in Figure 4.49. The web height is taken as equal to 1500 mm, which conforms to the recommended values $L/(7-9) = 12,000/(7-9) = 1714 - 1333$ mm. The web plate thickness is assumed to be 14 mm. The flange width is taken as equal to 360 mm, with a thickness of 24 mm. To classify the cross section chosen,

$$\varepsilon = \sqrt{\frac{235}{f_y}} = \sqrt{\frac{235}{275}} = 0.924$$

$C_1 = 165 \, mm$, $t_{fl} = 24$, $C_1/t_{fl} = 165/24 = 6.9 \leq 9 \times 0.924$
$= 8.316$ (Cross girder flange is class 1).

$C_2 = 1464 \, mm$, $t_w = 14$, $C_1/t_{fl} = 1464/14 = 104.6 < 124 \times 0.924$
$= 114.58$ (Cross girder web is class 3).

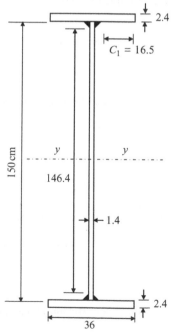

Figure 4.49 Welded plate girder section of cross girders.

To calculate the bending moment resistance, the elastic section modulus should be used:

$$\text{Inertia about } y\text{-}y = 1.4 \times 150^3/12 + 2 \times \left[36 \times 2.4^3/12 + 36 \times 2.4 \times 76.2^2\right]$$
$$= 1,397,185.8 \text{ cm}^4$$

$$W_{\text{eff,min}} = 1,397,185.8/77.4 = 18,051.5 \text{ cm}^3$$

Check of Bending Resistance

$$M_{c,\text{Rd}} = \frac{W_{\text{el,min}} \times f_y}{\gamma_{M0}} = \frac{18,051.5 \times 10^3 \times 275}{1.0} = 4,964,162,500 \text{ N mm}$$

$$= 4964.2 \text{ kN m} > M_{\text{Ed}} = 4677.3 \text{ kN m (Then O.K.)}$$

Check of Shear Resistance

$$V_{b,\text{Rd}} = V_{bw,\text{Rd}} + V_{bf,\text{Rd}} \leq \frac{\eta f_{yw} h_w t_w}{\sqrt{3}\gamma_{M1}}$$

By neglecting the flange contribution,

$$V_{b,\text{Rd}} = V_{bw,\text{Rd}} \leq \frac{1.2 \times 275 \times 1500 \times 14}{\sqrt{3} \times 1.1} = 3,637,306.7 \text{ N} = 3637.3 \text{ kN}$$

$$V_{bw,\text{Rd}} = \frac{\chi_w f_{yw} h_w t_w}{\sqrt{3}\gamma_{M1}}$$

$$\bar{\lambda}_w = 0.76\sqrt{\frac{f_{yw}}{\tau_{cr}}}, \quad \tau_{cr} = k_\tau \sigma_E$$

$$\sigma_E = 190,000(14/1500)^2 = 16.55 \text{ N/mm}^2$$

$$k_\tau = 5.34 + 4(1500/2000)^2 = 7.59$$

$$\bar{\lambda}_w = 0.76\sqrt{\frac{275}{7.59 \times 16.55}} = 1.125 > 1.08$$

$$\text{Then, } \chi_w = \frac{1.37}{0.7 + \bar{\lambda}_w} = \frac{1.37}{0.7 + 1.125} = 0.751$$

$$V_{bw,Rd} = \frac{0.751 \times 275 \times 1500 \times 14}{\sqrt{3} \times 1.1} = 2{,}276{,}347.8\,\text{N} = 2276.3\,\text{kN}$$
$$< 3637.3\,\text{kN}$$

$$\eta_3 = \frac{V_{Ed}}{V_{b,Rd}} = \frac{1600.8}{2276.3} = 0.703 < 1.0\,(\text{Then O.K.})$$

4.3.3 Calculation of Forces in Truss Members

4.3.3.1 General

To calculate the design forces in the truss members, we need to calculate the dead and live loads acting on the main truss in the longitudinal direction, which is addressed as follows.

Dead Loads

Weight of steel structure for part of bridge between the main trusses:

$$w_{s_1} = 1.75 + 0.04L + 0.0003L^2 \leq 3.5\,\text{kN/m}^2$$

$$w_{s_1} = 1.75 + 0.04 \times 60 + 0.0003 \times 60^2 = 5.23 > 3.5\,\text{kN/m}^2$$
$$\text{taken as } 3.5\,\text{kN/m}^2$$

Weight of steel structure for part of bridge outside the main trusses:

$$w_{s_2} = 1 + 0.03L\,\text{kN/m}^2$$

$$w_{s_2} = 1 + 0.03 \times 60 = 2.8\,\text{kN/m}^2$$

$$w_s = 3.5 \times 12/2 + 2.8 \times 2 = 26.6\,\text{kN/m}$$

Weight of reinforced concrete decks and haunches:

$$w_{RC} = (0.2 + 0.01) \times 25 \times 5 + (0.15 + 0.01) \times 25 \times 3 = 38.25\,\text{kN/m}$$

Weight of finishing (assume weight of finishing is 1.75 kN/m^2 for parts between sidewalks and 1.5 kN/m^2 for sidewalks):

$$w_F = 1.75 \times 5 + 1.5 \times 3 = 13.25\,\text{kN/m}$$

We can now calculate the total dead load acting on the main trusses in the longitudinal direction (see Figure 4.50) as follows:

$$w_{D.L.} = 26.6 + 38.25 + 13.25 = 78.1\,\text{kN/m}$$

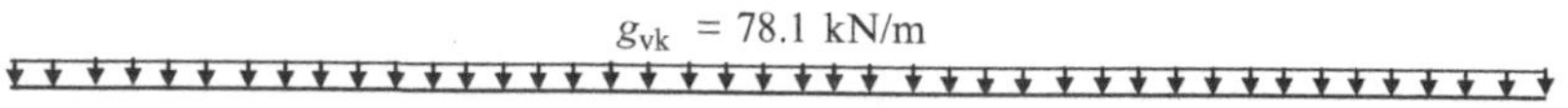

$$g_{vk} = 78.1\ \text{kN/m}$$

Figure 4.50 Dead loads acting on main trusses.

Live Loads

To determine the live loads acting on main trusses in the longitudinal directions, we can study different lateral sections through vehicles, distributed loads, and sidewalks of Load Model 1 acting on the bridge, as shown in Figure 4.51. From the lateral section shown in Figure 4.51, we can find that the maximum concentrated and distributed loads transferred to a main truss are 375 kN and 43.8 kN/m, respectively, as shown in Figure 4.52. We can also calculate the negative distributed reactions acting on a main truss in the longitudinal direction by investigating the case of loading shown in Figure 4.53. The negative distributed load acting on a main truss is 0.83 kN/m as shown in Figure 4.54. The calculated dead and live loads

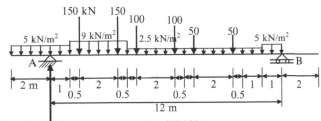

Reaction from concentrated loads = 375 kN

Reaction from distributed loads = 43.8 kN/m

Figure 4.51 Maximum reactions due to live loads transferred by cross girders on main trusses.

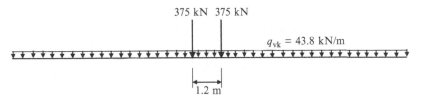

Figure 4.52 Live loads acting on main trusses.

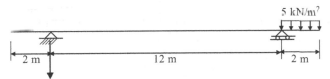

Negative reaction from distributed loads = −0.83 kN/m

Figure 4.53 Negative reactions due to live loads transferred by cross girders on main trusses.

$$q_{vk} = -0.83 \text{ kN/m}$$

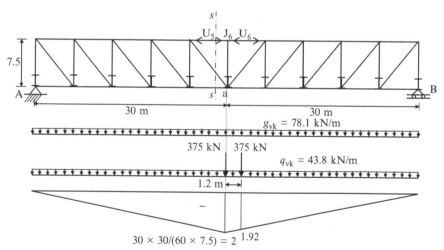

Figure 4.54 Negative distributed live loads acting on main trusses.

can be now used to determine the forces in the members of main trusses using the influence line method as shown in the coming sections.

4.3.3.2 Calculation of Force in the Upper Chord Member U_5

To determine the force in the upper chord truss member U_5 (see Figure 4.55) using the influence line method, we can follow the simple procedures of putting a unit concentrated moving load at midspan (point a), and using the sectioning method, we take a section s-s, as shown in Figure 4.55, and then take the moment at point a to calculate the force in the member due to the applied unit load. After that, we can put the previously calculated dead and live loads acting on a main truss in the longitudinal direction. The total force in the member will be the summation of the concentrated loads multiplied by the companion vertical coordinate in the influence line diagram and the summation of the distributed loads multiplied by the companion areas in the diagram. Hence, the forces due to the dead and live loads can be calculated as follows:

$$F_{D.L.}(U_5) = -0.5 \times 60 \times 2 \times 78.1 = -4686 \text{ kN}$$

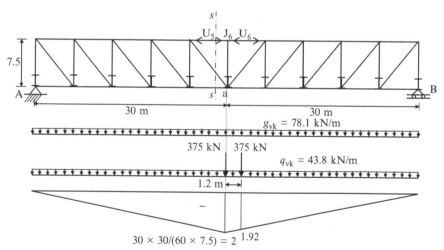

Figure 4.55 Determination of the compressive force in upper chord member U_5 using the influence line method.

$$F_{L.L.}(U_5) = -375 \times (2 + 1.92) - 0.5 \times 60 \times 2 \times 43.8 = -4098\,kN$$

$$F_{Ed}(U_5) = F_{D.L.} \times \gamma_g + F_{L.L.} \times \gamma_q$$

$$F_{Ed}(U_5) = -4686 \times 1.3 - 4098 \times 1.35$$
$$= -11624.1\,kN\,(\text{Compression force})$$

It should be noted that, from the equilibrium of joint J_6 (see Figure 4.55), the force in upper chord truss member U_5 is equal to that of U_6. It should also be noted that the negative distributed loads are not used since they will produce a small tensile force, which reduces the calculated compressive force.

4.3.3.3 Calculation of Force in the Lower Chord Member L_5

To determine the force in the lower chord truss member L_5 (see Figure 4.56) using the influence line method, we put the unit concentrated load at point a, and using the sectioning method, we take a section s-s, as shown in Figure 4.56, and then take the moment at point a. After that, we can put the previously calculated dead and live loads acting on a main truss in the longitudinal direction. The forces due to the dead and live loads can be calculated as follows:

$$F_{D.L.}(L_5) = 0.5 \times 60 \times 1.92 \times 78.1 = 4498.6\,kN$$

$$F_{L.L.}(L_5) = 375 \times (1.92 + 1.856) + 0.5 \times 60 \times 1.92 \times 43.8 = 3938.9\,kN$$

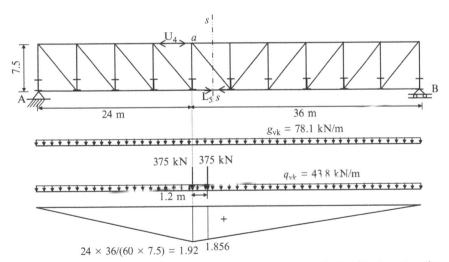

Figure 4.56 Determination of the tensile force in lower chord member L_5 using the influence line method.

$$F_{Ed}(L_5) = F_{D.L.} \times \gamma_g + F_{L.L.} \times \gamma_q$$

$$F_{Ed}(L_5)\text{maximum} = 4498.6 \times 1.3 + 3938.9 \times 1.35$$
$$= 11{,}165.7\,\text{kN (Tension force)}$$

Since this member is a tensile member, we should check the minimum force due to the negative distributed loads since they may change the force in the member to compression:

$$F_{L.L.}(L_5)(\text{negative}) = -0.5 \times 60 \times 1.92 \times 0.83 = -47.8\,\text{kN}$$

$$F_{Ed}(L_5)\text{minimum} = 4498.6 \times 1.3 - 47.8 \times 1.35$$
$$= 5783.7\,\text{kN (Tension force)}$$

It should be noted that, from the equilibrium of the truss (see Figure 4.56), the force in upper chord truss member U_4 is equal to that of the calculated lower chord member L_5 but with a negative sign (a compression force of 11,165.7 kN).

4.3.3.4 Calculation of Force in the Lower Chord Member L_4

We can repeat the earlier procedures now and change the pole where the moment is calculated to determine the force in the member, as shown in Figure 4.57. Hence, the forces due to the dead and live loads can be calculated as follows:

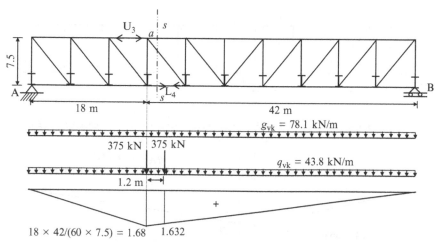

Figure 4.57 Determination of the tensile force in lower chord member L_4 using the influence line method.

$$F_{D.L.}(L_4) = 0.5 \times 60 \times 1.68 \times 78.1 = 3936.2\,\text{kN}$$

$$F_{L.L.}(L_4) = 375 \times (1.68 + 1.632) + 0.5 \times 60 \times 1.68 \times 43.8 = 3449.5\,\text{kN}$$

$$F_{Ed}(L_4) = F_{D.L.} \times \gamma_g + F_{L.L.} \times \gamma_q$$

$$F_{Ed}(L_4)\text{maximum} = 3936.2 \times 1.3 + 3449.5 \times 1.35$$
$$= 9773.9\,\text{kN (Tension force)}$$

$$F_{L.L.}(L_4)(\text{negative}) = -0.5 \times 60 \times 1.68 \times 0.83 = -41.8\,\text{kN}$$

$$F_{Ed}(L_4)\text{minimum} = 3936.2 \times 1.3 - 41.8 \times 1.35$$
$$= 5060.6\,\text{kN (Tension force)}$$

It should be noted that, from the equilibrium of the truss (see Figure 4.57), the force in upper chord truss member U_3 is equal to that of the calculated lower chord member L_4 but with a negative sign (a compression force of 9773.9 kN).

4.3.3.5 Calculation of Force in the Lower Chord Member L₃

The force in member L_3 due to the dead and live loads can be calculated, as shown in Figure 4.58, as follows:

$$F_{D.L.}(L_3) = 0.5 \times 60 \times 1.28 \times 78.1 = 2999\,\text{kN}$$

$$F_{L.L.}(L_3) = 375 \times (1.28 + 1.248) + 0.5 \times 60 \times 1.28 \times 43.8 = 2629.9\,\text{kN}$$

$$F_{Ed}(L_3) = F_{D.L.} \times \gamma_g + F_{L.L.} \times \gamma_q$$

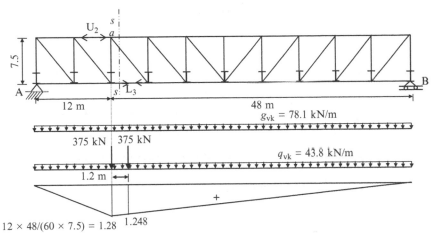

Figure 4.58 Determination of the tensile force in lower chord member L_3 using the influence line method.

$$F_{Ed}(L_3)\text{maximum} = 2999 \times 1.3 + 2629.9 \times 1.35$$
$$= 7449.1\,kN\,(\text{Tension force})$$

$$F_{L.L.}(L_3)(\text{negative}) = -0.5 \times 60 \times 1.28 \times 0.83 = -31.9\,kN$$

$$F_{Ed}(L_3)\text{minimum} = 2999 \times 1.3 - 31.9 \times 1.35$$
$$= 3855.6\,kN\,(\text{Tension force})$$

It should be noted that, from the equilibrium of the truss (see Figure 4.58), the force in upper chord truss member U_2 is equal to that of the calculated lower chord member L_3 but with a negative sign (a compression force of 7449.1 kN).

4.3.3.6 Calculation of Force in the Lower Chord Member L_2
The force in member L_2 due to the dead and live loads can be calculated, as shown in Figure 4.59, as follows:

$$F_{D.L.}(L_2) = 0.5 \times 60 \times 0.72 \times 78.1 = 1687\,kN$$

$$F_{L.L.}(L_2) = 375 \times (0.72 + 0.704) + 0.5 \times 60 \times 0.72 \times 43.8 = 1480.1\,kN$$

$$F_{Ed}(L_2) = F_{D.L.} \times \gamma_g + F_{L.L.} \times \gamma_q$$

$$F_{Ed}(L_2)\text{maximum} = 1687 \times 1.3 + 1480.1 \times 1.35$$
$$= 4191.2\,kN\,(\text{Tension force})$$

$$F_{L.L.}(L_2)(\text{negative}) = -0.5 \times 60 \times 0.72 \times 0.83 = -17.9\,kN$$

$$F_{Ed}(L_2)\text{minimum} = 1687 \times 1.3 - 17.9 \times 1.35$$
$$= 2168.9\,kN\,(\text{Tension force})$$

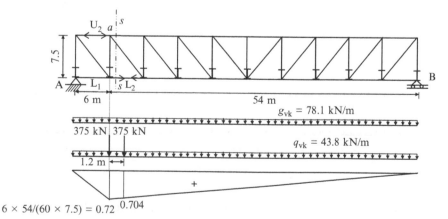

Figure 4.59 Determination of the tensile force in lower chord member L_2 using the influence line method.

It should be noted that, from the equilibrium of the truss (see Figure 4.59), the force in upper chord truss member U_1 is equal to that of the calculated lower chord member L_2 but with a negative sign (a compression force of 4191.2 kN). It should also be noted that the force in the lower chord member L_1 is zero under vertical loading.

4.3.3.7 Calculation of Force in the Diagonal Chord Member D_5

To determine the force in the diagonal chord truss member D_5 (see Figure 4.60) using the influence line method, we can follow the simple procedures of putting a unit concentrated moving load at point a adjacent to section s-s, shown in Figure 4.60, and study the equilibrium of the truss for the other side of section s-s to calculate the force in the member. Then, we put the unit concentrated moving load at point b adjacent to section s-s, shown in Figure 4.60, and study the equilibrium of the truss for the other side of section s-s to calculate the force in the member. The influence line of the diagonal member consists of two triangles as shown in Figure 4.60 having different signs. After that, we can put the previously calculated dead and live loads acting on a main truss in the longitudinal direction. It should be noted that the live loads can be put on the negative or positive triangle to

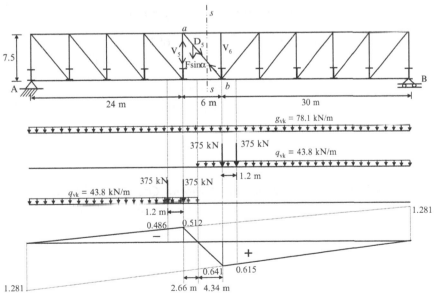

Figure 4.60 Determination of the force in diagonal member D_5 using the influence line method.

produce a compressive or tensile force, respectively, while the dead loads must be put on both triangles. Once again, the total force in the member will be the summation of concentrated loads multiplied by the companion vertical coordinate in the diagram and the summation of the distributed loads multiplied by the companion area in the diagram. Hence, the forces due to the dead and live loads can be calculated as follows:

$$A_{+ve}(D_5) = 0.5 \times 33.34 \times 0.641 = 10.69$$

$$A_{-ve}(D_5) = 0.5 \times 26.66 \times 0.512 = 6.82$$

$$A_{net}(D_5) = 10.69 - 6.82 = 3.87$$

$$F_{D.L.}(D_5) = 3.87 \times 78.1 = 302.2\,kN$$

$$F_{L.L.}(D_5)(positive) = 375 \times (0.641 + 0.615) + 10.69 \times 43.8 + 6.82 \times 0.83$$
$$= 944.9\,kN$$

$$F_{Ed}(D_5)maximum = F_{D.L.} \times \gamma_g + F_{L.L.} \times \gamma_q$$

$$F_{Ed}(D_5)maximum = 302.2 \times 1.3 + 944.9 \times 1.35$$
$$= 1668.5\,kN\,(Tension\,force)$$

$$F_{L.L.}(D_5)(negative) = -375 \times (0.512 + 0.486) - 6.82 \times 43.8 - 10.69$$
$$\times 0.83 = -681\,kN$$

$$F_{Ed}(D_5)minimum = F_{D.L.} \times \gamma_g + F_{L.L.} \times \gamma_q$$

$$F_{Ed}(D_5)minimum = 302.2 \times 1.3 - 681 \times 1.35$$
$$= -526.5\,kN\,(Compression\,force)$$

It should be noted that, from the equilibrium of joint J_5 (see Figure 4.60), the force in the vertical truss member V_5 is equal to that of D_5 multiplied by $\sin\alpha$ but with a negative sign (a compression force of $1668.5 \times \sin 51.34 = 1302.9$ kN).

4.3.3.8 Calculation of Force in the Diagonal Chord Member D_4

By repeating the procedures adopted for D_5, the force in the diagonal truss member D_4 can be calculated, as shown in Figure 4.61, as follows:

$$A_{+ve}(D_4) = 0.5 \times 40 \times 0.769 = 15.38$$

$$A_{-ve}(D_4) = 0.5 \times 20 \times 0.384 = 3.84$$

$$A_{net}(D_4) = 15.38 - 3.84 = 11.54$$

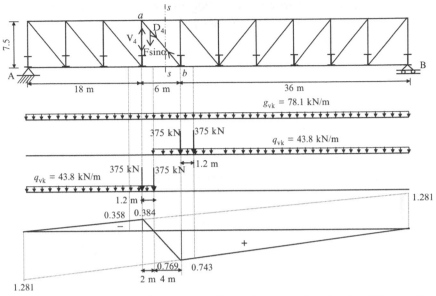

Figure 4.61 Determination of the force in diagonal member D_4 using the influence line method.

$$F_{D.L.}(D_4) = 11.54 \times 78.1 = 901.3 \, kN$$

$$F_{L.L.}(D_4)(\text{positive}) = 375 \times (0.769 + 0.743) + 15.38 \times 43.8 + 3.84 \times 0.83$$
$$= 1243.8 \, kN$$

$$F_{Ed}(D_4)\text{maximum} = F_{D.L.} \times \gamma_g + F_{L.L.} \times \gamma_q$$

$$F_{Ed}(D_4)\text{maximum} = 901.3 \times 1.3 + 1243.8 \times 1.35$$
$$= 2850.8 \, kN \, (\text{Tension force})$$

$$F_{L.L.}(D_4)(\text{negative}) = -375 \times (0.384 + 0.358) - 3.84 \times 43.8 - 15.38$$
$$\times 0.83 = -459.2 \, kN$$

$$F_{Ed}(D_4)\text{minimum} = F_{D.L.} \times \gamma_g + F_{L.L.} \times \gamma_q$$

$$F_{Ed}(D_4)\text{minimum} = 901.3 \times 1.3 - 459.2 \times 1.35$$
$$= 551.8 \, kN \, (\text{Tension force})$$

It should be noted that, from the equilibrium of joint J_4 (see Figure 4.61), the force in the vertical truss member V_4 is equal to that of D_4 multiplied by $\sin \alpha$ but with a negative sign (a compression force of $2850.8 \times \sin 51.34 = 2226.1 \, kN$).

4.3.3.9 Calculation of Force in the Diagonal Chord Member D₃

The force in the diagonal truss member D_3 can be calculated, as shown in Figure 4.62, as follows:

$$A_{+ve}(D_3) = 0.5 \times 46.67 \times 0.897 = 20.93$$

$$A_{-ve}(D_3) = 0.5 \times 13.33 \times 0.256 = 1.71$$

$$A_{net}(D_3) = 20.93 - 1.71 = 19.22$$

$$F_{D.L.}(D_3) = 19.22 \times 78.1 = 1501.1\,kN$$

$$F_{L.L.}(D_3)(positive) = 375 \times (0.897 + 0.871) + 20.93 \times 43.8 + 1.71 \times 0.83$$
$$= 1581.2\,kN$$

$$F_{Ed}(D_3)maximum = F_{D.L.} \times \gamma_g + F_{L.L.} \times \gamma_q$$

$$F_{Ed}(D_3)maximum = 1501.1 \times 1.3 + 1581.2 \times 1.35$$
$$= 4086.1\,kN\,(Tension\,force)$$

$$F_{L.L.}(D_3)(negative) = -375 \times (0.256 + 0.23) - 1.71 \times 43.8 - 20.93 \times 0.83$$
$$= -274.5\,kN$$

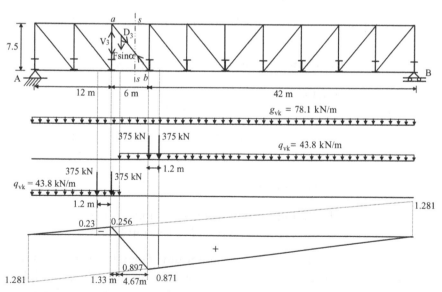

Figure 4.62 Determination of the force in diagonal member D_3 using the influence line method.

$$F_{Ed}(D_3)\text{minimum} = F_{D.L.} \times \gamma_g + F_{L.L.} \times \gamma_q$$

$$F_{Ed}(D_3)\text{minimum} = 1501.1 \times 1.3 - 274.5 \times 1.35$$
$$= 1580.9\,kN\,(\text{Tension force})$$

It should be noted that, from the equilibrium of joint J_3 (see Figure 4.62), the force in the vertical truss member V_3 is equal to that of D_3 multiplied by $\sin\alpha$ but with a negative sign (a compression force of $4086.1 \times \sin 51.34 = 3190.7$ kN).

4.3.3.10 Calculation of Force in the Diagonal Chord Member D_2

The force in the diagonal truss member D_2 can be calculated, as shown in Figure 4.63, as follows:

$$A_{+ve}(D_2) = 0.5 \times 53.33 \times 1.025 = 27.33$$
$$A_{-ve}(D_2) = 0.5 \times 6.67 \times 0.128 = 0.427$$
$$A_{net}(D_2) = 27.33 - 0.427 = 26.9$$

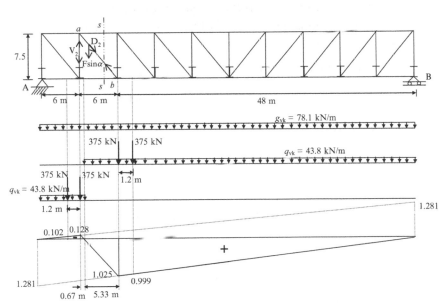

Figure 4.63 Determination of the force in diagonal member D_2 using the influence line method.

$$F_{D.L.}(D_2) = 26.9 \times 78.1 = 2100.9\,kN$$

$$F_{L.L.}(D_2)(positive) = 375 \times (1.025 + 0.999) + 27.33 \times 43.8 + 0.427 \times 0.83$$
$$= 1956.4\,kN$$

$$F_{Ed}(D_2)maximum = F_{D.L.} \times \gamma_g + F_{L.L.} \times \gamma_q$$

$$F_{Ed}(D_2)maximum = 2100.9 \times 1.3 + 1956.4 \times 1.35$$
$$= 5372.3\,kN\,(Tension\ force)$$

$$F_{L.L.}(D_2)(negative) = -375 \times (0.128 + 0.102) - 0.427 \times 43.8 - 27.33$$
$$\times 0.83 = -127.6\,kN$$

$$F_{Ed}(D_2)minimum = F_{D.L.} \times \gamma_g + F_{L.L.} \times \gamma_q$$

$$F_{Ed}(D_2)minimum = 2100.9 \times 1.3 - 127.6 \times 1.35$$
$$= 2558.9\,kN\,(Tension\ force)$$

It should be noted that, from the equilibrium of joint J_2 (see Figure 4.63), the force in the vertical truss member V_2 is equal to that of D_2 multiplied by $\sin\alpha$ but with a negative sign (a compression force of $5372.3 \times \sin 51.34 = 4195.1$ kN).

4.3.3.11 Calculation of Force in the Diagonal Chord Member D₁

The force in the diagonal truss member D_1 can be calculated, as shown in Figure 4.64, as follows:

$$A_{+ve}(D_1) = A_{net}(D_2) = 0.5 \times 60 \times 1.153 = 34.59$$

$$F_{D.L.}(D_1) = 34.59 \times 78.1 = 2701.5\,kN$$

$$F_{L.L.}(D_1)(positive) = 375 \times (1.153 + 1.127) + 34.59 \times 43.8 = 2370.0\,kN$$

$$F_{Ed}(D_1)maximum = F_{D.L.} \times \gamma_g + F_{L.L.} \times \gamma_q$$

$$F_{Ed}(D_1)maximum = 2701.5 \times 1.3 + 2370.0 \times 1.35$$
$$= 6711.5\,kN\,(Tension\ force)$$

$$F_{L.L.}(D_1)(negative) = -34.59 \times 0.83 = -28.7\,kN$$

$$F_{Ed}(D_2)minimum = F_{D.L.} \times \gamma_g + F_{L.L.} \times \gamma_q$$

$$F_{Ed}(D_2)minimum = 2100.9 \times 1.3 - 28.7 \times 1.35$$
$$= 2692.4\,kN\,(Tension\ force)$$

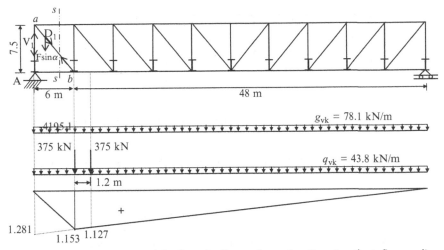

Figure 4.64 Determination of the force in diagonal member D_1 using the influence line method.

It should be noted that, from the equilibrium of joint J_1 (see Figure 4.64), the force in the vertical truss member V_1 is equal to that of D_1 multiplied by $\sin \alpha$ but with a negative sign (a compression force of $6711.5 \times \sin 51.34 = 5240.8$ kN).

4.3.3.12 Calculation of the Reactions at Supports
The reactions at supports can be also calculated using the influence line method, as shown in Figure 4.65, as follows:

$$A_{+ve}(R) = A_{net}(D_2) = 0.5 \times 60 \times 1.0 = 30.0$$

$$F_{D.L.}(R) = 30.0 \times 78.1 = 2343 \text{ kN}$$

$$F_{L.L.}(R)(\text{positive}) = 375 \times (1.0 + 0.98) + 30.0 \times 43.8 = 2056.5 \text{ kN}$$

$$F_{Ed}(R) = F_{D.L.} \times \gamma_g + F_{L.L.} \times \gamma_q$$

$$F_{Ed}(R)\text{maximum} = 2343 \times 1.3 + 2056.5 \times 1.35 = 5822.2 \text{ kN}$$

Figure 4.66 summarizes the calculated forces in the truss members and presents the commonly known distribution of forces in the N-shaped main truss under the dead and live cases of loading.

4.3.3.13 Design of the Maximum Compression Upper Chord Member U_5
After the calculation of the design forces in the main truss members, we can now design different members of the main truss. Let us start by designing the

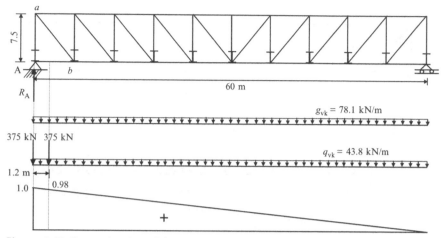

Figure 4.65 Determination of the reaction R_A using the influence line method.

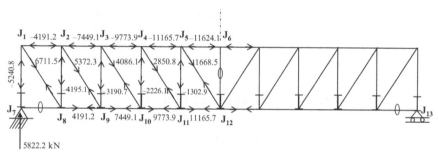

Figure 4.66 Distribution of forces in the N-shaped main truss under the dead and live cases of loading.

maximum compression upper chord member U_5, shown in Figure 4.67, carrying a compressive design force of $-11,624.1$ kN. It should be noted that box sections used with truss bridges may be bolted or welded. In bolted box sections, channels are commonly used in webs and connected to cover flange plates using bolts. However, bolted box sections require a lot of detailing and are time-consuming to fabricate. That is why welded box sections consisting of flange and web plates have been commonly used in bridges in the last decades, particularly for continuous chord members owing to the advanced techniques available nowadays for butt welding, while verticals and diagonals of truss bridges can be designed as bolted members since they are not continuous and can be assembled and erected in the construction field to avoid transportation problems. To assume a reasonable cross section

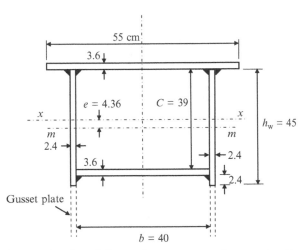

Figure 4.67 The cross section of the maximum compression member U₅ as well as that of the member U₄.

for the upper chord compression member, the following parameters can be considered:

$$h_w = \frac{a}{12-15} = \frac{6000}{12-15} = 500 - 400 \, mm, \text{ taken as } 450 \, mm.$$

$$b = (0.75 - 0.9), \quad h_w = (0.75 - 0.9) \times 450$$
$$= 337.5 - 405 \, mm, \text{ taken as } 400 \, mm.$$

It should be noted that the spacing between gusset plates (b) must be kept constant for the whole truss. Let us start by assuming the upper cover plate width of 550 mm, flange thickness of 36 mm, and web thickness of 24 mm. After that, we design the member and check the stresses. If the section is safe and economic, then the design is acceptable; otherwise, we change the dimensions accordingly and repeat the procedures. To classify the cross section chosen (see Figure 4.67),

$$\varepsilon = \sqrt{\frac{235}{f_y}} = \sqrt{\frac{235}{275}} = 0.924$$

$$b - 400 \, mm, \quad t_{fl} = 36, \quad b/t_{fl} = 400/36 = 11.1 < 30.5 \, (\text{Flange is Class 1})$$

$$C = 390 \, mm, \quad t_w = 24, \quad C/t_{fl} = 390/24 = 16.25 < 30.5 \, (\text{Web is class 1})$$

$$A = 55 \times 3.6 + 40 \times 3.6 + 2 \times 45 \times 2.4 = 558 \, cm^2$$

$$e = \frac{55 \times 3.6 \times 24.3 - 40 \times 3.6 \times 16.5}{558} = 4.36 \, cm$$

$$I_m = 2 \times 2.4 \times 45^3/12 + \left[55 \times 3.6^3/12 + 55 \times 3.6 \times 24.3^2\right]$$
$$+ \left[40 \times 3.6^3/12 + 40 \times 3.6 \times 16.5^2\right] = 192{,}940.4 \, \text{cm}^4$$

$$I_x = 192{,}940.4 - 558 \times 4.36^2 = 182{,}333 \, \text{cm}^4$$

$$I_y = 3.6 \times 55^3/12 + 3.6 \times 40^3/12 + 2 \times \left[45 \times 2.4^3/12 + 45 \times 2.4 \times 21.2^2\right]$$
$$= 166{,}295.2 \, \text{cm}^4$$

$$i_x = \sqrt{\frac{I_x}{A}} = \sqrt{\frac{182{,}333}{558}} = 18.08 \, \text{cm}$$

$$i_y = \sqrt{\frac{I_y}{A}} = \sqrt{\frac{166{,}295.2}{558}} = 17.26 \, \text{cm}$$

$$l_{bx} = l_{by} = 6000 \, \text{mm}$$

$$\bar{\lambda} = \frac{L_{cr}}{i} \frac{1}{\lambda_1}$$

$$\lambda_1 = 93.9 \times 0.924 = 86.7636$$

$$\bar{\lambda} = \frac{6000}{172.6} \times \frac{1}{86.7636} = 0.4$$

The axial compressive force in the upper chord member U_5 ($N_{Ed} = 11{,}624.1$ kN):

$$\frac{N_{Ed}}{N_{b,Rd}} \leq 1.0$$

where $N_{b,Rd} = \dfrac{\chi A f_y}{\gamma_{M1}}$

$$\chi = \frac{1}{\Phi + \sqrt{\Phi^2 - \bar{\lambda}^2}} \quad \text{but } \chi \leq 1.0$$

$$\Phi = 0.5\left[1 + \alpha\left(\bar{\lambda} - 0.2\right) + \bar{\lambda}^2\right]$$

$$\Phi = 0.5\left[1 + 0.34(0.4 - 0.2) + 0.4^2\right] = 0.614$$

$$\chi = \frac{1}{0.614 + \sqrt{0.614^2 - 0.4^2}} = 0.926 \quad \text{but } \chi \leq 1.0$$

Then, $N_{b,Rd} = \dfrac{0.926 \times 55{,}800 \times 275}{1.1} = 12{,}917{,}700 \, \text{N}$

$$N_{b,Rd} = 12{,}917.7 \, \text{kN} > N_{Ed} = 11{,}624.1 \, \text{kN} \, (\text{Then O.K.})$$

It should be noted that the design force of member U_4 is close to that of this designed member U_5; therefore, we can use the same cross section for U_4.

4.3.3.14 Design of the Compression Upper Chord Member U_3

Following the same procedures adopted for the compression member U_5, we can design the compression upper chord member U_3, shown in Figure 4.68, carrying a compressive design force of -9773.9 kN. To classify the cross section chosen (see Figure 4.67),

$$\varepsilon = \sqrt{\frac{235}{f_y}} = \sqrt{\frac{235}{275}} = 0.924$$

$$b = 400\,\text{mm}, \quad t_\text{fl} = 30, \quad b/t_\text{fl} = 400/33 = 13.3 < 30.5 \,(\text{Flange is Class 1})$$

$$C = 396\,\text{mm}, \quad t_\text{w} = 20, \quad C/t_\text{fl} = 396/20 = 19.8 < 30.5 \,(\text{Web is class 1})$$

$$A = 55 \times 3.0 + 40 \times 3.0 + 2 \times 45 \times 2.0 = 465\,\text{cm}^2$$

$$e = \frac{55 \times 3.0 \times 24.0 - 40 \times 3.0 \times 18.6}{465} = 3.72\,\text{cm}$$

$$I_\text{m} = 2 \times 2.0 \times 45^3/12 + \left[55 \times 3.0^3/12 + 55 \times 3.6 \times 24.0^2\right]$$
$$+ \left[40 \times 3.0^3/12 + 40 \times 3.0 \times 18.6^2\right] = 167{,}143.95\,\text{cm}^4$$

$$I_x = 167{,}143.95 - 565 \times 3.72^2 = 160{,}709.1\,\text{cm}^4$$

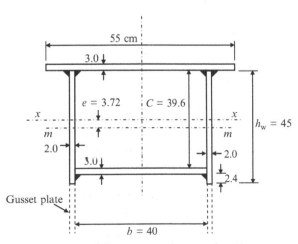

Figure 4.68 The cross section of the compression member U_3.

$$I_y = 3.0 \times 55^3/12 + 3.0 \times 40^3/12 + 2 \times \left[45 \times 2.0^3/12 + 45 \times 2.0 \times 21.0^2\right]$$
$$= 137{,}033.8 \text{ cm}^4$$

$$i_x = \sqrt{\frac{I_x}{A}} = \sqrt{\frac{160{,}709.1}{465}} = 18.59 \text{ cm}$$

$$i_y = \sqrt{\frac{I_y}{A}} = \sqrt{\frac{137{,}033.8}{465}} = 17.17 \text{ cm}$$

$$l_{bx} = l_{by} = 6000 \text{ mm}$$

$$\bar{\lambda} = \frac{L_{cr}}{i}\frac{1}{\lambda_1}$$

$$\lambda_1 = 93.9 \times 0.924 = 86.7636$$

$$\bar{\lambda} = \frac{6000}{171.7} \times \frac{1}{86.7636} = 0.4$$

The axial compressive force in the upper chord member U_3 ($N_{Ed} = 9773.9$ kN):

$$\frac{N_{Ed}}{N_{b,Rd}} \le 1.0$$

where $N_{b,Rd} = \dfrac{\chi A f_y}{\gamma_{M1}}$

$$\chi = \frac{1}{\Phi + \sqrt{\Phi^2 - \bar{\lambda}^2}} \quad \text{but } \chi \le 1.0$$

$$\Phi = 0.5\left[1 + \alpha(\bar{\lambda} - 0.2) + \bar{\lambda}^2\right]$$

$$\Phi = 0.5\left[1 + 0.34(0.4 - 0.2) + 0.4^2\right] = 0.614$$

$$\chi = \frac{1}{0.614 + \sqrt{0.614^2 - 0.4^2}} = 0.926 \quad \text{but } \chi \le 1.0$$

$$\text{Then, } N_{b,Rd} = \frac{0.926 \times 46{,}500 \times 275}{1.1} = 10{,}764{,}750 \text{ N}$$

$$N_{b,Rd} = 10{,}764.8 \text{ kN} > N_{Ed} = 9773.9 \text{ kN (Then O.K.)}$$

4.3.3.15 Design of the Compression Upper Chord Member U_2

The compression member U_2, shown in Figure 4.69, carrying a compressive design force of -7449.1 kN can be designed as follows. To classify the cross section chosen (see Figure 4.69),

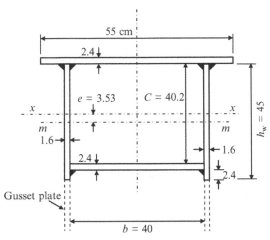

Figure 4.69 The cross section of the compression member U_2.

$$\varepsilon = \sqrt{\frac{235}{f_y}} = \sqrt{\frac{235}{275}} = 0.924$$

$$b = 400 \, \text{mm}, \quad t_{fl} = 24, \quad b/t_{fl} = 400/24 = 16.7 < 30.5 \, (\text{Flange is Class 1})$$

$$C = 402 \, \text{mm}, \quad t_w = 16, \quad C/t_{fl} = 402/16 = 25 < 30.5 \, (\text{Web is class 1})$$

$$A = 55 \times 2.4 + 40 \times 2.4 + 2 \times 45 \times 1.6 = 372 \, \text{cm}^2$$

$$e = \frac{55 \times 2.4 \times 23.7 - 40 \times 2.4 \times 18.9}{372} = 3.53 \, \text{cm}$$

$$I_m = 2 \times 1.6 \times 45^3/12 + \left[55 \times 2.4^3/12 + 55 \times 2.4 \times 23.7^2 \right]$$
$$+ \left[40 \times 2.4^3/12 + 40 \times 2.4 \times 18.9^2 \right] = 132,844.7 \, \text{cm}^4$$

$$I_x = 132,844.7 - 372 \times 3.53^2 = 128,209.2 \, \text{cm}^4$$

$$I_y = 2.4 \times 55^3/12 + 2.4 \times 40^3/12 + 2 \times \left[45 \times 1.6^3/12 + 45 \times 1.6 \times 20.8^2 \right]$$
$$= 108,405.9 \, \text{cm}^4$$

$$i_x = \sqrt{\frac{I_x}{A}} = \sqrt{\frac{128,209.2}{372}} = 18.56 \, \text{cm}$$

$$i_y = \sqrt{\frac{I_y}{A}} = \sqrt{\frac{108,405.9}{372}} = 17.07 \, \text{cm}$$

$$l_{bx} = l_{by} = 6000 \, \text{mm}$$

$$\bar{\lambda} = \frac{L_{cr}}{i} \frac{1}{\lambda_1}$$

$$\lambda_1 = 93.9 \times 0.924 = 86.7636$$

$$\bar{\lambda} = \frac{6000}{170.7} \times \frac{1}{86.7636} = 0.41$$

The axial compressive force in the upper chord member U_3 ($N_{Ed} = 7449.1$ kN):

$$\frac{N_{Ed}}{N_{b,Rd}} \leq 1.0$$

where $N_{b,Rd} = \dfrac{\chi A f_y}{\gamma_{M1}}$

$$\chi = \frac{1}{\varPhi + \sqrt{\varPhi^2 - \bar{\lambda}^2}} \quad \text{but } \chi \leq 1.0$$

$$\varPhi = 0.5 \left[1 + \alpha (\bar{\lambda} - 0.2) + \bar{\lambda}^2 \right]$$

$$\varPhi = 0.5 \left[1 + 0.34(0.41 - 0.2) + 0.41^2 \right] = 0.62$$

$$\chi = \frac{1}{0.62 + \sqrt{0.62^2 - 0.41^2}} = 0.922 \quad \text{but } \chi \leq 1.0$$

$$\text{Then, } N_{b,Rd} = \frac{0.922 \times 37200 \times 275}{1.1} = 8,574,600 \text{ N}$$

$$N_{b,Rd} = 8574.6 \text{ kN} > N_{Ed} = 7449.1 \text{ kN (Then O.K.)}$$

4.3.3.16 Design of the Compression Upper Chord Member U₁

The compression member U_1, shown in Figure 4.70, carrying a compressive design force of -4191.2 kN can be designed as follows. To classify the cross section chosen (see Figure 4.70),

$$\varepsilon = \sqrt{\frac{235}{f_y}} = \sqrt{\frac{235}{275}} = 0.924$$

$$b = 400 \text{ mm}, \quad t_{fl} = 12, \quad b/t_{fl} = 400/12 = 33.3 < 35.11 \text{ (Flange is Class 2)}$$

$$C = 414 \text{ mm}, \quad t_w = 12, \quad C/t_{fl} = 414/12 = 34.5 < 35.11 \text{ (Web is class 2)}$$

$$A = 55 \times 1.2 + 40 \times 1.2 + 2 \times 45 \times 1.2 = 222 \text{ cm}^2$$

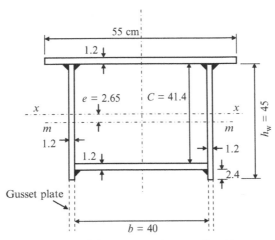

Figure 4.70 The cross section of the compression member U_1.

$$e = \frac{55 \times 1.2 \times 23.1 - 40 \times 1.2 \times 19.5}{222} = 2.65 \text{ cm}$$

$$I_m = 2 \times 1.2 \times 45^3/12 + \left[55 \times 1.2^3/12 + 55 \times 1.2 \times 23.1^2\right]$$
$$+ \left[40 \times 1.2^3/12 + 40 \times 1.2 \times 19.5^2\right] = 71{,}708.9 \text{ cm}^4$$

$$I_x = 71708.9 - 222 \times 2.65^2 = 70{,}149.9 \text{ cm}^4$$

$$I_y = 1.2 \times 55^3/12 + 1.2 \times 40^3/12 + 2 \times \left[45 \times 1.2^3/12 + 45 \times 1.2 \times 20.6^2\right]$$
$$= 68{,}881.3 \text{ cm}^4$$

$$i_x = \sqrt{\frac{I_x}{A}} = \sqrt{\frac{70149.9}{222}} = 17.78 \text{ cm}$$

$$i_y = \sqrt{\frac{I_y}{A}} = \sqrt{\frac{68881.3}{222}} = 17.61 \text{ cm}$$

$$l_{bx} = l_{by} = 6000 \text{ mm}$$

$$\bar{\lambda} = \frac{L_{cr}}{i}\frac{1}{\lambda_1}$$

$$\lambda_1 = 93.9 \times 0.924 = 86.7636$$

$$\bar{\lambda} = \frac{6000}{176.1} \times \frac{1}{86.7636} = 0.39$$

The axial compressive force in the upper chord member U_3 ($N_{Ed} = 4191.2$ kN):

$$\frac{N_{Ed}}{N_{b,Rd}} \leq 1.0$$

where $N_{b,Rd} = \dfrac{\chi A f_y}{\gamma_{M1}}$

$$\chi = \frac{1}{\Phi + \sqrt{\Phi^2 - \bar{\lambda}^2}} \quad \text{but } \chi \leq 1.0$$

$$\Phi = 0.5\left[1 + \alpha(\bar{\lambda} - 0.2) + \bar{\lambda}^2\right]$$

$$\Phi = 0.5\left[1 + 0.34(0.39 - 0.2) + 0.39^2\right] = 0.61$$

$$\chi = \frac{1}{0.61 + \sqrt{0.61^2 - 0.39^2}} = 0.928 \quad \text{but } \chi \leq 1.0$$

Then, $N_{b,Rd} = \dfrac{0.922 \times 22{,}200 \times 275}{1.1} = 5{,}150{,}400$ N

$$N_{b,Rd} = 5150.4\,\text{kN} > N_{Ed} = 4191.2\,\text{kN}\,(\text{Then O.K.})$$

4.3.3.17 Design of the Compression Vertical Member V_5

Let us now design the compression vertical member V_5, shown in Figure 4.71, carrying a compressive design force of -1302.9 kN. To assume a reasonable cross section for the compression vertical member, the following parameters can be considered:

$$d_1 = \frac{L}{15 - 22} = \frac{7500}{15 - 22} = 500 - 341\,\text{mm, taken as 350 mm.}$$

It should be noted that the vertical member must be inside the gusset plates spaced at a constant distance (b) of 400 mm. Let us start by assuming the flange thickness of 14 mm and web thickness of 10 mm. To classify the cross section chosen (see Figure 4.71),

$$\varepsilon = \sqrt{\frac{235}{f_y}} = \sqrt{\frac{235}{275}} = 0.924$$

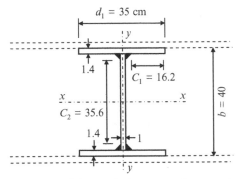

Figure 4.71 The cross section of the vertical compression member V_5.

$C_1 = 162\,\text{mm}, \quad t_{fl} = 14, \quad C_1/t_{fl} = 162/14 = 11.57 < 14 \times 0.924$
$= 12.9\,(\text{Flange is Class 3})$

$C_2 = 356\,\text{mm}, \quad t_w = 10, \quad C_2/t_{fl} = 356/10 = 35.6 < 33 \times 0.924$
$= 30.5\,(\text{Web is class 1})$

$$A = 2 \times 35 \times 1.4 + 37.2 \times 1.0 = 135.2\,\text{cm}^2$$

$$I_x = 1.0 \times 37.2^3/12 + 2 \times \left[35 \times 1.4^3/12 + 35 \times 1.4 \times 19.3^2\right] = 40{,}810\,\text{cm}^4$$

$$I_y = 37.2 \times 1^3/12 + 2 \times 1.4 \times 35^3/12 = 10{,}007.3\,\text{cm}^4$$

$$i_x = \sqrt{\frac{I_x}{A}} = \sqrt{\frac{40{,}810}{135.2}} = 17.37\,\text{cm}$$

$$i_y = \sqrt{\frac{I_y}{A}} = \sqrt{\frac{10{,}007.3}{135.2}} = 8.6\,\text{cm}$$

$$l_{by} = 6750\,\text{mm}$$

$$l_{bx} = 7500\,\text{mm}$$

$$\bar{\lambda} = \frac{L_{cr}}{i}\frac{1}{\lambda_1}$$

$$\lambda_1 = 93.9 \times 0.924 = 86.7636$$

$$\bar{\lambda} = \frac{6750}{86} \times \frac{1}{86.7636} = 0.905$$

where $N_{b,Rd} = \dfrac{\chi A f_y}{\gamma_{M1}}$

The axial compressive force in the vertical member V_5 ($N_{Ed} = 1302.9$ kN):

$$\frac{N_{Ed}}{N_{b,Rd}} \leq 1.0$$

$$\chi = \frac{1}{\Phi + \sqrt{\Phi^2 - \bar{\lambda}^2}} \quad \text{but } \chi \leq 1.0$$

$$\Phi = 0.5\left[1 + \alpha(\bar{\lambda} - 0.2) + \bar{\lambda}^2\right]$$

$$\Phi = 0.5\left[1 + 0.49(0.905 - 0.2) + 0.905^2\right] = 1.082$$

$$\chi = \frac{1}{1.082 + \sqrt{1.082^2 - 0.905^2}} = 0.597 \quad \text{but } \chi \leq 1.0$$

$$\text{Then, } N_{b,Rd} = \frac{0.597 \times 13520 \times 275}{1.1} = 2,017,860 \, \text{N}$$

$$N_{b,Rd} = 2017.9 \text{ kN} > N_{Ed} = 1302.9 \, \text{kN (Then O.K.)}$$

4.3.3.18 Design of the Compression Vertical Member V_4

Following the same procedures adopted for the design the compression vertical member V_5, we can design the vertical compression member, shown in Figure 4.72, carrying a compressive design force of -2226.1 kN, as follows. To classify the cross section chosen (see Figure 4.72),

$$\varepsilon = \sqrt{\frac{235}{f_y}} = \sqrt{\frac{235}{275}} = 0.924$$

$C_1 = 352$ mm, $t_{fl} = 12$, $C_1/t_{fl} = 352/12 = 29.3 < 30.5$ (Flange is Class 1)

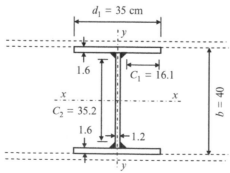

Figure 4.72 The cross section of the vertical compression member V_4.

$$C_2 = 163 \, \text{mm}, \quad t_w = 16, \quad C_2/t_{fl} = 163/16 = 10.2 < 30.5 \ (\text{Web is class 1})$$

$$A = 2 \times 35 \times 1.6 + 36.8 \times 1.2 = 156.2 \, \text{cm}^2$$

$$I_x = 1.2 \times 36.8^3/12 + 2 \times \left[35 \times 1.6^3/12 + 35 \times 1.6 \times 19.2^2\right]$$
$$= 46{,}295.2 \, \text{cm}^4$$

$$I_y = 36.8 \times 1.2^3/12 + 2 \times 1.6 \times 35^3/12 = 11{,}438.6 \, \text{cm}^4$$

$$i_x = \sqrt{\frac{I_x}{A}} = \sqrt{\frac{46295.2}{156.2}} = 17.22 \, \text{cm}$$

$$i_y = \sqrt{\frac{I_y}{A}} = \sqrt{\frac{11438.6}{156.2}} = 8.56 \, \text{cm}$$

$$l_{by} = 6750 \, \text{mm}$$

$$l_{bx} = 7500 \, \text{mm}$$

$$\bar{\lambda} = \frac{L_{cr}}{i} \frac{1}{\lambda_1}$$

$$\lambda_1 = 93.9 \times 0.924 = 86.7636$$

$$\bar{\lambda} = \frac{6750}{85.6} \times \frac{1}{86.7636} = 0.91$$

The axial compressive force in the vertical member V_5 ($N_{Ed} = 2226.1$ kN):

$$\frac{N_{Ed}}{N_{b,Rd}} \leq 1.0$$

where $N_{b,Rd} = \dfrac{\chi A f_y}{\gamma_{M1}}$

$$\chi = \frac{1}{\varPhi + \sqrt{\varPhi^2 - \bar{\lambda}^2}} \quad \text{but } \chi \leq 1.0$$

$$\varPhi = 0.5 \left[1 + \alpha(\bar{\lambda} - 0.2) + \bar{\lambda}^2\right]$$

$$\varPhi = 0.5 \left[1 + 0.49(0.91 - 0.2) + 0.91^2\right] = 1.088$$

$$\chi = \frac{1}{1.088 + \sqrt{1.088^2 - 0.91^2}} = 0.594 \quad \text{but } \chi \leq 1.0$$

Then, $N_{b,Rd} = \dfrac{0.594 \times 15620 \times 275}{1.1} = 2{,}319{,}570 \, \text{N}$

$$N_{b,Rd} = 2319.6 \, \text{kN} > N_{Ed} = 2226.1 \, \text{kN} \ (\text{Then O.K.})$$

4.3.3.19 Design of the Compression Vertical Member V₃

The compression vertical member V_3, shown in Figure 4.73, carrying a compressive design force of -3190.7 kN can be designed as follows. To classify the cross section chosen (see Figure 4.73),

$$\varepsilon = \sqrt{\frac{235}{f_y}} = \sqrt{\frac{235}{275}} = 0.924$$

$C_1 = 336\,\text{mm}, \quad t_{fl} = 16, \quad C_1/t_{fl} = 336/16 = 21.0 < 30.5\,(\text{Flange is Class 1})$

$C_2 = 159\,\text{mm}, \quad t_w = 24, \quad C_2/t_{fl} = 159/24 = 6.6 < 30.5\,(\text{Web is class 1})$

$$A = 2 \times 35 \times 2.4 + 35.2 \times 1.6 = 224.32\,\text{cm}^2$$

$$I_x = 1.6 \times 35.2^3/12 + 2 \times \left[35 \times 2.4^3/12 + 35 \times 2.4 \times 18.8^2\right]$$
$$= 65{,}273.8\,\text{cm}^4$$

$$I_y = 35.2 \times 1.6^3/12 + 2 \times 2.4 \times 35^3/12 = 17{,}162\,\text{cm}^4$$

$$i_x = \sqrt{\frac{I_x}{A}} = \sqrt{\frac{65{,}273.8}{224.32}} = 17.06\,\text{cm}$$

$$i_y = \sqrt{\frac{I_y}{A}} = \sqrt{\frac{17162}{224.32}} = 8.75\,\text{cm}$$

$$l_{by} = 6750\,\text{mm}$$

$$l_{bx} = 7500\,\text{mm}$$

$$\bar{\lambda} = \frac{L_{cr}}{i} \frac{1}{\lambda_1}$$

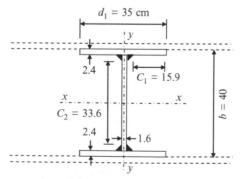

Figure 4.73 The cross section of the vertical compression member V_3.

$$\lambda_1 = 93.9 \times 0.924 = 86.7636$$

$$\bar{\lambda} = \frac{6750}{87.5} \times \frac{1}{86.7636} = 0.89$$

The axial compressive force in the vertical member V_5 ($N_{Ed} = 3190.7$ kN):

$$\frac{N_{Ed}}{N_{b,Rd}} \leq 1.0$$

where $N_{b,Rd} = \dfrac{\chi A f_y}{\gamma_{M1}}$

$$\chi = \frac{1}{\Phi + \sqrt{\Phi^2 - \bar{\lambda}^2}} \text{ but } \chi \leq 1.0$$

$$\Phi = 0.5 \left[1 + \alpha(\bar{\lambda} - 0.2) + \bar{\lambda}^2 \right]$$

$$\Phi = 0.5 \left[1 + 0.49(0.89 - 0.2) + 0.89^2 \right] = 1.065$$

$$\chi = \frac{1}{1.065 + \sqrt{1.065^2 - 0.89^2}} = 0.606 \text{ but } \chi \leq 1.0$$

Then, $N_{b,Rd} = \dfrac{0.606 \times 22{,}432 \times 275}{1.1} = 3{,}398{,}448$ N

$$N_{b,Rd} = 3398.4 \text{ kN} > N_{Ed} = 3190.7 \text{ kN (Then O.K.)}$$

4.3.3.20 Design of the Compression Vertical Member V_2

The compression vertical member V_2, shown in Figure 4.74, carrying a compressive design force of -4195.1 kN can be designed as follows. To classify the cross section chosen (see Figure 4.74),

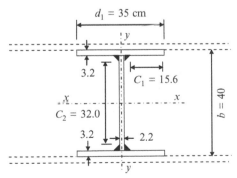

Figure 4.74 The cross section of the vertical compression member V_2.

$$\varepsilon = \sqrt{\frac{235}{f_y}} = \sqrt{\frac{235}{275}} = 0.924$$

$C_1 = 156 \, \text{mm}, \quad t_{fl} = 32, \quad C_1/t_{fl} = 156/16 = 4.9 < 30.5 \, \text{(Flange is Class 1)}$

$C_2 = 320 \, \text{mm}, \quad t_w = 22, \quad C_2/t_{fl} = 320/22 = 14.5 < 30.5 \, \text{(Web is class 1)}$

$$A = 2 \times 35 \times 3.2 + 33.6 \times 2.2 = 297.92 \, \text{cm}^2$$

$$I_x = 2.2 \times 33.6^3/12 + 2 \times \left[35 \times 3.2^3/12 + 35 \times 3.2 \times 18.4^2 \right] = 82{,}983 \, \text{cm}^4$$

$$I_y = 33.6 \times 2.2^3/12 + 2 \times 3.2 \times 35^3/12 = 22{,}896.5 \, \text{cm}^4$$

$$i_x = \sqrt{\frac{I_x}{A}} = \sqrt{\frac{82{,}983}{297.92}} = 16.69 \, \text{cm}$$

$$i_y = \sqrt{\frac{I_y}{A}} = \sqrt{\frac{22{,}896.5}{297.92}} = 8.77 \, \text{cm}$$

$$l_{by} = 6750 \, \text{mm}$$

$$l_{bx} = 7500 \, \text{mm}$$

$$\bar{\lambda} = \frac{L_{cr}}{i} \frac{1}{\lambda_1}$$

$$\lambda_1 = 93.9 \times 0.924 = 86.7636$$

$$\bar{\lambda} = \frac{6750}{87.7} \times \frac{1}{86.7636} = 0.89$$

The axial compressive force in the vertical member V_5 ($N_{Ed} = 4195.1 \, \text{kN}$):

$$\frac{N_{Ed}}{N_{b,Rd}} \leq 1.0$$

where $N_{b,Rd} = \dfrac{\chi A f_y}{\gamma_{M1}}$

$$\chi = \frac{1}{\Phi + \sqrt{\Phi^2 - \bar{\lambda}^2}} \quad \text{but } \chi \leq 1.0$$

$$\Phi = 0.5 \left[1 + \alpha(\bar{\lambda} - 0.2) + \bar{\lambda}^2 \right]$$

$$\Phi = 0.5 \left[1 + 0.49(0.89 - 0.2) + 0.89^2 \right] = 1.065$$

$$\chi = \frac{1}{1.065 + \sqrt{1.065^2 - 0.89^2}} = 0.606 \quad \text{but } \chi \leq 1.0$$

$$\text{Then, } N_{b,Rd} = \frac{0.606 \times 29,792 \times 275}{1.1} = 4,513,488 \, N$$

$$N_{b,Rd} = 4513.5 \, kN > N_{Ed} = 4195.1 \, kN \, (\text{Then O.K.})$$

4.3.3.21 Design of the Compression Vertical Member V_1

Let us now design the first vertical member V_1 shown in Figure 4.66. This member is of great importance for this through bridge since it carries not only the high compressive force coming from the supports but also a bending moment coming from the analysis of the end portal frame. The end portal frame is necessary for this through bridge to transfer wind load coming from the upper wind bracing to the bearings. The reactions coming from the upper wind bracing cannot be resisted by cross bracing since it will cause an obstacle to passing traffic. Therefore, these reactions can be transferred as shown in Figure 4.75 by an end frame action. The end frame consists of the first vertical members of the main trusses, the edge cross girder, and the edge member of the upper wind bracing, which has to be an I-shaped beam section. The end portal fame can be analyzed as shown in Figure 4.75 by assuming hinges at a distance varying from 1/3 to 1/2 of the depth of the frame. The first vertical member V_1 carries the compressive design force coming from the analysis of truss under vertical dead and live loads, which is equal to -5240.8 kN in addition to an added compressive force of -5240.8 kN and a bending moment of 2178.3 kN m coming from the analysis of the end portal frame. Let us assume the cross section shown in Figure 4.76, which is a compact class 1 cross section. The resistance to bending moment can be calculated as follows:

$$M_{c,Rd} = \frac{W_{pl} \times f_y}{\gamma_{M0}} \quad \text{for classes 1 and 2}$$

$$W_{pl} = 55 \times 30^2/4 - 2 \times 2.5 \times 30^2/4 - 40 \times 30^2/4 = 11,875 \, cm^3$$

$$M_{c,Rd} = \frac{W_{pl} \times f_y}{\gamma_{M0}} = \frac{11875 \times 10^3 \times 275}{1.0 \times 10^6} = 3265.6 \, kN \, m > 2178.3 \, kN \, m$$

On the other hand, the resistance to compressive forces can be calculated as follows:

$$A = 2 \times 55 \times 5 + 2 \times 30 \times 5 = 850 \, cm^2$$

$$I_x = 2 \times 5 \times 30^3/12 + 2 \times \left[55 \times 5^3/12 + 55 \times 5 \times 17.5^2\right] = 192,083.3 \, cm^4$$

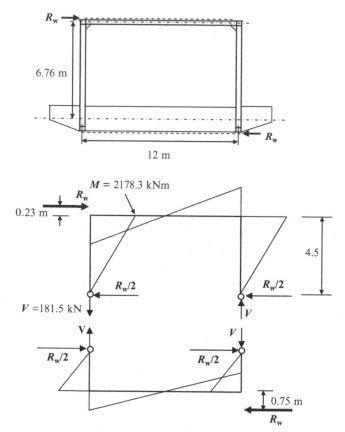

Figure 4.75 Analysis of forces on the end portal frame for the evaluation of axial force and bending moment on the vertical member V₁.

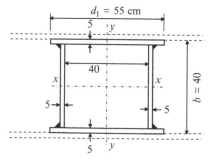

Figure 4.76 The cross section of the vertical compression member V₁.

$$I_y = 2 \times 5 \times 55^3/12 + 2 \times \left[30 \times 5^3/12 + 30 \times 5 \times 22.5^2\right] = 291,145.8 \, \text{cm}^4$$

$$i_x = \sqrt{\frac{I_x}{A}} = \sqrt{\frac{192,083.3}{850}} = 15.03 \, \text{cm}$$

$$i_y = \sqrt{\frac{I_y}{A}} = \sqrt{\frac{291,145.8}{850}} = 18.51 \, \text{cm}$$

$$l_{by} = 6750 \, \text{mm}$$

$$l_{bx} = 7500 \, \text{mm}$$

$$\bar{\lambda} = \frac{L_{cr}}{i} \frac{1}{\lambda_1}$$

$$\lambda_1 = 93.9 \times 0.924 = 86.7636$$

$$\bar{\lambda} = \frac{7500}{150.3} \times \frac{1}{86.7636} = 0.58$$

The axial compressive force in the vertical member V_5 ($N_{Ed} = 5422.3 \, \text{kN}$):

$$\frac{N_{Ed}}{N_{b,Rd}} \leq 1.0$$

where $N_{b,Rd} = \dfrac{\chi A f_y}{\gamma_{M1}}$

$$\chi = \frac{1}{\Phi + \sqrt{\Phi^2 - \bar{\lambda}^2}} \quad \text{but } \chi \leq 1.0$$

$$\Phi = 0.5\left[1 + \alpha(\bar{\lambda} - 0.2) + \bar{\lambda}^2\right]$$

$$\Phi = 0.5\left[1 + 0.49(0.58 - 0.2) + 0.58^2\right] = 0.733$$

$$\chi = \frac{1}{0.733 + \sqrt{0.733^2 - 0.58^2}} = 0.847 \quad \text{but } \chi \leq 1.0$$

Then, $N_{b,Rd} = \dfrac{0.847 \times 85,000 \times 275}{1.1} = 17,998,750 \, \text{N}$

$$N_{b,Rd} = 17,998.75 \, \text{kN} > N_{Ed} = 4195.1 \, \text{kN} \, (\text{Then O.K.})$$

Check of combined axial compression and bending moment can be done using the conservative interaction formula given in EC3 [1.27] as follows:

$$\frac{N_{Ed}}{N_{b,Rd}} + \frac{M_{Ed}}{M_{c,Rd}} = \frac{5422.3}{17,998.75} + \frac{2178.3}{3265.6} = 0.301 + 0.677 = 0.968$$

$$< 1.0 \, (\text{Then O.K.})$$

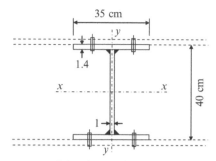

Figure 4.77 The cross section of the diagonal tension members D_5 and D_4.

4.3.3.22 Design of the Diagonal Member D_5

We can also design the diagonal member D_5, shown in Figure 4.77, carrying a maximum tensile design force of 1668.5 kN and a minimum compression force of −526.5 kN. We can use the same cross section used for V_5 and check the safety of the member against the tensile and compressive forces as follows.

Design as a Tension Member

The bolts used in connecting the member with gusset plates are M27 high-strength pretensioned bolts having a clearance of 3 mm (hole diameter $\varnothing = 30$ mm):

$$A = 35 \times 1.4 \times 2 + 37.2 \times 1 = 135.2\,\text{cm}^2$$

$$A_{\text{net}} = 135.2 - 4 \times 3.0 \times 1.4 = 118.4\,\text{cm}^2$$

$$N_{\text{pl,Rd}} = \frac{Af_y}{\gamma_{M0}} = \frac{135.2 \times 275 \times 100}{1.0} = 3{,}718{,}000\,\text{N} = 3718.0\,\text{kN} > N_{\text{Ed}}$$
$$= 1668.5\,\text{kN}$$

$$N_{\text{u,Rd}} = \frac{0.9 A_{\text{net}} f_u}{\gamma_{M2}} = \frac{0.9 \times 11{,}840 \times 430}{1.25} = 3{,}665{,}664\,\text{N} = 3665.7\,\text{kN}$$
$$> N_{\text{Ed}} = 1668.5\,\text{kN}$$

Design as a Compression Member

$$l_{bx} = 9600\,\text{mm}$$

$$l_{by} = 8640\,\text{mm}$$

$$\bar{\lambda} = \frac{L_{cr}}{i} \frac{1}{\lambda_1}$$

$$\lambda_1 = 93.9 \times 0.924 = 86.7636$$

$$\bar{\lambda} = \frac{8640}{86} \times \frac{1}{86.7636} = 1.158$$

The axial compressive force in the diagonal member D_5 ($N_{Ed} = 526.5$ kN):

$$\frac{N_{Ed}}{N_{b,Rd}} \leq 1.0$$

where $N_{b,Rd} = \dfrac{\chi A f_y}{\gamma_{M1}}$

$$\chi = \frac{1}{\Phi + \sqrt{\Phi^2 - \bar{\lambda}^2}} \quad \text{but } \chi \leq 1.0$$

$$\Phi = 0.5 \left[1 + \alpha(\bar{\lambda} - 0.2) + \bar{\lambda}^2 \right]$$

$$\Phi = 0.5 \left[1 + 0.49(1.158 - 0.2) + 1.158^2 \right] = 1.405$$

$$\chi = \frac{1}{1.405 + \sqrt{1.405^2 - 1.158^2}} = 0.454 \quad \text{but } \chi \leq 1.0$$

$$\text{Then, } N_{b,Rd} = \frac{0.454 \times 13,520 \times 275}{1.1} = 1,534,520 \, \text{N}$$

$$N_{b,Rd} = 1534.5 \, \text{kN} > N_{Ed} = 526.5 \, \text{kN (Then O.K.)}$$

It should be noted that the same cross section used for member D_5 can also be used for the diagonal tension member D_4.

4.3.3.23 Design of the Diagonal Tension Member D_3

The diagonal member D_3, shown in Figure 4.78, carrying a maximum tensile design force of 4086.1 kN can be designed adopting the same procedures used with D_5 as follows:

The bolts used in connecting the member with gusset plates are M27 high-strength pretensioned bolts having a clearance of 3 mm (hole diameter $\varnothing = 30$ mm):

$$A = 2 \times 35 \times 2 + 36 \times 1.2 = 183.2 \, \text{cm}^2$$

$$A_{net} = 183.2 - 8 \times 3.0 \times 2 = 135.2 \, \text{cm}^2$$

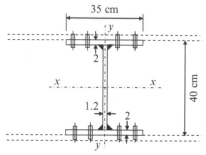

Figure 4.78 The cross section of the diagonal tension member D_3.

$$N_{pl,Rd} = \frac{Af_y}{\gamma_{M0}} = \frac{183.2 \times 275 \times 100}{1.0} = 5,038,000\,N = 5038\,kN > N_{Ed}$$
$$= 4086.1\,kN$$

$$N_{u,Rd} = \frac{0.9A_{net}f_u}{\gamma_{M2}} = \frac{0.9 \times 135.2 \times 430}{1.25} = 4,185,792\,N$$
$$= 4185.8\,kN > N_{Ed} = 4086.1\,kN$$

4.3.3.24 Design of the Diagonal Tension Member D₂

The diagonal member D_2, shown in Figure 4.79, carrying a maximum tensile design force of 5372.3 kN can be designed as follows:

The bolts used in connecting the member with gusset plates are M27 high-strength pretensioned bolts having a clearance of 3 mm (hole diameter $\varnothing = 30$ mm):

$$A = 2 \times 35 \times 2.8 + 34.4 \times 1.8 = 257.92\,cm^2$$

$$A_{net} = 257.92 - 8 \times 3.0 \times 2 = 190.72\,cm^2$$

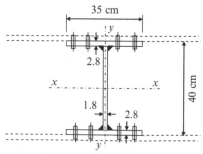

Figure 4.79 The cross section of the diagonal tension member D_2.

$$N_{pl,Rd} = \frac{Af_y}{\gamma_{M0}} = \frac{257.92 \times 275 \times 100}{1.0} = 7,092,800 \text{ N} = 7092.8 \text{ kN} > N_{Ed}$$
$$= 5372.3 \text{ kN}$$

$$N_{u,Rd} = \frac{0.9A_{net}f_u}{\gamma_{M2}} = \frac{0.9 \times 190.72 \times 100 \times 430}{1.25} = 5,904,691 \text{ N}$$
$$= 5904.7 \text{ kN} > N_{Ed} = 5372.3 \text{ kN}$$

4.3.3.25 Design of the Diagonal Tension Member D_1

The diagonal member D_1, shown in Figure 4.80, carrying a maximum tensile design force of 6711.5 kN can be designed as follows:

The bolts used in connecting the member with gusset plates are M27 high-strength pretensioned bolts having a clearance of 3 mm (hole diameter $\varnothing = 30$ mm):

$$A = 2 \times 35 \times 3.2 + 33.6 \times 2.4 = 304.64 \text{ cm}^2$$

$$A_{net} = 304.64 - 8 \times 3.0 \times 2 = 227.84 \text{ cm}^2$$

$$N_{pl,Rd} = \frac{Af_y}{\gamma_{M0}} = \frac{304.64 \times 275 \times 100}{1.0} = 8,377,600 \text{ N} = 8377.6 \text{ kN} > N_{Ed}$$
$$= 6711.5 \text{ kN}$$

$$N_{u,Rd} = \frac{0.9A_{net}f_u}{\gamma_{M2}} = \frac{0.9 \times 227.84 \times 100 \times 430}{1.25} = 7,053,926 \text{ N} = 7053.9 \text{ kN}$$
$$> N_{Ed} = 7053.9 \text{ kN}$$

4.3.3.26 Design of the Lower Chord Member L_5

Let us now design the tensile lower chord member L_5, shown in Figure 4.81, carrying a tensile design force of 11,165.7 kN. To assume a reasonable cross

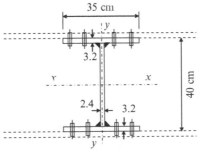

Figure 4.80 The cross section of the diagonal tension member D_1.

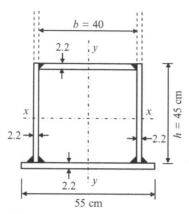

Figure 4.81 The cross section of the lower chord tension member L_5.

section for the lower chord tension members, the following parameters can be considered:

$$h = \frac{L}{15-30} = \frac{6000}{12-30} = 500 - 200 \text{ mm, taken as 450 mm.}$$

Once again, it should be noted that the gusset plates must be spaced at a constant distance (b) of 400 mm. Let us start by assuming the flange and web thicknesses of 22 mm (see Figure 4.71). It should also be noted that the gross and net cross-sectional areas of the lower chord members are the same since they are connected using butt weld. The design of section can be performed adopting the same procedures used for the diagonal tension members as follows:

$$A = A_{net} = 55 \times 2.2 + 40 \times 2.2 + 2 \times 45 \times 2.2 = 407 \text{ cm}^2$$

$$N_{pl,Rd} = \frac{Af_y}{\gamma_{M0}} = \frac{407 \times 275 \times 100}{1.0} = 11{,}192{,}500 \text{ N} = 11{,}192.5 \text{ kN} > N_{Ed}$$
$$= 11{,}165.7 \text{ kN}$$

$$N_{u,Rd} = \frac{0.9A_{net}f_u}{\gamma_{M2}} = \frac{0.9 \times 407 \times 100 \times 430}{1.25} = 12{,}600{,}720 \text{ N} = 12{,}600.7 \text{ kN}$$
$$> N_{Ed} = 11{,}165.7 \text{ kN}$$

4.3.3.27 Design of the Lower Chord Member L_4

Following the same procedures adopted for the design of the lower chord member L_5, we can design the tensile lower chord member L_4, shown in

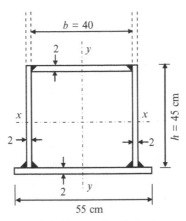

Figure 4.82 The cross section of the lower chord tension member L_4.

Figure 4.82, carrying a tensile design force of 9773.9 kN. The design of section can be as follows:

$$A = A_{net} = 55 \times 2.0 + 40 \times 2.0 + 2 \times 45 \times 2.0 = 370 \, cm^2$$

$$N_{pl,Rd} = \frac{A f_y}{\gamma_{M0}} = \frac{370 \times 275 \times 100}{1.0} = 10{,}175{,}000 \, N = 1017.5 \, kN > N_{Ed}$$
$$= 9773.9 \, kN$$

$$N_{u,Rd} = \frac{0.9 A_{net} f_u}{\gamma_{M2}} = \frac{0.9 \times 370 \times 100 \times 430}{1.25} = 11{,}455{,}200 \, N = 11{,}455.2 \, kN$$
$$> N_{Ed} = 9773.9 \, kN$$

4.3.3.28 Design of the Lower Chord Member L_3

The lower chord member L_3, shown in Figure 4.83, carrying a tensile design force of 7449.1 kN can be designed as follows:

$$A = A_{net} = 55 \times 1.6 + 40 \times 1.6 + 2 \times 45 \times 1.6 = 296 \, cm^2$$

$$N_{pl,Rd} = \frac{A f_y}{\gamma_{M0}} = \frac{296 \times 100 \times 275}{1.0} = 8{,}140{,}000 \, N = 8140 \, kN > N_{Ed}$$
$$= 7449.1 \, kN$$

$$N_{u,Rd} = \frac{0.9 A_{net} f_u}{\gamma_{M2}} = \frac{0.9 \times 296 \times 100 \times 430}{1.25} = 9{,}164{,}160 \, N$$
$$= 9164.2 \, kN > N_{Ed} = 7449.1 \, kN$$

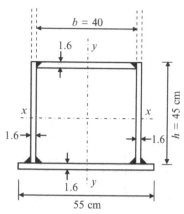

Figure 4.83 The cross section of the lower chord tension member L_3.

4.3.3.29 Design of the Lower Chord Member L_2

The lower chord member L_2, shown in Figure 4.84, carrying a tensile design force of 4191.2 kN can be designed as follows:

$$A = A_{net} = 55 \times 1.0 + 40 \times 1.0 + 2 \times 45 \times 1.0 = 185 \text{ cm}^2$$

$$N_{pl,Rd} = \frac{A f_y}{\gamma_{M0}} = \frac{185 \times 100 \times 275}{1.0} = 5,087,500 \text{ N} = 5087.5 \text{ kN} > N_{Ed}$$
$$= 4191.2 \text{ kN}$$

$$N_{u,Rd} = \frac{0.9 A_{net} f_u}{\gamma_{M2}} = \frac{0.9 \times 185 \times 100 \times 430}{1.25} = 5,727,600 \text{ N} = 5727.6 \text{ kN}$$
$$> N_{Ed} = 4191.2 \text{ kN}$$

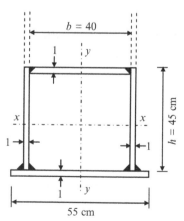

Figure 4.84 The cross section of the lower chord tension members L_2 and L_1.

It should be noted that the thickness used (1 cm) is the minimum thickness that can be used in bridges; therefore, this section will be used also for the zero member—under dead and live load cases of loading, lower chord member L_1.

4.3.3.30 Design of Stringer-Cross Girder Connection

The stringer is designed as a simply supported beam on cross girders; therefore, the connection is mainly transferring shear forces (maximum reaction from stringers of 604.5 kN) (see Figure 4.85). Using M27 high-strength pretensioned bolts of grade 8.8, having f_{ub} of 800 MPa, shear area A of 4.59 cm², and gross area A_g of 5.73 cm², we can determine the required number of bolts, following the rules specified in EC3 (BS EN 1993-1-8) [2.13], as follows:

$$F_{v,Rd} = \frac{\alpha_V f_{ub} A}{\gamma_{M2}}$$

$$F_{v,Rd} = \frac{0.6 \times 800 \times 459}{1.25} = 176,256\,N$$

Then, $F_{v,Rd}$ equals 176 kN (for bolts in single shear) and 353 kN (for bolts in double shear):

$$F_{s,Rd} = \frac{k_s n \mu}{\gamma_{M3}} F_{p,C}$$

$$F_{p,C} = 0.7 f_{ub} A_s = 0.7 \times 800 \times 573 = 320,880\,N$$

$$F_{s,Rd,ser} = \frac{1.0 \times 1.0 \times 0.4}{1.1} 320,880 = 116,683.6\,N.$$

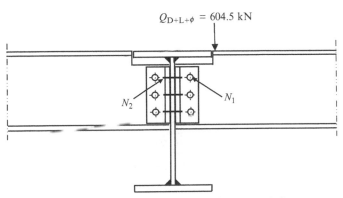

$Q_{D+L+\phi} = 604.5$ kN

N_2 N_1

Figure 4.85 The connection between a stringer and a cross girder.

Then, $F_{s,Rd}=117$ kN (for bolts in single shear at serviceability limit states) and $F_{s,Rd}=234$ kN (for bolts in double shear at serviceability limit states). At ultimate limit states, $F_{s,Rd,ult}$ can be calculated as follows:

$$F_{s,Rd,ult} = \frac{1.0 \times 1.0 \times 0.4}{1.25} 320{,}880 = 102{,}682 \, N.$$

Then, $F_{s,Rd}=103$ kN (for bolts in single shear at ultimate limit states) and $F_{s,Rd}=206$ kN (for bolts in double shear at ultimate limit states):

$$N_1 = \frac{604.5}{206} = 2.9 \text{ taken as 3 bolts,}$$

$$N_2 = \frac{604.5}{103} = 5.9 \text{ taken as 6 bolts}$$

4.3.3.31 Design of Cross Girder-Main Truss Connection

The cross girder is designed as a simply supported beam on main trusses; therefore, once again, the connection is mainly transferring shear forces (maximum reaction from cross girders of 1600.8 kN) (see Figure 4.86). We can determine the required number of bolts as follows:

$$N_3 = \frac{1600.8}{206} = 7.8 \text{ taken as 8 bolts,}$$

$$N_2 = \frac{1600.8}{103} = 15.5 \text{ taken as 16 bolts}$$

4.3.3.32 Design of Wind Bracings

Wind forces acting on the investigated through highway bridge (see Figure 4.87) as well as any other lateral forces directly applied to the bridge

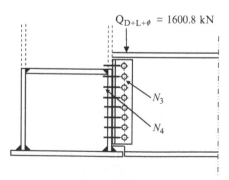

Figure 4.86 The connection between a cross girder and the main truss.

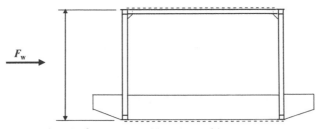

Figure 4.87 Design height for upper and lower wind bracings.

are transmitted to the bearings by systems of upper and lower wind bracings as well as end portal frames. The upper and lower wind bracings carry wind forces on the main truss as shown in Figure 4.87. Wind bracings are quite important to the lateral stability of the upper chord compression members since they define the buckling outside the plane of the truss, and therefore, wind forces applied to this bridge can be sufficiently estimated using the design rules specified in EC1 [3.2] as follows:

$$F_w = \frac{1}{2}\rho v_b^2 C A_{\text{ref},x}$$

$$v_b = c_{\text{dir}} \times c_{\text{season}} \times v_{b,0} = 1.0 \times 1.0 \times 26 = 26 \, \text{m/s}$$

$$A_{\text{ref},x} = 7.5 \times 60 = 450 \, \text{m}^2$$

$$F_w = \frac{1}{2} \times 1.25 \times 26^2 \times 5.7 \times 450 = 1{,}083{,}712.5 \, \text{N} = 1083.7 \, \text{kN}$$

Considering the structural analysis for the upper wind bracing system shown in Figure 4.88, the critical design wind force in the diagonal bracing members can be calculated as follows:

Distributed wind loads$(q_{\text{WL}}) = 1083.7 \times 0.5/60 = 9.03 \, \text{kN/m}$

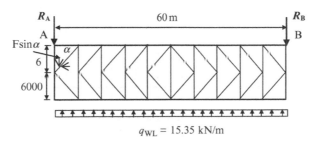

$q_{\text{WL}} = 15.35 \, \text{kN/m}$

Figure 4.88 Loads on the upper wind bracing.

Factored distributed wind loads $= q_{WL} \times \gamma_q = 9.03 \times 1.7 = 15.351 \, kN/m$

$$R_A = 15.351 \times 60/2 = 460.53 \, kN$$

$$\alpha = \tan^{-1}(6/6) = 45°$$

$$F_D = 460.53/(2 \times \sin 45) = 325.6 \, kN$$

The cross section of the bracing member (see Figure 4.28) can be determined as follows:

$$l_{bx} = 8490 \, mm, \quad l_{by} = 1.2 \times 8490 = 10,190 \, mm$$

Choose two angles back-to-back $150 \times 150 \times 15$, with 10 mm gusset plate between them (Figure 4.89):

$$A = 2 \times 43.2 = 86.4 \, cm^2, \quad i_x = 4.59 \, cm, \quad e = 4.26 \, cm,$$

$$i_y = \sqrt{4.59^2 + (4.26 + 1/2)^2} = 6.61 \, cm$$

$$\varepsilon = \sqrt{\frac{235}{275}} = 0.924$$

$$\bar{\lambda} = \frac{L_{cr}}{i} \frac{1}{\lambda_1}$$

$$\lambda_1 = 93.9 \times 0.924 = 86.7636$$

$$\bar{\lambda} = \frac{8490}{45.9} \frac{1}{86.7636} = 2.13$$

The axial compressive force in the diagonal bracing member ($N_{Ed} = 325.6 \, kN$):

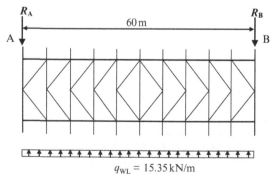

$$q_{WL} = 15.35 \, kN/m$$

Figure 4.89 Loads on the lower wind bracing.

$$\frac{N_{Ed}}{N_{b,Rd}} \leq 1.0$$

where $N_{b,Rd} = \dfrac{\chi A f_y}{\gamma_{M1}}$

$$A = 2 \times 43.2 = 86.4\,\text{cm}^2$$

$$\chi = \frac{1}{\Phi + \sqrt{\Phi^2 - \bar{\lambda}^2}} \quad \text{but } \chi \leq 1.0$$

$$\Phi = 0.5\left[1 + \alpha(\bar{\lambda} - 0.2) + \bar{\lambda}^2\right]$$

$$\Phi = 0.5\left[1 + 0.34(2.13 - 0.2) + 2.13^2\right] = 3.097$$

$$\chi = \frac{1}{3.097 + \sqrt{3.097^2 - 2.13^2}} = 0.187 \quad \text{but } \chi \leq 1.0$$

$$\text{Then, } N_{b,Rd} = \frac{0.187 \times 86.4 \times 100 \times 275}{1.1} = 403{,}920\,\text{N}$$

$$N_{b,Rd} = 403.9\,\text{kN} > N_{Ed} = 325.6\,\text{kN} \,(\text{Then O.K.})$$

4.3.3.33 Design of Roller Steel Fabricated Bearings

Let us now design the roller steel fabricated bearings shown in Figure 4.36 and detailed in Figure 4.90. The maximum vertical reaction at the supports of the main truss was previously calculated under dead and live loads with dynamic effect ($R_{D+L+\Phi}$), which was 5822.2 kN. The material of construction for the bearings is cast iron steel (ISO 3755) 340–550 having a yield stress of 340 MPa and an ultimate stress of 550 MPa.

Design of the Sole Plate

The reaction ($R_{D+L+\Phi}$) can be assumed as two equal concentrated loads at two points, which are the centers of gravity of half of the last vertical member V_{11} shown in Figure 4.90. To determine the centers of gravity (distance e), we can take the first area moment around the axis z-z, shown in Figure 4.90, as follows:

$$e = \frac{2 \times 27.5 \times 5 \times 13.75 + 30 \times 5 \times 2.5}{425} = 9.78\,\text{cm}$$

Assuming that the thickness of the sole plate is t_1, with detailed dimensions shown in Figure 4.90 based on the lower chord member L_1

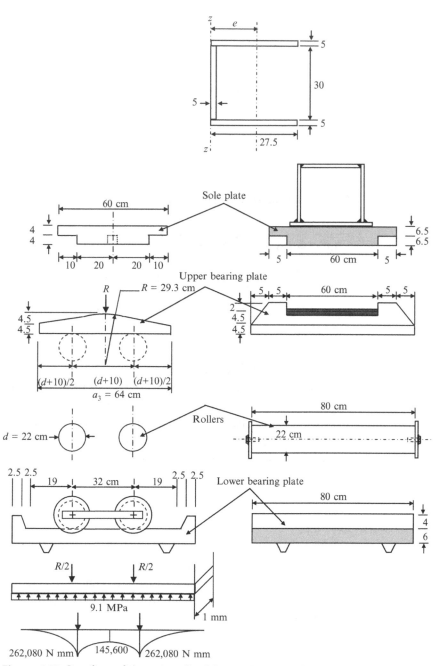

Figure 4.90 Detailing of the twin roller fabricated steel bridge bearings.

dimensions, we can determine the maximum moment applied to the sole plate (M) as follows:

$$M = R_{D+L+\Phi} \times e/2 = 5822.2 \times 10^3 \times 97.8/2 = 284,705,580 \, \text{N mm}.$$

$$\text{Section plastic modulus} \left(W_{pl} \right) = b_1 t_1^2/4 = 700 \times t_1^2/4 = 175 \times t_1^2$$

The plate thickness t_1 can be calculated now as follows:

$$\frac{M}{W_{pl}} = \frac{f_y}{\gamma_{M0}}$$

$$\frac{284,705,580}{175 \times t_1^2} = \frac{340}{1.0}$$

Then, $t_1 = 69.2$ mm, taken as 80 mm, as shown in Figure 4.90.

Design of the Rollers

The design of rollers requires determination of the diameter, length, and number of rollers to resist the vertical load as well as the arrangement and allowed movement in the direction of rollers. The design axial force per unit length of roller contact N'_{Sd} specified in BS EN 1337-1 [3.11] shall satisfy

$$N'_{Sd} \leq N'_{Rd}$$

where N'_{Rd} is the design value of resistance per unit length of roller contact, which is calculated as

$$N'_{Rd} = 23 \times R \times \frac{f_u^2}{E_d} \times \frac{1}{\gamma_m^2} = 23 \times R \times \frac{550^2}{210,000} \times \frac{1}{1} = 33.131 \times R$$

Assume the number of rollers is 2 and their length is 800 mm as shown in Figure 4.33:

$$N'_{Sd} = \frac{R_{D+L+\Phi}}{2 \times 800} = \frac{5822.2 \times 10^3}{1600} = 3638.9 \, \text{N/mm}$$

Then, the radius of rollers can be determined by equalizing N'_{Sd} with N'_{Rd} as follows:

$$3638.9 = 33.131 \times R$$

Then, $R = 109.8$ mm, taken as 110 mm and the diameter D is 220 mm.

Design of Upper Bearing Plate

The upper bearing plate is shown in Figure 4.90. The width and length of the plate are dependent on the spacing between rollers and the length of rollers as well as the allowed movement in the direction of rollers. The thickness of the upper bearing plate can be determined as follows:

$$M = \frac{R_{D+L+\Phi}}{2} \times \frac{(D+100)}{2} = \frac{5822.2 \times 10^3}{2} \times \frac{320}{2} = 465,776,000 \text{ N mm}.$$

$$W_{pl} = \frac{b_2 t_2^2}{4} = \frac{800 t_2^2}{4} = 200 t_2^2 \text{ mm}^3$$

The plate thickness t_2 can be calculated now as follows:

$$\frac{M}{W_{pl}} = \frac{f_y}{\gamma_{M0}}$$

$$\frac{465,776,000}{200 \times t_1^2} = \frac{340}{1.0}$$

Then, $t_1 = 82.8$ mm, taken as 90 mm, as shown in Figure 4.90.

The radius of the curved part of the upper bearing plate, which has a length of 600 mm as shown in Figure 4.90, can be determined the same way as that adopted for the design of the rollers:

$$N'_{Rd} = 23 \times R \times \frac{f_u^2}{E_d} \times \frac{1}{\gamma_m^2} = 23 \times R \times \frac{550^2}{210,000} \times \frac{1}{1} = 33.131 \times R$$

$$N'_{Sd} = \frac{R_{D+L+\Phi}}{600} = \frac{5822.2 \times 10^3}{600} = 97036.7 \text{ N/mm}$$

Then, the radius of rollers can be determined by equalizing N'_{Sd} with N'_{Rd} as follows:

$$97,036.7 = 33.131 \times R$$

Then, $R = 293$ mm.

Design of Lower Bearing Plate

The lower bearing plate is shown in Figure 4.90. The width and length of the plate are dependent on the strength of concrete and are dependent on the spacing between rollers and the length of rollers as well as the allowed movement in the direction of rollers. The thickness of the upper bearing plate can be determined as follows:

$$f_c = \frac{R_{D+L+\Phi}}{a_3 b_3} = \frac{5822.2 \times 10^3}{600 \times 800} = 9.1 \text{ MPa} < \frac{f_c}{\gamma_c} = \frac{40}{1.5}$$

$$= 26.7 \text{ MPa (for a typical concrete in bridges of C40/50 with } f_{ck})$$

The plate thickness t_3 can be calculated from the distribution of bending moment, caused by the pressure on the concrete foundation, as follows:

$M = 262,080 \, \text{N mm}$ per unit width of the plate.

$$W_{pl} = \frac{b_3 t_3^2}{4} = \frac{1 \times t_3^2}{4} = 0.25 \times t_2^2 \, \text{mm}^3$$

$$\frac{M}{W_{pl}} = \frac{f_y}{\gamma_{M0}}$$

$$\frac{262,080}{0.25 \times t_3^2} = \frac{340}{1.0}$$

Then, $t_3 = 55.5 \, \text{mm}$, taken as 60 mm, as shown in Figure 4.90.

4.3.3.34 Design of Hinged Line Rocker Steel Fabricated Bearings

Finally, we can now design the hinged line rocker steel fabricated bearings shown in Figure 4.36 and detailed in Figure 4.91. The maximum vertical reaction at the support of the main plate girder was previously calculated under dead and live loads with dynamic effect $(R_{D+L+\Phi})$, which was 5822.2 kN. The bearing is also subjected to a lateral force from the braking forces from traffic as well as subjected to a longitudinal force from the reactions of the upper and lower wind bracings, which cause moments around the longitudinal and lateral directions of the bearing base, respectively. Similar to the roller bearing, the material of construction for the bearings is cast iron steel (ISO 3755) 340-550 having a yield stress of 340 MPa and an ultimate stress of 550 MPa. It should be noted that the overall height of the hinged bearing must be exactly the same as that of the roller bearing. The general layout and assumed dimensions of the hinged line rocker bearing are shown in Figure 4.92. The braking Q_{lbk} forces can be calculated as follows:

$$Q_{lbk} = 360 + 2.7 \times L = 360 + 2.7 \times 60 = 522 \, \text{kN}, \quad \text{for Load model 1}$$

See Figure 4.91 for the direction of the forces. Also, the reactions from upper and lower wind bracings (R_{tot}) (see Figure 4.91 for the direction of the forces) were previously calculated as follows:

$$R_{tot} = 2 \times 460.53 = 921.06 \, \text{kN}$$

We can now determine the normal stress distribution due to the applied loads, shown in Figure 4.91, on the concrete foundation as follows:

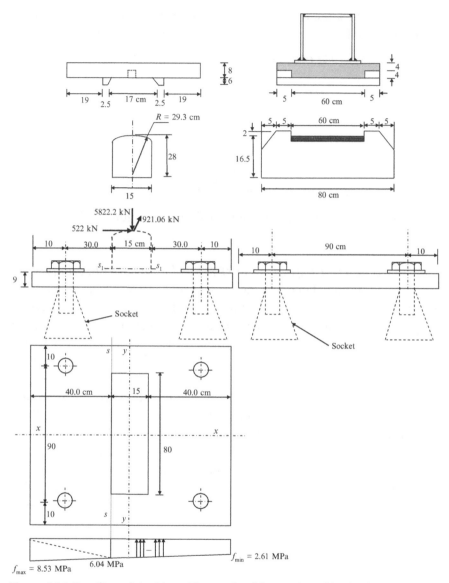

Figure 4.91 Detailing of the hinged line rocker fabricated steel bridge bearings.

$$f = -\frac{N}{A} \pm \frac{M_x}{I_x}y \pm \frac{M_y}{I_y}x$$

$$\frac{N}{A} = \frac{5822.2 \times 10^3}{950 \times 1100} = 5.57\,\text{MPa}$$

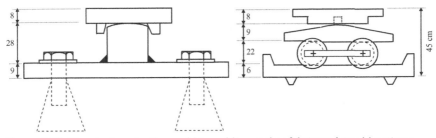

Figure 4.92 The designed roller and hinged line rocker fabricated steel bearings.

$$\frac{M_x}{I_x}y = \frac{921.06 \times 10^3 \times 370}{950 \times 1100^3/12}550 = 1.79\,\text{MPa}$$

$$\frac{M_y}{I_y}x = \frac{522 \times 10^3 \times 370}{1100 \times 950^3/12}475 = 1.17\,\text{MPa}$$

$$f_{\max} = -5.57 - 1.79 - 1.17 = -8.53\,\text{MPa}$$

$$f_{\min} = -5.57 + 1.79 + 1.17 = -2.61\,\text{MPa}$$

The critical bending moment on the base plate of the hinged bearing is at section s-s shown in Figure 4.91:

$$M = (0.5 \times 400 \times 6.04) \times 1100 \times 400/3 + (0.5 \times 400 \times 8.53) \times 1100$$
$$\times 400 \times 2/3 = 677,600,000\,\text{N mm}$$

$$W_{\text{pl}} = 1100 \times t_4^2/4 = 275t_4^2$$

$$\frac{M}{W_{\text{pl}}} = \frac{f_y}{\gamma_{M0}}$$

$$\frac{677,600,000}{275t_4^2} = \frac{340}{1.0}$$

Then, $t_4 = 85.1$ mm, taken as 90 mm.

The normal stresses at section s_1-s_1, shown in Figure 4.91, of the line rocker bearing can be checked as follows:

$$M_x = 921.06 \times 10^3 \times 280 = 257,896,800\,\text{N mm}.$$

$$M_y = 522 \times 10^3 \times 280 = 146,160,000\,\text{N mm}.$$

$$\frac{N}{A} = \frac{5822.2 \times 10^3}{150 \times 800} = 48.52\,\text{MPa}$$

$$\frac{M_x}{I_x}y = \frac{257{,}896{,}800}{150 \times 800^3/12}400 = 16.12\,\text{MPa}$$

$$\frac{M_y}{I_y}x = \frac{146{,}160{,}000}{800 \times 150^3/12}75 = 48.72\,\text{MPa}$$

$$f_{\text{max}} = -(48.52 + 16.12 + 48.72) = -113.4\,\text{MPa}$$
$$< 340\,\text{MPa}\,(\text{Then O.K.})$$

4.3.3.35 Design of Joint J_1

It is now possible to design the joints of the main trusses after designing all members and knowing all details regarding the joints. Let us start by designing joint J_1 (see Figure 4.93). For M27 high-strength pretensioned bolts used, the following design values are calculated:

$F_{v,Rd} = 176$ kN (single shear) and 353 kN (double shear)
$F_{s,Rd} = 117$ kN (single shear) and 234 kN (double shear)
$F_{s,ult} = 103$ kN (single shear) and 206 kN (double shear)

Number of Bolts for the Vertical Member V_1

$$N(V_1) = \frac{F_{\text{Ed}}}{F_{s,\text{ult}}} = \frac{5240.8}{103} = 50.9 \text{ bolts, taken as 54 bolts (27 bolts in each side}$$
acting in single shear)

Number of Bolts for the Diagonal Member D_1

$$N(D_1) = \frac{F_{\text{Ed}}}{F_{s,\text{ult}}} = \frac{6711.5}{206} = 32.6 \text{ bolts, taken as 40 bolts (20 bolts in each side}$$
acting in double shear)

4.3.3.36 Design of Joint J_2

Following the same procedures adopted for the design of joint J_1, we can design joint J_2 (see Figure 4.94) using the same M27 high-strength pretensioned bolts as follows.

Number of Bolts for the Vertical Member V_2

$$N(V_2) = \frac{F_{\text{Ed}}}{F_{s,\text{ult}}} = \frac{4195.1}{103} = 40.7 \text{ bolts, taken as 48 bolts (24 bolts in each side}$$
acting in single shear)

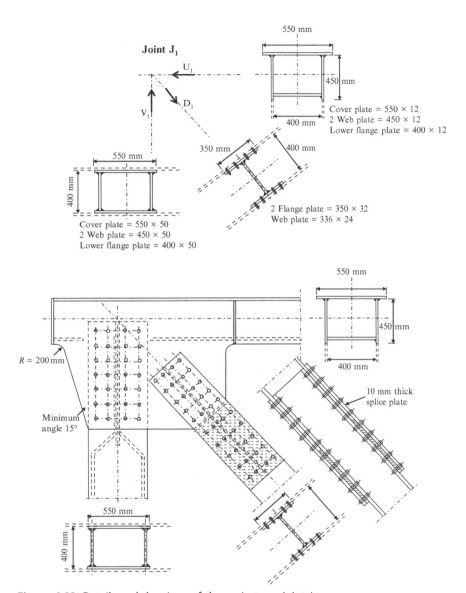

Figure 4.93 Details and drawings of the main truss joint J_1.

Number of Bolts for the Diagonal Member D_2

$$N(D_2) = \frac{F_{Ed}}{F_{s,ult}} = \frac{5372.3}{206} = 26 \text{ bolts, taken as 32 bolts (16 bolts in each side}$$

acting in double shear)

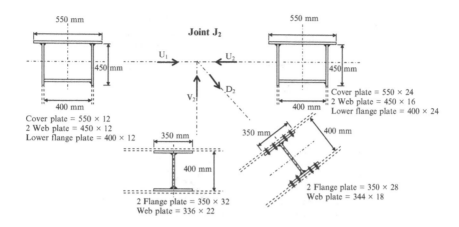

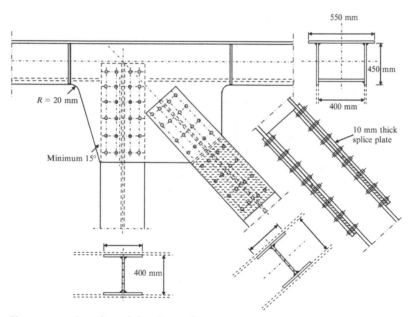

Figure 4.94 Details and drawings of the main truss joint J_2.

4.3.3.37 Design of Joint J_3

Joint J_3 (see Figure 4.95) can be designed using the same M27 high-strength pretensioned bolts as follows.

Number of Bolts for the Vertical Member V_3

$$N(V_3) = \frac{F_{Ed}}{F_{s,ult}} = \frac{3190.7}{103} = 31 \text{ bolts, taken as 32 bolts (16 bolts in each side}$$

acting in single shear)

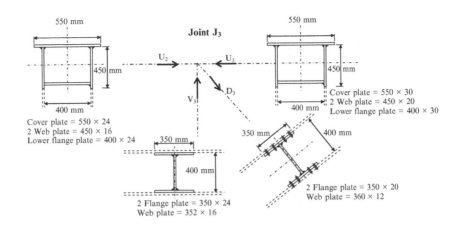

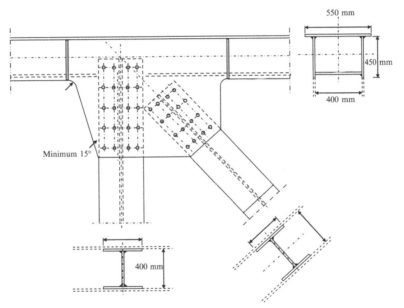

Figure 4.95 Details and drawings of the main truss joint J₃.

Number of Bolts for the Diagonal Member D₃

$$N(D_3) = \frac{F_{Ed}}{F_{s,ult}} = \frac{4086.1}{103} = 39.6 \text{ bolts, taken as 40 bolts (20 bolts in each side}$$

acting in single shear)

4.3.3.38 Design of Joint J₄

Joint J₄ (see Figure 4.96) can be designed using the same M27 high–strength pretensioned bolts as follows.

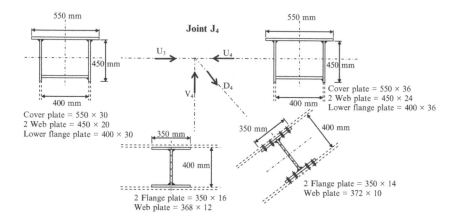

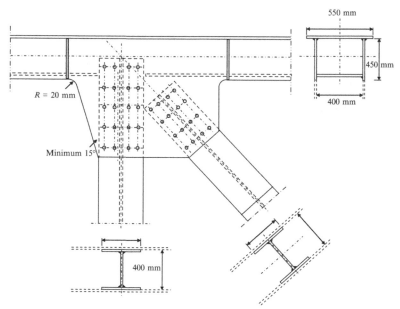

Figure 4.96 Details and drawings of the main truss joint J_4.

Number of Bolts for the Vertical Member V_4

$$N(V_4) = \frac{F_{Ed}}{F_{s,ult}} = \frac{2226.1}{103} = 21.6 \text{ bolts, taken as 24 bolts (12 bolts in each side}$$

acting in single shear)

Number of Bolts for the Diagonal Member D_4

$$N(D_4) = \frac{F_{Ed}}{F_{s,ult}} = \frac{2850.8}{103} = 27.8 \text{ bolts, taken as 32 bolts (16 bolts in each side}$$

acting in single shear)

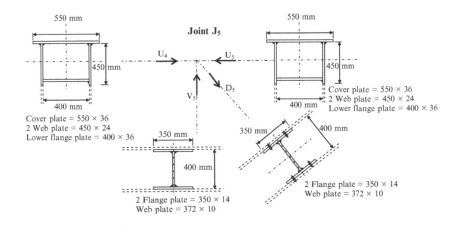

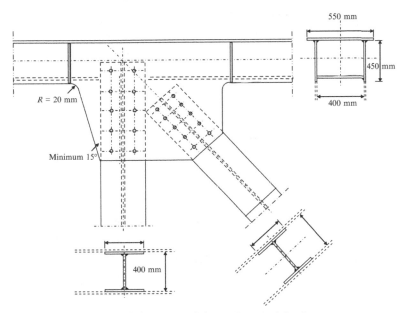

Figure 4.97 Details and drawings of the main truss joint J_5.

4.3.3.39 Design of Joint J_5

Joint J_5 (see Figure 4.97) can be designed using the same M27 high-strength pretensioned bolts as follows.

Number of Bolts for the Vertical Member V_5

$$N(V_5) = \frac{F_{Ed}}{F_{s,ult}} = \frac{1302.9}{103} = 12.6 \text{ bolts, taken as 16 bolts (8 bolts in each side}$$

acting in single shear)

Number of Bolts for the Diagonal Member D$_5$

$$N(D_5) = \frac{F_{Ed}}{F_{s,ult}} = \frac{1668.5}{103} = 16.2 \text{ bolts, taken as 20 bolts (10 bolts in each side}$$

acting in single shear)

4.3.3.40 Design of Joint J$_6$

Joint J$_6$ (see Figure 4.98) can be designed using the same M27 high-strength pretensioned bolts as follows.

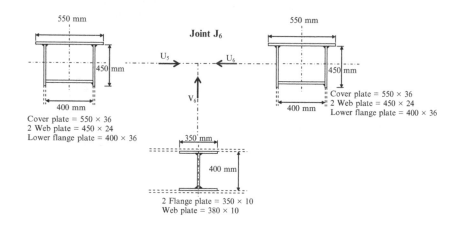

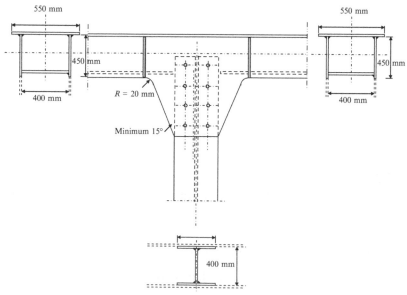

Figure 4.98 Details and drawings of the main truss joint J$_6$.

Number of Bolts for the Vertical Member V_6

The member is zero under the applied dead and live load cases of loading. The number of bolts can be taken as the minimum number based on the connection drawing. The number of connecting bolts of V_6 is taken as 16 bolts (8 bolts in each side acting in single shear).

4.3.3.41 Design of Joint J_7

Joint J_7 (see Figure 4.99) can be designed using the same M27 high–strength pretensioned bolts as follows.

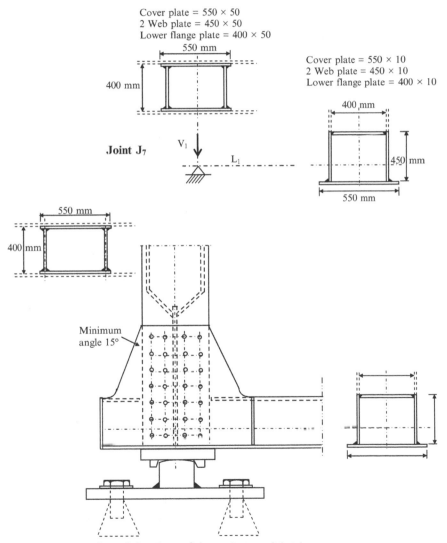

Figure 4.99 Details and drawings of the main truss joint J_7.

Number of Bolts for the Vertical Member V_1

$$N(V_1) = \frac{F_{Ed}}{F_{s,ult}} = \frac{5240.8}{103} = 50.9 \text{ bolts, taken as 54 bolts (27 bolts in each side}$$

acting in single shear)

4.3.3.42 Design of Joint J_8

Joint J_8 (see Figure 4.100) can be designed using the same M27 high-strength pretensioned bolts as follows.

Number of Bolts for the Vertical Member V_2

$$N(V_2) = \frac{F_{Ed}}{F_{s,ult}} = \frac{4195.1}{103} = 40.7 \text{ bolts, taken as 48 bolts (24 bolts in each side}$$

acting in single shear)

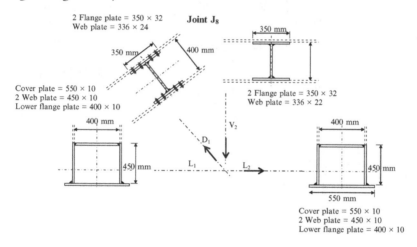

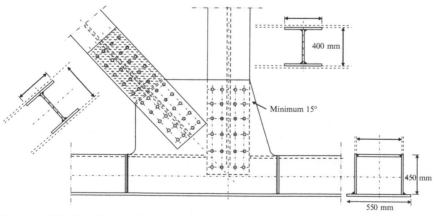

Figure 4.100 Details and drawings of the main truss joint J_8.

Number of Bolts for the Diagonal Member D_1

$$N(D_1) = \frac{F_{Ed}}{F_{s,ult}} = \frac{6711.5}{206} = 32.6 \text{ bolts, taken as 36 bolts (18 bolts in each side}$$

acting in double shear)

4.3.3.43 Design of Joint J_9

Joint J_9 (see Figure 4.101) can be designed using the same M27 high-strength pretensioned bolts as follows.

Number of Bolts for the Vertical Member V_3

$$N(V_3) = \frac{F_{Ed}}{F_{s,ult}} = \frac{3190.7}{103} = 31 \text{ bolts, taken as 32 bolts (16 bolts in each side}$$

acting in single shear)

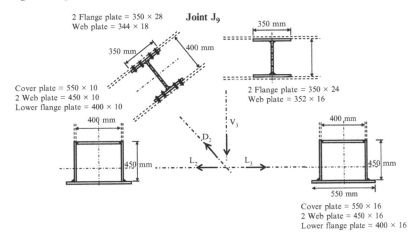

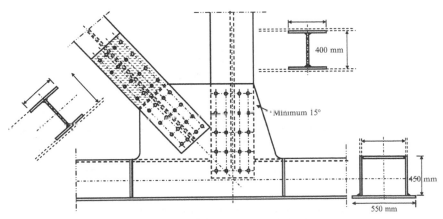

Figure 4.101 Details and drawings of the main truss joint J_9.

Number of Bolts for the Diagonal Member D_2

$$N(D_2) = \frac{F_{Ed}}{F_{s,ult}} = \frac{5372.3}{206} = 26 \text{ bolts, taken as 32 bolts (16 bolts in each side}$$

acting in double shear)

4.3.3.44 Design of Joint J₁₀

Joint J_{10} (see Figure 4.102) can be designed using the same M27 high-strength pretensioned bolts as follows.

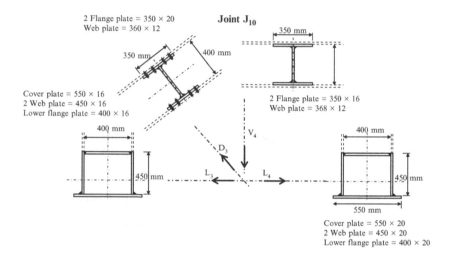

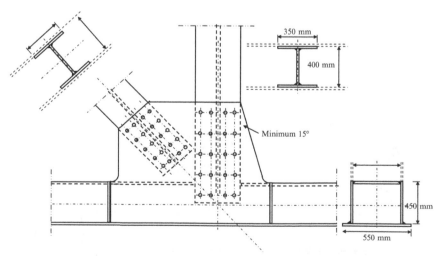

Figure 4.102 Details and drawings of the main truss joint J_{10}.

Number of Bolts for the Vertical Member V$_4$

$$N(V_4) = \frac{F_{Ed}}{F_{s,ult}} = \frac{2226.1}{103} = 21.6 \text{ bolts, taken as 24 bolts (12 bolts in each side}$$

acting in single shear)

Number of Bolts for the Diagonal Member D$_3$

$$N(D_3) = \frac{F_{Ed}}{F_{s,ult}} = \frac{4086.1}{103} = 39.7 \text{ bolts, taken as 40 bolts (20 bolts in each side}$$

acting in single shear)

4.3.3.45 Design of Joint J$_{11}$

Joint J$_{11}$ (see Figure 4.103) can be designed using the same M27 high-strength pretensioned bolts as follows.

Number of Bolts for the Vertical Member V$_5$

$$N(V_5) = \frac{F_{Ed}}{F_{s,ult}} = \frac{1302.9}{103} = 12.6 \text{ bolts, taken as 16 bolts (8 bolts in each side}$$

acting in single shear)

Number of Bolts for the Diagonal Member D$_4$

$$N(D_4) = \frac{F_{Ed}}{F_{s,ult}} = \frac{2850.8}{103} = 27.7 \text{ bolts, taken as 32 bolts (16 bolts in each side}$$

acting in single shear)

4.3.3.46 Design of Joint J$_{12}$

Joint J$_{12}$ (see Figure 4.104) can be designed using the same M27 high-strength pretensioned bolts as follows.

Number of Bolts for the Vertical Member V$_6$

The member is zero under the applied dead and live load cases of loading. The number of bolts can be taken as the minimum number based on the connection drawing. The number of connecting bolts of V$_6$ is taken as 16 bolts (8 bolts in each side acting in single shear).

Number of Bolts for the Diagonal Member D$_5$

$$N(D_5) = \frac{F_{Ed}}{F_{s,ult}} = \frac{1668.5}{103} = 16.2 \text{ bolts, taken as 16 bolts (8 bolts in each side}$$

acting in single shear)

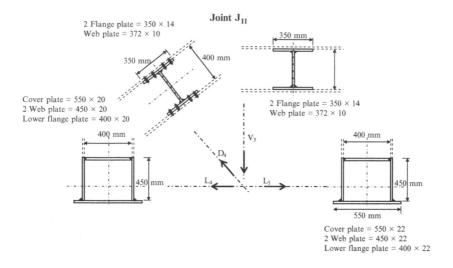

Joint J₁₁

2 Flange plate = 350 × 14
Web plate = 372 × 10

350 mm   400 mm

350 mm

350 mm

400 mm

Cover plate = 550 × 20
2 Web plate = 450 × 20
Lower flange plate = 400 × 20

2 Flange plate = 350 × 14
Web plate = 372 × 10

400 mm   400 mm

450 mm   450 mm

V₅
D₄
L₄   L₅

550 mm

Cover plate = 550 × 22
2 Web plate = 450 × 22
Lower flange plate = 400 × 22

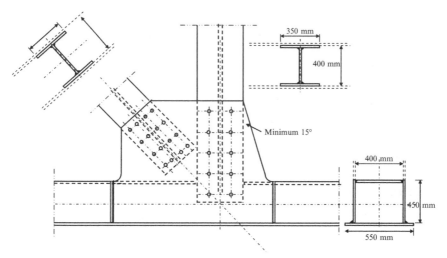

350 mm
400 mm

Minimum 15°

400 mm
450 mm
550 mm

Figure 4.103 Details and drawings of the main truss joint J₁₁.

4.3.3.47 Design of Joint J₁₃

Joint J₁₃ (see Figure 4.105) can be designed using the same M27 high-strength pretensioned bolts as follows.

Number of Bolts for the Vertical Member V₁₁

$$N(V_{11}) = \frac{F_{Ed}}{F_{s,ult}} = \frac{5240.8}{103} = 50.9 \text{ bolts, taken as } 54 \text{ bolts (27 bolts in each side}$$

acting in single shear)

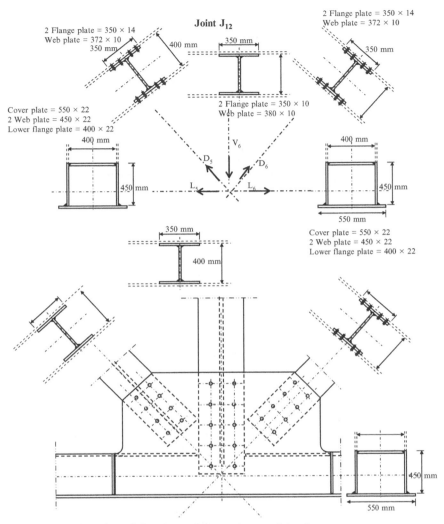

Figure 4.104 Details and drawings of the main truss joint J$_{12}$.

4.4 DESIGN EXAMPLE OF A HIGHWAY STEEL-CONCRETE COMPOSITE BRIDGE

The third design example presented in this chapter is for a highway steel-concrete composite bridge. The general layout of the through bridge is shown in Figures 4.106 and 4.107, with a brief introduction to the bridge components previously explained in Figure 1.22. The steel-concrete composite bridge has simply supported ends with a length between supports of

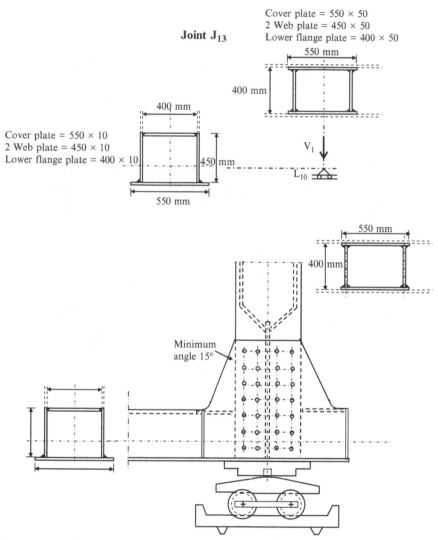

Figure 4.105 Details and drawings of the main truss joint J_{13}.

48 m and an overall length of 49 m. The overall width of the composite bridge is 13 m including two sidewalks of 1 m width each. The depth of reinforced concrete slab deck is 250 mm and the depth of the reinforced concrete haunch is 200 mm. The concrete slab decks are supported by five plate girders spaced at a distance of 2.5 m. The steel plate girder cross section in the middle 24 m consists of an upper flange plate of 700×38 mm^2, a web plate of 1724×16 mm^2, and a lower flange plate of 900×38 mm^2, as shown

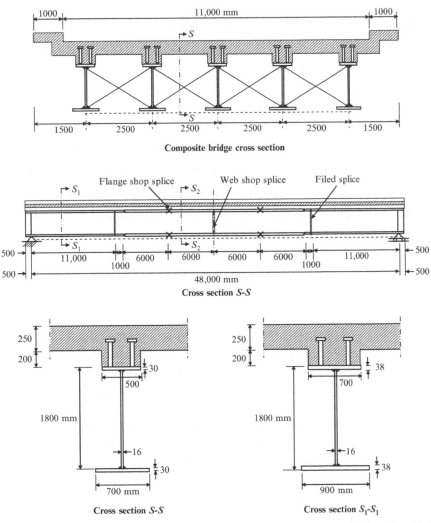

Figure 4.106 General layout of a highway steel-concrete composite bridge (the third design example).

in Figure 4.106, while the steel plate girder cross section in the remaining parts consists of an upper flange plate of 500×30 mm^2, a web plate of 1740×16 mm^2, and a lower flange plate of 700×30 mm^2, as shown in Figure 4.106. The web plate is stiffened by vertical stiffeners spaced at 1.5 m as shown in Figure 4.107. The steel material of construction of the double–track railway bridge conformed to standard steel grade EN 10025-2 (S 275) having a yield stress of 275 MPa and an ultimate strength of 430 MPa. The bridge has

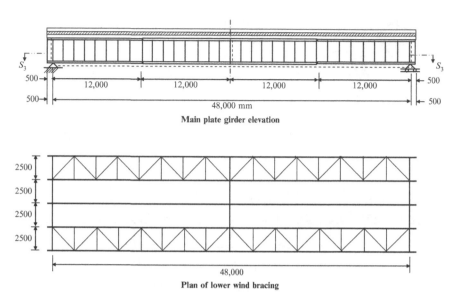

Main plate girder elevation

Plan of lower wind bracing

Figure 4.107 General layout of a highway steel-concrete composite bridge (the third design example).

lower wind K-shaped bracing as well as cross bracings of X-shaped truss members as shown in Figures 4.106 and 4.107. Figure 4.106 also shows the position of the flange and web shop splices as well as the positions of the field splices, while Figure 4.107 shows the stiffeners of the plate girder web. The composite action between the reinforced concrete slab deck and the steel plate girders was achieved via headed stud shear connectors having a diameter of 25 mm and an overall height of 300 mm. Two headed studs are welded on the top flanges of the steel plate girders as shown in Figure 4.106. The expected live loads on the bridge conforms to Load Model 1, which represents the static and dynamic effects of vertical loading due to normal road traffic as specified in EC1 [3.1]. The bolts used in different connections and field splices are M27 high-strength pretensioned bolts. Assume the unit weight of reinforced concrete slab decks is 25 kN/m^3. It is required to design the composite plate girder bridge adopting the design rules specified in EC4 [3.6]. It should be noted that composite slabs with metal decking (profiled steel sheeting) are commonly used nowadays in bridges owing to the elimination of formwork. However, they are quite costly compared with traditional haunched concrete slab decks. Designers therefore have to compare the cost of both constructions for the project under investigation. Chapter 2 of this book has detailed the shear connection with solid slabs, haunched solid slabs, and composite slabs with profiled steel sheeting, which is suggested by the reviewer, and shear connection in floors with precast hollow core slabs. The author has an already published paper dealing with the behavior of shear connection in composite slabs with profiled steel sheeting [2.71]. This design example presents the design of composite plate girders with haunched slab decks. The design example does not favor one construction technique over the other.

4.4.1 Calculation of Loads Acting on the Composite Bridge

To design the composite bridge, we need to calculate the dead and live loads acting on the bridge in the longitudinal direction, which is addressed as follows.

Dead Loads

Weight of steel structure for part of bridge between main trusses:

$$w_{s_1} = 1.75 + 0.04L + 0.0003L^2 \leq 3.5 \, \text{kN/m}^2$$

$$w_{s_1} = 1.75 + 0.04 \times 46 + 0.0003 \times 48^2 = 4.38$$
$$> 3.5 \, \text{kN/m}^2 \text{ taken as } 3.5 \, \text{kN/m}^2$$

Weight of steel structure for part of bridge outside main trusses:

$$w_{s_2} = 1 + 0.03L \, \text{kN/m}^2$$

$$w_{s_2} = 1 + 0.03 \times 48 = 2.44 \, \text{kN/m}^2$$

$$w_s = 3.5 \times 11/5 + 2.44 \times 2 \times 1/5 = 8.7 \, \text{kN/m}$$

Weight of reinforced concrete decks and haunches:

$$w_{\text{RC}} = (0.25 + 0.05) \times 25 \times 2.5 = 18.75 \, \text{kN/m}$$

Weight of finishing (assume weight of finishing is 1.75 kN/m²):

$$w_{\text{F}} = 1.75 \times 2.5 = 4.375 \, \text{kN/m}$$

We can now calculate the total dead load acting on an intermediate composite plate girder in the longitudinal direction (see Figure 4.108) as follows:

$$w_{\text{D.L.}} = 8.7 + 18.75 + 4.375 = 31.825 \, \text{kN/m}$$

Since the main composite plate girders are simply supported, we can calculate the maximum shear force and bending moment due to dead loads on an intermediate composite plate girder (see Figure 4.108) as follows:

$$Q_{\text{D.L.}} = g_{\text{vk}} \times L/2 = 31.825 \times 48/2 = 763.8 \, \text{kN}$$

$$M_{\text{D.L.}} = g_{\text{vk}} \times L^2/8 = 31.825 \times 48^2/8 = 9165.6 \, \text{kN m}$$

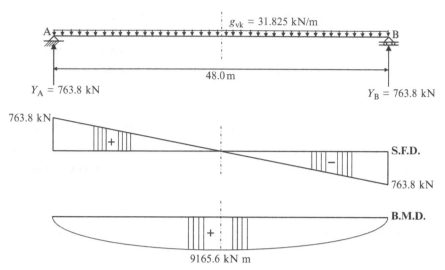

Figure 4.108 Straining actions from dead loads acting on one intermediate composite plate girder.

Live Loads

The live loads acting on the composite highway bridge conform to Load Model 1, which represents the static and dynamic effects of vertical loading due to normal road traffic as specified in EC1 [3.1]. To determine the worst cases of loading on an intermediate composite plate girder due to live loads, we can study a lateral section through vehicles and a lateral section through distributed loads of Load Model 1 acting on the bridge, as shown in Figure 4.109. From the section through vehicles, we can find that the maximum concentrated load transferred to the intermediate composite plate girder is 240 kN, while from the section through distributed loads, we can find that the maximum distributed load transferred to the composite plate girder is 17.3 kN/m. Therefore, the load distribution transferred to the composite plate girder in the longitudinal direction is as shown in Figure 4.110. From the previous analyses, we find that the worst case of loading for the evaluation of the maximum bending moment is that the centerline (midspan) of the composite plate girder divides the spacing between the resultant of the concentrated live loads and the closest load, with maximum bending moment located at the closest load (point a in Figure 4.111):

$$M_{L.L.} = 652.2 \times 23.7 - 17.3 \times 23.7^2/2 = 10598.5 \, kN \, m$$

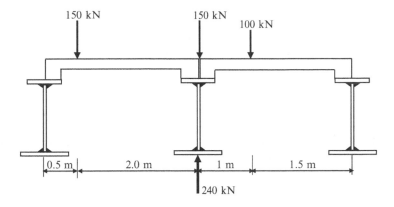

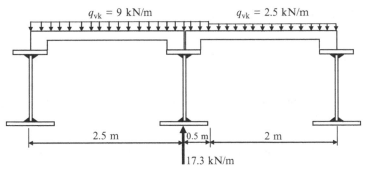

Figure 4.109 Calculation of straining actions from live loads transferred on intermediate composite plate girders.

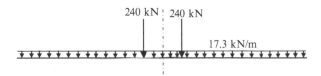

Figure 4.110 Transferred live loads on intermediate composite plate girders.

Also, from the previous analyses, we find that there is a single case of loading for the live loads to produce a maximum shear force at the supports of the intermediate composite plate girder, which is shown in Figure 4.112:

$$Q_{L.L.} = 889.2 \, kN$$

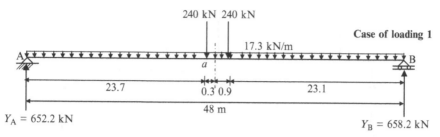

Figure 4.111 The critical case of loading for the maximum bending moment acting on an intermediate composite plate girder.

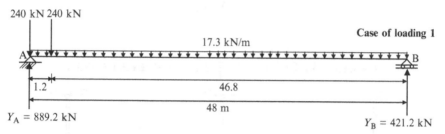

Figure 4.112 The critical case of loading for the maximum shearing force acting on an intermediate composite plate girder.

Bending Moment Due to Dead and Live Loads with Dynamic Effect Added (M_{Ed})

$$M_{Ed} = M_{D.L.} \times \gamma_g + M_{L.L.} \times \gamma_q = 9165.6 \times 1.3 + 10794.2 \times 1.35$$
$$= 26223.26 \, kN \, m$$

It should be noted that, according to EC0 (BS EN 1990) [3.4], the permanent actions of steel self-weight and superimposed load should be multiplied by 1.2, while the permanent actions of concrete weight should be multiplied by 1.35. Therefore, the total dead load is calibrated and multiplied by 1.3. On the other hand, variable actions comprising road traffic actions are multiplied by 1.35.

Shearing Force Due to Dead and Live Loads with Dynamic Effect Added (Q_{Ed})

$$Q_{Ed} = Q_{D.L.} \times \gamma_g + Q_{L.L.} \times \gamma_q = 763.8 \times 1.3 + 889.2 \times 1.35 = 2193.36 \, kN$$

Design Bending Moment (M_{Ed}) and Shear Force (Q_{Ed})

$$M_{Ed} = 26223.26 \, kN \, m$$
$$Q_{Ed} = 2193.36 \, kN$$

4.4.2 Design of the Composite Plate Girder Cross Section at Mid- and Quarter-Span

After the evaluation of the straining actions acting on the composite bridge, we can now design the critical cross sections of an intermediate composite plate girder as follows.

Design of the Intermediate Composite Plate Girder Cross Section at Midspan
Following the design rules specified in EC4 [3.6], the effective width of the concrete slab is 2.5 m and the cross-sectional dimensions are shown in Figure 4.113. Knowing that the design concrete strength f_{cd} is equal to f_{ck}/γ_c ($f_{cd}=40/1.5=26.67$ MPa), we can calculate the forces acting on the composite girder as shown in Figure 4.113. The position of the plastic neutral axis can be located from the equilibrium of these forces, assuming an initial position in the top flange of the steel plate girder, as follows:

$$14{,}168.4 + 3173.7 + x \times 700 \times 275/1000 = 9405 + 7585.6$$
$$+ 700 \times (38 - x) \times 275/1000$$

$$17{,}342.1 + 192.5 \times x = 16{,}990.6 - 192.5 \times x + 7315$$

$$2 \times 192.5 \times x = 6963.5$$

Then, $x=18.1$ mm.

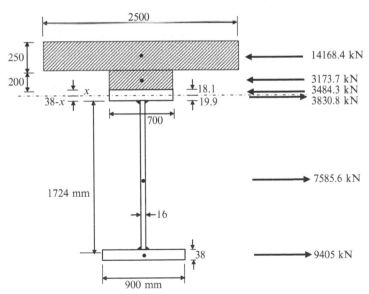

Figure 4.113 Calculation of bending resistance at the critical mid-span section of an intermediate composite plate girder.

We can now calculate the design plastic moment resistance as follows:

$$M_{pl,Rd} = 14{,}168.4 \times 343.1 + 3173.7 \times 118.1 + 3484.3 \times 18.1/2$$
$$+ 3830.8 \times 19.9/2 + 7585.6 \times 881.9 + 9405 \times 1762.9$$
$$= 28{,}575{,}456.5 \, kN \, mm = 28{,}575.5 \, kN \, mm > 26{,}223.26 \, kN \, m$$
(Then O.K.).

Design of the Intermediate Composite Plate Girder Cross Section at Quarter-Span

Since it is decided to reduce the cross section at quarter-span, as shown in Figure 4.106, we should check the safety of the proposed cross section against different stresses. Assuming the bending moment diagram is a second-degree parabola (see Figure 4.114), we can determine the bending moment at quarter-span as follows:

$$\frac{\Delta}{26{,}223.26} = \left(\frac{1}{2}\right)^2; \quad \text{then } \Delta = 6555.8 \, kN \, m$$

The design moment at quarter-span $(M_{Ed}) = 26{,}223.26 - 6555.8 = 19{,}667.5$ kNm. We can now repeat the previous procedures adopted for the heavier cross section for the design of the smaller steel plate girder cross section shown in Figure 4.115:

$$14{,}168.4 + 3173.7 + x \times 500 \times 275/1000$$
$$= 5775 + 7656 + 500 \times (30 - x) \times 275/1000$$

$$17{,}342.1 + 137.5 \times x = 13{,}431 - 137.5 \times x + 4125$$

$$2 \times 137.5 \times x = 213.9$$

Then, $x = 0.78$ mm.

The design plastic moment resistance can be calculated as follows:

$$M_{pl,Rd} = 14{,}168.4 \times 325.78 + 3173.7 \times 100.78 + 107.3 \times 0.78/2$$
$$+ 4017.8 \times 29.22/2 + 7656 \times 899.22 + 5775 \times 1784.22$$
$$= 22{,}182{,}667.7 \, kN \, mm = 22\,182.7 \, kN \, mm > 19{,}667.5 \, kN \, m$$
(Then O.K.).

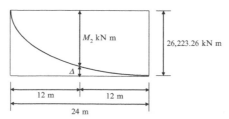

Figure 4.114 Calculation of bending moment acting at quarter span.

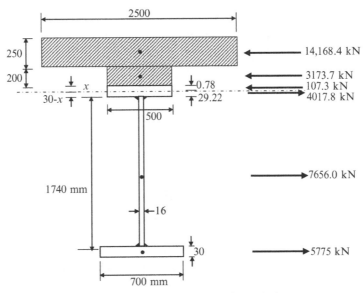

Figure 4.115 Calculation of bending resistance at the critical quarter-span section of an intermediate composite plate girder.

Check of Shear Forces

$$V_{pl,Rd} = \frac{A_v\left(f_y/\sqrt{3}\right)}{\gamma_{M0}} = \frac{(1740 \times 16) \times \left(275/\sqrt{3}\right)}{1.0} = 4,420,193.6\,N$$

$$= 4420.2\,kN > Q_{Ed} = 2193.36\,kN \text{ (Then O.K.)}$$

Design of Shear Connection

To ensure the transfer of shear stresses at the concrete slab deck–steel plate girder interface, the shear connection between the two components is designed in this section. This can be conducted from the elastic analysis of the cross section at supports shown in Figure 4.116. The maximum shear at supports (Q_{Ed}), previously calculated, is equal to 2193.36 kN. To calculate the elastic section properties, the reinforced concrete sections have to be transformed to equivalent steel sections, as shown in Figure 4.117, using the modular ratios of the two components. After the evaluation of the elastic section properties, the shear flow at the interface can be calculated and assessed against the shear resistances of the headed studs to determine the spacing between rows of shear connectors. These procedures can be performed as follows:

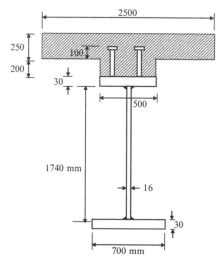

Figure 4.116 Cross section at supports of the composite plate girder.

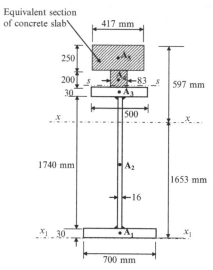

Figure 4.117 Calculation of shear forces at supports of the composite plate girder using elastic analysis.

$$E_s = 210,000\,\text{MPa}$$

$$E_{cm} = 35000\,\text{MPa}\,(\text{Short term})$$

$$\text{Modular ratio}\ \alpha_c = \frac{210,000}{35,000} = 6$$

$$\text{Width of slab} = 250/6 = 41.7 \, \text{cm}$$

$$\text{Width of haunch} = 50/6 = 8.3 \, \text{cm}$$

$$A_1 = 70 \times 3 = 210 \, \text{cm}^2$$

$$A_2 = 174 \times 1.6 = 278.4 \, \text{cm}^2$$

$$A_3 = 50 \times 3 = 150 \, \text{cm}^2$$

$$A_4 = 8.3 \times 20 = 166 \, \text{cm}^2$$

$$A_5 = 41.7 \times 25 = 1042.5 \, \text{cm}^2$$

$$A = 1846.9 \, \text{cm}^2$$

To determine the centroid, we can take the first area moment around the x_1-x_1 axis, as follows:

$$y_c = \frac{210 \times 1.5 + 278.4 \times 90 + 150 \times 178.5 + 166 \times 190 + 1042.5 \times 212.5}{1846.9}$$
$$= 165.3 \, \text{cm}$$

$$\begin{aligned}I_x = {} & [70 \times 3^3/12 + 210 \times 163.8^2] + [1.6 \times 174^3/12 + 278.4 \times 75.3^2] \\ & + [50 \times 3^3/12 + 150 \times 13.2^2] + [8.3 \times 20^3/12 + 166 \times 24.7^2] \\ & + [41.7 \times 25^3/12 + 1042.5 \times 47.2^2] = 10{,}425{,}383 \, \text{cm}^4\end{aligned}$$

Shear flow at section s-s:

$$q = \frac{Q_{Ed} \times S_{ss}}{I_x} = \frac{2193.36 \times 10^3 \times [166 \times 24.7 + 1042.5 \times 47.2] \times 10^3}{10{,}425{,}383 \times 10^4}$$
$$= 1121.5 \, \text{N/mm}$$

Maximum spacing between shear connectors in the longitudinal direction

$$(S_{max}) = 15t_f \sqrt{\frac{235}{f_y}} = 15 \times 30 \times \sqrt{\frac{235}{275}} = 416 \, \text{mm}$$

$$\text{Force per two headed studs} = S \times 1121.5 \, \text{N}$$

$$\text{Force per one headed stud} = S \times 560.75 \, \text{N}$$

Design resistance of headed studs can be calculated as follows:

$$P_{Rd} = \frac{0.8 f_u \pi d^2/4}{\gamma_v} = \frac{0.8 \times 430 \times 3.14 \times 25^2/4}{1.25} = 135{,}088 \, \text{N} = 135.1 \, \text{kN}$$

$$\alpha = 1$$

$$P_{Rd} = \frac{0.29 \alpha d^2 \sqrt{f_{ck} E_{cm}}}{\gamma_v} = \frac{0.29 \times 1.0 \times 25^2 \sqrt{40 \times 35,000}}{1.25} = 171,566.3 \, N$$
$$= 171.6 \, kN$$

Then, $P_{Rd} = 135.1$ kN.

Hence, the spacing between headed stud rows in the longitudinal direction can be calculated as follows:

$$P_{Rd} = \text{Force per one headed stud} = S \times 560.75 \, N$$

$$135.1 \times 10^3 = S \times 560.75$$

Then, $S = 240.9$ mm, taken as 240 mm < 416 mm (Then O.K.).

4.4.3 Design of Wind Bracings

Wind forces acting on the composite plate girder bridge (see Figure 4.118) are transmitted to the bearings by systems of cross and lower wind bracings. Wind forces applied to this bridge can be sufficiently estimated using the design rules specified in EC1 [3.2] as follows. The design rules specify a height of 2 m on top of the concrete slab deck to be used in the calculation of the area subjected to wind forces:

$$F_w = \frac{1}{2} \rho v_b^2 C A_{ref,x}$$

$$v_b = c_{dir} \times c_{season} \times v_{b,0} = 1.0 \times 1.0 \times 26 = 26 \, m/s$$

$$A_{ref,x} = 4.25 \times 49 = 208.25 \, m^2$$

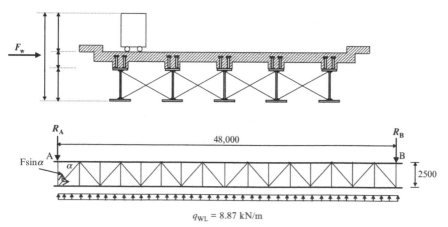

Figure 4.118 Loads on the lower wind bracing.

$$F_w = \frac{1}{2} \times 1.25 \times 26^2 \times 5.7 \times 208.25 = 501,518\,\text{N} = 501.5\,\text{kN}$$

Considering the structural analysis for the upper wind bracing system shown in Figure 4.118, the critical design wind force in the diagonal bracing members can be calculated as follows:

$$\text{Distributed wind loads}\,(q_{WL}) = 501.5 \times 0.5/48 = 5.22\,\text{kN/m}$$

$$\text{Factored distributed wind loads} = q_{WL} \times \gamma_q = 5.22 \times 1.7 = 8.87\,\text{kN/m}$$

$$R_A = 8.87 \times 24 = 212.9\,\text{kN}$$

$$\alpha = \tan^{-1}(2.5/3) = 39.8°$$

$$F_D = 212.9/(\sin 30.8) = 332.6\,\text{kN}$$

The cross section of the bracing member (see Figure 4.119) can be determined as follows:

$$l_{bx} = 3910\,\text{mm}, \quad l_{by} = 1.2 \times 3910 = 4690\,\text{mm}$$

Choose two angles back–to–back $100 \times 100 \times 10$, with 10 mm gusset plate between them:

$$A = 2 \times 19.2 = 38.4\,\text{cm}^2, \quad i_x = 3.05\,\text{cm}, \quad e = 2.83\,\text{cm},$$

$$i_y = \sqrt{3.05^2 + (2.83 + 0.5)^2} = 4.52\,\text{cm}$$

$$\varepsilon = \sqrt{\frac{235}{275}} = 0.924$$

$$\bar{\lambda} = \frac{L_{cr}}{i}\frac{1}{\lambda_1}$$

$$\lambda_1 = 93.9 \times 0.924 = 86.7636$$

$$\bar{\lambda} = \frac{3910}{30.5}\frac{1}{86.7636} = 1.478$$

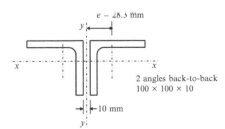

Figure 4.119 Lower wind bracing cross section.

The axial compressive force in the diagonal bracing member $(N_{Ed} = 332.6 \text{ kN})$:

$$\frac{N_{Ed}}{N_{b,Rd}} \leq 1.0$$

where $N_{b,Rd} = \dfrac{\chi A f_y}{\gamma_{M1}}$

$$A = 2 \times 19.2 = 38.4 \text{ cm}^2$$

$$\chi = \frac{1}{\Phi + \sqrt{\Phi^2 - \bar{\lambda}^2}} \quad \text{but } \chi \leq 1.0$$

$$\Phi = 0.5\left[1 + \alpha(\bar{\lambda} - 0.2) + \bar{\lambda}^2\right]$$

$$\Phi = 0.5\left[1 + 0.34(1.478 - 0.2) + 1.478^2\right] = 1.81$$

$$\chi = \frac{1}{1.81 + \sqrt{1.81^2 - 1.478^2}} = 0.35 \quad \text{but } \chi \leq 1.0$$

$$\text{Then, } N_{b,Rd} = \frac{0.35 \times 38.4 \times 100 \times 275}{1.1} = 336{,}000$$

$$N_{b,Rd} = 336 \text{ kN} > N_{Ed} = 332.6 \text{ kN (Then O.K.)}$$

4.4.4 Design of Web Stiffeners

There are two types of stiffeners used to strengthen the thin web plate of the main composite plate girder against buckling due to shear stresses, bending stresses, or both. The stiffeners at the supports are commonly known as load bearing stiffeners, while intermediate stiffeners are commonly known as stability stiffeners (intermediate transverse stiffeners). The design of the stiffeners can be performed as follows:

4.4.4.1 Load Bearing Stiffeners

To design the load bearing stiffener at supports (see Figure 4.120), we can also follow the design rules specified in EC3 [1.27, 2.11] for concentrically loaded compression members. The axial force in the stiffener is the maximum reaction at supports $(N_{Ed} = R_{D+L+\Phi})$, which is equal to 2193.36 kN. The design procedures can be performed as follows:

$$\frac{N_{Ed}}{N_{b,Rd}} \leq 1.0$$

where $N_{b,Rd} = \dfrac{\chi A f_y}{\gamma_{M1}}$

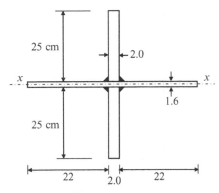

Figure 4.120 Load bearing web stiffeners at supports.

$$A = 2 \times 25 \times 2.0 + 46.0 \times 1.6 = 155.2 \, \text{cm}^2$$

$$I_x = 46 \times 1.6^3/12 + 2 \times \left[2 \times 25^3/12 + 50 \times 13.3^2\right] = 22{,}913 \, \text{cm}^4$$

$$\chi = \frac{1}{\Phi + \sqrt{\Phi^2 - \bar{\lambda}^2}} \quad \text{but } \chi \leq 1.0$$

$$\Phi = 0.5\left[1 + \alpha\left(\bar{\lambda} - 0.2\right) + \bar{\lambda}^2\right]$$

$$\bar{\lambda} = \sqrt{\frac{Af_y}{N_{cr}}}$$

$$N_{cr} = \frac{\pi^2 \times EI}{L^2} = \frac{3.14^2 \times 210{,}000 \times 22{,}913 \times 10^4}{3480^2} = 39{,}174{,}373 \, \text{N}$$

$$\bar{\lambda} = \sqrt{\frac{155.2 \times 100 \times 275}{39{,}174{,}373}} = 0.33$$

$$\Phi = 0.5\left[1 + 0.49(0.33 - 0.2) + 0.33^2\right] = 0.586$$

$$\chi = \frac{1}{0.586 + \sqrt{0.586^2 - 0.33^2}} = 0.934 \quad \text{but } \chi \leq 1.0$$

$$\text{Then, } N_{b,Rd} = \frac{0.934 \times 155.2 \times 100 \times 275}{1.1} = 3{,}623{,}920 \, \text{N}$$

$$N_{b,Rd} = 3623.9 \, \text{kN} > N_{Ed} = 2193.36 \, \text{kN} \ (\text{Then O.K.})$$

4.4.4.2 Intermediate Stiffeners

Intermediate stiffeners (see Figure 4.121) can be designed by choosing their dimensions such that

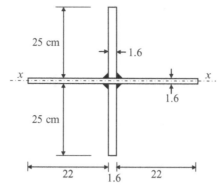

Figure 4.121 Intermediate stability web stiffeners.

$$\frac{a_1}{h_w} = \frac{1500}{1740} = 0.86 < \sqrt{2} = 1.414 \,(\text{Then O.K.})$$

$$\text{and } I_{st} \geq \frac{1.5 h_w^2 t_w^3}{a_1^2} = \frac{1.5 \times 174^3 \times 1.6^3}{150^2} = 1438.5 \,\text{cm}^4$$

$$I_{st} = 45.6 \times 1.6^3/12 + 2 \times \left[1.6 \times 25^3/12 + 40 \times 13.3^2\right] = 18333.4 \,\text{cm}^4$$
$$> 1438.5 \,\text{cm}^4 \,(\text{Then,O.K.})$$

4.4.5 Design of Field Splices

Figure 4.106 shows the locations of filed splices for the investigated bridge. Designing the splice requires determination of size of connecting plates as well as the number of bolts of the filed splice shown in Figure 4.122. The area of the upper flange plate equals to $50 \times 3 = 150$ cm^2; this can be compensated by three flange splice plates having cross-sectional area of 50×1.6 and $2 \times 23 \times 1.6$ cm^2 with a total area of 153.6 cm^2, which is greater than the original area. Also, the area of the lower flange plate equals to $70 \times 3 = 210$ cm^2; this can be compensated by three flange splice plates having cross-sectional area of 70×1.6 and $2 \times 33 \times 1.6$ cm^2 with a total area of 217.6 cm^2, which is greater than the original area, while the area of web plate $= 174 \times 1.6 = 278.4$ cm^2 can be compensated by two web splice plates having cross-sectional area of $2 \times 170 \times 1.0$ cm^2 with a total area of 340 cm^2, which is governed by the minimum thickness (10 mm) of plates used in steel bridges. The top row of bolts in the web (see Figure 4.122) is subjected to horizontal shear from the bending moment distribution, assuming the yield stress reached at the extreme and lower fibers of the flanges, and vertical shear from the applied loads. Using a spacing of 8.5 cm between two adjacent bolts, an edge spacing of 4.25 cm and a hole of 3 cm (2.7 cm bolt

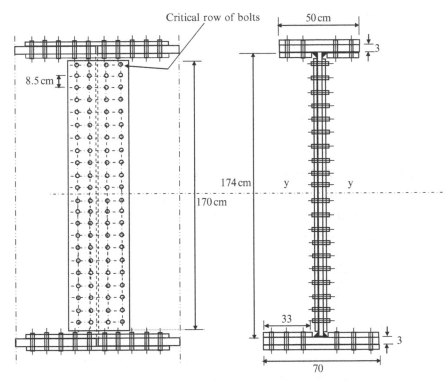

Figure 4.122 The field splice of the main plate girder.

diameter plus 0.3 cm clearance), we can determine the horizontal shear force (H) per bolt and the vertical shear per bolt (V) as follows:

$$H = \text{Area from centrelines between bolts} \times f_y$$

$$F_y = 275 \, \text{MPa}$$

$$H = (85 - 30) \times 16 \times 275/2 = 121,000 \text{N} = 121 \, \text{kN}$$

V = maximum shear resisted by web/total number of bolts.
Maximum shear resisted by web:

$$V_{b,Rd} = V_{bw,Rd} \leq \frac{1.2 \times 275 \times 1740 \times 16}{\sqrt{3} \times 1.1} = 5{,}304{,}232 \, \text{N} = 5304.2 \, \text{kN}$$

$$V = 5304.2/40 = 132.6 \, \text{kN}$$

The resultant of the forces per bolt (R) is equal to $\sqrt{121^2 + 132.6^2} = 179.5 \, \text{kN}$, which is less than 206 kN (the resistance of the bolt in double shear). Then O.K.

Flange Splices

Maximum force in the upper flange $= 150 \times 275 \times 100/1000 = 4125\,\text{kN}$

$N(\text{flange}) = 4125/206 = 20$ bolts (5 rows of four bolts in double shear)

Maximum force in the lower flange $= 210 \times 275 \times 100/1000 = 5775\,\text{kN}$

$N(\text{flange}) = 5775/206$

$\qquad = 28$ taken as 30 bolts (5 rows of six bolts in double shear)

4.4.6 Design of Roller Steel Fabricated Bearings

Let us now design the roller steel fabricated bearings shown in Figure 4.106 and detailed in Figure 4.123. The maximum vertical reaction at the supports of the main plate girder was previously calculated under dead and live loads with dynamic effect $(R_{D+L+\Phi})$, which was 2193.36 kN. The material of construction for the bearings is cast iron steel (ISO 3755) 340–550 having a yield stress of 340 MPa and an ultimate stress of 550 MPa.

Design of the Sole Plate

The reaction $(R_{D+L+\Phi})$ can be assumed as two equal concentrated loads at two points, which are the centers of gravity of half of the load bearing stiffener section shown in Figure 4.123. To determine the centers of gravity (distance e), we can take the first area moment around the axis z-z, shown in Figure 4.33, as follows:

$$e = \frac{2 \times 25 \times 1.0 \times 0.5 + 23.0 \times 1.6 \times 11.5}{2 \times 25 \times 1.0 + 23.0 \times 1.6} = 5.16\,\text{cm}$$

Assuming that the thickness of the sole plate is t_1, with detailed dimensions shown in Figure 4.33 based on the flange plate girder dimensions, we can determine the maximum moment applied to the sole plate (M) as follows:

$$M = R_{D+L+\Phi} \times e/2 = 2193.36 \times 10^3 \times 51.6/2 = 56{,}588{,}688\,\text{N\,mm}.$$

Section plastic modulus $(W_{pl}) = b_1 t_1^2/4 = 800 \times t_1^2/4 = 200 \times t_1^2$

The plate thickness t_1 can be calculated now as follows:

$$\frac{M}{W_{pl}} = \frac{f_y}{\gamma_{M0}}$$

$$\frac{56{,}588{,}688}{200 \times t_1^2} = \frac{340}{1.0}$$

Then, $t_1 = 28.8\,\text{mm}$, taken as 40 mm, as shown in Figure 4.123.

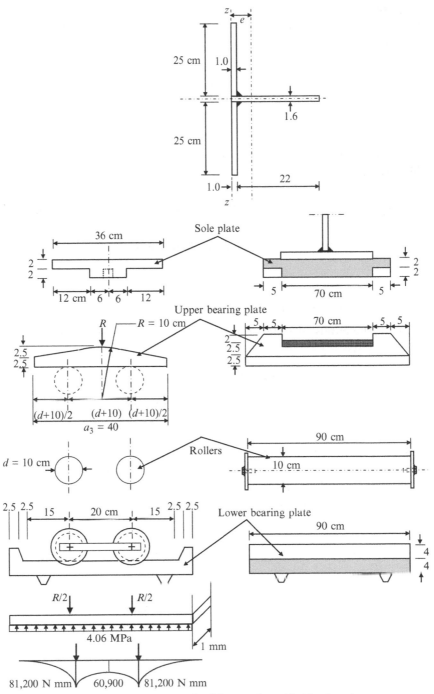

Figure 4.123 Detailing of the twin roller fabricated steel bridge bearings.

Design of the Rollers

The design of rollers requires determination of the diameter, length, and number of rollers to resist the vertical load as well as the arrangement and allowed movement in the direction of rollers. The design axial force per unit length of roller contact N'_{Sd} specified in BS EN 1337-1 [3.11] shall satisfy

$$N'_{Sd} \leq N'_{Rd}$$

where N'_{Rd} is the design value of resistance per unit length of roller contact, which is calculated as

$$N'_{Rd} = 23 \times R \times \frac{f_u^2}{E_d} \times \frac{1}{\gamma_m^2} = 23 \times R \times \frac{550^2}{210,000} \times \frac{1}{1} = 33.131 \times R$$

Assume the number of rollers is 2 and their length is 800 mm as shown in Figure 4.33:

$$N'_{Sd} = \frac{R_{D+L+\Phi}}{2 \times 900} = \frac{2193.36 \times 10^3}{1800} = 1218.5 \, \text{N/mm}$$

Then, the radius of rollers can be determined by equalizing N'_{Sd} with N'_{Rd} as follows:

$$1218.5 = 33.131 \times R$$

Then, $R = 37.7$ mm, taken as 50 mm and the diameter D is 100 mm.

Design of Upper Bearing Plate

The upper bearing plate is shown in Figure 4.123. The width and length of the plate are dependent on the spacing between rollers and the length of rollers as well as the allowed movement in the direction of rollers. The thickness of the upper bearing plate can be determined as follows:

$$M = \frac{R_{D+L+\Phi}}{2} \times \frac{(D+100)}{2} = \frac{2193.36 \times 10^3}{2} \times \frac{200}{2} = 109,668,000 \, \text{N mm}.$$

$$W_{pl} = \frac{b_2 t_2^2}{4} = \frac{900 t_2^2}{4} = 225 t_2^2 \, \text{mm}^3$$

The plate thickness t_2 can be calculated now as follows:

$$\frac{M}{W_{pl}} = \frac{f_y}{\gamma_{M0}}$$

$$\frac{109,668,000}{225 \times t_1^2} = \frac{340}{1.0}$$

Then, $t_1 = 37.8$ mm, taken as 50 mm, as shown in Figure 4.123.

The radius of the curved part of the upper bearing plate, which has a length of 700 mm as shown in Figure 4.123, can be determined the same way as that adopted for the design of the rollers:

$$N'_{Rd} = 23 \times R \times \frac{f_u^2}{E_d} \times \frac{1}{\gamma_m^2} = 23 \times R \times \frac{550^2}{210,000} \times \frac{1}{1} = 33.131 \times R$$

$$N'_{Sd} = \frac{R_{D+L+\Phi}}{700} = \frac{2193.36 \times 10^3}{700} = 3133.4\,\text{N/mm}$$

Then, the radius of rollers can be determined by equalizing N'_{Sd} with N'_{Rd} as follows:

$$3133.4 = 33.131 \times R$$

Then, $R = 94.6$ mm, taken as 100 mm.

Design of Lower Bearing Plate

The lower bearing plate is shown in Figure 4.123. The width and length of the plate are dependent on the strength of concrete and are dependent on the spacing between rollers and the length of rollers as well as the allowed movement in the direction of rollers. The thickness of the upper bearing plate can be determined as follows:

$$f_c = \frac{R_{D+L+\Phi}}{a_3 b_3} = \frac{2193.36 \times 10^3}{600 \times 900} = 4.06\,\text{MPa} < \frac{f_c}{\gamma_c} = \frac{40}{1.5} = 26.7\,\text{MPa (for a}$$

typical concrete in bridges of C40/50 with f_{ck})

The plate thickness t_3 can be calculated from the distribution of bending moment, caused by the pressure on the concrete foundation, as follows: $M = 81,200\,\text{N\,mm}$ per unit width of the plate:

$$W_{pl} = \frac{b_3 t_3^2}{4} = \frac{1 \times t_3^2}{4} = 0.25 \times t_2^2\,\text{mm}^3$$

$$\frac{M}{W_{pl}} = \frac{f_y}{\gamma_{M0}}$$

$$\frac{81200}{0.25 \times t_3^2} = \frac{340}{1.0}$$

Then, $t_3 = 30.9$ mm, taken as 40 mm, as shown in Figure 4.123.

4.4.7 Design of Hinged Line Rocker Steel Fabricated Bearings

We can also design the hinged line rocker steel fabricated bearings shown in Figure 4.106 and detailed in Figure 4.124. The maximum vertical reaction at the support of the main plate girder was previously calculated under dead and live loads with dynamic effect $(R_{D+L+\Phi})$, which was 2193.36 kN.

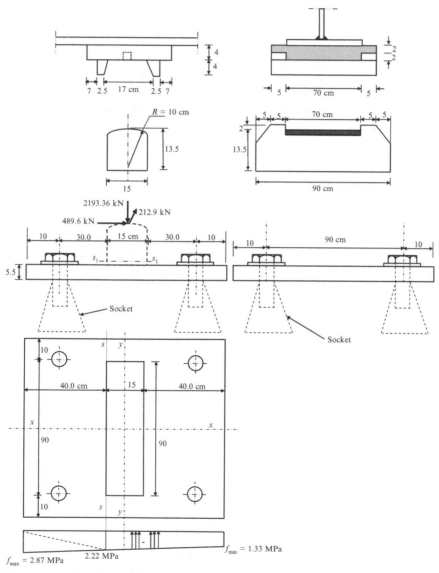

Figure 4.124 Detailing of the hinged line rocker fabricated steel bridge bearings.

The bearing is also subjected to a lateral force from the braking forces of moving traffic as well as subjected to a longitudinal force from the reactions of the lower wind bracings, which cause moments around longitudinal and lateral directions of the bearing base, respectively. Similar to the roller bearing, the material of construction for the bearings is cast iron steel (ISO 3755) 340-550 having a yield stress of 340 MPa and an ultimate stress of 550 MPa. It should be noted that the overall height of the hinged bearing must be exactly the same as that of the roller bearing. The general layout and assumed dimensions of the hinged line rocker bearing are shown in Figure 4.125. The braking Q_{lk} forces can be calculated as follows.

$Q_{lbk} = 360 + 2.7 \times 48 = 489.6$ (see Figure 4.124 for the direction of the forces). Also, the reactions from the lower wind bracings (R_{tot}) (see Figure 4.124 for the direction of the forces) were previously calculated as follows:

$$R_{tot} = 212.9\,kN$$

We can now determine the normal stress distribution due to the applied loads, shown in Figure 4.124, on the concrete foundation as follows:

$$f = -\frac{N}{A} \pm \frac{M_x}{I_x}y \pm \frac{M_y}{I_y}x$$

$$\frac{N}{A} = \frac{2193.36 \times 10^3}{950 \times 1100} = 2.098\,MPa$$

$$\frac{M_x}{I_x}y = \frac{212.9 \times 10^3 \times 190}{950 \times 1100^3/12}550 = 0.21\,MPa$$

$$\frac{M_y}{I_y}x = \frac{489.6 \times 10^3 \times 190}{1100 \times 950^3/12}475 = 0.56\,MPa$$

$$f_{max} = -2.098 - 0.21 - 0.56 = -2.87\,MPa$$

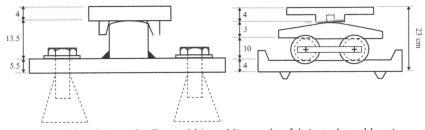

Figure 4.125 The designed roller and hinged line rocker fabricated steel bearings.

$$f_{\min} = -2.098 + 0.21 + 0.56 = -1.33\,\text{MPa}$$

The critical bending moment on the base plate of the hinged bearing is at section s-s shown in Figure 4.124:

$$M = (0.5 \times 400 \times 2.22) \times 1100 \times 400/3 + (0.5 \times 400 \times 2.87) \times 1100$$
$$\times 400 \times 2/3 = 233,493,333\,\text{N mm}$$

$$W_{\text{pl}} = 1100 \times t_4^2/4 = 275t_4^2$$

$$\frac{M}{W_{\text{pl}}} = \frac{f_y}{\gamma_{\text{M0}}}$$

$$\frac{233,493,333}{275t_4^2} = \frac{340}{1.0}$$

Then, $t_4 = 49.97$ mm, taken as 55 mm.

The normal stresses at section s_1-s_1, shown in Figure 4.124, of the line rocker bearing can be checked as follows:

$$M_x = 212.9 \times 10^3 \times 135 = 28,741,500\,\text{N mm}.$$

$$M_y = 489.6 \times 10^3 \times 135 = 66,096,000\,\text{N mm}.$$

$$\frac{N}{A} = \frac{2193.36 \times 10^3}{150 \times 900} = 16.25\,\text{MPa}$$

$$\frac{M_x}{I_x}y = \frac{28,741,500}{150 \times 900^3/12}450 = 1.42\,\text{MPa}$$

$$\frac{M_y}{I_y}x = \frac{66,096,000}{900 \times 150^3/12}75 = 19.56\,\text{MPa}$$

$$f_{\max} = -(16.25 + 1.42 + 19.56) = -37.23\,\text{MPa} < 340\,\text{MPa (Then O.K.)}$$

4.5 DESIGN EXAMPLE OF A DOUBLE-TRACK PLATE GIRDER PONY RAILWAY STEEL BRIDGE

The fourth design example presented in this chapter is for a double-track open-timber floor plate girder pony railway steel bridge. The general layout of the double-track pony bridge is shown in Figures 4.126 and 4.127. The bridge has simply supported ends, a length between supports of 27 m, and an overall length of 28 m. The width of the bridge (spacing between main

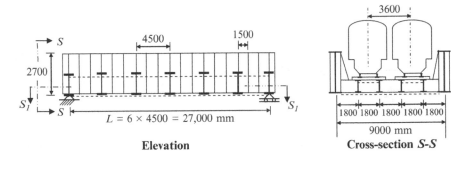

Elevation   **Cross-section *S-S***

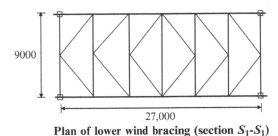

Plan of lower wind bracing (section S_1-S_1)

Figure 4.126 General layout of a double track open-timber floor plate girder pony railway steel bridge (the fourth design example).

plate girders) is 9 m as shown in Figure 4.126. It is required to design the bridge components adopting the design rules specified in EC3 [1.27]. The steel material of construction of the double-track railway bridge conformed to standard steel grade EN 10025-2 (S 275) having a yield stress of 275 MPa and an ultimate strength of 430 MPa. The bridge has only a lower wind bracing of K-shaped truss members as shown in Figure 4.126. In addition, the bridge has lateral shock (nosing force) bracing for the stringers as well as braking force bracing at the level of the lower wind bracing as shown in Figure 4.127. The lateral shock bracing eliminates bending moments around the vertical axis of the stringers, while braking force bracing eliminates bending moments around the vertical axis of the cross girder. The plate girder web is stiffened by vertical stiffeners, to safeguard against shear stresses and web buckling, spaced at a constant distance of 1.5 m. The expected live loads on the bridge conform to Load Model 71, which represents the static effect of vertical loading due to normal rail traffic as specified in EC1 [3.1]. The bolts used in connections and field splices are M27 high-strength pretensioned bolts of grade 8.8.

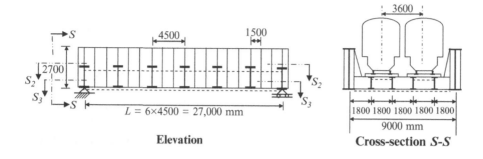

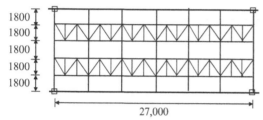

Plan of lateral shock (nosing force) bracing (section S_2-S_2)

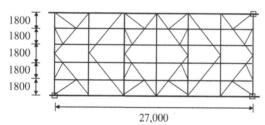

Plan of lower wind bracing with braking force bracing (section S_3-S_3)

Figure 4.127 General layout of a double track open-timber floor plate girder pony railway steel bridge (the fourth design example).

4.5.1 Design of the Stringers

Let us start by designing the stringers supporting the track as shown in Figure 4.126.

Dead Loads

$$\text{Half weight of the track load} = 3\,\text{kN/m}$$

$$\text{Weight of stringer bracing} = 0.3\,\text{kN/m}$$

$$\text{Own weight of stringer} = 1.5\,\text{kN/m}$$

$$\text{Total dead load} = g_{vk} = 4.8\,\text{kN/m}$$

Assuming the stringers are simply supported by the cross girders, we can calculate the maximum shear force and bending moment due to dead loads on a stringer (see Figure 4.128) as follows:

$$Q_{\text{D.L.}} = g_{vk} \times L/2 = 4.8 \times 4.5/2 = 10.8\,\text{kN}$$

$$M_{\text{D.L.}} = g_{vk} \times L^2/8 = 4.8 \times 4.5^2/8 = 12.15\,\text{kN m}$$

Live Loads

Considering the axle live loads on the bridge components according to Load Model 71, which represents the static effect of vertical loading due to normal rail traffic as specified in EC1 [3.1] (see Figure 4.129), three cases of loading for the evaluation of maximum bending moment due to the live loads on a stringer can be studied. The first case of loading is that the centerline at midspan of a stringer divides the spacing between the resultant of the concentrated live loads and the closest load, with maximum bending moment

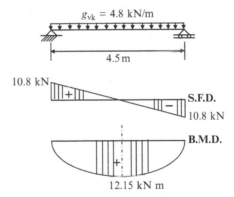

Figure 4.128 Straining actions from dead loads acting on a stringer.

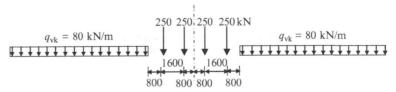

Figure 4.129 Axle live loads on the bridge conforming to Load Model 71 specified in EC1 [3.1].

located at the closest load (point a in Figure 4.130), while the second case of loading is that the centerline of the stringer is located under one of the intermediate concentrated loads, with maximum bending moment located at midspan, and finally, the third case of loading is that the stringer span is covered by the distributed live loads, with maximum bending moment located at midspan. The three cases of loading are shown in Figure 4.130:

$$M_{\text{L.L.}}(\text{case of loading 1}) = 220.83 \times 1.85 - 125 \times 1.6 = 208.54\,\text{kN m}$$

$$M_{\text{L.L.}}(\text{case of loading 2}) = 187.5 \times 2.25 - 125 \times 1.6 = 221.875\,\text{kN m}$$

$$M_{\text{L.L.}}(\text{case of loading 3}) = 40 \times 4.5^2/8 = 101.25\,\text{kN m}$$

Dynamic Factor Φ

Assuming a track with standard maintenance, therefore,

$$L_\Phi = 4.5 + 3 = 7.5\,\text{m}$$

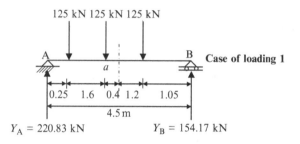

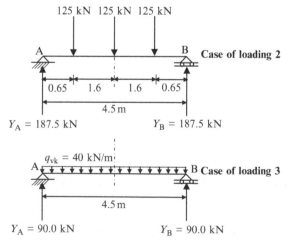

Figure 4.130 Cases of loading for the maximum bending moment acting on a stringer.

$$\Phi_3 = \frac{2.16}{\sqrt{7.5 - 0.2}} + 0.73 = 1.581, \quad \Phi_3 \geq 1.0 \text{ and } \leq 2.0.$$

Bending Moment Due to Dead and Live Loads with Dynamic Effect Added ($M_{D+L+\Phi}$)

$$M_{D+L+\Phi} = M_{D.L.} \times \gamma_g + \Phi \times M_{L.L.} \times \gamma_q$$
$$= 12.15 \times 1.2 + 1.581 \times 221.875 \times 1.45 = 523.2 \, kN \, m$$

It should be noted that the load factors adopted in this study are that of the ultimate limit state. This is attributed to the fact that the finite element models presented in Chapters 6 and 7 can be used to analyze the bridges and provide more accurate predictions for the deflections and other serviceability limit state cases of loading.

Shearing Force Due to Dead and Live Loads with Dynamic Effect Added ($Q_{D+L+\Phi}$)

There is only a single case of loading for live loads to produce a maximum shear force at the supports of the stringer, which is shown in Figure 4.131:

$$Q_{L.L.} = 241.7 \, kN$$

$$Q_{D+L+\Phi} = Q_{D.L.} \times \gamma_g + \Phi \times Q_{L.L.} \times \gamma_q$$
$$= 10.8 \times 1.2 + 1.581 \times 241.7 \times 1.45 = 567 \, kN$$

Design Bending Moment (M_{Ed}) and Shear Force (Q_{Ed})

$$M_{Ed} = M_{D+L+\Phi} = 523.2 \, kN \, m$$
$$Q_{Ed} = Q_{D+L+\Phi} = 567 \, kN$$

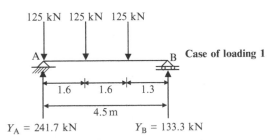

Figure 4.131 Cases of loading for the maximum shear force acting on a stringer.

Design of Stringer Cross Section

$$M_{c,Rd} = \frac{W_{pl} \times f_y}{\gamma_{M0}} \quad \text{for classes 1 and 2}$$

$$523.2 \times 10^6 = \frac{W_{pl} \times 275}{1.0}$$

$$W_{PL} = 1{,}902{,}545 \text{ mm}^3 = 1902.5 \text{ cm}^3$$

Choose UB 457 × 191 × 89 (equivalent to American W18 × 60), shown in Figure 4.132. W_{PL} around x–$x = 2014$ cm^3. To classify the cross section chosen,

$$\varepsilon = \sqrt{\frac{235}{f_y}} = \sqrt{\frac{235}{275}} = 0.924$$

$C_1 = 80.5$ mm, $t_{fl} = 17.7$, $C_1/t_{fl} = 80.5/17.7 = 4.5 \leq 9 \times 0.924$
$= 8.316$(Stringer flange is class 1)

$C_2 = 407.6$ mm, $t_w = 10.5$, $C_1/t_{fl} = 407.6/10.5 = 38.8 \leq 72 \times 0.924$
$= 66.5$(Stringer web is class 1)

Check of Bending Resistance

$$M_{c,Rd} = \frac{W_{pl} \times f_y}{\gamma_{M0}} = \frac{2014 \times 10^3 \times 275}{1.0} = 553{,}850{,}000 \text{ N mm}$$

$$= 553.85 \text{ kN m} > M_{Ed} = 523.2 \text{ kN m} \text{ (Then O.K.)}$$

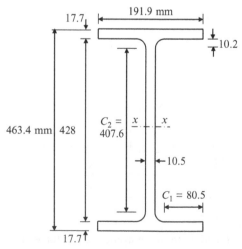

Figure 4.132 The cross-section of stringers (UB 457 × 191 × 89).

Check of Shear Resistance

$$V_{pl,Rd} = \frac{A_v\left(f_y/\sqrt{3}\right)}{\gamma_{M0}} = \frac{(428 \times 10.5) \times \left(275/\sqrt{3}\right)}{1.0} = 713,518.3\,\text{N}$$

$$= 713.5\,\text{kN} > Q_{Ed} = 567\,\text{kN}\,(\text{Then O.K.})$$

4.5.2 Design of the Cross Girders

The cross girders carry concentrated loads from the stringers as shown in Figure 4.133. Therefore, we can analyze an intermediate cross girder as follows:

Dead Loads

$$\text{Reaction from stringers due to dead loads} = 4.8 \times 4.5 = 21.6\,\text{kN}$$

$$\text{Own weight of cross girder} = 3.0\,\text{kN/m}$$

Assuming the cross girders are simply supported by the main plate girders, we can calculate the maximum shear force and bending moment

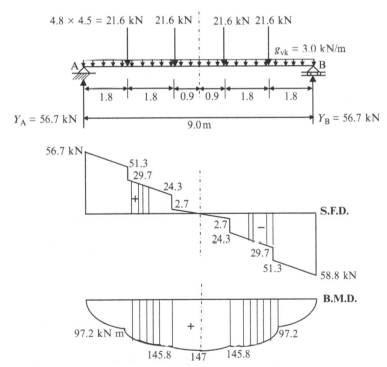

Figure 4.133 Straining actions from dead loads acting on an intermediate cross girder.

due to dead loads on an intermediate cross girder (see Figure 4.133) as follows:

$$Q_{D.L.} = 3 \times 9/2 + 2 \times 21.6 = 56.7 \, kN$$

$$M_{D.L.} = 3 \times 9^2/8 + 21.6 \times 1.8 + 21.6 \times 3.6 = 147 \, kN\,m$$

Live Loads

To determine the maximum reactions due to live loads transferred by the stringers to the cross girders, the case of loading shown in Figure 4.134 is studied. The maximum reaction $R_{L.L.}$ can be calculated as follows:

$$R_{L.L.} = 125 + 2 \times 125 \times (4.5 - 1.6)/5 + 125 \times (4.5 - 3.2)/5 + 40 \times 0.5$$
$$\times 0.25/4.5 + 40 \times 2.1 \times 1.05/4.5 = 342.9 \, kN$$

The maximum straining actions due to live loads on an intermediate cross girder can be then calculated (see Figure 4.135) as follows:

$$Q_{L.L.} = 2 \times 342.9 = 685.8 \, kN$$

$$M_{L.L.} = 342.8 \times 1.8 + 342.9 \times 3.6 = 1851.66 \, kN\,m$$

Dynamic Factor Φ

$$L_\Phi = 2 \times 9 = 18 \, m$$

$$\Phi_3 = \frac{2.16}{\sqrt{18 - 0.2}} + 0.73 = 1.264, \quad \Phi_3 \geq 1.0 \text{ and} \leq 2.0.$$

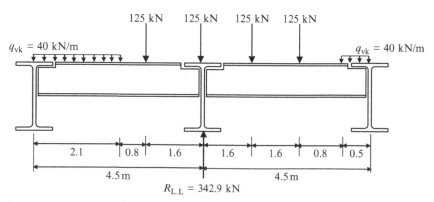

Figure 4.134 The case of loading producing maximum straining actions from live loads on an intermediate cross girder.

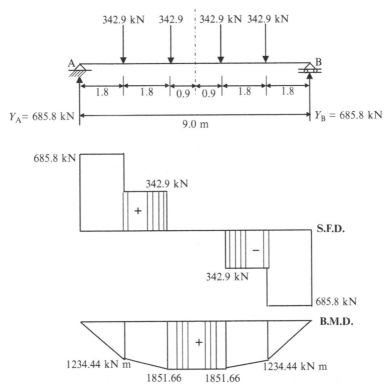

Figure 4.135 Straining actions from live loads acting on an intermediate cross girder.

Bending Moment Due to Dead and Live Loads with Dynamic Effect Added ($M_{D+L+\Phi}$)

$$M_{D+L+\Phi} = M_{D.L.} \times \gamma_g + \Phi \times M_{L.L.} \times \gamma_q$$
$$= 147 \times 1.2 + 1.264 \times 1851.66 \times 1.45 = 3570.1 \, \text{kN m}$$

Shearing Force Due to Dead and Live Loads with Dynamic Effect Added ($Q_{D+L+\Phi}$)

$$Q_{D+L+\Phi} = Q_{D.L.} \times \gamma_g + \Phi \times Q_{L.L.} \times \gamma_q$$
$$= 56.7 \times 1.2 + 1.264 \times 685.8 \times 1.45 = 1325 \, \text{kN}$$

Since the investigated bridge is a pony bridge, therefore, there is an additional bending moment, resulting from the flexibility of the U-frame, which must be added to the calculated bending moment due to the dead and live loads. The additional bending moment is equal to 1/100 the maximum force in the compression flange of the main plate girder (F_{max}) multiplied by the arm (r). The maximum force can be calculated as follows (Figure 4.136):

$$F_{max} = \frac{M_{Ed}(M.G.)}{y_{ct}} = \frac{20285.7}{0.99 \times 270} = 7589\,kN$$

Additional bending moment $= (7589/100) \times 2.2306 = 169.3\,kN\,m$

Design Bending Moment (M_{Ed}) and Shear Force (Q_{Ed})

$$M_{Ed} = M_{D+L+\Phi} = 3570.1 + 169.3 = 3739.4\,kN\,m$$

$$Q_{Ed} = Q_{D+L+\Phi} = 1325\,kN$$

Design of the Cross Girder Cross Section

$$M_{c,Rd} = \frac{W_{pl} \times f_y}{y_{M0}} \quad \text{for classes 1 and 2}$$

$$3739.4 \times 10^6 = \frac{W_{pl} \times 275}{1.0}$$

$$W_{PL} = 13,597,818\,mm^3 = 13,597.8\,cm^3$$

Choose UB $914 \times 419 \times 343$ (equivalent to American W36×230), shown in Figure 4.137. W_{PL} around x-$x = 15,480\,cm^3$. To classify the cross section chosen,

$$\varepsilon = \sqrt{\frac{235}{f_y}} = \sqrt{\frac{235}{275}} = 0.924$$

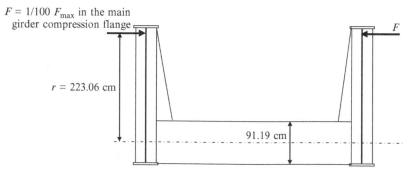

$F = 1/100\,F_{max}$ in the main girder compression flange

$r = 223.06\,cm$

$91.19\,cm$

F

Figure 4.136 Additional bending moment on cross girders due to flexibility of the U-frame.

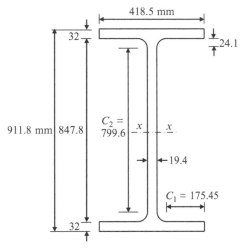

Figure 4.137 The cross-section of cross girders (UB 914 × 419 × 343).

$C_1 = 175.45\,\text{mm}$, $t_{fl} = 32$, $C_1/t_{fl} = 175.45/32 = 5.48 \leq 9 \times 0.924$
$= 8.316\,(\text{Cross girder flange is class 1})$

$C_2 = 799.6\,\text{mm}$, $t_w = 19.4$, $C_1/t_{fl} = 799.6/19.3 = 41.2 \leq 72 \times 0.924$
$= 66.5\,(\text{Cross girder web is class 1})$

Check of Bending Resistance

$$M_{c,Rd} = \frac{W_{pl} \times f_y}{\gamma_{M0}} = \frac{15,480 \times 10^3 \times 275}{1.0} = 4,257,000,000\,\text{N mm}$$
$$= 4257\,\text{kN m} > M_{Ed} = 3739.4\,\text{kN m}\,(\text{Then O.K.})$$

Check of Shear Resistance

$$V_{pl,Rd} = \frac{A_v\left(f_y/\sqrt{3}\right)}{\gamma_{M0}} = \frac{(847.8 \times 19.4) \times \left(275/\sqrt{3}\right)}{1.0} = 2,611,362.8\,\text{N}$$
$$- 2611.4\,\text{kN} > Q_{Ed} = 1325\,\text{kN}\,(\text{Then O.K.})$$

4.5.3 Design of the Main Plate Girders

Let us now design the main plate girders supporting the cross girders as shown in Figure 4.126. We can estimate the dead and live loads acting on a main plate girder as follows:

Dead Loads

$$\text{Weight of steel structure} = 11 + 0.5 \times 27 = 24.5\,\text{kN/m}$$

$$\text{Track load} = 6\,\text{kN/m}$$

$$\text{Total dead load} = g_{vk} = 1.8 \times 24.5/2 + 6 = 28.05\,\text{kN/m}$$

The main plate girders are simply supported; hence, we can calculate the maximum shear force and bending moment due to dead loads on a main plate girder (see Figure 4.138) as follows:

$$Q_{\text{D.L.}} = g_{vk} \times L/2 = 28.05 \times 27/2 = 378.7\,\text{kN}$$

$$M_{\text{D.L.}} = g_{vk} \times L^2/8 = 28.05 \times 27^2/8 = 2556.1\,\text{kN\,m}$$

Live Loads

Considering the axle loads on the bridge components according to Load Model 71 (see Figure 4.129), two cases of loading for the evaluation of maximum bending moment due to live loads on a main plate girder can be studied. The first case of loading is that the centerline of the main plate girder is located under one of the intermediate concentrated live loads, with maximum bending moment calculated at midspan (see Figure 4.139). On the other hand, the second case of loading is that the centerline of a main plate girder divides the spacing between the resultant of the concentrated live

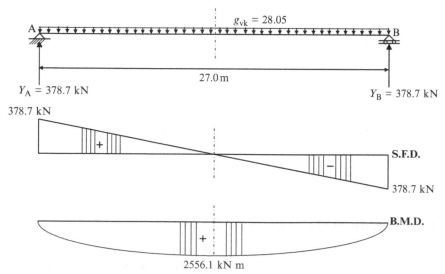

Figure 4.138 Straining actions from dead loads acting on one main plate girder.

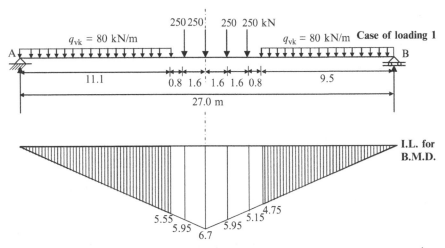

Figure 4.139 Determination of the maximum bending moment on one main plate girder due to live loads using the influence line method (case of loading 1).

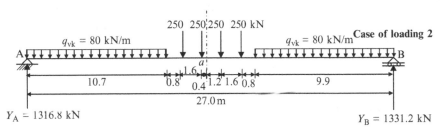

Figure 4.140 Determination of the maximum bending moment on one main plate girder due to live loads using the analytical method (case of loading 2).

loads and the closest load, with maximum bending moment located at the closest load (point a in Figure 4.140). The maximum bending moment under the first case of loading is calculated using the influence line method (by multiplying the concentrated loads by the companion coordinates on the bending moment diagram and by multiplying the distributed loads by the companion areas on the bending moment diagram), while that under the second case of loading is calculated analytically using structural analysis. Hence, the bending moments due to live loads can be calculated as follows:

$$M_{L.L.}(\text{case of loading 1}) = 250 \times [5.95 + 6.75 + 5.95 + 5.15] + 80 \times 0.5$$
$$\times 9.5 \times 4.75 + 80 \times 0.5 \times 11.1 \times 5.55$$
$$= 10{,}219.2 \, \text{kN m}$$

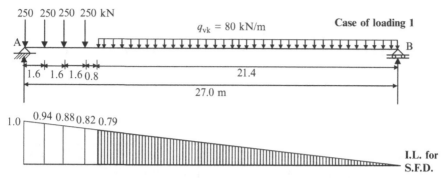

Figure 4.141 Determination of the maximum shear force on one main plate girder due to live loads using the influence line method (case of loading 1).

$$M_{L.L.}(\text{case of loading 2}) = 1316.8 \times 13.1 - 80 \times 10.7 \times 7.75 - 250 \times 1.6$$
$$= 10216.1\,\text{kN}\,\text{m}$$

There is only a single case of loading for the live loads to produce a maximum shear force at the supports of a main plate girder, which is shown in Figure 4.141. Once again, we can use the influence line method to calculate the maximum shear force due to this case of loading or analytically by taking moment at support B and evaluate the reaction at A:

$$Q_{L.L.} = 10216.1\,\text{kN}$$

Dynamic Factor Φ

$$L_{\Phi} = 27\,\text{m}$$

$$\Phi_3 = \frac{2.16}{\sqrt{27} - 0.2} + 0.73 = 1.162, \quad \Phi_3 \geq 1.0 \text{ and } \leq 2.0.$$

Bending Moment Due to Dead and Live Loads with Dynamic Effect Added ($M_{D+L+\Phi}$)

$$M_{D+L+\Phi} = M_{D.L.} \times \gamma_g + \Phi \times M_{L.L.} \times \gamma_q$$
$$= 2556.1 \times 1.2 + 1.162 \times 10219.2 \times 1.45 = 20285.7\,\text{kN}\,\text{m}$$

Shearing Force Due to Dead and Live Loads with Dynamic Effect Added ($Q_{D+L+\Phi}$)

$$Q_{D+L+\Phi} = Q_{D.L.} \times \gamma_g + \Phi \times Q_{L.L.} \times \gamma_q$$
$$= 378.7 \times 1.2 + 1.162 \times 1589.6 \times 1.45 = 3132.8\,\text{kN}$$

Design Bending Moment (M_{Ed}) and Shear Force (Q_{Ed})

$$M_{Ed} = M_{D+L+\Phi} = 20{,}285.7\,kN\,m$$

$$Q_{Ed} = Q_{D+L+\Phi} = 3132.8\,kN$$

Design of the Main Plate Girder Cross Section

Let us assume the main plate girder cross section shown in Figure 4.16. The cross section consists of two flange plates for the upper and lower flanges and a web plate. The web plate height is taken as equal to $L/10 = 27{,}000/10 = 2700$ mm, with a plate thickness of 16 mm. The width of the bottom plate of the upper and lower flanges of the cross section is taken as 0.2 the web height, which is equal to 540 mm, while the top plate width is taken as 500 mm, to allow for welding with the bottom flange plate. The flange plates have the same plate thickness of 30 mm. The choice of two flange plates for the upper and lower flanges is intended to curtail the top flange plate approximately at quarter-span as will be detailed in the coming sections. It should be noted that the web height value ($L/10$) is an acceptable recommended [1.9] value for railway steel bridges constructed in Great Britain and Europe. This value is an initial value for preliminary cross-sectional estimation. The cross section has to be checked, classified, designed, and assessed against deflection limits set by serviceability limit states. To classify the cross section chosen,

$$\varepsilon = \sqrt{\frac{235}{f_y}} = \sqrt{\frac{235}{275}} = 0.924$$

$C_1 = 254\,mm$, $t_{fl} = 60$, $C_1/t_{fl} = 254/60 = 3.2 \leq 9 \times 0.924$
$= 8.316$ (Main plate girder flange is class 1).

$C_2 = 2684\,mm$, $t_w = 16$, $C_1/t_{fl} = 2684/16 = 167.8 > 124 \times 0.924$
$= 114.58$ (Main plate girder web is class 4).

To calculate the bending moment resistance, the effective area should be used. Considering web plate buckling, the effective area of the web part in compression (see Figure 4.142) can be calculated as follows:

$$k_\sigma = 23.9$$

$$\bar{\lambda}_p = \frac{270/1.6}{28.4 \times 0.924 \times \sqrt{23.9}} = 1.315 > 0.673$$

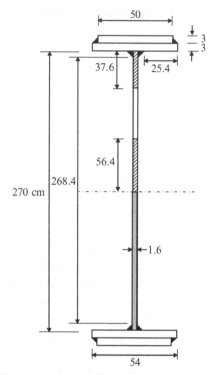

Figure 4.142 Reduced cross-section of plate girder.

$$\rho = \frac{1.315 - 0.055(3 - 1)}{1.315^2} = 0.697$$

$$b_{eff} = 0.633 \times 270/2 = 94\,\text{cm},$$

Then, $b_{eff1} = 0.6 \times 94 = 56.4$ cm and $b_{eff2} = 0.4 \times 94 = 37.4$ cm as shown in Figure 4.142.

To calculate the elastic section modulus, the elastic centroid of the section has to be located by taking the first area moment, as an example, around axis y_0-y_0 shown in Figure 4.143, as follows:

$$A = 54 \times 3 \times 2 + 50 \times 3 \times 2 + 191.4 \times 1.6 + 37.6 \times 1.6 = 990.4\,\text{cm}^2$$

$$y_c = \frac{\begin{bmatrix} 50 \times 3 \times 1.5 + 54 \times 3 \times 4.5 + 191.4 \times 1.6 \times 101.7 + 37.6 \\ \times 1.6 \times 257.2 + 54 \times 3 \times 277.5 + 50 \times 3 \times 280.5 \end{bmatrix}}{990.4}$$

$$y_c = 136\,\text{cm}$$

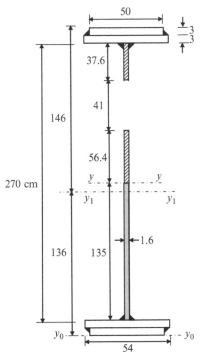

Figure 4.143 Calculation of properties of area for main plate girder.

Inertia about y_1-$y_1 = [50 \times 3^3/12 + 50 \times 3 \times 134.5^2]$

$$+ [54 \times 3^3/12 + 54 \times 3 \times 131.5^2]$$
$$+ [1.6 \times 191.4^3/12 + 1.6 \times 191.4 \times 34.3^2]$$
$$+ [1.6 \times 37.6^3/12 + 1.6 \times 37.6 \times 121.2^2]$$
$$+ [54 \times 3^3/12 + 54 \times 3 \times 141.5^2]$$
$$+ [50 \times 3^3/12 + 50 \times 3 \times 144.5^2] = 14{,}076{,}983.3 \, \text{cm}^4$$

$$W_{\text{eff,min}} = 14{,}076{,}983.3/146 = 96417.7 \, \text{cm}^3$$

Check of Bending Resistance

$$M_{\text{c,Rd}} = \frac{W_{\text{eff,min}} \times f_y}{\gamma_{\text{M0}}} = \frac{96{,}417.7 \times 10^3 \times 275}{1.0} = 26{,}514{,}900{,}000 \, \text{N mm}$$

$$= 26514.9 \, \text{kN m} > M_{\text{Ed}} = 20{,}285.7 \, \text{kN m} \, (\text{Then O.K.})$$

Check of Shear Resistance

$$V_{b,Rd} = V_{bw,Rd} + V_{bf,Rd} \leq \frac{\eta f_{yw} h_w t_w}{\sqrt{3}\gamma_{M1}}$$

By neglecting the flange contribution,

$$V_{b,Rd} = V_{bw,Rd} \leq \frac{1.2 \times 275 \times 2700 \times 16}{\sqrt{3} \times 1.1} = 7,482,459.6\,N$$

$$V_{bw,Rd} = \frac{\chi_w f_{yw} h_w t_w}{\sqrt{3}\gamma_{M1}}$$

$$\bar{\lambda}_w = 0.76\sqrt{\frac{f_{yw}}{\tau_{cr}}}, \quad \tau_{cr} = k_\tau \sigma_E$$

$$\sigma_E = 190,000(16/2700)^2 = 6.672\,N/mm^2$$

$$k_\tau = 4 + 5.34(2700/1500)^2 = 21.3$$

$$\bar{\lambda}_w = 0.76\sqrt{\frac{275}{21.3 \times 6.672}} = 1.057 > 1.08$$

Then, $\chi_w = \dfrac{0.83}{1.057} = 0.785$

$$V_{bw,Rd} = \frac{0.785 \times 275 \times 2700 \times 16}{\sqrt{3} \times 1.1} = 4,894,800\,N = 4894.8\,kN$$
$$< 7482.5\,kN$$

$$\eta_3 = \frac{V_{Ed}}{V_{b,Rd}} = \frac{3132.8}{4894.8} = 0.64 < 1.0\,(\text{Then O.K.})$$

It should be noted that for this type of bridges, it is recommended that further checks regarding the assessment of fatigue loading have to be performed. However, this can be done using advanced finite element modeling of the bridge.

4.5.4 Curtailment (Transition) of the Flange Plates of the Main Plate Girder

The critical cross section of the main plate girder at midspan, which is subjected to the maximum bending moment, was designed previously with two flange plates. Since the main plate girder is simply supported, the bending moment is decreased towards the supports. Therefore, we can stop the top flange plate at a certain distance to get the most benefit from the material.

This process is commonly called as curtailment (transition) of flange plates. It should be noted that, theoretically, curtailment (transition) of flange plates can be conducted by reducing the flange plate width, thickness, or both. However, in practice, fabricators prefer to keep the flange widths constant and vary the thickness because this option costs much less than reducing the flange width, which might require a very heavy grinding work. To avoid lateral torsional buckling of the compression top flange at the reduction zone, it is recommended practically to reduce the width or thickness by 40% of the original with a smooth transition zone sloping at 1 (vertical) to 10 (horizontal). It is also recommended that bridges with lengths of 20-30 m are curtailed in one step, while for bridges with spans greater than 30 m, two steps of curtailment (transition) are recommended. For the investigated design example, we can conduct one-step curtailment by reducing the top flange plate of the upper and lower flanges, as shown in Figure 4.144. To classify the reduced cross section,

$$\varepsilon = \sqrt{\frac{235}{f_y}} = \sqrt{\frac{235}{275}} = 0.924$$

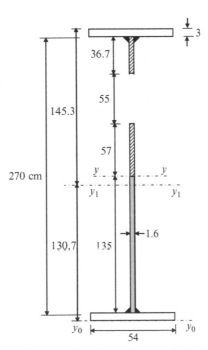

Figure 4.144 Calculation of properties of area for curtailed main plate girder.

$C_1 = 254\,\text{mm}, \quad t_{fl} = 30, \quad C_1/t_{fl} = 254/30 = 8.47\,(\text{Class 2}).$

$C_2 = 2684\,\text{mm}, \quad t_w = 16, \quad C_1/t_{fl} = 2684/16 = 167.75 > 124 \times 0.924$
$\quad = 114.58\,(\text{Class 4}).$

To calculate the bending moment resistance, the effective area should be used. Considering web plate buckling, the effective area of the part of web plate in compression (see Figure 4.144) can be calculated as follows:

$$k_\sigma = 23.9$$

$$\bar{\lambda}_p = \frac{270/1.6}{28.4 \times 0.924 \times \sqrt{23.9}} = 1.315 > 0.673$$

$$\rho = \frac{1.315 - 0.055(3-1)}{1.315^2} = 0.697$$

$$b_{\text{eff}} = 0.697 \times 270/2 = 94\,\text{cm},$$

Then, $b_{\text{eff1}} = 0.6 \times 94 = 56.4$ cm and $b_{\text{eff2}} = 0.4 \times 94 = 37.6$ cm as shown in Figure 4.144.

To calculate the elastic section modulus, the elastic centroid of the section has to be located by taking the first area moment, as an example, around axis y_0-y_0 shown in Figure 4.144, as follows:

$$A = 54 \times 3 \times 2 + 191.4 \times 1.6 + 37.6 \times 1.6 = 690.4\,\text{cm}^2$$

$$y_c = \frac{[54 \times 3 \times 1.5 + 191.4 \times 1.6 \times 98.7 + 37.6 \times 1.6 \times 254.2 + 54 \times 3 \times 274.5]}{690.4}$$

$$y_c = 130.7\,\text{cm}$$

Inertia about y_1-y_1 $= [54 \times 3^3/12 + 54 \times 3 \times 129.2^2]$
$\qquad + [54 \times 3^3/12 + 54 \times 3 \times 143.8^2]$
$\qquad + [1.6 \times 191.4^3/12 + 1.6 \times 191.4 \times 32^2]$
$\qquad + [1.6 \times 37.6^3/12 + 1.6 \times 37.6 \times 143.8^2]$
$\qquad = 8{,}227{,}509\,\text{cm}^4$

$$W_{\text{eff,min}} = 8{,}227{,}509/145.3 = 56{,}624.3\,\text{cm}^3$$

Bending Moment Resistance

$$M_{c,\text{Rd}} = \frac{W_{\text{eff,min}} \times f_y}{\gamma_{M0}} = \frac{56{,}624.3 \times 10^3 \times 275}{1.0} = 15{,}571{,}700{,}000\,\text{N\,mm}$$
$$= 15{,}571.7\,\text{kN\,m}$$

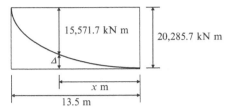

Figure 4.145 Calculation of curtailed flange plate lengths.

Length of Flange Plates

Assuming the overall bending moment diagram of the main plate girder is a second-degree parabola (see Figure 4.145), we can determine the length of the curtailed top flange plate of the upper and lower flanges as follows:

$$\left(\frac{x}{L/2}\right)^2 = \frac{20{,}285.7 - 15{,}571.7}{20{,}285.7} = \frac{4714}{20{,}285.7} = 0.23238$$

$$\frac{x}{13.5} = 0.482; \text{ then, } x = 6.5 \text{ m.}$$

Hence, the length of the smaller top plate is 13 m.

4.5.5 Design of the Fillet Weld Between Flange Plates and Web

To determine the size of fillet weld connecting the bottom flange plates of the upper and lower flanges with the web plate for the investigated bridge, we can calculate the maximum shear flow at the support for the reduced cross section, shown in Figure 4.146, as follows:

$$\text{Inertia about } y\text{-}y = 1.6 \times 270^3/12 + 2 \times \left[54 \times 3^3/12 + 54 \times 3 \times 136.5^2\right]$$
$$= 8{,}661{,}492 \text{ cm}^4$$

Shear flow at section s-s:

$$q = \frac{Q_{Ed} \times S_{ss}}{I_{yy}} = 2 \times a \times f_{vw,d}$$

$$f_{vw,d} = \frac{f_u/\sqrt{3}}{\beta_w \gamma_{M2}} = \frac{430/\sqrt{30}}{0.85 \times 1.25} = 233.7 \text{ N/mm}^2$$

$$q = \frac{3327.2 \times 10^3 \times (54 \times 3 \times 136.5) \times 10^3}{8{,}661{,}492 \times 10^4} = 2 \times a \times 233.7$$

Then, $a = 1.7$ mm, taken as 8 mm, which is the minimum size.

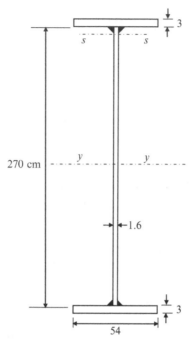

270 cm

1.6

54

Figure 4.146 Calculation of flange fillet weld size at supports.

4.5.6 Check of Lateral Torsional Buckling of the Plate Girder Compression Flange

To check the safety of the upper compression flange against lateral torsional buckling, we have to calculate the elastic critical moment for lateral torsional buckling (M_{cr}). However, to calculate M_{cr}, we have to evaluate the effective buckling length (unsupported length) of the compression flange of this investigated pony bridge (l_e), which can be calculated as follows:

$l_e = \dfrac{\pi}{\sqrt{2}}[EI_c a\delta]^{1/4}$, where I_c is the inertial of the compression flange (see Figure 4.147) about z-z axis, a is the spacing between cross girders, and δ is the flexibility of the U-frame reasonably assumed $= 0.00006$ mm/N for pony bridges.

$$I_c = 3 \times 50^3/12 + 3 \times 54^3/12 + 45 \times 1.6^3/12 = 70{,}631.4 \text{ cm}^4$$

$l_e = \dfrac{\pi}{\sqrt{2}}\left[210{,}000 \times 70{,}631.4 \times 10^4 \times 4500 \times 0.00006\right]^{1/4}$ (should be not less than $a = 450$ cm)

Considering the cross section at midspan shown in Figure 4.147, we can calculate M_{cr} as follows:

$$M_{cr} = C_b \frac{\pi}{kl_b}\sqrt{\left(\frac{\pi E_y}{kl_b}\right)^2 C_w I_z + E_z I_z G \mathcal{J}}$$

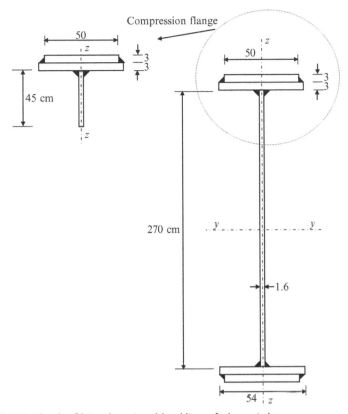

Figure 4.147 Check of lateral torsional buckling of plate girder.

Given, $C_b = 1.13, E = 210\,\text{GPa}, G = 81\,\text{GPa}, l_b = 5588.3\,\text{mm}$ and $k = 1$.

Inertia about $z\text{-}z(I_z) = 2 \times 3 \times 50^3/12 + 2 \times 3 \times 54^2/12 + 270 \times 1.6^3/12$
$$= 141,324\,\text{cm}^4$$

$$C_w = \frac{h^2 \times I_z}{4} = \frac{2820^2 \times 141,324 \times 10^4}{4} = 2.809662444 \times 10^{15}\,\text{mm}^6$$

$$j = \frac{1}{3}\left(2 \times 540 \times 30^3 + 2 \times 500 \times 30^3 + 2820 \times 16^3\right) = 22,570,240\,\text{mm}^2$$

$$M_{cr} = 1.13\frac{3.14}{5588.3}\sqrt{\left(\frac{3.14 \times 210}{5588.3}\right)^2 2.809662444 \times 10^{15} \times 141,324}$$

$$\sqrt{\times 10^4 + 210 \times 141,324 \times 10^4 \times 81 \times 22,570,240}$$

$$M_{cr} = 0.000634933\sqrt{5.528515455 \times 10^{22} + 5.425707933 \times 10^{20}}$$

$$M_{cr} = 150,021,280 \, \text{N mm} = 150,021.2 \, \text{kN m}$$

We can now check the safety against lateral torsional buckling following the rules specified in EC3 [1.27, 2.11] as follows:

$$\frac{M_{Ed}}{M_{b,Rd}} \leq 1.0$$

Given: $M_{Ed} = 20,285.7 \, \text{kN m}$, $W_y = 96,417.7 \, \text{cm}^3$

$$M_{b,Rd} = \chi_{LT} W_y \frac{f_y}{\gamma_{M1}}$$

$$\bar{\lambda}_{LT} = \sqrt{\frac{W_y f_y}{M_{cr}}} = \sqrt{\frac{26,514.9}{150,021.2}} = 0.42$$

$$\Phi_{LT} = 0.5\left[1 + \alpha_{LT}(\bar{\lambda}_{LT} - 0.2) + \bar{\lambda}_{LT}^2\right] = 0.5\left[1 + 0.76(0.42 - 0.2) + 0.42^2\right]$$
$$= 0.672$$

$$\chi_{LT} = \frac{1}{\Phi_{LT} + \sqrt{\Phi_{LT}^2 - \bar{\lambda}_{LT}^2}} \quad \text{but } \chi_{LT} \leq 1.0$$

$$\chi_{LT} = \frac{1}{0.672 + \sqrt{0.672^2 - 0.42^2}} \quad \text{but } \chi_{LT} \leq 1.0$$

$$\chi_{LT} = 0.836$$

$$M_{b,Rd} = \frac{0.836 \times 26,514.9}{1.0} = 22,166.5 \, \text{kN m} > 20,285.7 \, \text{kN m}$$

4.5.7 Design of Web Stiffeners

There are two types of stiffeners used to strengthen the thin web plate of the main plate girder against buckling due to shear stresses, bending stresses, or both. The stiffeners at the supports are commonly known as load bearing stiffeners, while intermediate stiffeners are commonly known as stability stiffeners (intermediate transverse stiffeners). The design of the stiffeners can be performed as follows:

4.5.7.1 Load Bearing Stiffeners

To design the load bearing stiffener at supports (see Figure 4.148), we can also follow the design rules specified in EC3 [1.27, 2.11] for concentrically loaded compression members. The axial force in the stiffener is the

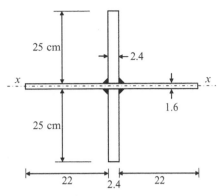

Figure 4.148 Load bearing web stiffeners at supports.

maximum reaction at supports ($N_{Ed} = R_{D+L+\Phi}$), which is equal to 3132.8 kN. The design procedures can be performed as follows:

$$\frac{N_{Ed}}{N_{b,Rd}} \leq 1.0$$

where $N_{b,Rd} = \dfrac{\chi A f_y}{\gamma_{M1}}$

$$A = 2 \times 25 \times 2.4 + 46.4 \times 1.6 = 194.24 \, \text{cm}^2$$

$$I_x = 46.4 \times 1.6^3/12 + 2 \times \left[2.4 \times 25^3/12 + 60 \times 13.3^2\right] = 27{,}492.6 \, \text{cm}^4$$

$$\chi = \frac{1}{\Phi + \sqrt{\Phi^2 - \bar{\lambda}^2}} \quad \text{but } \chi \leq 1.0$$

$$\Phi = 0.5\left[1 + \alpha(\bar{\lambda} - 0.2) + \bar{\lambda}^2\right]$$

$$\bar{\lambda} = \sqrt{\frac{A f_y}{N_{cr}}}$$

$$N_{cr} = \frac{\pi^2 \times EI}{L^2} = \frac{3.14^2 \times 210{,}000 \times 27{,}492.6 \times 10^4}{5400^2} = 19{,}521{,}217 \, \text{N}$$

$$\bar{\lambda} = \sqrt{\frac{194.24 \times 100 \times 275}{19{,}521{,}217}} = 0.523$$

$$\Phi = 0.5\left[1 + 0.49(0.523 - 0.2) + 0.523^2\right] = 0.716$$

$$\Phi = 0.5\left[1 + 0.49(0.523 - 0.2) + 0.523^2\right] = 0.716$$

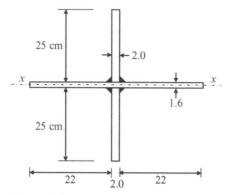

Figure 4.149 Intermediate stability web stiffeners.

$$\chi = \frac{1}{0.716 + \sqrt{0.716^2 - 0.523^2}} = 0.83 \text{ but } \chi \leq 1.0$$

$$\text{Then, } N_{b,Rd} = \frac{0.83 \times 194.26 \times 100 \times 275}{1.1} = 4,030,895 \text{ N}$$

$$N_{b,Rd} = 4030.9 \text{ kN} > N_{Ed} = 3132.8 \text{ kN (Then O.K.)}$$

4.5.7.2 Intermediate Stiffeners

Intermediate stiffeners (see Figure 4.149) can be designed by choosing its dimensions such that

$$\frac{a_1}{h_w} = \frac{1500}{2700} = 0.556 < \sqrt{2} = 1.414 \text{ (Then O.K.)}$$

and

$$I_{st} \geq \frac{1.5 h_w^2 t_w^3}{a_1^2} = \frac{1.5 \times 270^3 \times 1.6^3}{150^2} = 5374.8 \text{ cm}^4$$

$$I_{st} = 46 \times 1.6^3/12 + 2 \times \left[2 \times 25^3/12 + 50 \times 13.3^2 \right]$$

$$= 22,913 \text{ cm}^4 > 5374.8 \text{ cm}^4 \text{ (Then O.K.)}$$

4.5.8 Design of Stringer Bracing (Lateral Shock or Nosing Force Bracings)

The stringer bracing are subjected to lateral moving reversible force of 100 kN. The bracing members carry either tensile or compressive forces according to the changing direction of the lateral shock force (transverse horizontal force) (see Figure 4.150). The cross section of the bracing

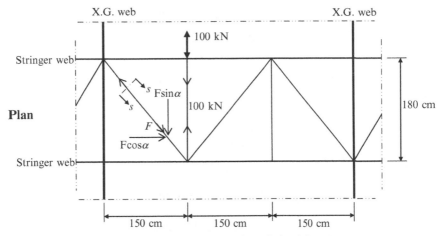

Figure 4.150 Analysis of forces acting on the lateral shock bracing.

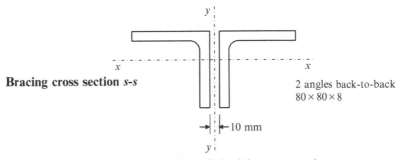

Figure 4.151 The cross section of the lateral shock bracing members.

member can be determined from designing the critical diagonal member for the compressive force as follows:

Assume the cross section of the stringer bracing as 2 angles back-to-back $80 \times 80 \times 8$ (see Figure 4.151); then,

$$\alpha = \tan^{-1}\frac{1.8}{1.5} = 50.2°$$

$$l_b = 2343\,\mathrm{mm}$$

$$\varepsilon = \sqrt{\frac{235}{275}} = 0.924$$

$$\bar{\lambda} = \frac{L_{cr}}{i}\frac{1}{\lambda_1}$$

$$\lambda_1 = 93.9 \times 0.924 = 86.7636$$

$$\bar{\lambda} = \frac{2343}{24.3} \frac{1}{86.7636} = 1.111$$

The axial compressive force in the diagonal bracing member ($N_{Ed} = 86.8$ kN):

$$\frac{N_{Ed}}{N_{b,Rd}} \le 1.0$$

where $N_{b,Rd} = \dfrac{\chi A f_y}{\gamma_{M1}}$

$$A = 2 \times 12.3 = 24.6 \text{ cm}^2$$

$$\chi = \frac{1}{\Phi + \sqrt{\Phi^2 - \bar{\lambda}^2}} \text{ but } \chi \le 1.0$$

$$\Phi = 0.5 \left[1 + \alpha(\bar{\lambda} - 0.2) + \bar{\lambda}^2 \right]$$

$$\Phi = 0.5 \left[1 + 0.34(1.111 - 0.2) + 1.111^2 \right] = 1.272$$

$$\chi = \frac{1}{1.272 + \sqrt{1.272^2 - 1.111^2}} = 0.529 \quad \chi \le 1.0$$

Then, $N_{b,Rd} = \dfrac{0.529 \times 24.6 \times 100 \times 275}{1.1} = 325{,}335 \text{ N}$

$$N_{b,Rd} = 325.3 \text{ kN} > N_{Ed} = 86.8 \text{ kN (Then O.K.)}$$

4.5.9 Design of Wind Bracings

Wind forces acting on the double-track railway bridge (see Figure 4.152) as well as any other lateral forces directly applied to the bridge are transmitted to the bearings by systems of wind bracing. For pony bridges, only the lower wind bracing carries wind forces on the moving train, wind forces on the main plate girder, and lateral shock (nosing force) applied to the tracks (see Figure 4.153). Wind forces applied to this bridge can be sufficiently estimated using the design rules specified in EC1 [3.2] as follows:

$$F_w = \frac{1}{2} \rho v_b^2 C A_{ref,x}$$

$$v_b = c_{dir} \times c_{season} \times v_{b,0} = 1.0 \times 1.0 \times 26 = 26 \text{ m/s}$$

$$A_{ref,x} = 4.9118 \times 28 = 137.53 \text{ m}^2$$

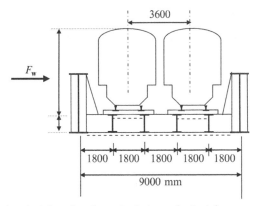

Figure 4.152 Design heights for the calculation of wind forces on the lower wind bracings.

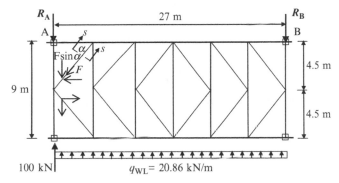

Figure 4.153 Loads on the lower wind bracing.

$$F_w = \frac{1}{2} \times 1.25 \times 26^2 \times 5.7 \times 137.53 = 331{,}208\,\text{N} = 331.2\,\text{kN}$$

Considering the structural analysis for the lower wind bracing system shown in Figure 4.153, the critical design wind force in the diagonal bracing members can be calculated as follows:

$$\text{Distributed wind loads}\,(q_{WL}) = 522.6 \times (5.5/7)/30 = 13.69\,\text{kN/m}$$

$$\text{Factored distributed wind loads} = q_{WL} \times \gamma_q = 13.69 \times 1.7 = 23.27\,\text{kN/m}$$

$$R_A = 100 + 20.86 \times 13.5 = 381.6\,\text{kN}$$

$$\alpha = \tan^{-1}(4.5/4.5) = 45°$$

$$F_D = 281.6/(2 \times \sin 45) = 199.1 \, \text{kN}$$

The cross section of the bracing member (see Figure 4.154) can be determined as follows:

$$l_{bx} = 6360 \, \text{mm}, \quad l_{by} = 1.2 \times 6360 = 7632 \, \text{mm}$$

Choose two angles back-to-back $150 \times 150 \times 15$, with 10 mm gusset plate between them:

$$A = 2 \times 43.2 = 86.4 \, \text{cm}^2, \quad i_x = 4.59 \, \text{cm}, \quad e = 4.26 \, \text{cm},$$

$$i_y = \sqrt{4.59^2 + (4.26 + 1/2)^2} = 6.61 \, \text{cm}$$

$$\varepsilon = \sqrt{\frac{235}{275}} = 0.924$$

$$\bar{\lambda} = \frac{L_{cr}}{i} \frac{1}{\lambda_1}$$

$$\lambda_1 = 93.9 \times 0.924 = 86.7636$$

$$\bar{\lambda} = \frac{6360}{45.9} \frac{1}{86.7636} = 1.597$$

The axial compressive force in the diagonal bracing member ($N_{Ed} = 298.7$ kN):

$$\frac{N_{Ed}}{N_{b,Rd}} \leq 1.0$$

where $N_{b,Rd} = \dfrac{\chi A f_y}{\gamma_{M1}}$

$$A = 2 \times 43.2 = 86.4 \, \text{cm}^2$$

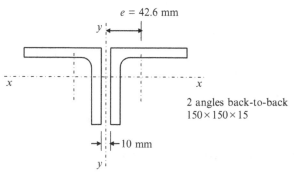

Figure 4.154 Upper wind bracing cross section s-s.

$$\chi = \frac{1}{\Phi + \sqrt{\Phi^2 - \bar{\lambda}^2}} \quad \text{but } \chi \le 1.0$$

$$\Phi = 0.5\left[1 + \alpha(\bar{\lambda} - 0.2) + \bar{\lambda}^2\right]$$

$$\Phi = 0.5\left[1 + 0.34(1.597 - 0.2) + 1.597^2\right] = 2.013$$

$$\chi = \frac{1}{2.013 + \sqrt{2.013^2 - 1.597^2}} = 0.309 \quad \text{but } \chi \le 1.0$$

$$\text{Then, } N_{b,Rd} = \frac{0.309 \times 86.4 \times 100 \times 275}{1.1} = 667,440\,\text{N}$$

$$N_{b,Rd} = 667.4\,\text{kN} > N_{Ed} = 199.1\,\text{kN} \,(\text{Then O.K.})$$

4.5.10 Design of Stringer-Cross Girder Connection

The stringer is designed as a simply supported beam on cross girders; therefore, the connection is mainly transferring shear forces (maximum reaction from stringers of 567 kN) (see Figure 4.155). Using M27 high-strength pretensioned bolts of grade 8.8, having f_{ub} of 800 MPa, shear area A of 4.59 cm^2, and gross area A_g of 5.73 cm^2, we can determine the required number of bolts, following the rules specified in EC3 (BS EN 1993-1-8) [2.13], as follows:

$$F_{v,Rd} = \frac{\alpha_V f_{ub} A}{\gamma_{M2}}$$

$$F_{v,Rd} = \frac{0.6 \times 800 \times 459}{1.25} = 176,256\,\text{N}$$

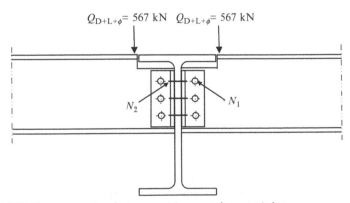

Figure 4.155 The connection between stringer and cross girder.

Then, $F_{v,Rd}$ equals 176 kN (for bolts in single shear) and 353 kN (for bolts in double shear):

$$F_{s,Rd} = \frac{k_s n \mu}{\gamma_{M3}} F_{p,C}$$

$$F_{p,C} = 0.7 f_{ub} A_s = 0.7 \times 800 \times 573 = 320{,}880\,N$$

$$F_{s,Rd,ser} = \frac{1.0 \times 1.0 \times 0.4}{1.1} 320{,}880 = 116{,}683.6\,N.$$

Then, $F_{s,Rd} = 117$ kN (for bolts in single shear at serviceability limit states) and $F_{s,Rd} = 234$ kN (for bolts in double shear at serviceability limit states). At ultimate limit states, $F_{s,Rd,ult}$ can be calculated as follows:

$$F_{s,Rd,ult} = \frac{1.0 \times 1.0 \times 0.4}{1.25} 320{,}880 = 102{,}682\,N.$$

Then, $F_{s,Rd} = 103$ kN (for bolts in single shear at ultimate limit states) and $F_{s,Rd} = 206$ kN (for bolts in double shear at ultimate limit states):

$$N_1 = \frac{567}{206} = 2.8 \quad \text{taken as 3 bolts,}$$

$$N_2 = \frac{567}{103} = 5.5 \quad \text{taken as 6 bolts}$$

4.5.11 Design of Cross Girder-Main Plate Girder Connection

The cross girder is designed as a simply supported beam on main plate girders; therefore, once again, the connection is mainly transferring shear forces (maximum reaction from cross girders of 1325 kN) (see Figure 4.156). We can determine the required number of bolts as follows:

$$N_3 = \frac{1325}{206} = 6.4 \quad \text{taken as 7 bolts,}$$

$$N_2 = \frac{1325}{103} = 13 \quad \text{taken as 14 bolts}$$

4.5.12 Design of Field Splices

Figure 4.157 shows the locations of filed splices for the investigated pony bridge. Designing the splice requires determination of size of connecting plates as well as the number of bolts of the filed splice shown in Figure 4.158. The area of the flange plate equals to $54 \times 3 = 162\ cm^2$; this can be

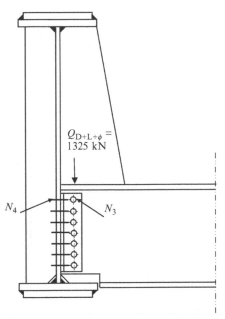

Figure 4.156 The connection between cross girder and main plate girder.

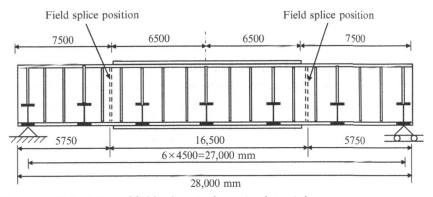

Figure 4.157 Positions of field splices in the main plate girder.

compensated by three flange splice plates having cross-sectional area of 54×1.6 and $2 \times 25 \times 1.6$ cm^2 with a total area of 166.4 cm^2, which is greater than the original area, while the area of web plate $= 270 \times 1.6 = 432$ cm^2 can be compensated by two web splice plates having cross-sectional area of $2 \times 260 \times 1.0$ cm^2 with a total area of 520 cm^2, which is governed by the minimum thickness (10 mm) of plates used in railway steel bridges. The top row of bolts in the web (see Figure 4.158) is

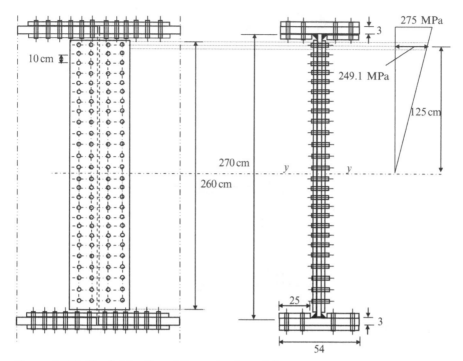

Figure 4.158 The field splice of the main plate girder.

subjected to horizontal shear from the bending moment distribution, assuming the yield stress reached at the extreme and lower fibers of the flanges, and vertical shear from the applied loads. Using a spacing of 10 cm between two adjacent bolts, an edge spacing of 5 cm, and a hole of 3 cm (2.7 cm bolt diameter plus 0.3 cm clearance), we can determine the horizontal shear force (H) per bolt and the vertical shear per bolt (V) as follows:

$$H = \text{Area from centrelines between bolts}$$
$$\times \text{ average stress at the bolt location}(f_{av})$$

$$f_{av} = 125 \times 275/138 = 249.1\,\text{MPa}$$

$$H = (100 - 30) \times 16 \times 249.1/2 = 139496\,\text{N} = 139.5\,\text{kN}$$

V = maximum shear resisted by web/total number of bolts.

Maximum shear resisted by web was previously calculated in the check of the safety of the plate girder against shear stresses and was 7482.5 kN. The total number of bolts in one side of the splice is 52:

$$V = 7482.5/52 = 143.9\,\text{kN}$$

The resultant of the forces per bolt (R) is equal to $\sqrt{139.5^2 + 143.9^2} = 200.4\,\mathrm{kN}$, which is less than 206 kN (the resistance of the bolt in double shear). Then O.K.

Flange Splices

Maximum force in the upper flange $= 162 \times 275 \times 100/1000 = 4455\,\mathrm{kN}$

$N(\text{flange}) = 4455/206 = 21.6\,\text{bolts}$ (6 rows of four bolts in double shear)

4.5.13 Design of Roller Steel Fabricated Bearings

Let us now design the roller steel fabricated bearings shown in Figure 4.126 and detailed in Figure 4.159. The maximum vertical reaction at the supports of the main plate girder was previously calculated under dead and live loads with dynamic effect $(R_{D+L+\Phi})$, which was 3132.8 kN. The material of construction for the bearings is cast iron steel (ISO 3755) 340-550 having a yield stress of 340 MPa and an ultimate stress of 550 MPa.

Design of the Sole Plate

The reaction $(R_{D+L+\Phi})$ can be assumed as two equal concentrated loads at two points, which are the centers of gravity of half of the load bearing stiffener section shown in Figure 4.159. To determine the centers of gravity (distance e), we can take the first area moment around the axis z-z, shown in Figure 4.159, as follows:

$$e = \frac{2 \times 25 \times 1.2 \times 0.6 + 23.2 \times 1.6 \times 11.6}{2 \times 25 \times 1.2 + 23.2 \times 1.6} = \frac{466.592}{97.12} = 4.8\,\mathrm{cm}$$

Assuming that the thickness of the sole plate is t_1, with detailed dimensions shown in Figure 4.159 based on the flange plate girder dimensions, we can determine the maximum moment applied to the sole plate (M) as follows:

$$M = R_{D+L+\Phi} \times e/2 = 3132.8 \times 10^3 \times 48/2 = 75{,}187{,}200\,\mathrm{N\,mm.}$$

Section plastic modulus $\left(W_{pl}\right) = b_1 t_1^2/4 = 700 \times t_1^2/4 = 175 \times t_1^2$

The plate thickness t_1 can be calculated now as follows:

$$\frac{M}{W_{pl}} = \frac{f_y}{\gamma_{M0}}$$

$$\frac{75{,}187{,}200}{175 \times t_1^2} = \frac{340}{1.0}$$

Then, $t_1 = 35.5$ mm, taken as 40 mm, as shown in Figure 4.159.

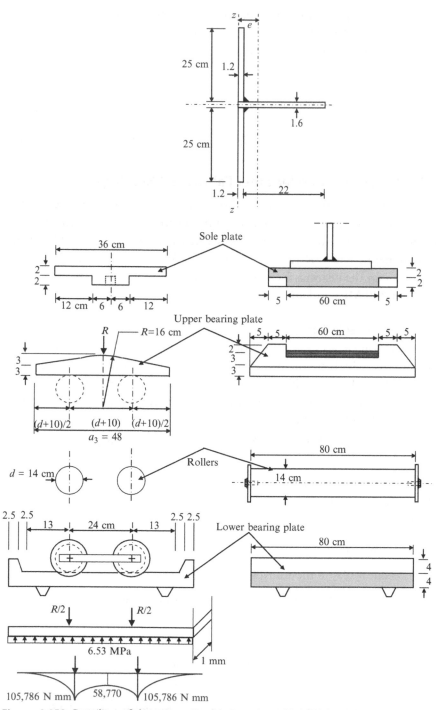

Figure 4.159 Detailing of the twin roller fabricated steel bridge bearings.

Design of the Rollers

The design of rollers requires determination of the diameter, length, and number of rollers to resist the vertical load as well as the arrangement and allowed movement in the direction of rollers. The design axial force per unit length of roller contact N_{Sd}' specified in BS EN 1337-1 [3.11] shall satisfy

$$N_{Sd}' \leq N_{Rd}'$$

where N_{Rd}' is the design value of resistance per unit length of roller contact, which is calculated as

$$N_{Rd}' = 23 \times R \times \frac{f_u^2}{E_d} \times \frac{1}{\gamma_m^2} = 23 \times R \times \frac{550^2}{210,000} \times \frac{1}{1} = 33.131 \times R$$

Assume the number of rollers is 2 and their length is 800 mm as shown in Figure 4.33:

$$N_{Sd}' = \frac{R_{D+L+\Phi}}{2 \times 800} = \frac{3132.8 \times 10^3}{1600} = 1958 \, \text{N/mm}$$

Then, the radius of rollers can be determined by equalizing N_{Sd}' with N_{Rd}' as follows:

$$1958 = 33.131 \times R$$

Then, $R = 59.1$ mm, taken as 70 mm and the diameter D is 140 mm.

Design of Upper Bearing Plate

The upper bearing plate is shown in Figure 4.159. The width and length of the plate are dependent on the spacing between rollers and the length of rollers as well as the allowed movement in the direction of rollers. The thickness of the upper bearing plate can be determined as follows:

$$M = \frac{R_{D+L+\Phi}}{2} \times \frac{(D+100)}{2} = \frac{3132.8 \times 10^3}{2} \times \frac{240}{2} = 187,968,000 \, \text{N mm}.$$

$$W_{pl} = \frac{b_2 t_2^2}{4} = \frac{800 t_2^2}{4} = 200 t_2^2 \, \text{mm}^3$$

The plate thickness t_2 can be calculated now as follows:

$$\frac{M}{W_{pl}} = \frac{f_y}{\gamma_{M0}}$$

$$\frac{187,968,000}{200 \times t_1^2} = \frac{340}{1.0}$$

Then, $t_1 = 52.6$ mm, taken as 60 mm, as shown in Figure 4.159.

The radius of the curved part of the upper bearing plate, which has a length of 600 mm as shown in Figure 4.159, can be determined the same way as that adopted for the design of the rollers:

$$N_{Rd}' = 23 \times R \times \frac{f_u^2}{E_d} \times \frac{1}{\gamma_m^2} = 23 \times R \times \frac{550^2}{210,000} \times \frac{1}{1} = 33.131 \times R$$

$$N_{Sd}' = \frac{R_{D+L+\Phi}}{600} = \frac{3,132,800}{600} = 5221.3\,\text{N/mm}$$

Then, the radius of rollers can be determined by equalizing N_{Sd}' with N_{Rd}' as follows:

$$5221.3 = 33.131 \times R$$

Then, $R = 157.6$ mm, taken as 160 mm.

Design of Lower Bearing Plate

The lower bearing plate is shown in Figure 4.159. The width and length of the plate are dependent on the strength of concrete and are dependent on the spacing between rollers and the length of rollers as well as the allowed movement in the direction of rollers. The thickness of the upper bearing plate can be determined as follows:

$$f_c = \frac{R_{D+L+\Phi}}{a_3 b_3} = \frac{3132.8 \times 10^3}{600 \times 800} = 6.53\,\text{MPa} < \frac{f_c}{\gamma_c} = \frac{40}{1.5} = 26.7\,\text{MPa}$$

$$\frac{f_c}{\gamma_c} = \frac{40}{1.5} = 26.7 \text{ (for a typical concrete in bridges of C40/50 with } f_{ck})$$

The plate thickness t_3 can be calculated from the distribution of bending moment, caused by the pressure on the concrete foundation, as follows: $M = 105,786\,\text{N\,mm}$ per unit width of the plate:

$$W_{pl} = \frac{b_3 t_3^2}{4} = \frac{1 \times t_3^2}{4} = 0.25 \times t_2^2\,\text{mm}^3$$

$$\frac{M}{W_{pl}} = \frac{f_y}{\gamma_{M0}}$$

$$\frac{105,786}{0.25 \times t_3^2} = \frac{340}{1.0}$$

Then, $t_3 = 35.3$ mm, taken as 40 mm, as shown in Figure 4.159.

4.5.14 Design of Hinged Line Rocker Steel Fabricated Bearings

Finally, we can now design the hinged line rocker steel fabricated bearings shown in Figure 4.126 and detailed in Figure 4.160. The maximum vertical

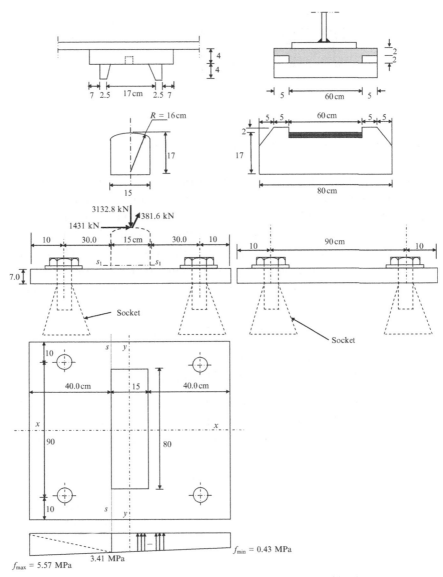

Figure 4.160 Detailing of the hinged line rocker fabricated steel bridge bearings.

reaction at the support of the main plate girder was previously calculated under dead and live loads with dynamic effect $(R_{D+L+\Phi})$, which was 3132.8 kN. The bearing is also subjected to a lateral force from the braking and traction forces from tracks as well as subjected to a longitudinal force from the reactions of the lower wind bracing, which cause moments around

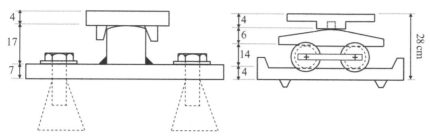

Figure 4.161 The designed roller and hinged line rocker fabricated steel bearings.

longitudinal and lateral directions of the bearing base, respectively. Similar to the roller bearing, the material of construction for the bearings is cast iron steel (ISO 3755) 340-550 having a yield stress of 340 MPa and an ultimate stress of 550 MPa. It should be noted that the overall height of the hinged bearing must be exactly the same as that of the roller bearing. The general layout and assumed dimensions of the hinged line rocker bearing are shown in Figure 4.161. The traction Q_{lak} and braking Q_{lbk} forces can be calculated as follows:

$$Q_{lak} = 33 \times L_{a,b} = 33 \times 27 = 991 \, kN \leq 1000 [kN], \text{ for Load Models 71}$$

$$Q_{lbk} = 20 \times L_{a,b} = 20 \times 27 = 540$$
$$\leq 6000 [kN], \text{ for Load Models 71,SW/0,SW/2 and HSLM}$$

Total the braking and traction forces $(Q_{tot}) = 1431$ kN (see Figure 4.160 for the direction of the forces). Also, the reaction from the lower wind bracings (R) (see Figure 4.160 for the direction of the forces) was previously calculated as follows:

$$R_{tot} = 381.6 \, kN$$

We can now determine the normal stress distribution due to the applied loads, shown in Figure 4.160, on the concrete foundation as follows:

$$f = -\frac{N}{A} \pm \frac{M_x}{I_x} y \pm \frac{M_y}{I_y} x$$

$$\frac{N}{A} = \frac{3,132,800}{950 \times 1100} = 3.0 \, MPa$$

$$\frac{M_x}{I_x} y = \frac{381.6 \times 10^3 \times 240}{950 \times 1100^3 / 12} 550 = 0.49 \, MPa$$

$$\frac{M_y}{I_y} x = \frac{1431 \times 10^3 \times 240}{1100 \times 950^3 / 12} 475 = 2.08 \, MPa$$

$$f_{max} = -3.0 - 0.49 - 2.08 = -5.57 \, MPa$$

$$f_{min} = -3.0 + 0.49 + 2.08 = -0.43 \, MPa$$

The critical bending moment on the base plate of the hinged bearing is at section s-s shown in Figure 4.160:

$$M = (0.5 \times 400 \times 3.41) \times 1100 \times 400/3 + (0.5 \times 400 \times 5.57) \times 1100$$
$$\times 400 \times 2/3 = 426{,}800{,}000 \, N \, mm$$

$$W_{pl} = 1100 \times t_4^2/4 = 275 t_4^2$$

$$\frac{M}{W_{pl}} = \frac{f_y}{\gamma_{M0}}$$

$$\frac{426{,}800{,}000}{275 t_4^2} = \frac{340}{1.0}$$

Then, $t_4 = 67.6$ mm, taken as 70 mm.

The normal stresses at section s_1-s_1, shown in Figure 4.160, of the line rocker bearing can be checked as follows:

$$M_x = 381.6 \times 10^3 \times 170 = 64{,}872{,}000 \, N \, mm.$$

$$M_y = 1431 \times 10^3 \times 170 = 243{,}270{,}000 \, N \, mm.$$

$$\frac{N}{A} = \frac{3{,}132{,}800}{150 \times 800} = 26.11 \, MPa$$

$$\frac{M_x}{I_x} y = \frac{64{,}872{,}000}{150 \times 800^3/12} 400 = 4.05 \, MPa$$

$$\frac{M_y}{I_y} x = \frac{243{,}270{,}000}{800 \times 150^3/12} 75 = 81.09 \, MPa$$

$$f_{max} = -(26.11 + 4.05 + 81.09) = -111.25 \, MPa$$
$$< 340 \, MPa \, (Then \, O.K.)$$

4.6 DESIGN EXAMPLE OF A DECK TRUSS HIGHWAY STEEL BRIDGE

The fifth design example presented in this chapter is for a deck truss highway steel bridge. The general layout of the through bridge is shown in Figures 4.162 and 4.163. The truss bridge has simply supported ends with a length between supports of 40 m. The truss bridge has a Warren truss with 8 equal panels of 5 m. It is required to design the bridge adopting the design

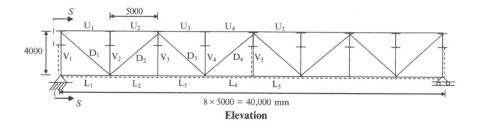

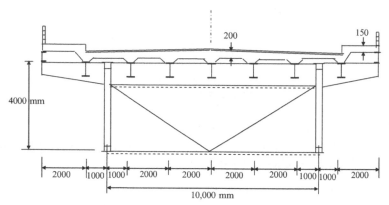

Figure 4.162 General layout of a deck truss highway steel bridge (the fifth design example).

rules specified in EC3 [1.27]. The steel material of construction of the bridge conformed to standard steel grade EN 10025-2 (S 275) having a yield stress of 275 MPa and an ultimate strength of 430 MPa. The dimensions and general layout of the bridge are shown in Figures 4.162 and 4.63. The bridge has upper and lower wind bracings of K-shaped truss members as well as cross bracing of K-shaped truss. The expected live loads on the highway bridge conform to Load Model 1, which represents the static and dynamic effects of vertical loading due to normal road traffic as specified in EC1 [3.1]. The bolts used in connections and field splices are M27 high-strength pretensioned bolts. The unit weight of reinforced concrete slab decks used is 25 kN/m³.

4.6.1 Design of the Stringers

Let us start by designing the stringers, the longitudinal steel beams, supporting the reinforced concrete slab deck as shown in Figure 4.162.

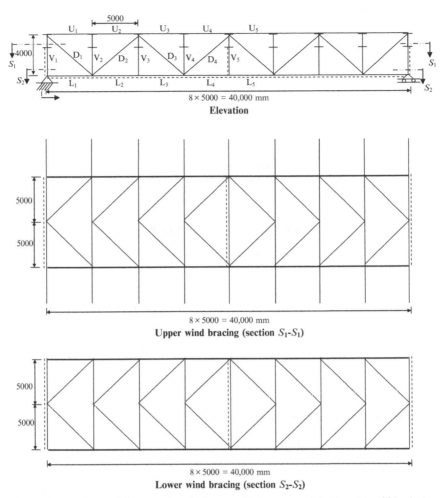

Figure 4.163 General layout of a deck truss highway steel bridge (the fifth design example).

Dead Loads

The general layout of an intermediate stringer is shown in Figure 4.164. The dead loads acting on an intermediate stringer can be calculated as follows:

$$\text{Flooring}(1.75 \, \text{kN/m}^2) = 1.75 \times 2 = 3.5 \, \text{kN/m}$$

$$\text{Reinforced concrete slab deck}(0.2 \, \text{m thickness}) = 5 \times 2 = 10 \, \text{kN/m}$$

$$\text{Haunch}(\text{Equivalent to 1 cm slab thickness}) = 0.25 \times 2 = 0.5 \, \text{kN/m}$$

$$\text{Own weight of stringer} = 1.5 \, \text{kN/m}$$

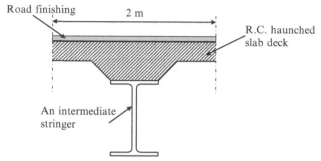

Figure 4.164 General layout of a an intermediate stringer.

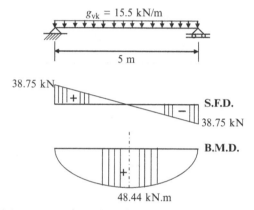

Figure 4.165 Straining actions from dead loads acting on an intermediate stringer.

$$\text{Total dead load} = g_{vk} = 15.5 \, \text{kN/m}$$

Assuming the stringers are simply supported by the cross girders, we can calculate the maximum shear force and bending moment due to dead loads on an intermediate stringer (see Figure 4.165) as follows:

$$Q_{D.L.} = g_{vk} \times L/2 = 15.5 \times 5/2 = 38.75 \, \text{kN}$$

$$M_{D.L.} = g_{vk} \times L^2/8 = 15.5 \times 5^2/8 = 48.44 \, \text{kN m}$$

Live Loads

The live loads acting on the highway bridge conform to Load Model 1, which represents the static and dynamic effects of vertical loading due to normal road traffic as specified in EC1 [3.1]. To determine the worst cases of loading on an intermediate stringer due to live loads, we can study a lateral section through vehicles and a lateral section through distributed loads of Load Model 1 acting on the bridge, as shown in Figure 4.166. From the section through vehicles, we find that the maximum concentrated load transferred to the stringer is 200 kN. While from the section through distributed loads, we find that the maximum distributed load transferred to the stringer is 14.34 kN/m. Therefore, the load distribution transferred to the stringer in the longitudinal direction is as shown in Figure 4.167. Two cases of loading for the evaluation of maximum bending moment due to live loads on a stringer can be studied. The first case of loading

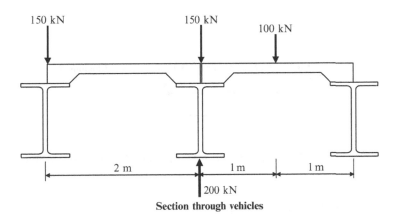

Section through vehicles

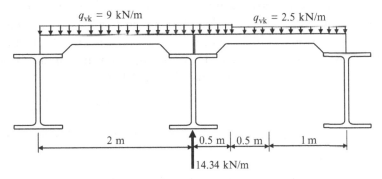

Section through distributed loads

Figure 4.166 Calculation of straining actions from live loads transferred on an intermediate stringer

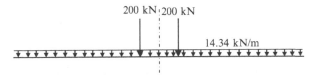

Figure 4.167 Transferred live loads on an intermediate stringer.

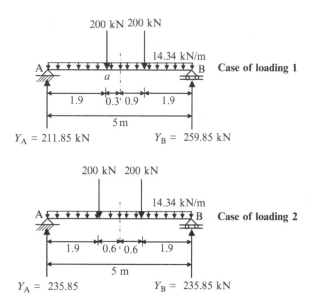

Figure 4.168 Cases of loading for the maximum bending moment acting on an intermediate stringer.

is that the centerline of the stringer divides the spacing between the resultant of the concentrated live loads and the closest load, with maximum bending moment calculated at the closest load (point a in Figure 4.168), while the second case of loading is that the centerline of the stringer is located in the middle between the concentrated live loads, with maximum bending moment located at midspan as shown in Figure 4.168:

$$M_{L.L.}(\text{case of loading}\,1) = 211.85 \times 2.2 - 14.34 \times 2.2^2/2 = 431.37\,\text{kN m}$$

$$M_{L.L.}(\text{case of loading}\,2) = 200 \times 1.9 + 14.34 \times 5^2/8 = 424.81\,\text{kN m}$$

There is a single case of loading for live loads to produce a maximum shear force at the supports of the stringer, which is shown in Figure 4.169:

$$Q_{L.L.} = 387.85\,\text{kN}$$

Bending Moment Due to Dead and Live Loads with Dynamic Effect Added (M_{Ed})

$$M_{Ed} = M_{D.L.} \times \gamma_g + M_{L.L.} \times \gamma_q = 48.44 \times 1.3 + 431.37 \times 1.35$$
$$= 645.3\,\text{kN m}$$

It should be noted that, according to EC0 (BS EN 1990) [3.4], the permanent actions of steel self-weight and superimposed load should be

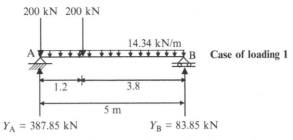

$Y_A = 387.85$ kN $Y_B = 83.85$ kN

Figure 4.169 Cases of loading for the maximum shear force acting on a stringer.

multiplied by 1.2 at the ultimate limit state, while the permanent actions of concrete weight should be multiplied by 1.35. Therefore, the total dead load is calibrated and multiplied by 1.3. On the other hand, variable actions comprising road traffic actions are multiplied by 1.35 at the ultimate limit state. Once again, it should be noted that the load factors adopted in this study are that of the ultimate limit state. This is attributed to the fact that the finite element models presented in Chapters 6 and 7 can be used to analyze the bridges and provide more accurate predictions for the deflections and other serviceability limit state cases of loading.

Shearing Force Due to Dead and Live Loads with Dynamic Effect Added (Q_{Ed})

$$Q_{Ed} = Q_{D.I.} \times \gamma_g + Q_{L.L.} \times \gamma_q = 38.75 \times 1.3 + 387.85 \times 1.35 = 574\,\text{kN}$$

Design Bending Moment (M_{Ed}) and Shear Force (Q_{Ed})

$$M_{Ed} = 645.3\,\text{kN m}$$

$$Q_{Ed} = 574\,\text{kN}$$

Design of Stringer Cross Section

$$M_{c,Rd} = \frac{W_{pl} \times f_y}{\gamma_{M0}} \quad \text{for classes 1 and 2}$$

$$645.3 \times 10^6 = \frac{W_{pl} \times 275}{1.0}$$

$$W_{PL} = 2{,}346{,}545.5\,\text{mm}^3 = 2346.5\,\text{cm}^3$$

Choose UB 533 × 210 × 92 (equivalent to American W21 × 62), shown in Figure 4.170. W_{PL} around $x\text{-}x = 2360$ cm³. To classify the cross section chosen,

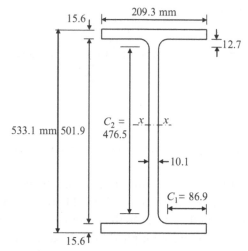

Figure 4.170 The cross-section of stringers (UB 533 × 210 × 92).

$$\varepsilon = \sqrt{\frac{235}{f_y}} = \sqrt{\frac{235}{275}} = 0.924$$

$C_1 = 86.9 \, \text{mm}, \quad t_{fl} = 15.6, \quad C_1/t_{fl} = 86.9/15.6 = 5.57 \le 9 \times 0.924$
$= 8.316 \, (\text{Stringer flange is class 1})$

$C_2 = 476.5 \, \text{mm}, \quad t_w = 10.1, \quad C_1/t_{fl} = 476.5/10.1 = 47.2 \le 72 \times 0.924$
$= 66.5 \, (\text{Stringer web is class 1})$

Check of Bending Resistance

$$M_{c,Rd} = \frac{W_{pl} \times f_y}{\gamma_{M0}} = \frac{2360 \times 10^3 \times 275}{1.0} = 649,000,000 \, \text{N mm}$$
$$= 649 \, \text{kN m} > M_{Ed} = 645.3 \, \text{kN m} \, (\text{Then O.K.})$$

Check of Shear Resistance

$$V_{pl,Rd} = \frac{A_v \left(f_y / \sqrt{3} \right)}{\gamma_{M0}} = \frac{(501.9 \times 10.1) \times \left(275 / \sqrt{3} \right)}{1.0} = 804,800 \, \text{N}$$
$$= 804.8 \, \text{kN} > Q_{Ed} = 574 \, \text{kN} \, (\text{Then O.K.})$$

4.6.2 Design of the Cross Girders

The cross girders (the lateral steel beams) carry concentrated loads from the stringers as shown in Figure 4.171. Therefore, the dead and live loads acting on an intermediate cross girder can be calculated as follows:

Dead Loads

Intermediate reactions from stringers due to dead loads $= 15.5 \times 5$
$= 77.5\,kN$

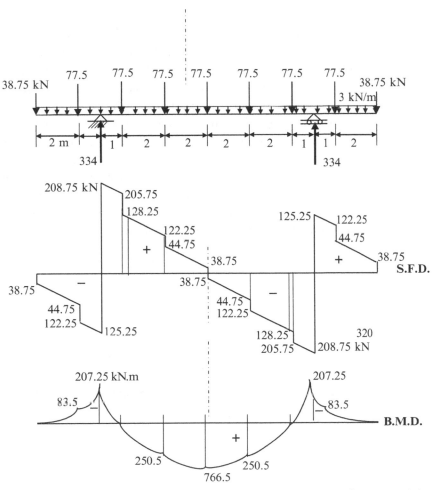

Figure 4.171 Straining actions from dead loads acting on an intermediate cross girder.

Reactions from stringers at edges due to dead loads $= 77.5/2 = 38.75\,\text{kN}$

Own weight of cross girder $= 3.0\,\text{kN/m}$

Assuming the cross girders are simply supported by the main plate girders, we find that the maximum shear force and bending moment due to dead loads on an intermediate cross girder (see Figure 4.171) are as follows:

$$Q_{\text{D.L.}} = 208.75\,\text{kN}$$

$$M_{\text{D.L.}} = 334\,\text{kN m}$$

Live Loads
To determine the worst cases of loading on an intermediate cross girder due to live loads, we can study different longitudinal sections through vehicles, distributed loads, and sidewalks of Load Model 1 acting on the bridge, as shown in Figure 4.172. From the different sections, we can find that the maximum concentrated and distributed loads transferred to the intermediate cross girder are as shown in Figure 4.172. The case of loading for the evaluation of maximum positive bending moment due to live loads on an intermediate cross girder can be studied, as shown in Figure 4.172. The case of loading is that the larger concentrated load from vehicles transferred is located at the centerline (midspan) of an intermediate cross girder, with maximum bending moment located at midspan as shown in Figure 4.172. The maximum positive bending moment is calculated as follows:

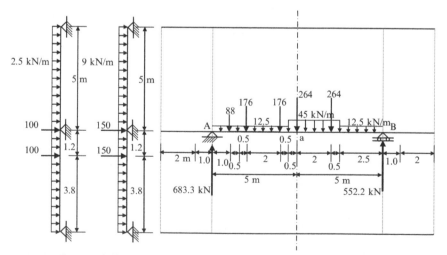

Figure 4.172 Case of loading for maximum positive bending moment from live loads acting on an intermediate cross girder.

$M_{L.L.}$(maximum positive bending moment)
$$= 552.2 \times 5 - 264 \times 2 - 45 \times 2.5 \times 1.25 - 12.5 \times 2.5 \times 3.75$$
$$= 1975.2 \, kN \, m$$

The case of loading for the evaluation of maximum negative bending moment due to live loads on an intermediate cross girder can be also studied, as shown in Figure 4.173. The maximum negative bending moment is calculated as follows:

$M_{L.L.}$(maximum negative bending moment)
$$= 264 \times 0.5 + 45 \times 1 \times 0.5 + 25 \times 2 \times 2 = 254.5 \, kN \, m$$

The case of loading for live loads to produce a maximum shear force at the supports of an intermediate cross girder is shown in Figure 4.174. It should be noted that for this deck bridge, cars are allowed to go on top of the supports, which are the main trusses, as shown in Figure 4.174.

$$Q_{L.L.} = 661.25 \, kN$$

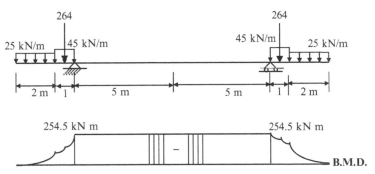

Figure 4.173 Case of loading for maximum negative bending moment from live loads acting on an intermediate cross girder.

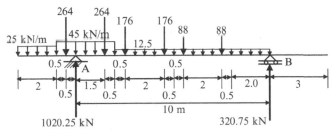

Figure 4.174 Case of loading for maximum shearing force from live loads acting on an intermediate cross girder.

Bending Moment Due to Dead and Live Loads with Dynamic Effect Added (M_{Ed})

$$M_{Ed} = M_{D.L.} \times \gamma_g + M_{L.L.} \times \gamma_q = 334 \times 1.3 + 1975.2 \times 1.35 = 3100.7 \, kN \, m$$

Shearing Force Due to Dead and Live Loads with Dynamic Effect Added (Q_{Ed})

$$Q_{Ed} = Q_{D.L.} \times \gamma_g + Q_{L.L.} \times \gamma_q = 208.75 \times 1.3 + 661.25 \times 1.35 = 1164.1 \, kN$$

Design Bending Moment (M_{Ed}) and Shear Force (Q_{Ed})

$$M_{Ed} = 3100.7 \, kN \, m$$

$$Q_{Ed} = 1164.1 \, kN$$

Design of the Cross Girder Cross Section

The cross girder is designed as a welded plate girder as shown in Figure 4.175. The web height is taken as equal to 1200 mm, which conforms to the

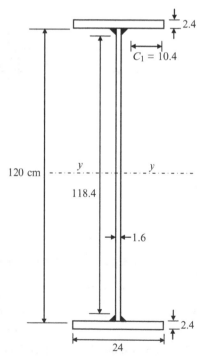

Figure 4.175 Welded plate girder section of cross girders.

recommended values $L/(7-9)=12,000/(7-9)=1429-1111$ mm. The web plate thickness is assumed to be 16 mm. The flange width is taken as equal to 240 mm, with a thickness of 24 mm. To classify the cross section chosen,

$$\varepsilon = \sqrt{\frac{235}{f_y}} = \sqrt{\frac{235}{275}} = 0.924$$

$C_1 = 104$ mm, $t_{fl} = 24$, $C_1/t_{fl} = 104/24 = 4.3 \leq 9 \times 0.924$
  $= 8.316$ (Cross girder flange is class 1).

$C_2 = 1184$ mm, $t_w = 16$, $C_1/t_{fl} = 1184/16 = 74 < 83 \times 0.924$
  $= 76.7$ (Cross girder web is class 1).

To calculate the bending moment resistance, the plastic section modulus should be used:

$$W_{pl} = 24 \times 124.8^3/4 - 2 \times 11.2 \times 120^3/4 = 12,810.2 \, \text{cm}^3$$

Check of Bending Resistance

$$M_{c,Rd} = \frac{W_{el,min} \times f_y}{\gamma_{M0}} = \frac{12,810.2 \times 10^3 \times 275}{1.0} = 3,522,800,000 \, \text{N mm}$$
$$= 3522.8 \, \text{kN m} > M_{Ed} = 3100.7 \, \text{kN m (Then O.K.)}$$

Check of Shear Resistance

$$V_{pl,Rd} = \frac{A_v\left(f_y/\sqrt{3}\right)}{\gamma_{M0}} = \frac{(1200 \times 16) \times \left(275/\sqrt{3}\right)}{1.0} = 3,048,409 \, \text{N}$$
$$= 3048.4 \, \text{kN} > Q_{Ed} = 1164.1 \, \text{kN (Then O.K.)}$$

4.6.3 Calculation of Forces in Truss Members

4.6.3.1 General

To calculate the design forces in the truss members, we need to calculate the dead and live loads acting on the main truss in the longitudinal direction, which is addressed as follows.

Dead Loads

Weight of steel structure for part of bridge between main trusses:

$$w_{s_1} = 1.75 + 0.04L + 0.0003L^2 \leq 3.5\,\text{kN/m}^2$$

$$w_{s_1} = 1.75 + 0.04 \times 40 + 0.0003 \times 40^2 = 3.83$$
$$> 3.5\,\text{kN/m}^2 \text{ taken as } 3.5\,\text{kN/m}^2$$

Weight of steel structure for part of bridge outside main trusses:

$$w_{s_2} = 1 + 0.03L\,\text{kN/m}^2$$

$$w_{s_2} = 1 + 0.03 \times 40 = 2.2\,\text{kN/m}^2$$

$$w_s = 3.5 \times 10/2 + 2.2 \times 3 = 24.1\,\text{kN/m}$$

Weight of reinforced concrete decks and haunches:

$$w_{RC} = (0.2 + 0.01) \times 25 \times 6 + (0.15 + 0.01) \times 25 \times 2 = 39.5\,\text{kN/m}$$

Weight of finishing (assume weight of finishing is 1.75 kN/m^2 for parts between sidewalks and 1.5 kN/m^2 for sidewalks):

$$w_F = 1.75 \times 6 + 1.5 \times 2 = 13.5\,\text{kN/m}$$

We can now calculate the total dead load acting on main trusses in the longitudinal direction (see Figure 4.176) as follows:

$$w_{D.L.} = 24.1 + 39.5 + 13.5 = 77.1\,\text{kN/m}$$

Live Loads

To determine the live loads acting on main trusses in the longitudinal directions, we can study different lateral sections through vehicles, distributed loads, and sidewalks of Load Model 1 acting on the bridge, as shown in Figure 4.177. From the lateral section shown in Figure 4.177, we can find that the maximum concentrated and distributed loads transferred to a main truss are 450 kN and 45.65 kN/m, respectively, as shown in Figure 4.178. We can also calculate the negative distributed reactions acting on a main truss in the longitudinal by investigating the case of loading shown in Figure 4.179. The negative concentrated and distributed loads acting on a main truss are 7.5 kN and 2.45 kN/m, respectively, as shown in Figure 4.180. The calculated dead and live loads can be now used to

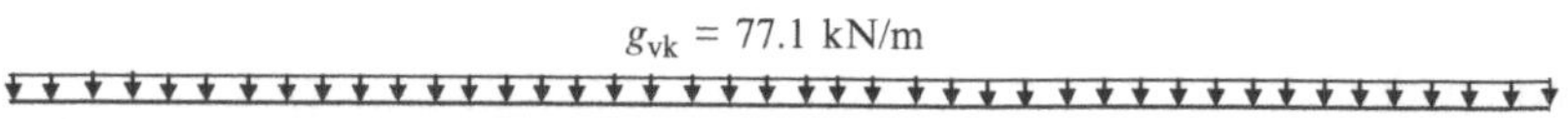

$$g_{vk} = 77.1\,\text{kN/m}$$

Figure 4.176 Dead loads acting on main trusses.

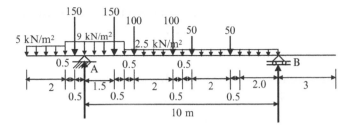

Reaction from concentrated loads = 450 kN

Reaction from distributed loads = 45.65 kN/m

Figure 4.177 Maximum reactions due to live loads transferred by cross girders on main trusses.

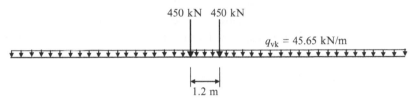

Figure 4.178 Live loads acting on main trusses.

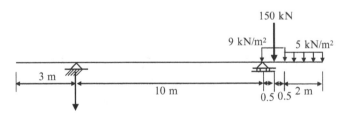

Negative reaction from concentrated loads = −7.5 kN

Negative reaction from distributed loads = −2.45 kN/m

Figure 4.179 Negative reactions due to live loads transferred by cross girders on main trusses.

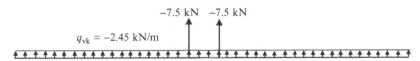

Figure 4.180 Negative concentrated and distributed live loads acting on main trusses.

determine the forces in the members of main trusses using the influence line method as shown in the coming sections.

4.6.3.2 Calculation of Force in the Upper Chord Member L_4

To determine the force in the lower chord truss member L_4 (see Figure 4.181) using the influence line method, we can follow the simple procedures of putting a unit concentrated moving load at midspan (point a), and using the sectioning method, we take a section s-s, as shown in Figure 4.181, and then take the moment at point a to calculate the force in the member due to the applied unit load. After that, we can put the previously calculated dead and live loads acting on a main truss in the longitudinal direction. The total force in the member will be the summation of the concentrated loads multiplied by the companion vertical coordinate in the influence line diagram and the summation of the distributed loads multiplied by the companion areas in the diagram. Hence, the forces due to the dead and live loads can be calculated as follows:

$$F_{D.L.}(L_4) = 0.5 \times 40 \times 2.5 \times 77.1 = 3855 \, kN$$

$$F_{L.L.}(L_4)positive = 450 \times (2.5 + 2.35) + 0.5 \times 40 \times 2.5 \times 45.65 = 4465 \, kN$$

$$F_{L.L.}(L_4)negative = -7.5 \times (2.5 + 2.35) - 0.5 \times 40 \times 2.5 \times 2.45$$
$$= -158.9 \, kN$$

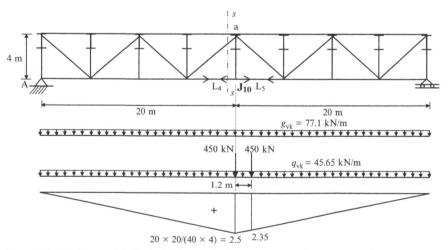

Figure 4.181 Determination of the tensile force in lower chord member L_4 using the influence line method.

$$F_{Ed}(L_4)\text{maximum} = F_{D.L.} \times \gamma_g + F_{L.L.} \times \gamma_q$$

$$F_{Ed}(L_4)\text{maximum} = 3855 \times 1.3 + 4465 \times 1.35$$
$$= 11{,}039.3\,kN\,(\text{Tension force})$$

$$F_{Ed}(L_4)\text{minimum} = 3855 \times 1.3 - 158.9 \times 1.35 = 4797\,kN\,(\text{Tension force})$$

It should be noted that, from the equilibrium of joint J_{10} (see Figure 4.181), the force in lower chord truss member L_4 is equal to that of L_5.

4.6.3.3 Calculation of Force in the Lower Chord Member L_3

To determine the force in the lower chord truss member L_3 (see Figure 4.182) using the influence line method, we can follow the same procedures adopting for member L_4. Hence, the forces due to the dead and live loads can be calculated as follows:

$$F_{D.L.}(L_3) = 0.5 \times 40 \times 1.875 \times 77.1 = 2891.3\,kN$$

$$F_{L.L.}(L_3)\text{positive} = 450 \times (1.875 + 1.8) + 0.5 \times 40 \times 1.875 \times 45.65$$
$$= 3365.6\,kN$$

$$F_{L.L.}(L_3)\text{negative} = -7.5 \times (1.875 + 1.8) - 0.5 \times 40 \times 1.875 \times 2.45$$
$$= -119.4\,kN$$

$$F_{Ed}(L_3)\text{maximum} = F_{D.L.} \times \gamma_g + F_{L.L.} \times \gamma_q$$

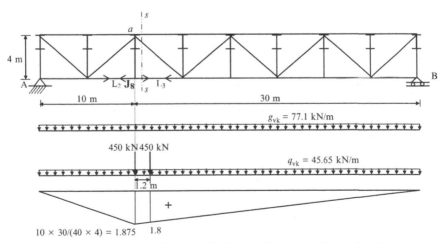

Figure 4.182 Determination of the tensile force in lower chord member L_3 using the influence line method.

$$F_{Ed}(L_3)\text{maximum} = 28{,}991.3 \times 1.3 + 3365.6 \times 1.35$$
$$= 8302.3\,kN\ (\text{Tension force})$$

$$F_{Ed}(L_3)\text{minimum} = 2891.3 \times 1.3 - 119.4 \times 1.35$$
$$= 3597.5\,kN\ (\text{Tension force})$$

It should be noted that, from the equilibrium of joint J_8 (see Figure 4.182), the force in lower chord truss member L_2 is equal to that of L_3. It should also be noted that, under the dead and live cases of loading, the force in the lower chord member L_1 is zero.

4.6.3.4 Calculation of Force in the Upper Chord Member U₄

We can repeat the previous procedures now and change the pole where the moment is calculated to determine the force in the upper chord member U_4, as shown in Figure 4.183. Hence, the forces due to the dead and live loads can be calculated as follows:

$$F_{D.L.}(U_4) = -0.5 \times 40 \times 2.3475 \times 77.1 = -3614.1\,kN$$

$$F_{L.L.}(U_4) = -450 \times (2.3475 + 2.23125) - 0.5 \times 40 \times 2.3475 \times 45.65$$
$$= -4198.6\,kN$$

$$F_{Ed}(U_4) = F_{D.L.} \times \gamma_g + F_{L.L.} \times \gamma_q$$

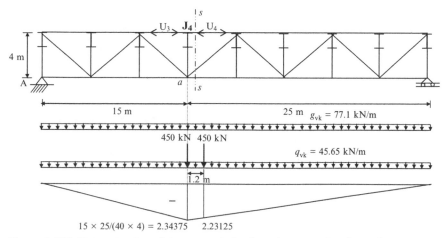

Figure 4.183 Determination of the compressive force in upper chord member U_4 using the influence line method.

$$F_{Ed}(U_4) = -3614.1 \times 1.3 - 4198.6 \times 1.35$$
$$= -10,366.4 \, kN \, (Compression \, force)$$

It should be noted that, from the equilibrium of the truss (see Figure 4.183), the force in upper chord truss member U_3 is equal to that of the calculated upper chord member U_4.

4.6.3.5 Calculation of Force in the Lower Chord Member U_2

The force in member U_2 due to the dead and live loads can be calculated, as shown in Figure 4.184, as follows:

$$F_{D.L.}(U_2) = -0.5 \times 40 \times 1.09375 \times 77.1 = -1686.6 \, kN$$

$$F_{L.L.}(U_2) = -450 \times (1.09375 + 1.05625) - 0.5 \times 40 \times 1.09375 \times 45.65$$
$$= -1966.1 \, kN$$

$$F_{Ed}(U_2) = F_{D.L.} \times \gamma_g + F_{L.L.} \times \gamma_q$$

$$F_{Ed}(U_2) = -1686.6 \times 1.3 - 1966.1 \times 1.35$$
$$= -4846.8 \, kN \, (Compression \, force)$$

It should be noted that, from the equilibrium of the truss (see Figure 4.184), the force in upper chord truss member U_2 is equal to that of the upper chord member U_1.

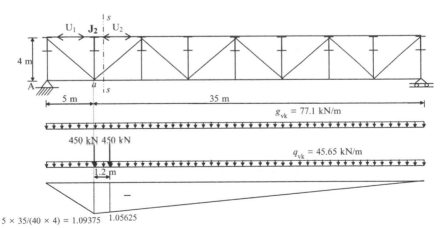

Figure 4.184 Determination of the tensile force in lower chord member L_3 using the influence line method.

4.6.3.6 Calculation of Force in the Diagonal Chord Member D₄

To determine the force in the diagonal chord truss member D_4 (see Figure 4.185) using the influence line method, we can follow the simple procedures of putting a unit concentrated moving load at point a adjacent to section $s-s$ shown in Figure 4.185 and study the equilibrium of the truss for the other side of section $s-s$ to calculate the force in the member. Then, we put the unit concentrated moving load at point b adjacent to section $s-s$ shown in Figure 4.185 and study the equilibrium of the truss for the other side of section $s-s$ to calculate the force in the member. The influence line of the diagonal member consists of two triangles as shown in Figure 4.185 having different signs. After that, we can put the previously calculated dead and live loads acting on a main truss in the longitudinal direction. It should be noted that the live loads can be put on the negative or positive triangle to produce a compressive or tensile force, respectively, while the dead loads must be put on both triangles. Once again, the total force in the member will be the summation of concentrated loads multiplied by the companion vertical coordinate in the diagram and the summation of the distributed loads multiplied by the companion area in the diagram. Hence, the forces due to the dead and live loads can be calculated as follows:

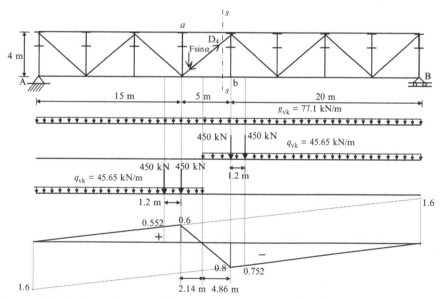

Figure 4.185 Determination of the force in diagonal member D_4 using the influence line method.

$$A_{-ve}(D_4) = -0.5 \times 22.86 \times 0.8 = -9.14$$

$$A_{+ve}(D_4) = 0.5 \times 17.14 \times 0.6 = 5.14$$

$$A_{net}(D_4) = -9.14 + 5.14 = -4.0$$

$$F_{D.L.}(D_4) = -4 \times 77.1 = -308.4\,kN$$

$$F_{L.L.}(D_5)(negative) = -450 \times (0.8 + 0.752) - 9.14 \times 45.65 - 7.5$$
$$\times (0.6 + 0.552) - 2.45 \times 5.14 = -1136.9\,kN$$

$$F_{L.L.}(D_4)(positive) = 450 \times (0.6 + 0.552) + 5.14 \times 45.65 + 7.5$$
$$\times (0.8 + 0.752) + 2.45 \times 9.14 = 787.1\,kN$$

$$F_{Ed}(D_4)maximum = F_{D.L.} \times \gamma_g + F_{L.L.} \times \gamma_q$$

$$F_{Ed}(D_4)\,maximum = -308.4 \times 1.3 - 1136.9 \times 1.35$$
$$= -1935.7\,kN\,(Compression\ force)$$

$$F_{Ed}(D_4)minimum = F_{D.L.} \times \gamma_g + F_{L.L.} \times \gamma_q$$

$$F_{Ed}(D_4)minimum = -308.4 \times 1.3 + 787.1 \times 1.35$$
$$= +661.7\,kN\,(Tension\ force)$$

4.6.3.7 Calculation of Force in the Diagonal Chord Member D₃

By repeating the procedures adopted for D_4, the force in the diagonal truss member D_3 can be calculated, as shown in Figure 4.186, as follows:

$$A_{+ve}(D_3) = 0.5 \times 28.57 \times 1 = 14.29$$

$$A_{-ve}(D_3) = -0.5 \times 11.43 \times 0.4 = -2.29$$

$$A_{net}(D_3) = 14.29 - 2.29 = 12.0$$

$$F_{D.L.}(D_3) = 12 \times 77.1 = +925.1\,kN$$

$$F_{L.L.}(D_3)(positive) = 450 \times (1 + 0.952) + 14.29 \times 45.65 + 7.5$$
$$\times (0.4 + 0.352) + 2.29 \times 2.45 = 1542\,kN$$

$$F_{L.L.}(D_3)(negative) = -450 \times (0.4 + 0.352) - 2.29 \times 45.65 - 7.5$$
$$\times (1 + 0.952) - 14.29 \times 2.45 = -492.6\,kN$$

$$F_{Ed}(D_3)maximum = F_{D.L.} \times \gamma_g + F_{L.L.} \times \gamma_q$$

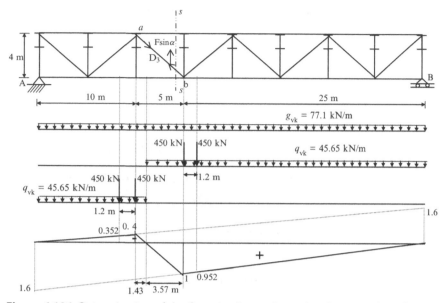

Figure 4.186 Determination of the force in diagonal member D_3 using the influence line method.

$$F_{Ed}(D_3)\text{maximum} = 925.1 \times 1.3 + 1542 \times 1.35$$
$$= 3284.3\,\text{kN (Tension force)}$$

$$F_{Ed}(D_3)\text{minimum} = F_{D.L.} \times \gamma_g + F_{L.L.} \times \gamma_q$$

$$F_{Ed}(D_3)\text{minimum} = 925.1 \times 1.3 - 492.6 \times 1.35$$
$$= 537.6\,\text{kN (Tension force)}$$

4.6.3.8 Calculation of Force in the Diagonal Chord Member D_2

The force in the diagonal truss member D_2 can be calculated, as shown in Figure 4.187, as follows:

$$A_{-ve}(D_2) = -0.5 \times 34.29 \times 1.2 = -20.57$$

$$A_{+ve}(D_2) = 0.5 \times 5.71 \times 0.2 = 0.57$$

$$A_{net}(D_2) = -20.57 + 0.57 = -20.0$$

$$F_{D.L.}(D_2) = -20 \times 77.1 = -1542\,\text{kN}$$

$$F_{L.L.}(D_2)(\text{negative}) = -450 \times (1.2 + 1.152) - 20.57 \times 45.65 - 7.5$$
$$\times (0.2 + 0.152) - 0.57 \times 2.45 = -2001.5\,\text{kN}$$

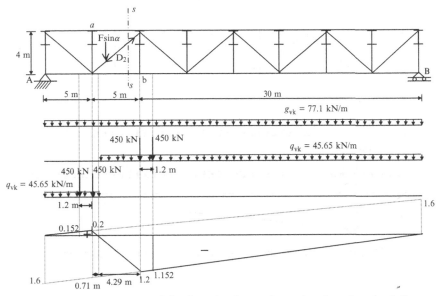

Figure 4.187 Determination of the force in diagonal member D_2 using the influence line method.

$$F_{L.L.}(D_2)(\text{positive}) = 450 \times (0.2 + 0.152) + 0.57 \times 45.65 + 20.57$$
$$\times 2.45 + 7.5 \times (1.2 + 1.152) = 252.5\,\text{kN}$$

$$F_{Ed}(D_2)\text{maximum} = F_{D.L.} \times \gamma_g + F_{L.L.} \times \gamma_q$$

$$F_{Ed}(D_2)\text{maximum} = -1542 \times 1.3 - 2001.5 \times 1.35$$
$$= -4706.6\,\text{kN}\,(\text{Compression force})$$

$$F_{Ed}(D_2)\text{minimum} = F_{D.L.} \times \gamma_g + F_{L.L.} \times \gamma_q$$

$$F_{Ed}(D_2)\text{minimum} = -1542 \times 1.3 + 252.5 \times 1.35$$
$$= -1663.7\,\text{kN}\,(\text{Compression force})$$

4.6.3.9 Calculation of Force in the Diagonal Chord Member D_1

The force in the diagonal truss member D_1 can be calculated, as shown in Figure 4.188, as follows:

$$A_{+\text{ve}}(D_1) = 0.5 \times 40 \times 1.4 = 28$$
$$F_{D.L.}(D_1) = 28 \times 77.1 = 2158.8\,\text{kN}$$

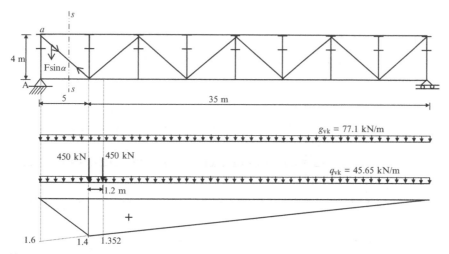

Figure 4.188 Determination of the force in diagonal member D_1 using the influence line method.

$$F_{L.L.}(D_1)(positive) = 450 \times (1.4 + 1.352) + 28 \times 45.65 = 2516.6\,kN$$

$$F_{L.L.}(D_2)(negative) = -7.5 \times (1.4 + 1.352) - 28 \times 2.45 = -89.24\,kN$$

$$F_{Ed}(D_1)maximum = F_{D.L.} \times \gamma_g + F_{L.L.} \times \gamma_q$$

$$F_{Ed}(D_1)maximum = 2158.8 \times 1.3 + 2516.6 \times 1.35$$
$$= 6203.9\,kN\,(Tension\ force)$$

$$F_{Ed}(D_1)minimum = F_{D.L.} \times \gamma_g + F_{L.L.} \times \gamma_q$$

$$F_{Ed}(D_1)minimum = 2158.8 \times 1.3 - 89.24 \times 1.35$$
$$= 2684.9\,kN\,(Tension\ force)$$

4.6.3.10 Calculation of the Reactions at Supports

The reactions at supports can be also calculated using the influence line method, as shown in Figure 4.189, as follows:

$$A_{+ve}(R) = A_{net}(D_2) = 0.5 \times 40 \times 1.0 = 20.0$$

$$F_{D.L.}(R) = 20.0 \times 77.1 = 1542\,kN$$

$$F_{L.L.}(R)(positive) = 450 \times (1.0 + 0.97) + 20.0 \times 45.65 = 1799.5\,kN$$

$$F_{Ed}(R) = F_{D.L.} \times \gamma_g + F_{L.L.} \times \gamma_q$$

$$F_{Ed}(R)maximum = 1542 \times 1.3 + 1799.5 \times 1.35 = 4433.9\,kN$$

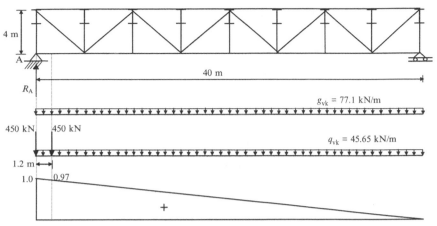

Figure 4.189 Determination of the reaction R_A using the influence line method.

4.6.3.11 Calculation of Force in the Vertical Members

The force in the vertical truss members V_2 and V_4 can be calculated, as shown in Figure 4.190, as follows:

$$A_{-ve}(V_4) = -0.5 \times 10 \times 1.0 = -5$$

$$F_{D.L.}(V_4) = -5 \times 77.1 = -385.5\,\text{kN}$$

$$F_{L.L.}(V_4)(\text{negative}) = -450 \times (1.0 + 0.76) + 5 \times 45.65 = -1020.3\,\text{kN}$$

$$F_{Ed}(D_1)\text{maximum} = F_{D.L.} \times \gamma_g + F_{L.L.} \times \gamma_q$$

$$F_{Ed}(D_1)\text{maximum} = -385.5 \times 1.3 - 1020.3 \times 1.35$$
$$= -1878.6\,\text{kN}\,(\text{Compression force}).$$

It should be noted that from the truss equilibrium, the forces in members V_3 and V_5 are zero under the applied dead and live loads, while the force in member V_1 is equal to the reaction at supports.

Figure 4.191 summarizes the calculated forces in the truss members and presents the commonly known distribution of forces in the Warren main truss under the dead and live cases of loading.

4.6.3.12 Design of the Maximum Compression Upper Chord Members U_4 and U_3

After the calculation of the design forces in the main truss members, we can now design different members of the main truss. Let us start by designing the maximum compression upper chord members U_4 and U_3, shown in Figure 4.192, carrying a compressive design force of $-10{,}366.4\,\text{kN}$.

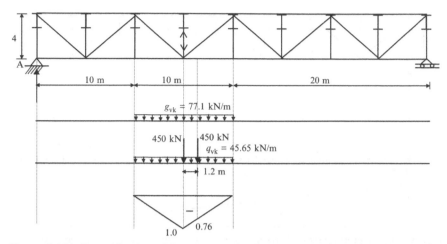

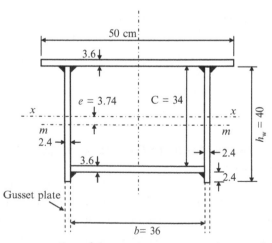

Figure 4.190 Determination of the compression force in vertical member V_4 using the influence line method.

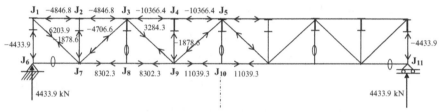

Figure 4.191 Distribution of forces in the W-shaped main truss under the dead and live cases of loading.

Figure 4.192 The cross section of the maximum compression members U_4 and U_3.

To assume a reasonable cross section for the upper chord compression member, the following parameters can be considered:

$$h_w = \frac{a}{12-15} = \frac{5000}{12-15} = 417-333 \text{ mm, taken as 400 mm.}$$

$$b = (0.75 - 0.9)h_w = (0.75 - 0.9) \times 400$$
$$= 300 - 360 \text{ mm, taken as 360 mm.}$$

It should be noted that the spacing between gusset plates (b) must be kept constant for the whole truss. Let us start by assuming the upper cover plate width of 500 mm, flange thickness of 36 mm, and web thickness of 24 mm. After that, we design the member and check the stresses. If the section is safe and economic, then the design is acceptable; otherwise, we change the dimensions accordingly and repeat the procedures. To classify the cross section chosen (see Figure 4.192),

$$\varepsilon = \sqrt{\frac{235}{f_y}} = \sqrt{\frac{235}{275}} = 0.924$$

$$b = 360 \text{ mm}, \quad t_{fl} = 36, \quad b/t_{fl} = 360/36 = 10 < 30.5 \text{ (Flange is Class 1)}$$

$$C = 340 \text{ mm}, \quad t_w = 24, \quad C/t_{fl} = 340/24 = 14.2 < 30.5 \text{ (Web is Class 1)}$$

$$A = 50 \times 3.6 + 36 \times 3.6 + 2 \times 40 \times 2.4 = 501.6 \text{ cm}^2$$

$$e = \frac{50 \times 3.6 \times 21.8 - 36 \times 3.6 \times 16.8}{501.6} = 3.74 \text{ cm}$$

$$I_m = 2 \times 2.4 \times 40^3/12 + [50 \times 3.6^3/12 + 50 \times 3.6 \times 21.8^2]$$
$$+ [36 \times 3.6^3/12 + 36 \times 3.6 \times 15.8^2] = 143831 \text{ cm}^4$$

$$I_x = 143831 - 501 \times 3.74^2 = 136815 \text{ cm}^4$$

$$I_y = 3.6 \times 50^3/12 + 3.6 \times 36^3/12 + 2 \times [40 \times 2.4^3/12 + 40 \times 2.4 \times 19.2^2]$$
$$= 122368 \text{ cm}^4$$

$$i_x = \sqrt{\frac{I_x}{A}} = \sqrt{\frac{136815}{501.6}} = 16.52 \text{ cm}$$

$$i_y = \sqrt{\frac{I_y}{A}} = \sqrt{\frac{122368}{501.6}} = 15.62 \text{ cm}$$

$$l_{bx} = l_{by} = 5000 \text{ mm}$$

$$\bar{\lambda} = \frac{L_{cr}}{i} \frac{1}{\lambda_1}$$

$$\lambda_1 = 93.9 \times 0.924 = 86.7636$$

$$\bar{\lambda} = \frac{5000}{156.2} \times \frac{1}{86.7636} = 0.369$$

The axial compressive force in the upper chord member U_4 ($N_{Ed} = 10{,}366.4$ kN):

$$\frac{N_{Ed}}{N_{b,Rd}} \leq 1.0$$

where $N_{b,Rd} = \dfrac{\chi A f_y}{\gamma_{M1}}$

$$\chi = \frac{1}{\Phi + \sqrt{\Phi^2 - \bar{\lambda}^2}} \quad \text{but } \chi \leq 1.0$$

$$\Phi = 0.5\left[1 + \alpha\left(\bar{\lambda} - 0.2\right) + \bar{\lambda}^2\right]$$

$$\Phi = 0.5\left[1 + 0.34(0.369 - 0.2) + 0.369^2\right] = 0.597$$

$$\chi = \frac{1}{0.597 + \sqrt{0.597^2 - 0.369^2}} = 0.938 \quad \text{but } \chi \leq 1.0$$

Then, $N_{b,Rd} = \dfrac{0.938 \times 501{,}600 \times 275}{1.1} = 11{,}762{,}520 \text{ N}$

$$N_{b,Rd} = 11762.5 \text{ kN} > N_{Ed} = 10366.4 \text{ kN (Then O.K.)}$$

4.6.3.13 Design of the Compression Upper Chord Members U_2 and U_1

Following the same procedures adopted for the compression members U_4 and U_3, we can design the compression upper chord members U_2 and U_1, shown in Figure 4.193, carrying a compressive design force of -4846.8 kN. To classify the cross section chosen (see Figure 4.193),

$$\varepsilon = \sqrt{\frac{235}{f_y}} = \sqrt{\frac{235}{275}} = 0.924$$

$b = 360$ mm, $t_{fl} = 14$, $b/t_{fl} = 360/14 = 25.7 < 30.5$ (Flange is Class 1)

$C = 362$ mm, $t_w = 14$, $C/t_{fl} = 362/14 = 25.9 < 30.5$ (Web is Class 1)

$$A = 55 \times 1.4 + 36 \times 1.4 + 2 \times 40 \times 1.4 = 232.4 \text{ cm}^2$$

$$e = \frac{50 \times 1.4 \times 20.7 - 36 \times 1.4 \times 16.9}{232.4} = 2.57 \text{ cm}$$

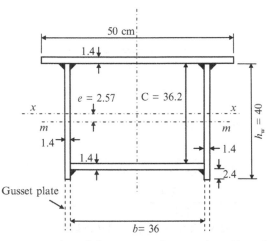

Figure 4.193 The cross section of the compression members U₁ and U₂.

$$I_m = 2 \times 1.4 \times 40^3/12 + \left[50 \times 1.4^3/12 + 50 \times 1.4 \times 20.7^2\right]$$
$$+ \left[36 \times 1.4^3/12 + 36 \times 1.4 \times 16.9^2\right] = 59342\,\text{cm}^4$$

$$I_x = 59342 - 232.4 \times 2.57^2 = 57807\,\text{cm}^4$$

$$I_y = 1.4 \times 50^3/12 + 1.4 \times 36^3/12 + 2 \times \left[40 \times 1.4^3/12 + 40 \times 1.4 \times 18.7^2\right]$$
$$= 59210\,\text{cm}^4$$

$$i_x = \sqrt{\frac{I_x}{A}} = \sqrt{\frac{57807}{232.4}} = 15.77\,\text{cm}$$

$$i_y = \sqrt{\frac{I_y}{A}} = \sqrt{\frac{59210}{232.4}} = 15.96\,\text{cm}$$

$$l_{bx} = l_{by} = 5000\,\text{mm}$$

$$\bar{\lambda} = \frac{L_{cr}}{i} \frac{1}{\lambda_1}$$

$$\lambda_1 = 93.9 \times 0.924 = 86.7636$$

$$\bar{\lambda} = \frac{5000}{157.7} \times \frac{1}{86.7636} = 0.365$$

The axial compressive force in the upper chord member U₂ ($N_{Ed} = 4846.8\,\text{kN}$):

$$\frac{N_{Ed}}{N_{b,Rd}} \leq 1.0$$

where $N_{b,Rd} = \frac{\chi A f_y}{\gamma_{M1}}$

$$\chi = \frac{1}{\Phi + \sqrt{\Phi^2 - \bar{\lambda}^2}} \quad \text{but } \chi \le 1.0$$

$$\Phi = 0.5\left[1 + \alpha(\bar{\lambda} - 0.2) + \bar{\lambda}^2\right]$$

$$\Phi = 0.5\left[1 + 0.34(0.365 - 0.2) + 0.365^2\right] = 0.595$$

$$\chi = \frac{1}{0.595 + \sqrt{0.595^2 - 0.365^2}} = 0.939 \quad \text{but } \chi \le 1.0$$

$$\text{Then, } N_{b,Rd} = \frac{0.939 \times 232.4 \times 275}{1.1} = 5,455,590 \, \text{N}$$

$$N_{b,Rd} = 5455.6 \, \text{kN} > N_{Ed} = 4846.8 \, \text{kN (Then O.K.)}$$

4.6.3.14 Design of the Lower Chord Member L_4 and L_5

Let us now design the tensile lower chord members L_4 and L_5, shown in Figure 4.194, carrying a tensile design force of 11,039.3 kN. To assume a reasonable cross section for the lower chord tension members, the following parameters can be considered:

$$h = \frac{L}{12 - 30} = \frac{5000}{12 - 30} = 417 - 166 \, \text{mm}, \quad \text{taken as 400 mm.}$$

Once again, it should be noted that the gusset plates must be spaced at a constant distance (b) of 400 mm. Let us start by assuming the flange and web thickness of 26 mm (see Figure 4.194). It should also be noted that the gross

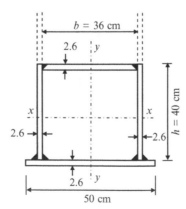

Figure 4.194 The cross section of the lower chord tension members L_4 and L_5.

and net cross-sectional areas of the lower chord members are the same since they are connected using butt weld. The design of section can be performed as follows:

$$A = A_{net} = 50 \times 2.6 + 36 \times 2.2 + 2 \times 40 \times 2.6 = 431.6 \, \text{cm}^2$$

$$N_{pl,Rd} = \frac{Af_y}{\gamma_{M0}} = \frac{431.6 \times 275 \times 100}{1.0} = 11,869,000 \, \text{N} = 11869 \, \text{kN} > N_{Ed}$$
$$= 11039.3 \, \text{kN}$$

$$N_{u,Rd} = \frac{0.9A_{net}f_u}{\gamma_{M2}} = \frac{0.9 \times 431.6 \times 100 \times 430}{1.25} = 13,362,336 \, \text{N}$$
$$= 13,362.3 \, \text{kN} > N_{Ed} = 11,039.3 \, \text{kN}$$

4.6.3.15 Design of the Lower Chord Members L_3 and L_2

Following the same procedures adopted for the design of the lower chord members L_5 and L_4, we can design the tensile lower chord members L_3 and L_2, shown in Figure 4.195, carrying a tensile design force of 8302.3 kN. The design of section can be as follows:

$$A = A_{net} = 50 \times 2.0 + 36 \times 2.0 + 2 \times 40 \times 2.0 = 332 \, \text{cm}^2$$

$$N_{pl,Rd} = \frac{Af_y}{\gamma_{M0}} = \frac{332 \times 275 \times 100}{1.0} = 9,130,000 \, \text{N} = 9130 \, \text{kN} > N_{Ed}$$
$$= 8302.3 \, \text{kN}$$

$$N_{u,Rd} = \frac{0.9A_{net}f_u}{\gamma_{M2}} = \frac{0.9 \times 332 \times 100 \times 430}{1.25} = 10,278,720 \, \text{N} = 10,278.7 \, \text{kN}$$
$$> N_{Ed} = 8302 \, \text{kN}$$

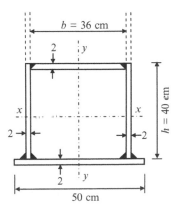

Figure 4.195 The cross section of the lower chord tension member L_3 and L_2.

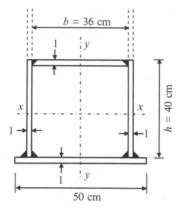

Figure 4.196 The cross section of the lower chord member L_1.

It should be noted that member L_1 having a force of zero under the dead and live cases of loading can be taken with the same cover plate width and the same web height but with a minimum thickness of 1 cm as shown in Figure 4.196.

4.6.3.16 Design of the Compression Vertical Member V_1

Let us now design the compression vertical member V_1, shown in Figure 4.197, carrying a compressive design force of -4433.9 kN. To assume a reasonable cross section for the compression vertical member, the following parameters can be considered:

$$d_1 = \frac{L}{15 - 22} = \frac{4000}{15 - 22} = 26.7 - 181.8\,\text{mm} \quad \text{taken as 260 mm.}$$

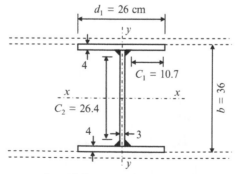

Figure 4.197 The cross section of the vertical compression member V_1.

It should be noted that the vertical member must be inside the gusset plates spaced at a constant distance (b) of 400 mm. Let us start by assuming the flange thickness of 40 mm and web thickness of 30 mm. To classify the cross section chosen (see Figure 4.197),

$$\varepsilon = \sqrt{\frac{235}{f_y}} = \sqrt{\frac{235}{275}} = 0.924$$

$$C_1 = 107 \, \text{mm}, \quad t_{fl} = 40, \quad C_1/t_{fl} = 107/40 = 2.8 < (\text{Flange is Class 1})$$

$$C_2 = 264 \, \text{mm}, \quad t_w = 30, \quad C_2/t_{fl} = 264/30 = 8.8 < 33 \times 0.924$$
$$= 30.5 (\text{Web is Class 1})$$

$$A = 2 \times 26 \times 4 + 28 \times 3.0 = 292 \, \text{cm}^2$$

$$I_x = 3.0 \times 28^3/12 + 2 \times \left[26 \times 4^3/12 + 26 \times 4 \times 16^2\right] = 65,189 \, \text{cm}^4$$

$$I_y = 28 \times 3^3/12 + 2 \times 4 \times 26^3/12 = 11,780 \, \text{cm}^4$$

$$i_x = \sqrt{\frac{I_x}{A}} = \sqrt{\frac{65,189}{292}} = 14.94 \, \text{cm}$$

$$i_y = \sqrt{\frac{I_y}{A}} = \sqrt{\frac{11,780}{292}} = 6.35 \, \text{cm}$$

$$l_{by} = 3600 \, \text{mm}$$

$$l_{bx} = 4000 \, \text{mm}$$

$$\bar{\lambda} = \frac{L_{cr}}{i} \frac{1}{\lambda_1}$$

$$\lambda_1 = 93.9 \times 0.924 = 86.7636$$

$$\bar{\lambda} = \frac{3600}{63.5} \times \frac{1}{86.7636} = 0.653$$

The axial compressive force in the vertical member V_1 ($N_{Ed} = 4433.9$ kN):

$$\frac{N_{Ed}}{N_{b,Rd}} \leq 1.0$$

where $N_{b,Rd} = \dfrac{\chi A f_y}{\gamma_{M1}}$

$$\chi = \frac{1}{\Phi + \sqrt{\Phi^2 - \bar{\lambda}^2}} \quad \text{but } \chi \leq 1.0$$

$$\Phi = 0.5\left[1 + \alpha(\bar{\lambda} - 0.2) + \bar{\lambda}^2\right]$$

$$\Phi = 0.5\left[1 + 0.49(0.653 - 0.2) + 0.653^2\right] = 0.824$$

$$\chi = \frac{1}{0.824 + \sqrt{0.824^2 - 0.653^2}} = 0.754 \text{ but } \chi \leq 1.0$$

Then, $N_{b,Rd} = \dfrac{0.754 \times 29,200 \times 275}{1.1} = 5,504,200\,N$

$N_{b,Rd} = 5504.2\,kN > N_{Ed} = 4433.9\,kN$ (Then O.K.)

4.6.3.17 Design of the Compression Vertical Members V_2 and V_4

Following the same procedures adopted for the design the compression vertical member V_1, we can design the vertical compression members V_2 and V_4 shown in Figure 4.198, carrying a compressive design force of $-1878.6\,kN$, as follows. To classify the cross section chosen (see Figure 4.198),

$$\varepsilon = \sqrt{\frac{235}{f_y}} = \sqrt{\frac{235}{275}} = 0.924$$

$C_1 = 117\,mm, \quad t_{fl} = 20, \quad C_1/t_{fl} = 117/20 = 5.85 < 30.5$ (Flange is Class 1)

$C_2 = 304\,mm, \quad t_w = 10, \quad C_2/t_{fl} = 304/10 = 30.4 < 30.5$ (Web is Class 1)

$$A = 2 \times 26 \times 2 + 32 \times 1.0 = 136\,cm^2$$

$$I_x = 1.0 \times 32^3/12 + 2 \times \left[26 \times 2^3/12 + 26 \times 2 \times 17^2\right] = 32821\,cm^4$$

$$I_y = 32 \times 1.0^3/12 + 2 \times 2 \times 26^3/12 = 5861.3\,cm^4$$

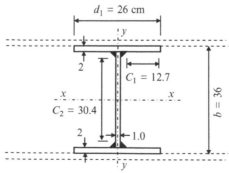

Figure 4.198 The cross section of the vertical compression members V_2 and V_4.

$$i_x = \sqrt{\frac{I_x}{A}} = \sqrt{\frac{32{,}821}{136}} = 15.53\,\text{cm}$$

$$i_y = \sqrt{\frac{I_y}{A}} = \sqrt{\frac{58{,}61.3}{136}} = 6.56\,\text{cm}$$

$$l_{by} = 3600\,\text{mm}$$

$$l_{bx} = 4000\,\text{mm}$$

$$\bar{\lambda} = \frac{L_{cr}}{i}\frac{1}{\lambda_1}$$

$$\lambda_1 = 93.9 \times 0.924 = 86.7636$$

$$\bar{\lambda} = \frac{3600}{65.6} \times \frac{1}{86.7636} = 0.633$$

The axial compressive force in the vertical member V_2 ($N_{Ed} = 1878.6$ kN):

$$\frac{N_{Ed}}{N_{b,Rd}} \le 1.0$$

where $N_{b,Rd} = \dfrac{\chi A f_y}{\gamma_{M1}}$

$$\chi = \frac{1}{\Phi + \sqrt{\Phi^2 - \bar{\lambda}^2}} \quad \text{but } \chi \le 1.0$$

$$\Phi = 0.5\left[1 + \alpha\left(\bar{\lambda} - 0.2\right) + \bar{\lambda}^2\right]$$

$$\Phi = 0.5\left[1 + 0.49(0.633 - 0.2) + 0.633^2\right] = 0.806$$

$$\chi = \frac{1}{0.806 + \sqrt{0.806^2 - 0.633^2}} = 0.766 \quad \text{but } \chi \le 1.0$$

Then, $N_{b,Rd} = \dfrac{0.766 \times 13600 \times 275}{1.1} = 2604400\,\text{N}$

$$N_{b,Rd} = 2604.4\,\text{kN} > N_{Ed} = 1878.6\,\text{kN (Then O.K.)}$$

It should be noted that members V_3 and V_5 having forces of zero under the dead and live cases of loading can be taken with the same flange plate width and the same web height but with a minimum thickness of 1 cm as shown in Figure 4.199.

4.6.3.18 Design of the Diagonal Member D_1

We can also design the diagonal member D_1, shown in Figure 4.200, carrying a maximum tensile design force of 6203.9 kN. The bolts used in

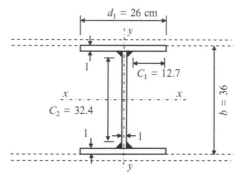

Figure 4.199 The cross section of the vertical members V_3 and V_5.

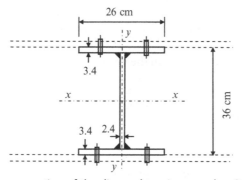

Figure 4.200 The cross section of the diagonal tension member D_1.

connecting the member with gusset plates are M27 high-strength pretensioned bolts having a clearance of 3 mm (hole diameter $\varnothing = 30$ mm). The
member can be designed as follows:

$$A_{net} = 246.88 - 4 \times 3.4 \times 3 = 206.08 \, \text{cm}^2$$

$$N_{pl,Rd} = \frac{A f_y}{\gamma_{M0}} = \frac{246.88 \times 275 \times 100}{1.0} = 6,789,200 \, \text{N} = 6789.2 \, \text{kN} > N_{Ed}$$
$$= 6203.9 \, \text{kN}$$

$$N_{u,Rd} = \frac{0.9 A_{net} f_u}{\gamma_{M2}} = \frac{0.9 \times 20,608 \times 430}{1.25} = 6,380,236.8 \, \text{N} = 6380.2 \, \text{kN}$$
$$> N_{Ed} = 6203.9 \, \text{kN}$$

4.6.3.19 Design of the Diagonal Tension Member D₃

The diagonal member D_3, shown in Figure 4.201, carrying a maximum tensile design force of 3284.3 kN can be designed adopting the same procedures
used with D_1 as follows:

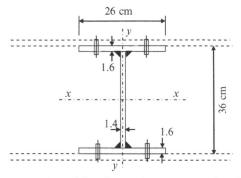

Figure 4.201 The cross section of the diagonal tension member D_3.

The bolts used in connecting the member with gusset plates are M27 high-strength pretensioned bolts having a clearance of 3 mm (hole diameter $\emptyset = 30$ mm):

$$A = 2 \times 26 \times 1.6 + 32.8 \times 1.4 = 129.12\,\text{cm}^2$$

$$A_{\text{net}} = 129.12 - 4 \times 3.0 \times 1.6 = 109.92\,\text{cm}^2$$

$$N_{\text{pl,Rd}} = \frac{A f_y}{\gamma_{M0}} = \frac{129.12 \times 275 \times 100}{1.0} = 3{,}550{,}800\,\text{N} = 3550.8\,\text{kN} > N_{\text{Ed}}$$
$$= 3284.3\,\text{kN}$$

$$N_{\text{u,Rd}} = \frac{0.9 A_{\text{net}} f_u}{\gamma_{M2}} = \frac{0.9 \times 10{,}992 \times 430}{1.25} = 3{,}403{,}123\,\text{N} = 3403.1\,\text{kN}$$
$$> N_{\text{Ed}} = 3284.3\,\text{kN}$$

4.6.3.20 Design of the Compression Diagonal Member D_2

The diagonal compression members V_2, shown in Figure 4.202, carrying a compressive design force of -4706.6 kN can be designed as follows. To classify the cross section chosen (see Figure 4.202),

$$\varepsilon = \sqrt{\frac{235}{f_y}} = \sqrt{\frac{235}{275}} \llcorner 0.924$$

$$C_1 = 180\,\text{mm}, \quad t_{\text{fl}} = 30, \quad C_1/t_{\text{fl}} = 180/30 = 6.0 < 30.5\,(\text{Flange is Class 1})$$

$$C_2 = 300\,\text{mm}, t_{\text{w}} = 20, \quad C_2/t_{\text{fl}} = 300/20 = 15 < 30.5\,(\text{Web is Class 1})$$

$$A = 2 \times 26 \times 3 + 30 \times 2 \times 2 = 276\,\text{cm}^2$$

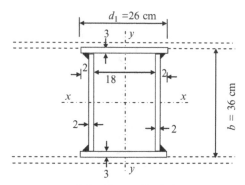

Figure 4.202 The cross section of the diagonal compression member D_2.

$$I_x = 2 \times 2 \times 30^3/12 + 2 \times \left[26 \times 3^3/12 + 26 \times 3 \times 16.5^2 \right] = 51,588 \, \text{cm}^4$$

$$I_y = 2 \times 3 \times 26^3/12 + 2 \times \left[30 \times 2^3/12 + 30 \times 2 \times 10^2 \right] = 20,828 \, \text{cm}^4$$

$$i_x = \sqrt{\frac{I_x}{A}} = \sqrt{\frac{51,588}{276}} = 13.67 \, \text{cm}$$

$$i_y = \sqrt{\frac{I_y}{A}} = \sqrt{\frac{20,828}{276}} = 8.69 \, \text{cm}$$

$$l_{by} = 5760 \, \text{mm}$$

$$l_{bx} = 6400 \, \text{mm}$$

$$\bar{\lambda} = \frac{L_{cr}}{i} \frac{1}{\lambda_1}$$

$$\lambda_1 = 93.9 \times 0.924 = 86.7636$$

$$\bar{\lambda} = \frac{5760}{86.9} \times \frac{1}{86.7636} = 0.764$$

The axial compressive force in the vertical member D_2 ($N_{Ed} = 4706.6$ kN):

$$\frac{N_{Ed}}{N_{b,Rd}} \leq 1.0$$

where $N_{b,Rd} = \dfrac{\chi A f_y}{\gamma_{M1}}$

$$\chi = \frac{1}{\Phi + \sqrt{\Phi^2 - \bar{\lambda}^2}} \quad \text{but } \chi \leq 1.0$$

$$\Phi = 0.5\left[1 + \alpha(\overline{\lambda} - 0.2) + \overline{\lambda}^2\right]$$

$$\Phi = 0.5\left[1 + 0.49(0.764 - 0.2) + 0.764^2\right] = 0.888$$

$$\chi = \frac{1}{0.888 + \sqrt{0.888^2 - 0.764^2}} = 0.746 \text{ but } \chi \leq 1.0$$

Then, $N_{b,Rd} = \dfrac{0.746 \times 27,600 \times 275}{1.1} = 5,147,400 \text{ N}$

$$N_{b,Rd} = 5147.4 \text{ kN} > N_{Ed} = 4706.6 \text{ kN (Then O.K.)}$$

4.6.3.21 Design of the Compression Diagonal Member D_4

The diagonal compression members D_4, shown in Figure 4.203, carrying a compressive design force of -1935.7 kN can be designed as follows. To classify the cross section chosen (see Figure 4.203),

$$\varepsilon = \sqrt{\frac{235}{f_y}} = \sqrt{\frac{235}{275}} = 0.924$$

$$C_1 = 115 \text{ mm}, t_{fl} = 26, C_1/t_{fl} = 115/26 = 4.4 < 30.5 \text{ (Flange is Class 1)}$$

$$C_2 = 292 \text{ mm}, t_w = 14, C_2/t_{fl} = 292/14 = 20.9 < 30.5 \text{ (Web is Class 1)}$$

$$A = 2 \times 26 \times 2.6 + 30.8 \times 1.4 = 178.32 \text{ cm}^2$$

$$I_x = 1.4 \times 30.8^3/12 + 2 \times \left[2.6 \times 26^3/12 + 2.6 \times 26 \times 16.7^2\right] = 48,731 \text{ cm}^4$$

$$I_y = 2 \times 2.6 \times 26^3/12 + 30.8 \times 1.4^3/12 = 7623 \text{ cm}^4$$

$$i_x = \sqrt{\frac{I_x}{A}} = \sqrt{\frac{48,731}{178.32}} = 16.53 \text{ cm}$$

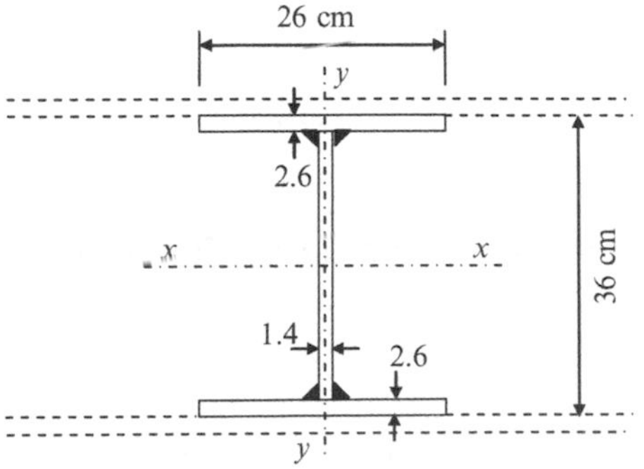

Figure 4.203 The cross section of the diagonal compression member D_4.

$$i_y = \sqrt{\frac{I_y}{A}} = \sqrt{\frac{7623}{178.32}} = 6.54 \, \text{cm}$$

$$l_{by} = 5760 \, \text{mm}$$

$$l_{bx} = 6400 \, \text{mm}$$

$$\bar{\lambda} = \frac{L_{cr}}{i} \frac{1}{\lambda_1}$$

$$\lambda_1 = 93.9 \times 0.924 = 86.7636$$

$$\bar{\lambda} = \frac{5760}{65.4} \times \frac{1}{86.7636} = 1.015$$

The axial compressive force in the vertical member D_4 ($N_{Ed} = 1935.7 \, \text{kN}$):

$$\frac{N_{Ed}}{N_{b,Rd}} \leq 1.0$$

where $N_{b,Rd} = \dfrac{\chi A f_y}{\gamma_{M1}}$

$$\chi = \frac{1}{\Phi + \sqrt{\Phi^2 - \bar{\lambda}^2}} \quad \text{but } \chi \leq 1.0$$

$$\Phi = 0.5 \left[1 + \alpha(\bar{\lambda} - 0.2) + \bar{\lambda}^2 \right]$$

$$\Phi = 0.5 \left[1 + 0.49(1.015 - 0.2) + 1.015^2 \right] = 1.219$$

$$\chi = \frac{1}{1.219 + \sqrt{1.219^2 - 1.015^2}} = 0.528 \quad \text{but } \chi \leq 1.0$$

Then, $N_{b,Rd} = \dfrac{0.528 \times 17832 \times 275}{1.1} = 2{,}353{,}824 \, \text{N}$

$$N_{b,Rd} = 2353.8 \, \text{kN} > N_{Ed} = 1935.7 \, \text{kN} \, (\text{Then O.K.})$$

4.6.3.22 Design of Stringer-Cross Girder Connection

The stringer is designed as a simply supported beam on cross girders; therefore, the connection is mainly transferring shear forces (maximum reaction from stringers of 574 kN) (see Figure 4.204). Using M27 high-strength pretensioned bolts of grade 8.8, having f_{ub} of 800 MPa, shear area A of 4.59 cm^2, and gross area A_g of 5.73 cm^2, we can determine the

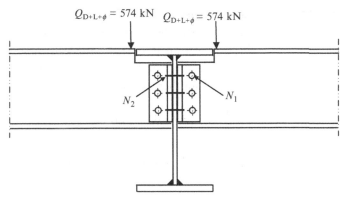

Figure 4.204 The connection between a stringer and a cross girder.

required number of bolts, following the rules specified in EC3 (BS EN 1993-1-8) [2.13], as follows:

$$F_{v,Rd} = \frac{\alpha_V f_{ub} A}{\gamma_{M2}}$$

$$F_{v,Rd} = \frac{0.6 \times 800 \times 459}{1.25} = 1,76,256\,N$$

Then, $F_{v,Rd}$ equals 176 kN (for bolts in single shear) and 353 kN (for bolts in double shear):

$$F_{s,Rd} = \frac{k_s n \mu}{\gamma_{M3}} F_{p,C}$$

$$F_{p,C} = 0.7 f_{ub} A_s = 0.7 \times 800 \times 573 = 320,880\,N$$

$$F_{s,Rd,ser} = \frac{1.0 \times 1.0 \times 0.4}{1.1} 320,880 = 116,683.6\,N.$$

Then, $F_{s,Rd} = 117$ kN (for bolts in single shear at serviceability limit states) and $F_{s,Rd} = 234$ kN (for bolts in double shear at serviceability limit states). At ultimate limit states, $F_{s,Rd,ult}$ can be calculated as follows:

$$F_{s,Rd,ult} = \frac{1.0 \times 1.0 \times 0.4}{1.25} 320,880 = 102,682\,N.$$

Then, $F_{s,Rd} = 103$ kN (for bolts in single shear at ultimate limit states) and $F_{s,Rd} = 206$ kN (for bolts in double shear at ultimate limit states):

$$N_1 = \frac{574}{206} = 2.8 \text{ taken as 3 bolts,}$$

$$N_2 = \frac{574}{103} = 5.6 \text{ taken as 6 bolts}$$

4.6.3.23 Design of Cross Girder-Main Truss Connection

The cross girder is designed as a simply supported beam on main trusses; therefore, once again, the connection is mainly transferring shear forces (maximum reaction from cross girders of 1600.8 kN) (see Figure 4.205). We can determine the required number of bolts as follows:

$$N_3 = \frac{1164.1}{206} = 5.6 \text{ taken as 6 bolts,}$$

$$N_2 = \frac{1164.1}{103} = 11.3 \text{ taken as 12 bolts}$$

4.6.3.24 Design of Wind Bracings

Wind forces acting on the investigated deck highway bridge (see Figure 4.206) as well as any other lateral forces directly applied to the bridge are transmitted to the bearings by systems of upper and lower wind bracings as well as cross bracings. The upper and lower wind bracings carry wind forces on the main truss as shown in Figure 4.206. Wind bracings are quite important to the lateral stability of the upper chord compression members since they define the buckling outside the plane of the truss, and therefore, wind forces applied to this bridge can be sufficiently estimated using the design rules specified in EC1 [3.2] as follows:

$$F_w = \frac{1}{2} \rho v_b^2 C A_{ref,x}$$

$$v_b = c_{dir} \times c_{season} \times v_{b,0} = 1.0 \times 1.0 \times 26 = 26 \, \text{m/s}$$

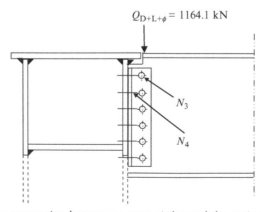

Figure 4.205 The connection between a cross girder and the main truss.

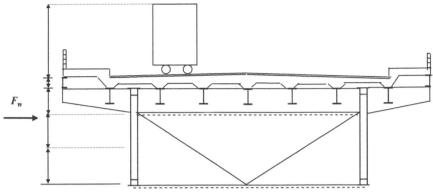

Figure 4.206 Design height for the calculation of wind forces on the upper and lower wind bracings.

$$A_{ref,x} = 6 \times 40 = 240\,\text{m}^2$$

$$F_w = \frac{1}{2} \times 1.25 \times 26^2 \times 5.7 \times 240 = 578{,}000\,\text{N} = 578\,\text{kN}$$

Consider the structural analysis for the upper wind bracing system shown in Figure 4.207. Assume that the upper wind bracing carries wind forces on the moving traffic, upper part of truss, and forces on the midspacing between upper and lower wind bracings. The critical design wind force in the diagonal bracing members can be calculated as follows:

Distributed wind loads $(q_{WL}) = 578 \times (4.6/6)/40 = 11.1\,\text{kN/m}$

Factored distributed wind loads $= q_{WL} \times \gamma_q = 11.1 \times 1.7 = 18.9\,\text{kN/m}$

$$R_A = 18.9 \times 20 = 378\,\text{kN}$$

$$\alpha = \tan^{-1}(6/6) = 45°$$

$$F_D = 378/(2 \times \sin 45) = 267.3\,\text{kN}$$

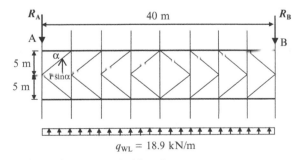

$q_{WL} = 18.9$ kN/m

Figure 4.207 Loads on the upper wind bracing.

The cross section of the bracing member (see Figure 4.28) can be determined as follows:

$$l_{bx} = 7070 \, \text{mm}, \quad l_{by} = 1.2 \times 7070 = 8484 \, \text{mm}$$

Choose two angles back-to-back $150 \times 150 \times 15$, with 10 mm gusset plate between them:

$$A = 2 \times 43.2 = 86.4 \, \text{cm}^2, \quad i_x = 4.59 \, \text{cm}, \quad e = 4.26 \, \text{cm},$$

$$i_y = \sqrt{4.59^2 + (4.26 + 1/2)^2} = 6.61 \, \text{cm}$$

$$\varepsilon = \sqrt{\frac{235}{275}} = 0.924$$

$$\bar{\lambda} = \frac{L_{cr}}{i} \frac{1}{\lambda_1}$$

$$\lambda_1 = 93.9 \times 0.924 = 86.7636$$

$$\bar{\lambda} = \frac{7070}{45.9} \frac{1}{86.7636} = 1.775$$

The axial compressive force in the diagonal bracing member ($N_{Ed} = 267.3 \, \text{kN}$):

$$\frac{N_{Ed}}{N_{b,Rd}} \leq 1.0$$

where $N_{b,Rd} = \dfrac{\chi A f_y}{\gamma_{M1}}$

$$A = 2 \times 43.2 = 86.4 \, \text{cm}^2$$

$$\chi = \frac{1}{\Phi + \sqrt{\Phi^2 - \bar{\lambda}^2}} \quad \text{but } \chi \leq 1.0$$

$$\Phi = 0.5\left[1 + \alpha(\bar{\lambda} - 0.2) + \bar{\lambda}^2\right]$$

$$\Phi = 0.5\left[1 + 0.34(1.775 - 0.2) + 1.775^2\right] = 2.343$$

$$\chi = \frac{1}{2.343 + \sqrt{2.343^2 - 1.775^2}} = 0.258 \quad \text{but } \chi \leq 1.0$$

Then, $N_{b,Rd} = \dfrac{0.258 \times 86.4 \times 100 \times 275}{1.1} = 557280 \, \text{N}$

$$N_{b,Rd} = 557.3 \, \text{kN} > N_{Ed} = 267.3 \, \text{kN} \, (\text{Then O.K.})$$

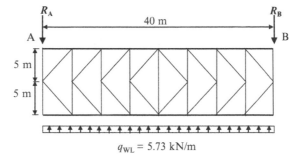

$q_{WL} = 5.73$ kN/m

Figure 4.208 Loads on the lower wind bracing.

Similar to the upper wind bracing analysis, the forces in the members of the lower wind bracing can be calculated with the loads acting on the lower wind bracing as shown in Figure 4.208. The lower wind bracing carries the wind forces acting on the midspacing between the upper and lower wind bracings. It should be noted that the level of the upper wind bracing is taken at the bottom flange of the cross girders. Since the cross girders are connected to the main upper chord member, therefore, it is assumed that the upper chord members are restrained to buckle outside the plane of the main truss at the interactions with the cross girders.

4.6.3.25 Design of Roller Steel Fabricated Bearings

Let us now design the roller steel fabricated bearings shown in Figure 4.162 and detailed in Figure 4.209. The maximum vertical reaction at the supports of the main truss was previously calculated under dead and live loads with dynamic effect $(R_{D+L+\Phi})$, which was 4433.9 kN. The material of construction for the bearings is cast iron steel (ISO 3755) 340–550 having a yield stress of 340 MPa and an ultimate stress of 550 MPa.

Design of the Sole Plate

The reaction $(R_{D+L+\Phi})$ can be assumed as two equal concentrated loads at two points, which are the centers of gravity of half of the last vertical member V_9 shown in Figure 4.209. To determine the centers of gravity (distance e), we can take the first area moment around the axis z-z, shown in Figure 4.209, as follows:

$$e = \frac{2 \times 4 \times 13 \times 6.5 + 28 \times 1.5 \times 0.75}{(2 \times 4 \times 13 + 28 \times 1.5)} = 4.85 \text{ cm}$$

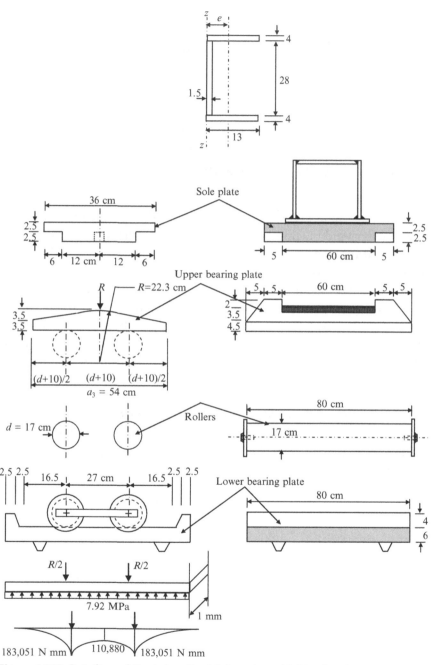

Figure 4.209 Detailing of the twin roller fabricated steel bridge bearings.

Assuming that the thickness of the sole plate is t_1, with detailed dimensions shown in Figure 4.209 based on the lower chord member L_1 dimensions, we can determine the maximum moment applied to the sole plate (M) as follows:

$$M = R_{D+L+\Phi} \times \frac{e}{2} = 4433.9 \times 10^3 \times \frac{48.5}{2} = 107{,}522{,}075 \text{ N mm.}$$

Section plastic modulus $(W_{pl}) = b_1 t_1^2/4 = 700 \times t_1^2/4 = 175 \times t_1^2$
The plate thickness t_1 can be calculated now as follows:

$$\frac{M}{W_{pl}} = \frac{f_y}{\gamma_{M0}}$$

$$\frac{107{,}522{,}075}{175 \times t_1^2} = \frac{340}{1.0}$$

Then, $t_1 = 42.5$ mm, taken as 50 mm, as shown in Figure 4.209.

Design of the Rollers

The design of rollers requires determination of the diameter, length, and number of rollers to resist the vertical load as well as the arrangement and allowed movement in the direction of rollers. The design axial force per unit length of roller contact N'_{Sd} specified in BS EN 1337-1 [3.11] shall satisfy

$$N'_{Sd} \leq N'_{Rd}$$

where N'_{Rd} is the design value of resistance per unit length of roller contact, which is calculated as

$$N'_{Rd} = 23 \times R \times \frac{f_u^2}{E_d} \times \frac{1}{\gamma_m^2} = 23 \times R \times \frac{550^2}{210{,}000} \times \frac{1}{1} = 33.131 \times R$$

Assume the number of rollers is 2 and their length is 800 mm as shown in Figure 4.33:

$$N'_{Sd} = \frac{R_{D+L+\Phi}}{2 \times 800} = \frac{4433.9 \times 10^3}{1600} = 2771.2 \text{ N/mm}$$

Then, the radius of rollers can be determined by equalizing N'_{Sd} with N'_{Rd} as follows:

$$2771.2 = 33.131 \times R$$

Then, $R = 83.6$ mm, taken as 85 mm and the diameter D is 170 mm.

Design of Upper Bearing Plate

The upper bearing plate is shown in Figure 4.209. The width and length of the plate are dependent on the spacing between rollers and the length of rollers as well as the allowed movement in the direction of rollers. The thickness of the upper bearing plate can be determined as follows:

$$M = \frac{R_{D+L+\Phi}}{2} \times \frac{(D+100)}{2} = \frac{4433.9 \times 10^3}{2} \times \frac{270}{2} = 299,288,250 \text{ N mm.}$$

$$W_{pl} = \frac{b_2 t_2^2}{4} = \frac{800 t_2^2}{4} = 200 t_2^2 \text{ mm}^3$$

The plate thickness t_2 can be calculated now as follows:

$$\frac{M}{W_{pl}} = \frac{f_y}{\gamma_{M0}}$$

$$\frac{299,288,250}{200 \times t_1^2} = \frac{340}{1.0}$$

Then, $t_1 = 66.3$ mm, taken as 70 mm, as shown in Figure 4.209.

The radius of the curved part of the upper bearing plate, which has a length of 600 mm as shown in Figure 4.209, can be determined the same way as that adopted for the design of the rollers:

$$N'_{Rd} = 23 \times R \times \frac{f_u^2}{E_d} \times \frac{1}{\gamma_m^2} = 23 \times R \times \frac{550^2}{210,000} \times \frac{1}{1} = 33.131 \times R$$

$$N'_{Sd} = \frac{R_{D+L+\Phi}}{600} = \frac{4433.9 \times 10^3}{600} = 7389.8 \text{ N/mm}$$

Then, the radius of rollers can be determined by equalizing N'_{Sd} with N'_{Rd} as follows:

$$7389.8 = 33.131 \times R$$

Then, $R = 223$ mm.

Design of Lower Bearing Plate

The lower bearing plate is shown in Figure 4.209. The width and length of the plate are dependent on the strength of concrete and are dependent on the spacing between rollers and the length of rollers as well as the allowed movement in the direction of rollers. The thickness of the upper bearing plate can be determined as follows:

$$f_c = \frac{R_{D+L+\Phi}}{a_3 b_3} = \frac{4433.9 \times 10^3}{700 \times 800}$$

$$= 7.92 \, \text{MPa} < \frac{f_c}{\gamma_c} = \frac{40}{1.5} = 26.7 \, \text{MPa} < \frac{f_c}{\gamma_c} = \frac{40}{1.5} = 26.7 \, \text{MPa}$$

(for a typical concrete in bridges of C40/50 with f_{ck})

The plate thickness t_3 can be calculated from the distribution of bending moment, caused by the pressure on the concrete foundation, as follows:
$M = 183,051 \, \text{N} \, \text{mm}$ per unit width of the plate:

$$W_{pl} = \frac{b_3 t_3^2}{4} = \frac{1 \times t_3^2}{4} = 0.25 \times t_3^2 \, \text{mm}^3$$

$$\frac{M}{W_{pl}} = \frac{f_y}{\gamma_{M0}}$$

$$\frac{183,051}{0.25 \times t_3^2} = \frac{340}{1.0}$$

Then, $t_3 = 46.4$ mm, taken as 50 mm, as shown in Figure 4.209.

4.6.3.26 Design of Hinged Line Rocker Steel Fabricated Bearings

Finally, we can now design the hinged line rocker steel fabricated bearings shown in Figure 4.162 and detailed in Figure 4.210. The maximum vertical reaction at the support of the main plate girder was previously calculated under dead and live loads with dynamic effect ($R_{D+L+\Phi}$), which was 4433.9 kN. The bearing is also subjected to a lateral force from the braking forces from traffic as well as subjected to a longitudinal force from the reactions of the upper and lower wind bracings, which cause moments around longitudinal and lateral directions of the bearing base, respectively. Similar to the roller bearing, the material of construction for the bearings is cast iron steel (ISO 3755) 340–550 having a yield stress of 340 MPa and an ultimate stress of 550 MPa. It should be noted that the overall height of the hinged bearing must be exactly the same as that of the roller bearing. The general layout and assumed dimensions of the hinged line rocker bearing are shown in Figure 4.211. The braking Q_{lbk} forces can be calculated as follows:

$$Q_{lbk} = 360 + 2.7 \times L = 360 + 2.7 \times 40 = 468 \, \text{kN}, \quad \text{for Load model 1}$$

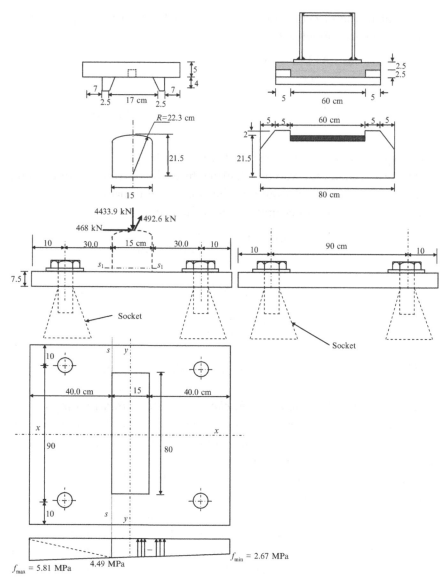

Figure 4.210 Detailing of the hinged line rocker fabricated steel bridge bearings.

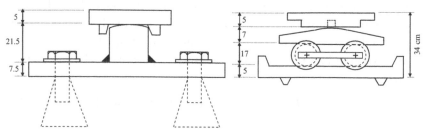

Figure 4.211 The designed roller and hinged line rocker fabricated steel bearings.

See Figure 4.210 for the direction of the forces. Also, the reactions from upper and lower wind bracings (R_{tot}) (see Figure 4.210 for the direction of the forces) were previously calculated as follows:

$$R_{tot} = 492.6\,kN$$

We can now determine the normal stress distribution due to the applied loads, shown in Figure 4.210, on the concrete foundation as follows:

$$f = -\frac{N}{A} \pm \frac{M_x}{I_x}y \pm \frac{M_y}{I_y}x$$

$$\frac{N}{A} = \frac{4433.9 \times 10^3}{950 \times 1100} = 4.24\,MPa$$

$$\frac{M_x}{I_x}y = \frac{492.6 \times 10^3 \times 290}{950 \times 1100^3/12}550 = 0.75\,MPa$$

$$\frac{M_y}{I_y}x = \frac{468 \times 10^3 \times 290}{1100 \times 950^3/12}475 = 0.82\,MPa$$

$$f_{max} = -4.24 - 0.75 - 0.82 = -5.81\,MPa$$

$$f_{min} = -4.24 + 0.75 + 0.82 = -2.67\,MPa$$

The critical bending moment on the base plate of the hinged bearing is at section s-s shown in Figure 4.210:

$$M = (0.5 \times 400 \times 4.49) \times 1100 \times 400/3 + (0.5 \times 400 \times 5.81) \times 1100 \\ \times 400 \times 2/3 = 472,560,000\,N\,mm$$

$$W_{pl} = 1100 \times \frac{t_4^2}{4} = 275t_4^2$$

$$\frac{M}{W_{pl}} = \frac{f_y}{\gamma_{M0}}$$

$$\frac{472,560,000}{275t_4^2} = \frac{340}{1.0}$$

Then, $t_4 = 71.1$ mm, taken as 75 mm.

The normal stresses at section s_1-s_1, shown in Figure 4.210, of the line rocker bearing can be checked as follows:

$$M_x = 492.6 \times 10^3 \times 215 = 105,909,000\,N\,mm.$$

$$M_y = 468 \times 10^3 \times 215 = 100,620,000\,N\,mm.$$

$$\frac{N}{A} = \frac{4433.9 \times 10^3}{150 \times 800} = 36.95\,MPa$$

$$\frac{M_x}{I_x}y = \frac{105,909,000}{150 \times 800^3/12}400 = 6.62\,\text{MPa}$$

$$\frac{M_y}{I_y}x = \frac{100,620,000}{800 \times 150^3/12}75 = 33.54\,\text{MPa}$$

$$f_{max} = -(36.95 + 6.62 + 33.54) = -77.11\,\text{MPa} < 340\,\text{MPa}\,(\text{Then O.K.})$$

4.6.3.27 Design of Joint J_1

It is now possible to design the joints of the main trusses after designing all members and knowing all details regarding the joints. Let us start by designing joint J_1 (see Figure 4.212). For M27 high-strength pretensioned bolts used, the following design values are calculated:

$F_{v,Rd} = 176$ kN (single shear) and 353 kN (double shear)

$F_{s,Rd} = 117$ kN (single shear) and 234 kN (double shear)

$F_{s,ult} = 103$ kN (single shear) and 206 kN (double shear)

Number of Bolts for the Vertical Member V_1

$$N(V_1) = \frac{F_{Ed}}{F_{s,ult}} = \frac{4433.9}{206} = 21.5\text{ bolts, taken as 24 bolts (12 bolts in each side}$$

acting in double shear)

Number of Bolts for the Diagonal Member D_1

$$N(D_1) = \frac{F_{Ed}}{F_{s,ult}} = \frac{6203.9}{206} = 30.1\text{ bolts, taken as 32 bolts (16 bolts in each side}$$

acting in double shear)

4.6.3.28 Design of Joint J_2

Following the same procedures adopted for the design of joint J_1, we can design joint J_2 (see Figure 4.213) using the same M27 high-strength pretensioned bolts as follows.

Number of Bolts for the Vertical Member V_2

$$N(V_2) = \frac{F_{Ed}}{F_{s,ult}} = \frac{1878.6}{103} = 18.2\text{ bolts, taken as 20 bolts (10 bolts in each side}$$

acting in single shear)

4.6.3.29 Design of Joint J_3

Joint J_3 (see Figure 4.214) can be designed using the same M27 high-strength pretensioned bolts as follows.

Number of Bolts for the Vertical Member D_2

$$N(D_2) = \frac{F_{Ed}}{F_{s,ult}} = \frac{4706.6}{206} = 22.8\text{ bolts, taken as 24 bolts (12 bolts in each side}$$

acting in double shear)

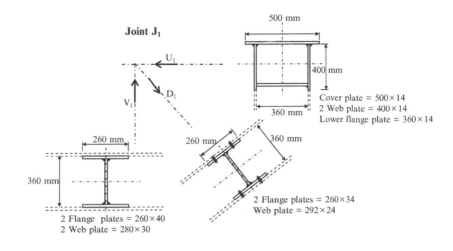

Joint J$_1$

Cover plate = 500×14
2 Web plate = 400×14
Lower flange plate = 360×14

500 mm
400 mm
360 mm

260 mm   260 mm   360 mm

360 mm

2 Flange plates = 260×34
Web plate = 292×24

2 Flange plates = 260×40
2 Web plate = 280×30

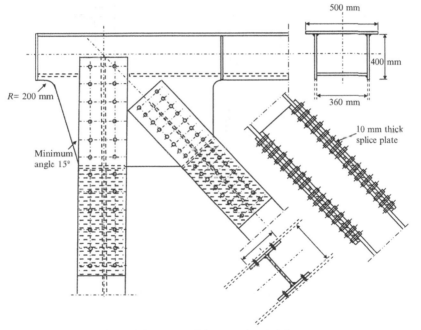

R= 200 mm

Minimum
angle 15°

500 mm
400 mm
360 mm

10 mm thick
splice plate

Figure 4.212 Details and drawings of the main truss joint J$_1$.

Number of Bolts for the Vertical Member V$_3$

The member is zero under the applied dead and live load cases of loading. The number of bolts can be taken as the minimum number based on the connection drawing. The number of connecting bolts of V$_3$ is taken as 16 bolts (8 bolts in each side acting in single shear).

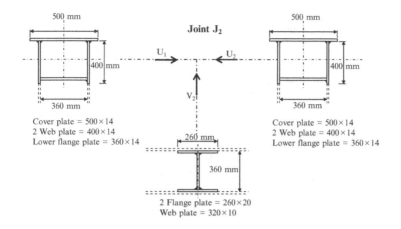

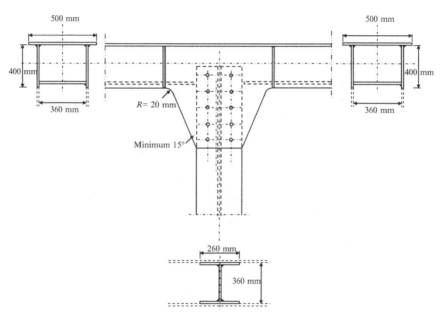

Figure 4.213 Details and drawings of the main truss joint J_2.

Number of Bolts for the Diagonal Member D_3

$$N(D_3) = \frac{F_{Ed}}{F_{s,ult}} = \frac{3284.3}{206} = 15.9 \text{ bolts, taken as 16 bolts (8 bolts in each side}$$

acting in double shear)

4.6.3.30 Design of Joint J_4

Joint J_4 (see Figure 4.215) can be designed using the same M27 high-strength pretensioned bolts as follows.

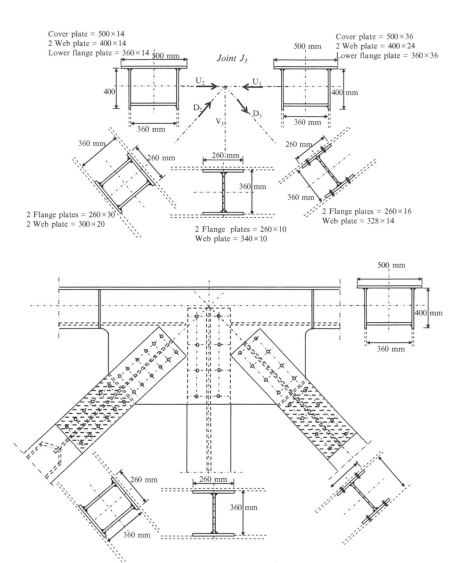

Figure 4.214 Details and drawings of the main truss joint J_3.

Number of Bolts for the Vertical Member V_4

$$N(V_4) - \frac{F_{Fd}}{F_{s,ult}} = \frac{1878.6}{103} = 18.2 \text{ bolts, taken as 20 bolts (10 bolts in each side}$$

acting in single shear)

4.6.3.31 Design of Joint J_5

Joint J_5 (see Figure 4.216) can be designed using the same M27 high–strength pretensioned bolts as follows.

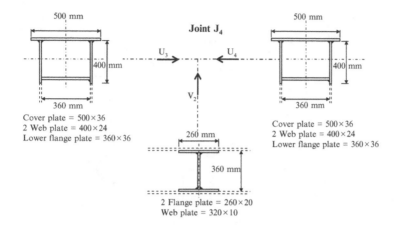

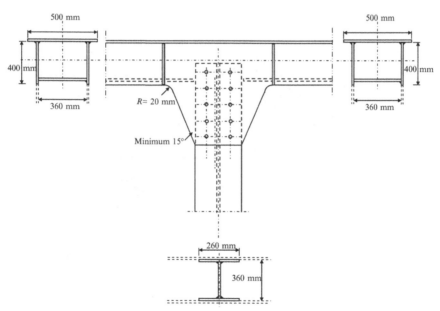

Figure 4.215 Details and drawings of the main truss joint J_4.

Number of Bolts for the Vertical Member D_4

$$N(D_4) = \frac{F_{Ed}}{F_{s,ult}} = \frac{1935.7}{103} = 18.8 \text{ bolts, taken as 20 bolts (10 bolts in each side}$$

acting in single shear)

Number of Bolts for the Vertical Member V_5

The member is zero under the applied dead and live load cases of loading. The number of bolts can be taken as the minimum number based on the

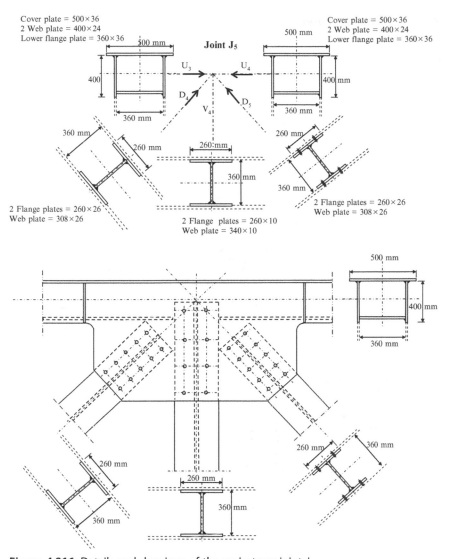

Figure 4.216 Details and drawings of the main truss joint J₅.

connection drawing. The number of connecting bolts of V_5 is taken as 16 bolts (8 bolts in each side acting in single shear).

Number of Bolts for the Diagonal Member D_5

$$N(D_5) = \frac{F_{Ed}}{F_{s,ult}} = \frac{1935.7}{103} = 18.8 \text{ bolts, taken as 20 bolts (10 bolts in each side}$$

acting in single shear)

4.6.3.32 Design of Joint J_6

Joint J_7 (see Figure 4.217) can be designed using the same M27 high-strength pretensioned bolts as follows.

Number of Bolts for the Vertical Member V_1

$$N(V_1) = \frac{F_{Ed}}{F_{s,ult}} = \frac{4433.9}{206} = 21.5 \text{ bolts, taken as 24 bolts (12 bolts in each side}$$

acting in double shear)

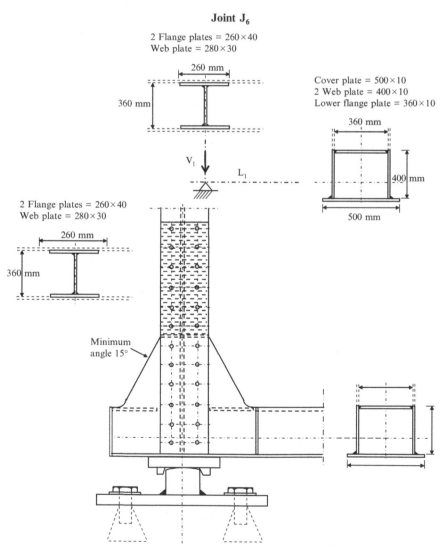

Joint J_6

2 Flange plates = 260×40
Web plate = 280×30

260 mm

360 mm

Cover plate = 500×10
2 Web plate = 400×10
Lower flange plate = 360×10

360 mm

400 mm

500 mm

2 Flange plates = 260×40
Web plate = 280×30

260 mm

360 mm

V_1

L_1

Minimum angle 15°

Figure 4.217 Details and drawings of the main truss joint J_6.

4.6.3.33 Design of Joint J₇

Joint J₇ (see Figure 4.218) can be designed using the same M27 high-strength pretensioned bolts as follows.

Number of Bolts for the Vertical Member D₁

$$N(D_1) = \frac{F_{Ed}}{F_{s,ult}} = \frac{6203.9}{206} = 30.1 \text{ bolts, taken as 32 bolts (16 bolts in each side}$$

acting in double shear)

Number of Bolts for the Vertical Member V₂

$$N(V_2) = \frac{F_{Ed}}{F_{s,ult}} = \frac{1878.6}{103} = 18.2 \text{ bolts, taken as 20 bolts (10 bolts in each side}$$

acting in single shear)

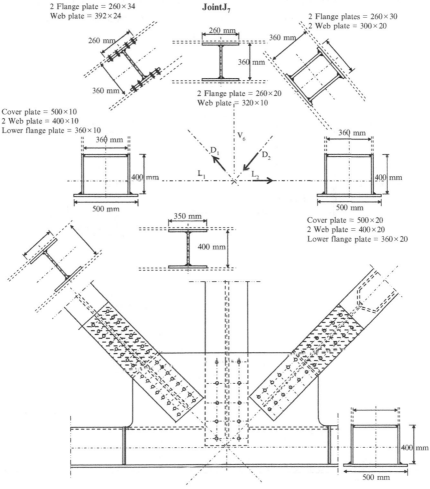

Figure 4.218 Details and drawings of the main truss joint J₇.

Number of Bolts for the Diagonal Member D_2

$$N(D_2) = \frac{F_{Ed}}{F_{s,ult}} = \frac{4706.6}{206} = 22.8 \text{ bolts, taken as 24 bolts (12 bolts in each side}$$

acting in double shear)

4.6.3.34 Design of Joint J_8

Joint J_8 (see Figure 4.219) can be designed using the same M27 high-strength pretensioned bolts as follows.

Number of Bolts for the Vertical Member V_3

The member is zero under the applied dead and live load cases of loading. The number of bolts can be taken as the minimum number based on the connection drawing. The number of connecting bolts of V_3 is taken as 16 bolts (8 bolts in each side acting in single shear).

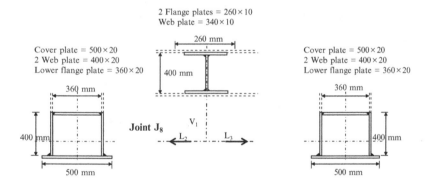

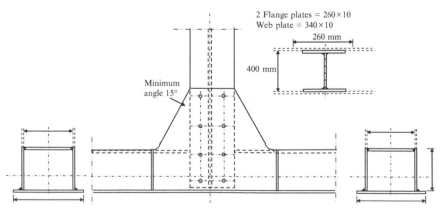

Figure 4.219 Details and drawings of the main truss joint J_8.

4.6.3.35 Design of Joint J_9

Joint J_9 (see Figure 4.220) can be designed using the same M27 high-strength pretensioned bolts as follows.

Number of Bolts for the Vertical Member D_3

$$N(D_3) = \frac{F_{Ed}}{F_{s,ult}} = \frac{3284.3}{206} = 15.9 \text{ bolts, taken as 16 bolts (8 bolts in each side}$$

acting in double shear)

Number of Bolts for the Vertical Member V_4

$$N(V_4) = \frac{F_{Ed}}{F_{s,ult}} = \frac{1878.6}{103} = 18.2 \text{ bolts, taken as 20 bolts (10 bolts in each side}$$

acting in single shear)

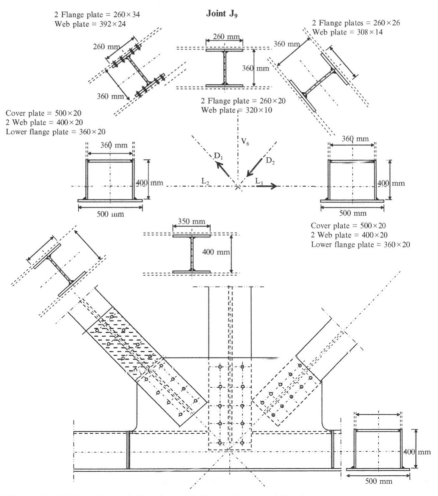

Figure 4.220 Details and drawings of the main truss joint J_9.

Number of Bolts for the Diagonal Member D₄

$$N(D_4) = \frac{F_{Ed}}{F_{s,ult}} = \frac{1935.7}{103} = 18.8 \text{ bolts, taken as 20 bolts (10 bolts in each side}$$

acting in single shear)

4.6.3.36 Design of Joint J₁₀

Joint J_{10} (see Figure 4.221) can be designed using the same M27 high-strength pretensioned bolts as follows.

Number of Bolts for the Vertical Member V₅

The member is zero under the applied dead and live load cases of loading. The number of bolts can be taken as the minimum number based on the connection drawing. The number of connecting bolts of V_5 is taken as 16 bolts (8 bolts in each side acting in single shear).

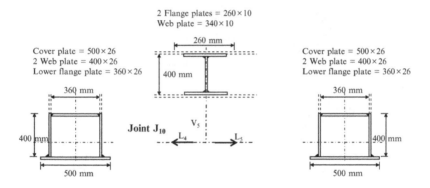

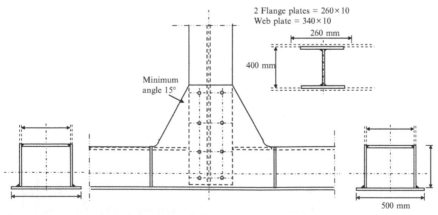

Figure 4.221 Details and drawings of the main truss joint J_{10}.

4.6.3.37 Design of Joint J₁₁

Joint J_{11} (see Figure 4.222) can be designed using the same M27 high-strength pretensioned bolts as follows.

Number of Bolts for the Vertical Member V_{11}

$$N(V_{11}) = \frac{F_{Ed}}{F_{s,ult}} = \frac{4433.9}{206} = 21.5 \text{ bolts, taken as } 24 \text{ bolts (12 bolts in each side}$$

acting in double shear)

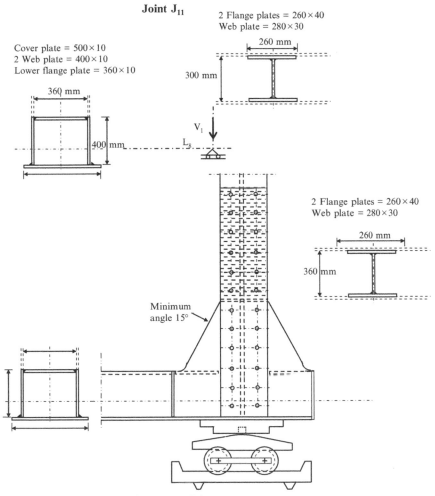

Figure 4.222 Details and drawings of the main truss joint J_{11}.

CHAPTER 5

Finite Element Analysis of Steel and Steel-Concrete Composite Bridges

5.1 GENERAL REMARKS

Chapters 1–4 have provided literature review and the required background regarding the general layout, material behavior of components and loads, stability of bridges and bridge components, and design of steel and steel-concrete composite bridges. It is now possible to present the main parameters affecting finite element analysis and modeling of the bridges. This chapter presents the more commonly used finite elements in meshing and modeling of the bridges. The chapter highlights the choice of correct finite element types and mesh size that can accurately simulate the complicated behavior of the different components of steel and steel-concrete composite bridges. The chapter provides a brief revision for the linear and nonlinear analyses required to study the stability of the bridges and bridge components. Also, the chapter details how to incorporate the nonlinear material properties of the different components, previously presented in Chapter 2, in the finite element analyses. In addition, this chapter details modeling of shear connection, previously presented in Chapters 2 and 3, for steel-concrete composite bridges. Furthermore, the chapter presents the application of different loads and boundary conditions, previously presented in Chapter 3, on the bridges. It should be noted that this chapter focuses on the finite element modeling using any software or finite element package, as an example in this book, the use of ABAQUS [1.29] software in finite element modeling.

The author aims that this chapter provides useful guidelines to readers, students, researchers, academics, designers, and practitioners on how to choose the best finite element type and mesh to represent different components of steel and steel-concrete composite bridges. There are many parameters that control the choice of finite element type and mesh such as the geometry, cross-section classification, and loading and boundary conditions of the bridge components. The aforementioned issues are addressed in this chapter. Accurate finite element modeling depends on the efficiency in simulating the nonlinear material behavior of the bridge components.

Finite Element Analysis and Design of Steel and Steel–Concrete Composite Bridges
469

This chapter shows how to correctly represent different linear and nonlinear regions in the stress-strain curves of the bridge components. Most of bridge components have initial local and overall geometric imperfections as well as residual stresses as a result of the manufacturing process. Ignoring the simulation of these initial imperfections and residual stresses would result in poor finite element models that are unable to describe the performance of the bridge components and the bridges as a whole. The correct simulation of different initial geometric imperfections and residual stresses available in the cross sections of the bridge components is addressed in this chapter. Steel and steel-concrete composite bridges are subjected to moving live loads and different boundary conditions, which differs from that commonly applied to buildings. Improper simulation of the applied loads and boundary conditions on a bridge component would not provide an accurate finite element model. Therefore, correct simulation of different loads and boundary conditions that are commonly associated with the bridge is the main focus of this chapter.

5.2 CHOICE OF FINITE ELEMENT TYPES FOR STEEL AND STEEL-CONCRETE COMPOSITE BRIDGES

To explain how to choose the best finite element type to simulate the behavior of a bridge, let us look at the components of the typical steel and steel-concrete composite bridges shown in Figures 1.20–1.22. We can see that there are main bridge components comprising the structural steel members and reinforced concrete deck slabs. On the other hand, there are also connecting bridge components comprising shear connectors, bolts, welds, bracing members, bearings, etc. To simulate the behavior of the main and connecting components of a bridge, we can use the different continuum, structural, and special-purpose finite elements provided in most available general-purpose finite element software, with ABAQUS [1.29] element library presented as an example in this book. General-purpose continuum elements cover all types of solid 1D, 2D, and 3D elements, while structural elements cover most elements used in structural engineering such as membrane, truss, beam, and shell elements. Finally, special-purpose finite elements cover elements used to simulate a special connecting element such as springs and joint elements. There are also elements used to model the interactions and contact behavior among main and connecting bridge components, which are mainly general contact, contact pair, and interface elements. The aforementioned elements will be highlighted in the coming section to help readers to choose the best finite elements to simulate the different components of steel and steel-concrete composite bridges.

5.2.1 Main Continuum, Structural, and Special-Purpose Finite Elements

To choose the best finite element for the structural steel members, we have to look into the classification of the cross section, which is normally specified in all current codes of practice. There are three commonly known cross-section classifications that are compact, noncompact, and slender sections. Compact sections have a thick plate thickness and can develop their plastic moment resistance without the occurrence of local buckling. Noncompact sections are sections in which the stress in the extreme fibers can reach the yield stress, but local buckling is liable to prevent the development of the plastic moment resistance. Finally, slender sections are those sections in which local buckling will occur in one or more parts of the cross section before reaching the yield strength. Compact sections in 3D can be modeled using either solid elements or shell elements that are able to model thick sections. However, noncompact and slender sections are only modeled using shell elements that are able to model thin sections. It should be noted that many general-purpose programs have shell elements that are used to simulate thin and thick sections.

Let us now look in more detail and classify shell elements commonly used in modeling noncompact and slender structural members. There are two main shell element categories known as *conventional and continuum shell elements*, examples shown in Figure 5.1. *Conventional shell elements* cover elements used for 3D shell geometries, elements used for axisymmetric geometries, and elements used for stress/displacement analysis. The conventional shell elements can be classified as thick shell elements, thin shell elements, and general-purpose shell elements that can be used for the analysis of thick or thin shells. Conventional shell elements have 6 degrees of freedom per node; however, it is possible to have shells with 5 degrees of freedom per node. Numerical integration is normally used to predict the behavior within the shell element. Conventional shell elements can use *full or reduced numerical integration*, as shown in Figure 5.2. Reduced-integration shell elements use lower-order integration to form the element stiffness. However, the mass matrix and distributed loadings are still integrated exactly. Reduced integration usually provides accurate results provided that the elements are not distorted or loaded in in-plane bending. Reduced integration significantly reduces running time, especially in three dimensions. Shell elements are

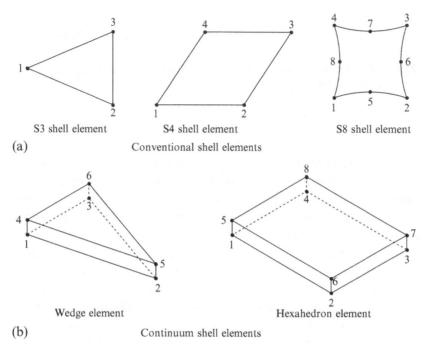

S3 shell element   S4 shell element   S8 shell element

(a)   Conventional shell elements

Wedge element   Hexahedron element

(b)   Continuum shell elements

Figure 5.1 Shell element types.

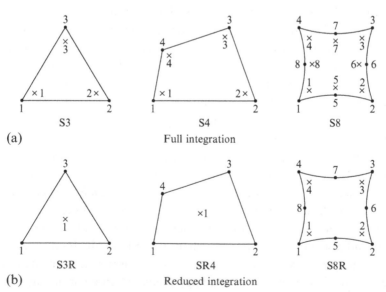

S3   S4   S8

(a)   Full integration

S3R   SR4   S8R

(b)   Reduced integration

Figure 5.2 Full and reduced integration of shell elements.

commonly identified based on the number of element nodes and the integration type. Hence, a shell element S8 means a stress-displacement shell having eight nodes with full integration, while a shell element S8R means a stress-displacement shell having eight nodes with reduced integration. On the other hand, *continuum shell elements* are general-purpose shells that allow finite membrane deformation and large rotations and, thus, are suitable for nonlinear geometric analysis. These elements include the effects of transverse shear deformation and thickness change. Continuum shell elements employ first-order layer-wise composite theory and estimate through-thickness section forces from the initial elastic moduli. Unlike conventional shells, continuum shell elements can be stacked to provide more refined through-thickness response. Stacking continuum shell elements allows for a richer transverse shear stress and force prediction. It should be noted that most metal structures are modeled using conventional shell elements, and hence, they are detailed in this book.

General-purpose conventional shell elements allow transverse shear deformation. They use thick shell theory as the shell thickness increases and become discrete Kirchhoff thin shell elements as the thickness decreases. The transverse shear deformation becomes very small as the shell thickness decreases. Examples of these elements are S3, S3R, S4, and S4R shells. Thick shells are needed in cases where transverse shear flexibility is important and second-order interpolation is desired. When a shell is made of the same material throughout its thickness, this occurs when the thickness is more than about 1/15 of a characteristic length on the surface of the shell, such as the distance between supports. An example of thick elements is S8R. Thin shells are needed in cases where transverse shear flexibility is negligible and the Kirchhoff constraint must be satisfied accurately (i.e., the shell normal remains orthogonal to the shell reference surface). For homogeneous shells, this occurs when the thickness is less than about 1/15 of a characteristic length on the surface of the shell, such as the distance between supports. However, the thickness may be larger than 1/15 of the element length.

Conventional shell elements can also be classified as *finite-strain and small-strain shell elements*. Element types S3, S3R, S4, and S4R account for finite membrane strains and arbitrarily large rotations; therefore, they are suitable for large-strain analysis. On the other hand, small-strain shell elements such as S8R shell elements are used for not only arbitrarily large rotations but only small strains. The change in thickness with deformation is ignored in these elements. For conventional shell elements used in ABAQUS [1.29], we must specify a section Poisson's ratio as part of the shell section definition to allow for the shell thickness in finite-strain elements to change as a function of the membrane strain. If the section Poisson's ratio is defined as zero, the shell thickness will remain constant and the elements are, therefore, suited for small-strain, large-rotation analysis. The change in thickness is ignored for the small-strain shell elements in ABAQUS [1.29].

Conventional reduced-integration shell elements can be also classified based on the number of degrees of freedom per node. Hence, there are two types of conventional reduced-integration shell elements known as *five-degree-of-freedom and six-degree-of-freedom shells*. Five-degree-of-freedom conventional shells have 5 degrees of freedom per node, which are three translational displacement components and two in-plane rotation components. On the other hand, six-degree-of-freedom shells have 6 degrees of freedom per node (three translational displacement components and three rotation components). The number of degrees of freedom per node is commonly denoted in the shell name by adding digit 5 or 6 at the end of the reduced-integration shell element name. Therefore, reduced-integration shell elements S4R5 and S4R6 have 5 and 6 degrees of freedom per node, respectively. The elements that use 5 degrees of freedom per node such as S4R5 and S8R5 can be more economical. However, they are suitable only for thin shells and they cannot be used for thick shells. The elements that use 5 degrees of freedom per node cannot be used for finite-strain applications, although they model large rotations with small strains accurately.

There are a number of issues that must be considered when using shell elements. Both S3 and S3R refer to the same three-node triangular shell element. This element is a degenerated version of S4R that is fully compatible with S4 and S4R elements. S3 and S3R provide accurate results in most loading situations. However, because of their constant bending and membrane strain approximations, high mesh refinement may be required to capture pure bending deformations or solutions to problems involving high strain gradients. Curved elements such as S8R5 shell elements are preferable for modeling bending of a thin curved shell. Element type S8R5 may give inaccurate results for buckling problems of doubly curved shells due to the fact that the internally defined integration point may not be positioned on the actual shell surface. Element type S4 is a fully integrated, general-purpose, finite-membrane-strain shell element. Element type S4 has four integration locations per element compared with one integration location for S4R, which makes the element computation more expensive. S4 is compatible with both S4R and S3R. S4 can be used in areas where greater solution accuracy is required or for problems where in-plane bending is expected. In all of these situations, S4 will outperform element type S4R. Further details regarding the elements used in modeling steel and metal structures are found in Ellobody et al. [5.1].

As mentioned earlier, steel and steel-concrete composite bridges that are composed of compact steel sections can be modeled using solid elements. In addition, reinforced concrete deck slabs in steel-concrete composite bridges can be modeled using either using thick shell elements, previously highlighted, or more commonly using solid or continuum elements. Modeling of concrete deck slabs using solid elements has the edge over modeling the slabs using thick shell elements since reinforcement bars or prestressing tendons used with the slabs can be accurately represented. Solid or continuum elements are volume elements that do not include structural elements such as beams, shells, and trusses. The elements can be composed of a single homogeneous material or can include several layers of different materials for the analysis of laminated composite solids. The naming conventions for solid elements depend on the element dimensionality, number of nodes in the element, and integration type. For example, C3D8R elements are continuum elements (C) having 3D eight nodes (8) with reduced integration (R). Solid elements provide accurate results if not distorted, particularly for quadrilaterals and hexahedra, as shown in Figure 5.3. The triangular and tetrahedral elements are less sensitive to distortion. Solid elements can be used for linear analysis and for complex nonlinear analyses involving stress, plasticity, and large deformations. Solid element library includes first-order (linear) interpolation elements and second-order (quadratic) interpolation elements commonly in three dimensions. Tetrahedra, triangular prisms, and hexahedra (bricks) are very common 3D elements, as shown in Figure 5.3. Modified second-order triangular and tetrahedral elements as well as reduced-integration solid elements can be also used. First-order plane-strain, axisymmetric quadrilateral and hexahedral solid elements provide constant volumetric strain throughout the element, whereas second-order elements provide higher accuracy than first-order elements for smooth problems that do not involve severe element distortions. They capture stress concentrations more effectively and are better for modeling geometric features. They can model a curved surface with fewer elements. Finally, second-order elements are very effective in bending-dominated problems. First-order triangular and tetrahedral elements should be avoided as much as possible in stress analysis problems; the elements are overly stiff

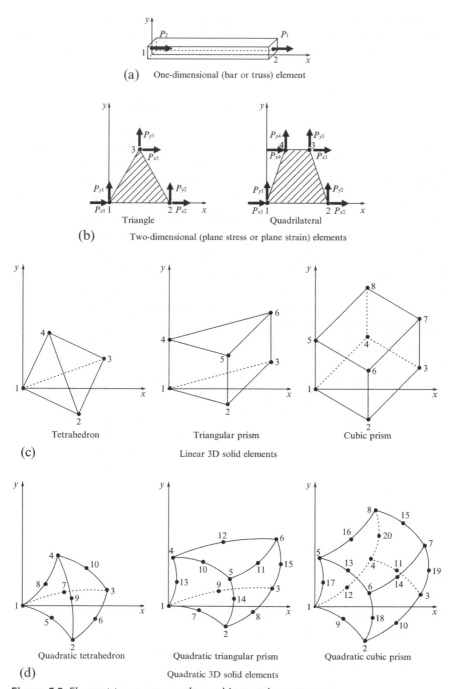

(a)   One-dimensional (bar or truss) element

(b)   Two-dimensional (plane stress or plane strain) elements

(c)   Linear 3D solid elements

(d)   Quadratic 3D solid elements

Figure 5.3 Element types commonly used in metal structures.

and exhibit slow convergence with mesh refinement, which is especially a problem with first-order tetrahedral elements. If they are required, an extremely fine mesh may be needed to obtain results with sufficient accuracy.

Similar to the behavior of shells, reduced integration can be used with solid elements to form the element stiffness. The mass matrix and distributed loadings use full integration. Reduced integration reduces running time, especially in three dimensions. For example, element type C3D20 has 27 integration points, while C3D20R has eight integration points only. Therefore, element assembly is approximately 3.5 × more costly for C3D20 than for C3D20R. Second-order reduced-integration elements generally provide accurate results than the corresponding fully integrated elements. However, for first-order elements, the accuracy achieved with full versus reduced integration is largely dependent on the nature of the problem. Triangular and tetrahedral elements are geometrically flexible and can be used in many models. It is very convenient to mesh a complex shape with triangular or tetrahedral elements. A good mesh of hexahedral elements usually provides a solution with equivalent accuracy at less cost. Quadrilateral and hexahedral elements have a better convergence rate than triangular and tetrahedral elements. However, triangular and tetrahedral elements are less sensitive to initial element shape, whereas first-order quadrilateral and hexahedral elements perform better if their shape is approximately rectangular. First-order triangular and tetrahedral elements are usually overly stiff, and fine meshes are required to obtain accurate results. For stress/displacement analyses, the first-order tetrahedral element C3D4 is a constant stress tetrahedron, which should be avoided as much as possible. The element exhibits slow convergence with mesh refinement. This element provides accurate results only in general cases with very fine meshing. Therefore, C3D4 is recommended only for filling in regions of low stress gradient to replace the C3D8 or C3D8R elements, when the geometry precludes the use of C3D8 or C3D8R elements throughout the model. For tetrahedral element meshes, the second-order or the modified tetrahedral elements such as C3D10 should be used. Similarly, the linear version of the wedge element C3D6 should generally be used only when necessary to complete a mesh, and, even then, the element should be far from any area where accurate results are needed. This element provides accurate results only with very fine meshing. A solid section definition is used to define the section properties of solid elements. A material definition must be defined with the solid section definition, which is assigned to a region in the finite element model.

Plane-stress and plane-strain structures can be modeled using 2D solid elements. The naming conventions for the elements depend on the element type (PE or PS) for (plane strain or plane stress), respectively, and number of nodes in the element. For example, CPE3 elements are continuum (C), plane-strain (PE) linear elements having three (3) nodes, as shown in Figure 5.3. The elements have 2 active degrees of freedom per node in the element plane. Quadratic 2D elements are suitable for curved geometry of structures. Structural metallic link members and metallic truss members can be modeled using 1D solid elements. The naming conventions for 1D solid elements depend on the number of nodes in the element. For example, C1D3 elements are continuum (C) elements having three (3) nodes. The elements have 1 active degree of freedom per node.

Axisymmetric solid elements are 3D elements that are used to model metal structures that have axisymmetric geometry. The element nodes are commonly using cylindrical coordinates (r, θ, z), where r is the radius from origin (coordinate 1), θ is the angle in degrees measured from horizontal axis (coordinate 2), and z is the perpendicular dimension (coordinate 3) as shown in Figure 5.4. Coordinate 1 must be greater than or equal to zero. Degree of freedom 1 is the translational displacement along the radius (u_r) and degree of freedom 2 is the translational displacement along the perpendicular direction (u_z). The naming conventions for axisymmetric solid elements with non-linear asymmetric deformation depend on the number of nodes in the element and integration type. For example, CAXA8R elements are continuum (C) elements and axisymmetric solid elements with nonlinear asymmetric

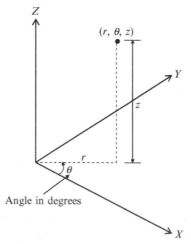

Figure 5.4 Cylindrical coordinates for axisymmetric solid elements.

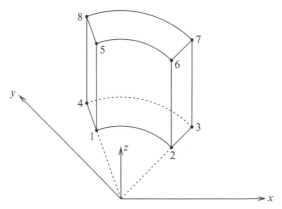

Figure 5.5 CAXA8R axisymmetric solid elements.

deformation (AXA) having eight (8) nodes with reduced (R) integration as shown in Figure 5.5. Stress/displacement axisymmetric solid elements without twist have two active degrees of freedom per node.

Reinforcement bars and prestressing tendons are commonly modeled in general-purpose software as "embedded" uniaxial (1D) finite elements in the form of individual bars or smeared layers. As an example in ABAQUS [1.29], reinforcement bars and prestressing tendons are included in the "host" elements (concrete elements) using the REBAR option. The option is used to define layers of uniaxial reinforcement in membrane, shell, and solid elements. Such layers are treated as a smeared layer with a constant thickness equal to the area of each reinforcing bar divided by the reinforcing bar spacing. The option also can be used to add additional stiffness, volume, and mass to the model. In additions, it can be used to add discrete axial reinforcement in beam elements. Embedded rebars can have material properties that are distinct from those of the underlying or host element. To define a rebar layer, modelers can specify one or multiple layers of reinforcement in membrane, shell, or solid elements. For each layer, modelers can specify the rebar properties by including the rebar layer name; the cross-sectional area of each rebar; the rebar spacing in the plane of the membrane, shell, or solid element; the position of the rebars in the thickness direction (for shell elements only), measured from the midsurface of the shell (positive in the direction of the positive normal to the shell); the rebar material name; the initial angular orientation, in degrees, measured relative to the local 1-direction; and the isoparametric direction from which the rebar angle output will be measured. Figures 5.6–5.8 show examples of reinforcement bars imbedded in different finite elements as presented in ABAQUS [1.29].

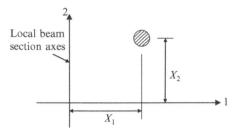

Figure 5.6 Rebar location in a beam section as presented in ABAQUS [1.29].

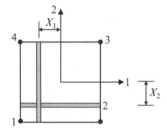

Figure 5.7 Rebar location in a 3D shell or membrane element as presented in ABAQUS [1.29].

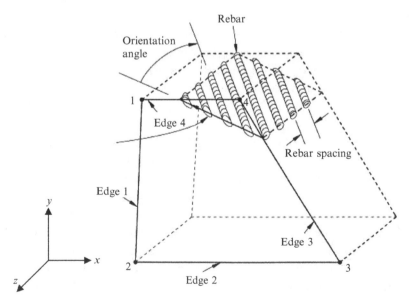

Figure 5.8 Rebar orientation in a 3D solid element as presented in ABAQUS [1.29].

For prestressed concrete structures, prestress forces in the tendons can be defined using INITIAL CONDITION option, which can add prestress forces in the rebars. If prestress is defined in the rebars and unless the prestress is held fixed, it will be allowed to change during an equilibrating static analysis step. This is a result of the straining of the structure as the self-equilibrating stress state establishes itself. An example is the pretension type of concrete prestressing in which reinforcing tendons are initially stretched to a desired tension before being covered by concrete. After the concrete cures and bonds to the rebar, the release of the initial rebar tension transfers load to the concrete, introducing compressive stresses in the concrete. The resulting deformation in the concrete reduces the stress in the rebar. Alternatively, modelers can keep the initial stress defined in some or all of the rebars constant during the initial equilibrium solution. An example is the posttension type of concrete prestressing; the rebars are allowed to slide through the concrete (normally they are in conduits), and the prestress loading is maintained by some external source (prestressing jacks). The magnitude of the prestress in the rebar is normally part of the design requirements and must not be reduced as the concrete compresses under the loading of the prestressing. Normally, the prestress is held constant only in the first step of an analysis. This is generally the more common assumption for prestressing. If the prestress is not held constant in analysis steps following the step in which it is held constant, the stress in the rebar will change due to additional deformation in the concrete. If there is no additional deformation, the stress in the rebar will remain at the level set by the initial conditions. If the loading history is such that no plastic deformation is induced in the concrete or rebar in steps subsequent to the steps in which the prestress is held constant, the stress in the rebar will return to the level set by the initial conditions upon the removal of the loading applied in those steps.

Rebar force output detailed in ABAQUS [1.29] is available at the rebar integration locations. The rebar force is equal to the rebar stress times the current rebar cross-sectional area. The current cross-sectional area of the rebar is calculated by assuming the rebar is made of an incompressible material, regardless of the actual material definition. For rebars in membrane, shell, or solid elements (see Figures 5.6–5.8), output variables identify the current orientation of rebar within the element and the relative rotation of the rebar as a result of finite deformation. These quantities are measured with respect to the user-specified isoparametric direction in the element, not the default local element system or the orientation-defined system. The output quantities of rebar angles can be measured from either of the isoparametric directions in the plane of the membrane, shell, or solid elements. Modelers can specify the desired isoparametric direction from which the rebar angle will be measured (1 or 2). The rebar angle is measured from the isoparametric direction to the rebar with a positive angle defined as a counterclockwise rotation around the element's normal direction. The default direction is the first isoparametric direction.

Bolted connections and shear connectors can be modeled using combinations of spring elements. Spring elements available in ABAQUS [1.29] can couple a force with a relative displacement and can couple a moment with a relative rotation. Spring elements can be linear or nonlinear. Figure 5.9 shows the definition of symbols for spring elements. The terms "force" and "displacement" are used throughout the description of spring elements. When the spring is associated with displacement degrees of freedom, these variables are the force and relative displacement in the spring. If the springs are associated with rotational degrees of freedom, they are torsional springs. In this case, the variables will be the moment transmitted by the spring and the relative rotation across the spring. Spring elements are used to model actual physical springs as well as idealizations of axial or torsional components. They can also model restraints to prevent rigid-body motion. They are also used to represent structural dampers by specifying structural damping factors to form the imaginary part of the spring stiffness. SPRING1, SPRING2, and SPRINGA elements are available in ABAQUS [1.29]. SPRING1 is between a node and ground, acting in a fixed direction. SPRING2 is between two nodes, acting in a fixed direction. While, SPRINGA acts between two nodes, with its line of action being the line joining the two nodes, so that this line of action can rotate in large-displacement analysis. The spring behavior can be linear or nonlinear in any of the spring elements. Element types SPRING1 and SPRING2 can be associated with displacement or rotational degrees of freedom.

The relative displacement definition depends on the element type. The relative displacement across a SPRING1 element is the ith component of displacement of the spring's node, with i defined as shown in Figure 5.9. The relative displacement across a SPRING2 element is the difference

Figure 5.9 Definition of symbols for spring elements.

between the ith component of displacement of the spring's first node and the jth component of displacement of the spring's second node, with i and j defined as shown in Figure 5.9. For a SPRINGA element, the relative displacement in a geometrically linear analysis is measured along the direction of the SPRINGA element. While in geometrically nonlinear analysis, the relative displacement across a SPRINGA element is the change in length in the spring between the initial and the current configuration.

The spring behavior can be linear or nonlinear. In either case, modelers must associate the spring behavior with a region of their model. Modelers can define linear spring behavior by specifying a constant spring stiffness (force per relative displacement). The spring stiffness can depend on temperature and field variables. For direct-solution steady-state dynamic analysis, the spring stiffness can depend on frequency, as well as on temperature and field variables. On the other hand, modelers can define nonlinear spring behavior by giving pairs of force-relative displacement values. These values should be given in ascending order of relative displacement and should be provided over a sufficiently wide range of relative displacement values so that the behavior is defined correctly. ABAQUS [1.29] assumes that the force remains constant (which results in zero stiffness) outside the range given; see Figure 5.10. Initial forces in nonlinear springs should be defined by giving a nonzero force at zero relative displacement. The spring stiffness can depend on temperature and field variables. Modelers can define the direction of action for SPRING1 and SPRING2 elements by giving the

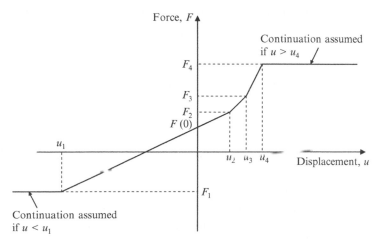

Figure 5.10 Nonlinear spring force-relative displacement relationship according to ABAQUS [1.29].

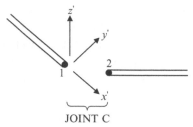

JOINT C
(Local system attached to node 1)

Figure 5.11 JOINTC elements available in ABAQUS [1.29].

degree of freedom at each node of the element. This degree of freedom may be in a local coordinate system. The local system is assumed to be fixed; even in large-displacement analysis, SPRING1 and SPRING2 elements act in a fixed direction throughout the analysis.

ABAQUS [1.29] also provides the capability of modeling flexible joints using JOINTC elements; see Figure 5.11. The elements are used to model joint interactions and are made up of translational and rotational springs and parallel dashpots, which are elements that can couple a force with a relative velocity, in a local, corotational coordinate system. The JOINTC element is provided to model the interaction between two nodes that are (almost) coincident geometrically and that represent a joint with internal stiffness and/or damping so that the second node of the joint can displace and rotate slightly with respect to the first node. Similar functionality is available using connector elements. The joint behavior consists of linear or nonlinear springs and dashpots in parallel, coupling the corresponding components of relative displacement and of relative rotation in the joint. Each spring or dashpot definition defines the behavior for one of the six local directions; up to six spring definitions and six dashpot definitions can be included. If no specification is given for a particular local relative motion in the joint, the joint is assumed to have no stiffness with respect to that component. The joint behavior can be defined in a local coordinate system that rotates with the motion of the first node of the element. If a local coordinate system is not defined, the global system is used. In large-displacement analysis, the formulation for the relationship between moments and rotations limits the usefulness of these elements to small relative rotations. The relative rotation across a JOINTC element should be of a magnitude to qualify as a small rotation. ABAQUS [1.29] also offers different connector types and connector elements to model the behavior of connectors. The analyst is often faced with modeling problems in which two different parts are connected in some way. Sometimes, connections are simple, such as two panels of sheet metal spot welded together or a link member connected to a frame with a hinge. Connector elements can account for internal friction, such as the lateral force or moments on a bolt generating friction in the translation of the bolt along a slot. Failure conditions are reached, where excess force or displacement inside the connection causes the entire connection or a single component of relative motion to break free. Connector elements can provide an easy and versatile way to model many types of physical mechanisms whose geometry is discrete (i.e., node to node), yet the kinematic and kinetic relationships describing the connection are complex.

5.2.2 Contact and Interaction Elements

5.2.2.1 General

Contact between two bodies can be modeled using the contact analysis capabilities in ABAQUS [1.29]. The software provides more than one approach for defining contact comprising general contact (Figure 5.12), contact pairs (Figure 5.13), and contact elements. Each approach has unique advantages and limitations. A contact simulation using contact pairs or general contact is defined by specifying surface definitions for the bodies that could potentially be in contact. The surfaces will interact with one another (the contact interactions). The mechanical and thermal contact property models can be specified in these types of contact, such as the pressure-overclosure relationship, the friction coefficient, or the contact conduction

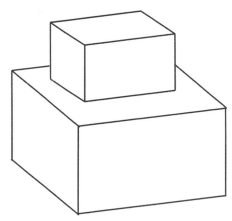

Figure 5.12 An example of general contact approach with feature edges at the perimeter of an active contact region as given in ABAQUS [1.29].

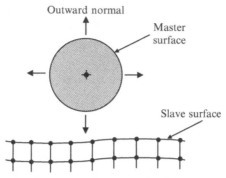

Figure 5.13 An example of contact pair approach with master and slave surfaces as given in ABAQUS [1.29].

coefficient. Surfaces can be defined at the beginning of a simulation or upon restart as part of the model definition. ABAQUS [1.29] has four classifications of contact surfaces comprising element-based deformable and rigid surfaces, node-based deformable and rigid surfaces, analytical rigid surfaces, and Eulerian material surfaces.

Contact interactions for contact pairs and general contact are defined in ABAQUS [1.29] by specifying surface pairings and self-contact surfaces. General contact interactions typically are defined by specifying self-contact for the default surface, which allows an easy, yet powerful, definition of contact. Self-contact for a surface that spans multiple bodies implies self-contact for each body as well as contact between the bodies. At least one surface in an interaction must be a non-node-based surface, and at least one surface in an interaction must be a nonanalytical rigid surface. Surface properties can be defined for particular surfaces in a contact model. Contact interactions in a model can refer to a contact property definition, in much the same way that elements refer to an element property definition. By default, the surfaces interact (have constraints) only in the normal direction to resist penetration. The other mechanical contact interaction models available depend on the contact algorithm. Some of the available models are softened contact, contact damping, friction, and spot welds bonding two surfaces together until the welds fail.

According to ABAQUS [1.29], contact pairs and general contact combine to provide the capability of modeling contact between two deformable bodies, with the structures being either 2D or 3D, and they can undergo either small or finite sliding. They can model contact between a rigid surface and a deformable body, with the structures can be either 2D or 3D, and they can undergo either small or finite sliding. They can also model finite-sliding self-contact of a single deformable body. In addition, they can model small-sliding or finite-sliding interaction between a set of points and a rigid surface. These models can be either 2D or 3D. Also, they can model contact between a set of points and a deformable surface. These models can be either 2D or 3D. They can also model problems where two separate surfaces need to be "tied" together so that there is no relative motion between them. This modeling technique allows for joining dissimilar meshes. Furthermore, contact pairs and general contact can model coupled thermal-mechanical interaction between deformable bodies with finite relative motion.

For most contact problems, modelers can choose whether to define contact interactions using general contact or contact pairs. The distinction between general contact and contact pairs lies primarily in the user interface, the default numerical settings, and the available options. The general contact and contact pair implementations share many underlying algorithms. The contact interaction domain, contact properties, and surface attributes are specified independently for general contact, offering a more flexible way to add detail incrementally to a model. The simple interface for specifying general contact allows for a highly automated contact definition; however, it is also possible to define contact with the general contact interface to mimic traditional contact pairs. Conversely, specifying self-contact of a surface spanning multiple bodies with the contact pair user interface (if the surface-to-surface formulation is used) mimics the highly automated approach often used for general contact. Pairwise specifications of contact interactions will often result in more efficient or robust analyses as compared to an all-inclusive self-contact approach to defining contact. General contact uses the finite-sliding, surface-to-surface formulation. Contact pairs use the finite-sliding, node-to-surface formulation by default.

General contact automatically accounts for thickness and offsets associated with shell-like surfaces. Contact pairs that use the finite-sliding, node-to-surface formulation do not account for shell thickness and offsets. General contact uses the penalty method to enforce the contact constraints by default. Contact pairs that use the finite-sliding, node-to-surface formulation use a Lagrange multiplier method to enforce contact constraints by default in most cases. General contact automatically assigns pure master and slave roles for most contact interactions and automatically assigns balanced master-slave roles to other contact interactions. Contact pairs have unique capabilities, which are not available for general contact such as modeling contact involving node-based surfaces or surfaces on 3D beam elements; small-sliding contact and tied contact; the finite-sliding, node-to-surface contact formulation; debonding and cohesive contact behavior; and surface interactions in analyses without displacement degrees of freedom, such as pure heat transfer and pressure-penetration loading. Surface-based contact methods associated with general contact and contact pairs cannot be used for certain classes of problems such as contact interaction between two pipelines or tubes modeled with pipe, beam, or truss elements where one pipe lies inside the other; contact between two nodes along a fixed direction in space; simulations using axisymmetric elements with asymmetric deformations; and heat transfer analyses where the heat flow is 1D. These situations require defining a contact simulation using contact elements.

5.2.2.2 Defining General Contact Interactions

ABAQUS [1.29] provides two algorithms for modeling contact and interaction problems, which are the general contact algorithm and the contact pair algorithm. The general contact algorithm is specified as part of the model definition. It allows very simple definitions of contact with very few restrictions on the types of surfaces involved. Also, it uses sophisticated tracking algorithms to ensure that proper contact conditions are enforced efficiently. It can be used simultaneously with the contact pair algorithm (i.e., some interactions can be modeled with the general contact algorithm, while others are modeled with the contact pair algorithm). In addition, it can be used with 2D or 3D surfaces. Furthermore, it uses the finite-sliding, surface-to-surface contact formulation.

The definition of a general contact interaction consists of specifying the general contact algorithm and defining the contact domain (i.e., the surfaces that interact with one another), the contact surface properties, and the mechanical contact property models. The general contact algorithm allows for quite general characteristics in the surfaces that it uses. A convenient method of specifying the contact domain is using cropped surfaces. Such surfaces can be used to perform "contact in a box" by using a contact domain that is enclosed in a specified rectangular box in the original configuration. The general contact algorithm uses the surface-to-surface contact formulation as the primary formulation and can use the edge-to-surface contact formulation as a supplementary formulation. The general contact algorithm does not consider contact involving analytical surfaces or node-based surfaces, although these surface types can be included in contact pairs in analyses that also use general contact. The general contact algorithm can consider 3D edge-to-surface contact, which is more effective at resolving some interactions than the surface-to-surface contact formulation. The edge-to-surface contact formulation is primarily intended to avoid localized penetration of a feature's edge of one surface into a relatively smooth portion of another surface when the normal directions of the respective surface facets in the active contact region form an oblique angle.

When a surface is used in a general contact interaction, all applicable facets are included in the contact definition; however, modelers can specify which edges to consider for edge-to-surface contact. The contact area associated with a feature edge depends on the mesh size; therefore, contact pressures (in units of force per area) associated with edge-to-surface contact are mesh-dependent. Both surface-to-surface and edge-to-surface contact constraints may be active at the same nodes. To help avoid numerical overconstraint issues, edge-to-surface contact constraints are always enforced with a penalty method. General contact is defined at the beginning of an analysis. Only one general contact definition can be specified, and this definition is in effect for every step of the analysis. Modelers can specify the regions of the model that can potentially come into contact with each other by defining general contact inclusions and exclusions. Only one contact inclusions definition and one contact exclusions definition are allowed in the model definition. All contact inclusions in an analysis are applied first, and then, all contact exclusions are applied, regardless of the order in which they are specified. The contact exclusions take precedence over the contact inclusions. The general contact algorithm will consider only those interactions specified by the contact inclusions definition and not specified by the contact exclusions definition. General contact interactions typically are defined by specifying self-contact for the default automatically generated surface provided by ABAQUS [1.29]. All surfaces used in the general contact algorithm can span multiple unattached bodies, so self-contact in this algorithm is not limited to contact of a single body with itself. For example, self-contact of a surface that spans two bodies implies contact between the bodies as well as contact of each body with itself.

Defining contact inclusions means specifying the regions of the model that should be considered for contact purposes. Modelers can specify self-contact for a default unnamed, all-inclusive surface defined automatically by ABAQUS. This default surface contains, with the exceptions noted later, all exterior element faces. This is the simplest way to define the contact domain. On the other hand, modelers can refine the contact domain definition by specifying the regions of the model to exclude from contact. Possible motivations for specifying contact exclusions include (1) avoiding physically unreasonable contact interactions; (2) improving computational performance by excluding parts of the model that are not likely to interact; (3) ignoring contact for all the surface pairings specified, even if these interactions are specified directly or indirectly in the contact inclusions definition; and finally (4) excluding multiple surface pairings from the contact domain. All of the surfaces specified must be element-based surfaces. Keep in mind that surfaces can be defined to span multiple unattached bodies, so self-contact exclusions are not limited to exclusions of single-body contact. When you specify pure master-slave contact surface weighting for a particular general contact surface pair, contact exclusions are generated automatically for the master-slave orientation opposite to that specified. ABAQUS [1.29] assigns default pure master-slave roles for contact involving disconnected bodies within the general contact domain, and contact exclusions are generated by default for the opposite master-slave orientations.

5.2.2.3 Defining Contact Pair Interactions

Contact pairs in ABAQUS [1.29] can be used to define interactions between bodies in mechanical, coupled temperature-displacement, and heat transfer simulations. Contact pairs can be formed using a pair of rigid or deformable surfaces or a single deformable surface. Modelers do not have to use surfaces with matching meshes. Contact pairs cannot be formed with one 2D surface and one 3D surface. Modelers can define contact in terms of two surfaces that may interact with each other as a "contact pair" or in terms of a single surface that may interact with itself in "self-contact." ABAQUS enforces contact conditions by forming equations involving groups of nearby nodes from the respective surfaces or, in the case of self-contact, from separate regions of the same surface. To define a contact pair, modelers must indicate which pairs of surfaces may interact with one another or which surfaces may interact with themselves. Contact surfaces should extend far enough to include all regions that may come into contact during an analysis; however, including additional surface nodes and faces that never experience contact may result in significant extra computational cost (e.g., extending a slave surface such that it includes many nodes that remain separated from the master surface throughout an analysis can significantly increase memory usage unless penalty contact enforcement is used). Every contact pair is assigned a contact formulation and must refer to an interaction property. Contact formulations are based on whether the tracking approach assumes finite- or small-sliding and whether the contact discretization is based on a node-to-surface or surface-to-surface approach. When a contact pair contains two surfaces, the two surfaces are not allowed to include any of the same nodes, and modelers must choose which surface will be the slave and which will be the master. The larger of the two surfaces should act as the master surface. If the surfaces are of comparable size, the surface on the stiffer body should act as the master surface. If the surfaces are of comparable size and stiffness, the surface with the coarser mesh should act as the master surface.

For node-to-surface contact, it is possible for master surface nodes to penetrate the slave surface without resistance. This penetration tends to occur if the master surface is more refined than the slave surface or a large contact pressure develops between soft bodies. Refining the slave surface mesh often minimizes the penetration of the master surface nodes. If the refinement technique does not work or is not practical, a symmetric master-slave method can be used if both surfaces are element-based surfaces with deformable or deformable-made-rigid parent elements. To use this method, modelers can define two contact pairs using the same two surfaces, but they have to switch the roles of master and slave surfaces for the two contact pairs. Using symmetric master-slave contact pairs can lead to over-constraint problems when very stiff or "hard" contact conditions are enforced. For softened contact conditions, use of symmetric master-slave contact pairs will cause deviations from the specified pressure-versus-overclosure behavior, because both contact pairs contribute to the overall interface stress without accounting for one another. Likewise, use of symmetric master-slave contact pairs will cause deviations from the friction model if an optional shear stress limit is specified, because the contact stresses observed by each contact pair will be approximately one-half of the total interface stress. Similarly, it can be difficult to interpret the results at the interface for symmetric master-slave contact pairs. In this case, both surfaces at the interface act as slave surfaces, so each has contact constraint values associated with it. The constraint values that represent contact pressures are not independent of each other.

ABAQUS [1.29] requires master contact surfaces to be single-sided for node-to-surface contact and for some surface-to-surface contact formulations. This requires that modelers consider the proper orientation for master surfaces defined on elements, such as shells and membranes, that have positive and negative directions. For node-to-surface contact, the orientation of slave surface normals is irrelevant, but for surface-to-surface contact, the orientation of single-sided slave surfaces is taken into consideration. Double-sided element-based surfaces are allowed for the default surface-to-surface contact formulations, although they are not always appropriate for cases with deep initial penetrations. If the master and slave surfaces are both double-sided, the positive or negative orientation of the contact normal direction will be chosen such as to minimize (or avoid) penetrations for each contact constraint. If either or both of the surfaces are single-sided, the positive or negative orientation of the contact normal direction will be determined from the single-sided surface normals rather than the relative positions of the surfaces.

When the orientation of a contact surface is relevant to the contact formulation, modelers must consider the following aspects for surfaces on structural (beam and shell), membrane, truss, or rigid elements:

(1) Adjacent surface faces must have consistent normal directions.
(2) The slave surface should be on the same side of the master surface as the outward normal. If, in the initial configuration, the slave surface is on the opposite side of the master surface as the outward normal, ABAQUS [1.29] will detect overclosure of the surfaces and may have difficulty finding an initial solution if the overclosure is severe. An improper specification of the outward normal will often cause an analysis to immediately fail to converge.
(3) Contact will be ignored with surface-to-surface discretization if single-sided slave and master surfaces have normal directions that are in approximately the same direction. It should be noted that discontinuous contact surfaces are allowed in many cases, but the master surface for finite-sliding, node-to-surface contact cannot be made up of two or more disconnected regions (they must be continuous across element edges in 3D models or across nodes in 2D models). ABAQUS [1.29] cannot use 3D beams or trusses to form a master surface because the elements do not have enough information to create unique surface normals. However, these elements can be used to define a slave surface. 2D beams and trusses can be used to form both master and slave surfaces. For small-sliding contact problems, the contact area is calculated in the input file preprocessor from the undeformed shape of the model; thus, it does not change throughout the analysis, and contact pressures for small-sliding contact are calculated according to this invariant contact area. This behavior is different from that in finite-sliding contact problems, where the contact area and contact pressures are calculated according to the deformed shape of the model.

5.2.2.4 Defining Contact with Contact Elements

As mentioned earlier, some contact problems cannot be modeled using general contact and contact pair formulations. Therefore, ABAQUS [1.29] offers a variety of contact elements that can be used when contact between two bodies cannot be simulated with the surface-based contact approach. These elements include the following:

(1) Gap contact elements, which can be used for mechanical and thermal contact between two nodes. For example, these elements can be used to model the contact between a piping system and its supports. They can also be used to model an inextensible cable that supports only tensile loads.

(2) Tube-to-tube contact elements, which can be used to model contact between two pipes or tubes in conjunction with slide lines.

(3) Slide line contact elements, which can be used to model finite-sliding contact between two axisymmetric structures that may undergo asymmetric deformations in conjunction with slide lines. Slide line elements can, for example, be used to model threaded connectors.

(4) Rigid surface contact elements can be used to model contact between an analytical rigid surface and an axisymmetric deformable body that may undergo asymmetric deformations. For example, rigid surface contact elements might be used to model the contact between a rubber seal and a much stiffer structure.

5.2.2.5 Frictional Behavior

When surfaces are in contact, they usually transmit shear as well as normal forces across their interface. There is generally a relationship between these two force components. The relationship, known as the friction between the contacting bodies, is usually expressed in terms of the stresses at the interface of the bodies. The friction models available in ABAQUS [1.29] can include the classical isotropic Coulomb friction model, which in its general form allows the friction coefficient to be defined in terms of slip rate, contact pressure, average surface temperature at the contact point, and field variables; see Figure 5.14. Coulomb friction model provides the option for modelers to define a static friction coefficient and a kinetic friction coefficient with a smooth transition zone defined by an exponential curve. Coulomb friction model also allows the introduction of a shear stress limit, which is the maximum value of shear stress that can be carried by the interface before the surfaces begin to slide. ABAQUS assumes by default that the interaction between contacting bodies is frictionless. Modelers can include a friction model in a contact property definition for both surface-based contact and element-based contact. The basic concept of the Coulomb friction model is to relate the maximum allowable frictional (shear) stress across an interface

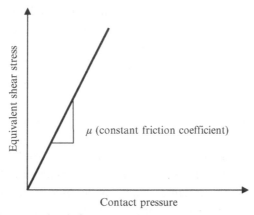

Figure 5.14 The basic Coulomb friction model as given in ABAQUS [1.29].

to the contact pressure between the contacting bodies. In the basic form of the Coulomb friction model, two contacting surfaces can carry shear stresses up to a certain magnitude across their interface before they start sliding relative to one another; this state is known as sticking. The Coulomb friction model defines this critical shear stress, τ_{crit}, at which sliding of the surfaces starts as a fraction of the contact pressure, p, between the surfaces ($\tau_{crit} = \mu p$). The stick/slip calculations determine when a point transitions from sticking to slipping or from slipping to sticking. The fraction, μ, is known as the coefficient of friction. For the case when the slave surface consists of a node-based surface, the contact pressure is equal to the normal contact force divided by the cross-sectional area at the contact node. In ABAQUS [1.29], the default cross-sectional area is 1.0, and modelers can specify a cross-sectional area associated with every node in the node-based surface when the surface is defined or, alternatively, assign the same area to every node through the contact property definition. The basic friction model assumes that it is the same in all directions (isotropic friction). For a 3D simulation, there are two orthogonal components of shear stress along the interface between the two bodies. These components act in the slip directions for the contact surfaces or contact elements.

There are two ways to define the basic Coulomb friction model in ABA-QUS [1.29]. In the default model, the friction coefficient is defined as a function of the equivalent slip rate and contact pressure. The coefficient of friction can be set to any nonnegative value. A zero friction coefficient means that no shear forces will develop and the contact surfaces are free to slide. You do not need to define a friction model for such a case.

Experimental data show that the friction coefficient that opposes the initiation of slipping from a sticking condition is different from the friction coefficient that opposes established slipping. The former is typically referred to as the "static" friction coefficient, and the latter is referred to as the "kinetic" friction coefficient. Typically, the static friction coefficient is higher than the kinetic friction coefficient. In the default model, the static friction coefficient corresponds to the value given at zero slip rate, and the kinetic friction coefficient corresponds to the value given at the highest slip rate. The transition between static friction and kinetic friction is defined by the values given at intermediate slip rates. In this model, the static and kinetic friction coefficients can be functions of contact pressure, temperature, and field variables. ABAQUS [1.29] also provides a model to specify a static friction coefficient and a kinetic friction coefficient directly. In this model, it is assumed that the friction coefficient decays exponentially from the static value to the kinetic value according to a formula given in the software. Modelers can specify an optional equivalent shear stress limit, τ_{max}, so that, regardless of the magnitude of the contact pressure stress, sliding will occur if the magnitude of the equivalent shear stress reaches this value; see Figure 5.15. A value of zero is not allowed. This shear stress limit is typically introduced in cases when the contact pressure stress may become very large (as can happen in some manufacturing processes), causing the Coulomb theory to provide a critical shear stress at the interface that exceeds the yield stress in the material beneath the contact surface. A reasonable upper bound

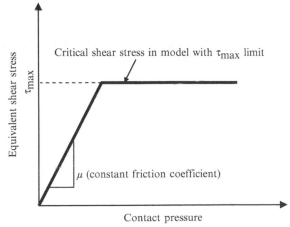

Figure 5.15 Friction model with a limit on the critical shear stress as given in ABAQUS [1.29].

estimate for τ_{max} is $\frac{\sigma_y}{\sqrt{3}}$, where σ_y is the Mises yield stress of the material adjacent to the surface. It should be noted that ABAQUS [1.29] offers the option
of specifying an infinite coefficient of friction ($\mu = \infty$). This type of surface
interaction is called "rough" friction, and with it, all relative sliding motion
between two contacting surfaces is prevented (except for the possibility of
"elastic slip" associated with penalty enforcement) as long as the corresponding normal-direction contact constraints are active. Rough friction is
intended for nonintermittent contact; once surfaces close and undergo
rough friction, they should remain closed. Convergence difficulties may
arise in ABAQUS if a closed contact interface with rough friction opens,
especially if large shear stresses have developed. The rough friction model
is typically used in conjunction with the no separation contact pressure-
overclosure relationship for motions normal to the surfaces, which prohibits
separation of the surfaces once they are closed. It should also be noted that in
ABAQUS [1.29], the sticking constraints at an interface between two surfaces can be enforced exactly by using the Lagrange multiplier implementation. With this method, there is no relative motion between two closed
surfaces until $\tau = \tau_{crit}$. However, the Lagrange multipliers increase the computational cost of the analysis by adding more degrees of freedom to the
model and often by increasing the number of iterations required to obtain
a converged solution. The Lagrange multiplier formulation may even prevent convergence of the solution, especially if many points are iterating
between sticking and slipping conditions. This effect can occur particularly
if locally, there is a strong interaction between slipping/sticking conditions
and contact stresses.

5.3 CHOICE OF FINITE ELEMENT MESH FOR THE BRIDGES AND BRIDGE COMPONENTS

The brief survey of the different finite elements mentioned earlier, available
in ABAQUS [1.29] element library, provided a useful background to help
beginners to choose the best finite element types to represent the different
components of steel and steel-concrete composite bridges. After choosing
the best finite element type, we need to look into the geometry of the bridge
and the bridge components to decide the best finite element mesh. Now, we
need to differentiate between modeling the individual bridge components
and modeling the whole bridge. To make it clear for readers, if we, for
example, model the shear connection of headed stud shear connectors in
solid concrete slabs, we will include the exact dimensions of the connection

components comprising the headed studs, surrounding concrete, steel beam, and interfaces among the aforementioned components. In this case, the finite element mesh will be simulating the shear connection behavior and the finite element model developed can be used to predict the local behavior of the shear connection. The finite element model in this case can evaluate the shear connection capacity, failure mode, and load-slip characteristic of the headed stud. This has been reported previously by the author in [2.68, 2.69]. However, if we model full-scale composite girders with solid slabs having numerous headed stud shear connectors, then the finite element mesh used to model the local shear connection behavior will not be used. This is attributed to the fact that including all the details of every connection around every headed stud will result in a huge finite element mesh for the composite girder that may be impossible to be analyzed. In this case, we can incorporate the local behavior of the headed stud in a shear connection to the overall composite beam behavior using springs or JOINTC elements. In this case, we can study the overall behavior of the composite beam using the developed finite element model. In this case, the finite element mesh of the composite girder will be reasonable in size, and the finite element model can evaluate the moment resistance of the composite beam, load-displacement relationships, failure modes, etc. This was also previously reported by the author in [2.68, 5.2]. Using the same approach, we can develop a finite element mesh to study the whole bridge behavior, and in parallel, we can develop other local finite element models to study the behavior of the individual components and incorporate them in the whole bridge model. In this book, Chapters 6 and 7 will include finite element models developed for the individual bridge components as well as for the whole bridges.

Structural steel members have flat and curved regions. Therefore, the finite element mesh of the individual members has to cover both flat and curved regions. Also, most structural steel members have short dimensions, which are commonly the lateral dimensions of the cross section, and long dimensions, which are the longitudinal axial dimension of the structural member that defines the structural member length. Therefore, the finite element mesh has to cover both lateral and longitudinal regions of the structural steel member. To mesh the structural steel member correctly using shell elements, we have to start with a short dimension for the chosen shell element and decide the best *aspect ratio*. The aspect ratio is defined as the ratio of the longest dimension to the shortest dimension of a quadrilateral finite element. As the aspect ratio is increased, the accuracy of the results is decreased. The aspect ratio should be kept approximately constant for all finite element

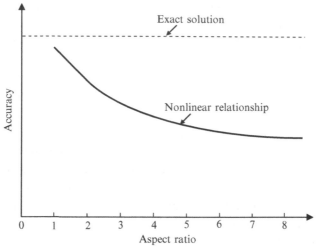

Figure 5.16 Effect of aspect ratio of finite elements on the accuracy of results.

analyses performed on the member. Therefore, most general-purpose finite element computer programs specify a maximum value for the aspect ratio that should not be exceeded; otherwise, the results will be inaccurate. Figure 5.16 presents a schematic diagram showing the effect of aspect ratio on the accuracy of results. The best aspect ratio is one and the maximum value, as an example the value recommended by ABAQUS [1.29], is five. It should be noted that the smaller the aspect ratio, the larger the number of elements and the longer the computational time. Hence, it is recommended to start with an aspect ratio of one and mesh the structural steel member and compare the numerical results against test results or exact closed-form solutions. Then, we can repeat the procedure using aspect ratios of two and three and plot the three numerical results against test results or exact solutions. After that, we can go back and choose different short dimensions smaller or larger than that initially chosen for the shell finite element and repeat the aforementioned procedures and again plot the results against test results or exact closed-form solutions. Plotting the results will determine the best finite element mesh that provides accurate results with less computational time. The studies we conduct to choose the best finite element mesh are commonly called as *convergence studies*. It should be noted that in regions of the structural member where the stress gradient is small, aspect ratios higher than five can be used and still can produce satisfactory results.

Similar to structural steel members, concrete slab decks can be meshed using the same approach. If we use solid elements to model the concrete slab deck, then we keep an aspect ratio of approximately 1:1:1 (width × height × length) and we start through the depth of the concrete slab and extend towards the length. The maximum aspect ratio is 1:1:5 in regions with high stress gradient. Structural steel members and concrete slab decks having cross sections that are symmetric about one or two axes can be modeled by cutting half or quarter of the member, respectively, owing to symmetry. Use of symmetry reduces the size of the finite element mesh considerably and consequently reduces the computational time significantly. Detailed discussions on how symmetry can be efficiently used in finite element modeling are presented in [5.1, 5.3]. However, researchers and modelers have to be very careful when using symmetry to reduce the mesh size of the individual components of bridges and the whole bridges. This is attributed to the fact that most structural steel members that have slender cross sections can fail owing to local buckling or local yielding. Failure due to local buckling or local yielding can occur in any region of the structural steel member due to initial local and overall geometric imperfections. Therefore, the whole structural steel members have to be modeled even if the cross section is symmetric about the two axes. In addition, symmetries have to be in loading, boundary conditions, geometry, and material properties. If the cross section is symmetric but the bridge component or the whole bridge is subjected to different loading along the length or width of the bridge or the boundary conditions are not the same at both ends, the whole bridge component or the whole bridge has to be modeled. Therefore, it is better to define symmetry in this book as correspondence in size, shape, and position of loads; material properties; boundary conditions; residual stresses due to processing; and initial local and overall geometric imperfections that are on opposite sides of a dividing line or plane. The posttensioned concrete slabs tested and modeled by Ellobody and Bailey [5.4] are examples of how using symmetry can considerably limit the size of the mesh and result in a significant saving in the solution time (see Figures 5.17 and 5.18). A combination of C3D8 and C3D6 elements available within the ABAQUS [1.29] element library was used to model the concrete slab, tendon, and anchorage elements. The slab had dimensions of 4300 × 1600 × 160 mm (length × width × depth). Due to symmetry, only one quarter of the slab was modeled (Figure 5.17) and the total number of elements used in the model was 6414, including the interface elements. A sensitivity study was carried out and it was found that a mesh size of 24 mm (width), 23.25 mm (depth), and 100 mm (length), for most of the elements, achieved accurate results. All nodes at symmetry surfaces (1) and (2) were prevented to

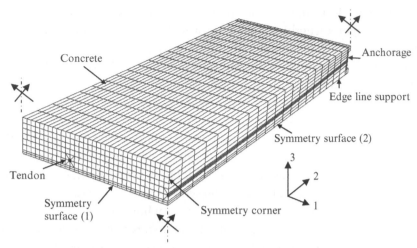

Figure 5.17 Using symmetry to model posttensioned concrete slabs tested and modeled by Ellobody and Bailey [5.4].

Figure 5.18 Full-scale specimens of posttensioned concrete slabs tested and modeled by Ellobody and Bailey [5.4]. (For the color version of this figure, the reader is referred to the online version of this chapter.)

displace in 2-direction and 1-direction, respectively. All nodes at the corner location were prevented to displace in 1-direction and 2-direction.

It should be noted most current efficient general-purpose finite element computer programs have the ability to perform meshing of the metal structures automatically. However, in many cases, the resulting finite element meshes may be very fine so that it takes huge time in the analysis process.

Therefore, it is recommended in this book to use guided meshing where the modelers apply the aforementioned fundamentals in developing the finite element mesh using current software. In this case, automatic meshing software can be of great benefit for modelers.

5.4 MATERIAL MODELING OF THE BRIDGE COMPONENTS

5.4.1 General

As mentioned previously, if we classify steel and steel-concrete composite bridges according to the materials of construction, we will find mainly two materials. The two materials are the structural steel and concrete, which are briefly highlighted in Chapter 2 based on the information presented in current codes of practice. In this section, it is important to detail how the stress–strain curves of the two main components are accurately incorporated in finite element modeling. It should be noted that there are other steels used in bridges, which may have higher ultimate stresses compared with structural steels such as steels used in bearings, shear connectors, bolts, prestressing tendons, and reinforcement bars. However, these can be incorporated in the same way as structural steels. Also, it should be noted that bridges may have different other materials such as stainless steels, aluminum, and cold-formed steel. These also can be treated following the same approach as the structural steel as previously presented in [5.1]. However, it is recommended in this book that other materials used in bridges of special nature should be investigated individually and incorporated differently in the model such as advanced composite laminates, materials used in bearings (elastomeric materials), and dampers. In the coming sections, the author aims to provide a good insight regarding material modeling of structural steel and concrete as adopted in most available general-purpose computer programs, with ABAQUS [1.29] presented as an example.

5.4.2 Material Modeling of Structural Steel

Structural steel members have linear-nonlinear stress–strain curves, as shown in Figure 2.3. The engineering stress–strain curves are determined from tensile coupon tests or stub column tests specified in most current international specifications. Although the testing procedures of tensile coupon tests and stub column tests are outside the scope of this book, it is important in this chapter to detail how the linear and nonlinear regions of the stress–strain curves are incorporated in the finite element models. The main important parameters needed from the stress–strain curve are the measured initial Young's modulus (E_o), the measured proportional limit stress (σ_p), the measured static yield stress (σ_y) that is commonly taken as the 0.1% or 0.2% proof stress $(\sigma_{0.1}$ or $\sigma_{0.2})$ for materials having a rounded stress–strain curve with no distinct yield plateau, the measured ultimate tensile strength (σ_u), and the measured elongation after fracture (ε_f). It should be noted that structural steel members used in bridges undergo large inelastic strains. Therefore, the engineering stress–strain curves must be converted to true stress-logarithmic plastic true strain curves. The true stress (σ_{true}) and plastic true strain $(\varepsilon_{true}^{pl})$ were calculated using Equations (5.1) and (5.2) as given in ABAQUS [1.29]:

$$\sigma_{true} = \sigma(1 + \varepsilon) \qquad (5.1)$$

$$\varepsilon_{true}^{pl} = \ln(1 + \varepsilon) - \sigma_{true}/E_o \qquad (5.2)$$

where E_o is the initial Young's modulus σ and ε are the measured nominal (engineering) stress and strain values, respectively.

The initial part of the stress-strain curve from origin to the proportional limit stress can be represented based on linear elastic model as given in ABAQUS [1.29]. The linear elastic model can define isotropic, orthotropic, or anisotropic material behavior and is valid for small elastic strains (normally less than 5%). Depending on the number of symmetry planes for the elastic properties, a material can be classified as either isotropic (an infinite number of symmetry planes passing through every point) or anisotropic (no symmetry planes). Some materials have a restricted number of symmetry planes passing through every point; for example, orthotropic materials have two orthogonal symmetry planes for the elastic properties. The number of independent components of the elasticity tensor depends on such symmetry properties. The simplest form of linear elasticity is the isotropic case. The elastic properties are completely defined by giving the Young's modulus (E_o) and the Poisson's ratio (υ). The shear modulus (G) can be expressed in terms of E_o. Values of Poisson's ratio approaching 0.5 result in nearly incompressible behavior.

The nonlinear part of the curve passed the proportional limit stress can be represented based on classical plasticity model as given in ABAQUS [1.29]. The model allows the input of a nonlinear curve by giving tabular values of stresses and strains. When performing an elastic-plastic analysis at finite strains, it is assumed that the plastic strains dominate the deformation and that the elastic strains are small. It is justified because structural steels used in bridges have a well-defined yield stress.

The classical metal plasticity models use Mises or Hill yield surfaces with associated plastic flow, which allow for isotropic and anisotropic yield, respectively. The models assume perfect plasticity or isotropic hardening behavior. Perfect plasticity means that the yield stress does not change with plastic strain. Isotropic hardening means that the yield surface changes size uniformly in all directions such that the yield stress increases (or decreases) in all stress directions as plastic straining occurs. Associated plastic flow means that as the material yields, the inelastic deformation rate is in the direction of the normal to the yield surface (the plastic deformation is volume-invariant). This assumption is generally acceptable for most calculations with metal. The classical metal plasticity models can be used in any procedure that uses elements with displacement degrees of freedom. The Mises and Hill yield surfaces assume that yielding of the metal is independent of the equivalent pressure stress. The Mises yield surface is used to define isotropic yielding. It is defined by giving the value of the uniaxial yield stress as a function of uniaxial equivalent plastic strain as mentioned previously. The Hill yield surface allows anisotropic yielding to be modeled. Further details regarding the modeling of different metals are found in [5.1].

5.4.3 Material Modeling of Concrete

5.4.3.1 General

There are mainly two material modeling approaches for concrete in ABAQUS [1.29], which are concrete smeared cracking and concrete damaged plasticity. Both models can be used to model plain and reinforced concrete. The reinforcement bars can be used with both models as previously highlighted in Section 5.2. In the coming sections, the two modeling approaches are briefly highlighted to enable modelers to choose the appropriate approach.

5.4.3.2 Concrete Smeared Cracking

Concrete smeared cracking model in ABAQUS [1.29] provides a general capability for modeling concrete in all types of structures, including beams, trusses, shells, and solids. The model can be used for plain concrete, even though it is intended primarily for the analysis of reinforced concrete structures. Also, the model can be used with rebar to model concrete reinforcement. In addition, concrete smeared cracking model is designed for applications in which the concrete is subjected to essentially monotonic straining at low confining pressures. The model consists of an isotropically hardening yield surface that is active when the stress is dominantly compressive and

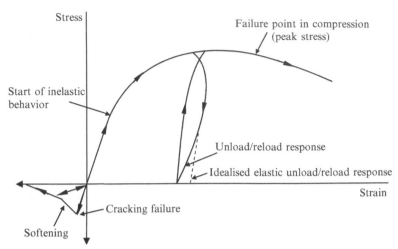

Figure 5.19 Uniaxial behavior of plain concrete as given in ABAQUS [1.29].

an independent "crack detection surface" that determines if a point fails by cracking. The model uses oriented damaged elasticity concepts (smeared cracking) to describe the reversible part of the material's response after cracking failure. The model requires that the linear elastic material model be used to define elastic properties and cannot be used with local orientations; see Figure 5.19.

Reinforcement in concrete structures is typically provided by means of rebars, which are 1D strain theory elements (rods) that can be defined singly or embedded in oriented surfaces. Rebars are typically used with metal plasticity models to describe the behavior of the rebar material and are superposed on a mesh of standard element types used to model the concrete. With concrete smeared cracking modeling approach, the concrete behavior is considered independently of the rebar. Effects associated with the rebar/concrete interface, such as bond slip and dowel action, are modeled approximately by introducing some "tension stiffening" into the concrete modeling to simulate load transfer across cracks through the rebar. Defining the rebar can be tedious in complex problems, but it is important that this be done accurately since it may cause an analysis to fail due to lack of reinforcement in key regions of a model.

Concrete smeared cracking model is intended as a model of concrete behavior for relatively monotonic loadings under fairly low confining pressures (less than 4–5 × the magnitude of the largest stress that can be carried by the concrete in uniaxial compression); see Figure 5.20. Cracking is assumed

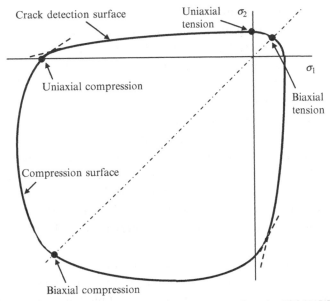

Figure 5.20 Yield and failure surfaces in plane stress as given in ABAQUS [1.29].

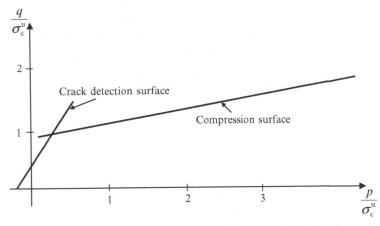

Figure 5.21 Yield and failure surfaces in the (p-q) plane as given in ABAQUS [1.29].

to be the most important aspect of the behavior, and representation of cracking and of postcracking behavior dominates the modeling. Cracking is assumed to occur when the stress reaches a failure surface that is called the "crack detection surface." This failure surface is a linear relationship between the equivalent pressure stress, p, and the Mises equivalent deviatoric stress, q; see Figure 5.21. When a crack has been detected, its orientation is stored for subsequent calculations. Subsequent cracking at the same point is restricted to being orthogonal to this direction since stress components associated with an open crack are not included in the definition of the failure surface used for detecting the additional cracks. Cracks are irrecoverable: they remain for the rest of the calculation (but may open and close). No more than three cracks can occur at any point (two in a plane-stress case and one in a uniaxial stress case). Following crack detection, the crack affects the calculations because a damaged elasticity model is used. Concrete smeared cracking model is a smeared crack model in the sense that it does not track individual "macrocracks." Constitutive calculations are performed independently at each integration point of the finite element model. The presence of cracks enters into these calculations by the way in which the cracks affect the stress and material stiffness associated with the integration point.

The postfailure behavior for direct straining across cracks is modeled with tension stiffening, which allows modelers to define the strain-softening behavior for cracked concrete; see Figure 5.22. This behavior also allows for the effects of the reinforcement interaction with concrete to be simulated in a simple manner. Tension stiffening is required in the concrete smeared cracking model. Modelers can specify tension stiffening by means of a postfailure stress-strain relation or by applying a fracture energy cracking criterion; see Figure 5.23. Specification of strain-softening behavior in reinforced concrete generally means specifying the postfailure stress as a function of strain across the crack. In cases with little or no reinforcement, this

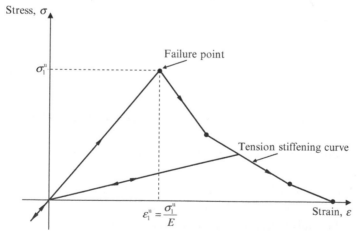

Figure 5.22 Tension-stiffening model as given in ABAQUS [1.29].

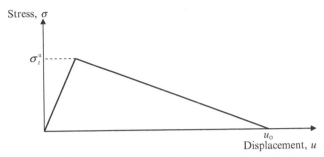

Figure 5.23 Fracture energy cracking model as given in ABAQUS [1.29].

specification often introduces mesh sensitivity in the analysis results in the sense that the finite element predictions do not converge to a unique solution as the mesh is refined because mesh refinement leads to narrower crack bands. This problem typically occurs if only a few discrete cracks form in the structure and mesh refinement does not result in the formation of additional cracks. If cracks are evenly distributed (either due to the effect of rebar or due to the presence of stabilizing elastic material, as in the case of plate bending), mesh sensitivity is less of a concern.

In practical calculations for reinforced concrete, the mesh is usually such that each element contains rebars. The interaction between the rebars and the concrete tends to reduce the mesh sensitivity, provided that a reasonable amount of tension stiffening is introduced in the concrete model to simulate this interaction; see Figure 5.22. The tension-stiffening effect must be estimated. It depends on such factors as the density of reinforcement, the quality of the bond between the rebar and the concrete, the relative size of the concrete aggregate compared to the rebar diameter, and the mesh. A reasonable starting point for relatively heavily reinforced concrete modeled with a fairly detailed mesh is to assume that the strain softening after failure reduces the stress linearly to zero at a total strain of about $10\times$ the strain at failure. The strain at failure in standard concretes is typically 10^{-4}, which suggests that tension stiffening that reduces the stress to zero at a total strain of about 10^{-3} is reasonable. This parameter should be calibrated to a particular case. The choice of tension-stiffening parameters is important in ABAQUS [1.29] since, generally, more tension stiffening makes it easier to obtain numerical solutions. Too little tension stiffening will cause the local cracking failure in the concrete to introduce temporarily unstable behavior in the overall response of the model. Few practical designs exhibit such behavior, so that the presence of this type of response in the analysis model usually indicates that the tension stiffening is unreasonably low.

As the concrete cracks, its shear stiffness is diminished. This effect is defined by specifying the reduction in the shear modulus as a function of the opening strain across the crack. You can also specify a reduced shear modulus for closed cracks. This reduced shear modulus will also have an effect when the normal stress across a crack becomes compressive. The new shear stiffness will have been degraded by the presence of the crack. When the principal stress components are dominantly compressive, the response of the concrete is modeled by an elastic–plastic theory using a simple form of yield surface written in terms of the equivalent pressure stress, p, and the Mises equivalent deviatoric stress, q; this surface is illustrated in Figure 5.21. Associated flow and isotropic hardening are used. This model significantly simplifies the actual behavior. The associated flow assumption generally overpredicts the inelastic volume strain. The yield surface cannot be matched accurately to data in triaxial tension and triaxial compression tests because of the omission of third stress invariant dependence. When the concrete is strained beyond the ultimate stress point, the assumption that the elastic response is not affected by the inelastic deformation is not realistic. In addition, when concrete is subjected to very high pressure stress, it exhibits inelastic response: no attempt has been made to build this behavior into the model. Modelers can define the stress–strain behavior of plain concrete in uniaxial compression outside the elastic range. Compressive stress data are provided as a tabular function of plastic strain and, if desired, temperature and field variables. Positive (absolute) values should be given for the compressive stress and strain. The stress–strain curve can be defined beyond the ultimate stress, into the strain-softening regime.

The cracking and compressive responses of concrete that are incorporated in the concrete model are illustrated by the uniaxial response of a specimen shown in Figure 5.19. When concrete is loaded in compression, it initially exhibits elastic response. As the stress is increased, some nonrecoverable (inelastic) straining occurs and the response of the material softens. An ultimate stress is reached, after which the material loses strength until it can no longer carry any stress. If the load is removed at some point after inelastic straining has occurred, the unloading response is softer than the initial elastic response: the elasticity has been damaged. This effect is ignored in the model, since we assume that the applications involve primarily monotonic straining, with only occasional, minor unloadings. When a uniaxial concrete specimen is loaded in tension, it responds elastically until, at a stress that is typically 7-10% of the ultimate compressive stress, cracks form. Cracks form so quickly that, even in the stiffest testing machines available, it is very difficult to observe the actual behavior. The model assumes that cracking causes damage, in the sense that open cracks can be represented by a loss of elastic stiffness. It is also assumed that there is no permanent strain associated with cracking. This will allow cracks to close completely if the stress across them becomes compressive.

In multiracial stress states, these observations are generalized through the concept of surfaces of failure and flow in stress space. These surfaces are fitted to experimental data. The surfaces used are shown in Figures 5.20 and 5.21. Modelers can specify failure ratios to define the shape of the failure surface (possibly as a function of temperature and predefined field variables). Four failure ratios can be specified, which are (1) the ratio of the ultimate biaxial compressive stress to the ultimate uniaxial compressive stress; (2) the absolute value of the ratio of the uniaxial tensile stress at failure to the ultimate uniaxial compressive stress; (3) the ratio of the magnitude of a principal component of plastic strain at ultimate stress in biaxial compression to the plastic strain at ultimate stress in uniaxial compression; and finally (4) the ratio of the tensile principal stress at cracking, in plane stress, when the other principal stress is at the ultimate compressive value, to the tensile cracking stress under uniaxial tension. It should be noted that, because the model is intended for application to problems involving relatively monotonic straining, no attempt is made to include the prediction of cyclic response or of the reduction in the elastic stiffness caused by inelastic straining under predominantly compressive stress. Nevertheless, it is likely that, even in those applications for which the model is designed, the strain trajectories will not be entirely radial, so that the model should predict the response to occasional strain reversals and strain trajectory direction changes in a reasonable way. Isotropic hardening of the "compressive" yield surface forms the basis of this aspect of the model's inelastic response prediction when the principal stresses are dominantly compressive.

5.4.3.3 Concrete Damaged Plasticity

Concrete damaged plasticity model in ABAQUS [1.29] provides a general capability for modeling concrete and other quasibrittle materials in all types of structures (beams, trusses, shells, and solids). The model uses concepts of isotropic damaged elasticity in combination with isotropic tensile and compressive plasticity to represent the inelastic behavior of concrete. Also, the model can be used for plain concrete, even though it is intended primarily for the analysis of reinforced concrete structures, and can be used with rebar to model concrete reinforcement. In addition, the model is designed for applications in which concrete is subjected to monotonic, cyclic, and/ or dynamic loading under low confining pressures. The model consists of the combination of nonassociated multihardening plasticity and scalar (isotropic) damaged elasticity to describe the irreversible damage that occurs during the fracturing process. Furthermore, the model allows user control of stiffness recovery effects during cyclic load reversals and can be defined to be sensitive to the rate of straining. The model can be used in conjunction with a viscoplastic regularization of the constitutive equations in ABAQUS to improve the convergence rate in the softening regime.

Concrete damaged plasticity model requires that the elastic behavior of the material be isotropic and linear. The model is a continuum, plasticity-based, damage model for concrete. It assumes that the main two failure mechanisms are tensile cracking and compressive crushing of the concrete material. The evolution of the yield (or failure) surface is controlled by two hardening variables, ε_t^{pl} and ε_c^{pl}, linked to failure mechanisms under tension and compression loading, respectively. ABAQUS refers to ε_t^{pl} and ε_c^{pl} as tensile and compressive equivalent plastic strains, respectively. The model assumes that the uniaxial tensile and compressive response of concrete is characterized by damaged plasticity, as shown in Figure 5.24. Under uniaxial tension, the stress–strain response follows a linear elastic relationship until the value of the failure stress, σ_{to}, is reached. The failure stress corresponds to the onset of microcracking in the concrete material. Beyond the failure stress, the formation of microcracks is represented macroscopically with a softening stress–strain response, which induces strain localization in the concrete structure. Under uniaxial compression, the response is linear until the value of initial yield, σ_{co}. In the plastic regime, the response is typically characterized by stress hardening followed by strain softening beyond the ultimate stress, σ_{cu}. This representation, although somewhat simplified, captures the main features of the response of concrete. It is assumed that the uniaxial stress-strain curves can be converted into stress versus plastic-strain curves. As shown in Figure 5.25, when the concrete specimen is unloaded from any point on the strain-softening branch of the stress–strain curves, the unloading response is weakened: the elastic stiffness of the material appears to be damaged (or degraded). The degradation of the elastic stiffness is characterized by two damage variables, d_t and d_c, which are assumed to be functions of the plastic strains, temperature, and field variables.

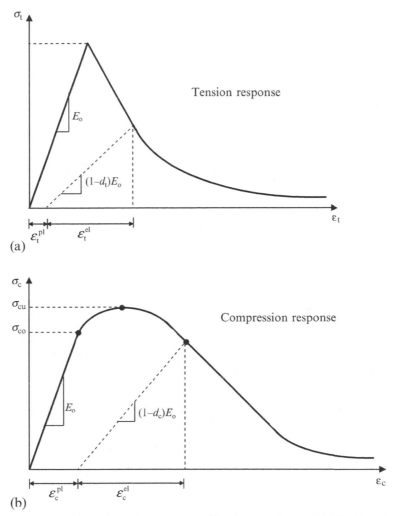

Figure 5.24 Response of concrete to uniaxial loading as given in ABAQUS [1.29].

In ABAQUS [1.29], reinforcement in concrete structures is typically provided by means of rebars, which are 1D rods that can be defined singly or embedded in oriented surfaces. Rebars are typically used with metal plasticity models to describe the behavior of the rebar material and are superposed on a mesh of standard element types used to model the concrete. With this modeling approach, the concrete behavior is considered independently of the rebar. Effects associated with the rebar/concrete interface, such as bond slip and dowel action, are modeled approximately by introducing

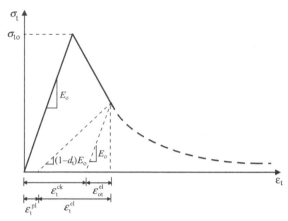

Figure 5.25 Illustration of the definition of the cracking strain ε_t^{ck} used for the definition of tension-stiffening data as given in ABAQUS [1.29].

some "tension stiffening" into the concrete modeling to simulate load transfer across cracks through the rebar. Defining the rebar can be tedious in complex problems, but it is important that this be done accurately since it may cause an analysis to fail due to the lack of reinforcement in key regions of a model. The postfailure behavior for direct straining is modeled with tension stiffening, which allows modelers to define the strain-softening behavior for cracked concrete. This behavior also allows for the effects of the reinforcement interaction with concrete to be simulated in a simple manner. Tension stiffening is required in the concrete damaged plasticity model. Modelers can specify tension stiffening by means of a postfailure stress–strain relation or by applying a fracture energy cracking criterion.

In reinforced concrete, the specification of postfailure behavior generally means giving the postfailure stress as a function of cracking strain, ε_t^{ck}. The cracking strain is defined as the total strain minus the elastic strain corresponding to the undamaged material as illustrated in Figure 5.25. To avoid potential numerical problems, ABAQUS [1.29] enforces a lower limit on the postfailure stress equal to 100 of the initial failure stress ($\sigma_{to}/100$). Tension-stiffening data are given in terms of the cracking strain, ε_t^{ck}. When unloading data are available, the data are provided to ABAQUS in terms of tensile damage curves, and the software automatically converts the cracking strain values to plastic-strain values. In cases with little or no reinforcement, the specification of a postfailure stress–strain relation introduces mesh sensitivity in the results, in the sense that the finite element predictions do not converge to a unique solution as the mesh is refined because mesh refinement leads to narrower crack bands. This problem typically occurs if cracking failure occurs only at localized regions in the structure and mesh refinement does not result in the formation of additional cracks. If cracking failure is distributed evenly (either due to the effect of rebar or due to the presence of stabilizing elastic material, as in the case of plate bending), mesh sensitivity is less of a concern.

In practical calculations for reinforced concrete, the mesh is usually such that each element contains rebars. The interaction between the rebars and the concrete tends to reduce the mesh sensitivity, provided that a reasonable amount of tension stiffening is introduced in the concrete model to simulate this interaction. This requires an estimate of the tension–stiffening effect, which depends on such factors as the density of reinforcement, the quality of the bond between the rebar and the concrete, the relative size of the concrete aggregate compared to the rebar diameter, and the mesh. A reasonable starting point for relatively heavily reinforced concrete modeled with a fairly detailed mesh is to assume that the strain softening after failure reduces the stress linearly to zero at a total strain of about $10 \times$ the strain at failure. The strain at failure in standard concretes is typically 10^{-4}, which suggests that tension stiffening that reduces the stress to zero at a total strain of about 10^{-3} is reasonable. This parameter should be calibrated to a particular case. The choice of tension–stiffening parameters is important since, generally, more tension stiffening makes it easier to obtain numerical solutions. Too little tension stiffening will cause the local cracking failure in the concrete to introduce temporarily unstable behavior in the overall response of the model. Few practical designs exhibit such behavior, so that the presence of this type of response in the analysis model usually indicates that the tension stiffening is unreasonably low.

When there is no reinforcement in significant regions of the model, the tension-stiffening approach described earlier will introduce unreasonable mesh sensitivity into the results. However, it is generally accepted that Hillerborg's [5.5] fracture energy proposal is adequate to allay the concern for many practical purposes. Hillerborg defines the energy required to open a unit area of crack, G_f, as a material parameter, using brittle fracture concepts. With this approach, the concrete's brittle behavior is characterized by a stress–displacement response rather than a stress–strain response; see Figure 5.26. Under tension, a concrete specimen will crack across some section. After it has been pulled apart sufficiently for most of the stress to be removed (so that the undamaged elastic strain is small), its length will be determined primarily by the opening at the crack. The opening does not

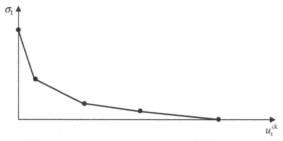

Figure 5.26 Postfailure stress-displacement curve as given in ABAQUS [1.29].

depend on the specimen's length. This fracture energy cracking model can be invoked by specifying the postfailure stress as a tabular function of cracking displacement, as shown in Figure 5.26. Alternatively, the fracture energy, G_f, can be specified directly as a material property and, in this case, define the failure stress, σ_{to}, as a tabular function of the associated fracture energy. This model assumes a linear loss of strength after cracking, as shown in Figure 5.27. The cracking displacement at which complete loss of strength takes place is, therefore, $(u_{to} = 2G_f/\sigma_{to})$. Typical values of G_f range from 40 (0.22 lb/in) for a typical construction concrete (with a compressive strength of approximately 20 MPa, 2850 lb/in.^{2}) to 120 N/m (0.67 lb/in) for a high-strength concrete (with a compressive strength of approximately 40 MPa, 5700 lb/in.^{2}). If tensile damage, d_t, is specified, ABAQUS automatically converts the cracking displacement values to "plastic" displacement values. The implementation of this stress-displacement concept in a finite element model requires the definition of a characteristic length associated with an integration point. The characteristic crack length is based on the element geometry and formulation. This definition of the characteristic crack length is used because the direction in which cracking occurs is not known in advance. Therefore, elements with large aspect ratios will have rather different behaviors depending on the direction in which they

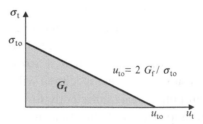

Figure 5.27 Postfailure stress-fracture energy curve as given in ABAQUS [1.29].

crack: some mesh sensitivity remains because of this effect, and elements that have aspect ratios close to one are recommended.

Modelers can define the stress-strain behavior of plain concrete in uniaxial compression outside the elastic range. Compressive stress data are provided as a tabular function of inelastic (or crushing) strain, ε_c^{in}, and, if desired, strain rate, temperature, and field variables. Positive (absolute) values should be given for the compressive stress and strain. The stress-strain curve can be defined beyond the ultimate stress, into the strain-softening regime. Hardening data are given in terms of an inelastic strain, ε_c^{in}, instead of plastic strain, ε_c^{pl}. The compressive inelastic strain is defined as the total strain minus the elastic strain corresponding to the undamaged material, $\varepsilon_c^{in} = \varepsilon_c - \varepsilon_{oc}^{el}$, where $\varepsilon_{oc}^{el} = \sigma_c/E_o$ as illustrated in Figure 5.28. Damage, d_t and/or d_c, can be specified in tabular form. (If damage is not specified, the model behaves as a plasticity model; consequently, $\varepsilon_t^{pl} = \varepsilon_t^{ck}$, $\varepsilon_c^{pl} = \varepsilon_c^{in}$.) In ABAQUS [1.29], the damage variables are treated as nondecreasing material point quantities. At any increment during the analysis, the new value of each damage variable is obtained as the maximum between the value at the end of the previous increment and the value corresponding to the current state (interpolated from the user-specified tabular data). The choice of the damage properties is important since, generally, excessive damage may have a critical effect on the rate of convergence. It is recommended to avoid using values of the damage variables above 0.99, which corresponds to a 99% reduction of the stiffness. Also, modelers can define the uniaxial tension damage variable, d_t, as a tabular function of either cracking strain or cracking displacement and

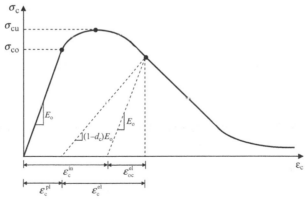

Figure 5.28 Definition of the compressive inelastic strain used for the definition of compression hardening data as given in ABAQUS [1.29].

the uniaxial compression damage variable, d_c, as a tabular function of inelastic (crushing) strain. Modelers also can define flow potential, yield surface, and ABAQUS viscosity parameters for the concrete damaged plasticity material model. The concrete damaged plasticity model assumes nonassociated potential plastic flow. The flow potential used for this model is the Drucker-Prager hyperbolic approach. This flow potential, which is continuous and smooth, ensures that the flow direction is always uniquely defined. The function approaches the linear Drucker-Prager flow potential asymptotically at high confining pressure stress and intersects the hydrostatic pressure axis at 90°. The model makes use of the yield function of Lubliner et al. [5.6], with the modifications proposed by Lee and Fenves [5.7] to account for different evolution of strength under tension and compression. The evolution of the yield surface is controlled by the hardening variables, ε_t^{pl} and ε_c^{pl}.

Unlike concrete models based on the smeared crack approach, the concrete damaged plasticity model does not have the notion of cracks developing at the material integration point. However, it is possible to introduce the concept of an effective crack direction with the purpose of obtaining a graphic visualization of the cracking patterns in the concrete structure. Different criteria can be adopted within the framework of scalar-damage plasticity for the definition of the direction of cracking. Following Lubliner et al. [5.6], ABAQUS [1.29] assumes that cracking initiates at points where the tensile equivalent plastic strain is greater than zero, $\varepsilon_t^{pl} > 0$, and the maximum principal plastic strain is positive. The direction of the vector normal to the crack plane is assumed to be parallel to the direction of the maximum principal plastic strain. ABAQUS offers a variety of elements for use with the concrete damaged plasticity model: truss, shell, plane-stress, plane-strain, generalized plane-strain, axisymmetric, and 3D elements. Most beam elements can be use. For general shell analysis, more than the default number of five integration points through the thickness of the shell should be used; nine thickness integration points are commonly used to model progressive failure of the concrete through the thickness with acceptable accuracy.

5.5 LINEAR AND NONLINEAR ANALYSES OF THE BRIDGES AND BRIDGE COMPONENTS

5.5.1 General

In *linear analyses*, it is assumed that the displacements of the finite element model are infinitesimally small and that the material is linearly elastic. In addition, it was assumed that the boundary conditions remain unchanged during the application of loading on the finite element model. The finite element equations correspond to linear analysis of a structural problem because the displacement response is a linear function of the applied force vector. This means that if the forces are increased with a constant factor, the corresponding displacements will be increased with the same factor. On the other hand, in nonlinear analysis, the aforementioned assumptions are not valid. The assumption is that the displacement must be small and the finite element stiffness matrix and force vector are constant and independent on the element displacements, because all integrations have been performed over the original volume of the finite elements and the strain–displacement relationships. The assumption of a linear elastic material was implemented in the use of constant stress–strain relationships. Finally, the assumption that the boundary conditions remain unchanged was reflected in the use of constant restraint relations for the finite element equilibrium equation.

Recognizing the previous discussion, we can define three main nonlinear analyses commonly known as *materially nonlinear analysis, geometrically (large displacement and large rotation) nonlinear analysis*, and *materially and geometrically nonlinear analysis*. In materially nonlinear analysis, the nonlinear effect lies in the nonlinear stress–strain relationship, with the displacements and strains infinitesimally small. Therefore, the usual engineering stress and strain measurements can be employed. In geometrically nonlinear analysis, the structure undergoes large rigid-body displacements and rotations. Majority of geometrically nonlinear analyses were based on von Karman nonlinear equations such as the analyses presented in [5.8–5.15]. The equations allow coupling between bending and membrane behavior with the retention of Kirchhoff normality constraint [1.16]. Finally, materially and geometrically nonlinear analysis combines both nonlinear stress–strain relationship and large displacements and rotations experienced by the structure.

Most available general-purpose finite element computer program divides the problem history (overall finite element analysis) into different steps as shown in Figure 5.29 An *analysis procedure* can be specified for each step, with prescribing loads, boundary conditions, and output requests specified for each step. A step is a phase of the problem history, and in its simplest form, a step can be just a static analysis of a load changing from one magnitude to another. For each step, modelers can choose an analysis procedure. This choice defines the type of analysis to be performed during the step such as static stress analysis, eigenvalue buckling analysis, or any other types of analyses. Static analyses are used when inertia effects can be neglected.

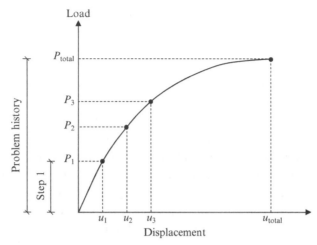

Figure 5.29 Load-displacement history in a nonlinear analysis.

The analyses can be linear or nonlinear and assume that time-dependent material effects, such as creep, are negligible. Linear static analysis involves the specification of load cases and appropriate boundary conditions. If all or part of a structure has linear response, substructuring is a powerful capability for reducing the computational cost of large analyses. Static nonlinear analyses can also involve geometric nonlinearity and/or material nonlinearity effects. If geometrically nonlinear behavior is expected in a step, the large-displacement formulation should be used. Only one procedure is allowed per step and any combination of available procedures can be used from step to step. However, information from a previous step can be imported to the current step by calling the results from the previous step. The loads, boundary conditions, and output requests can be inserted in any step.

Most available general-purpose finite element computer programs classify the steps into two main kinds of steps, which are commonly named as *general analysis steps* and *linear perturbation steps*. General analysis steps can be used to analyze linear or nonlinear response. On the other hand, linear perturbation steps can be used only to analyze linear problems. Linear analysis is always considered to be linear perturbation analysis about the state at the time when the linear analysis procedure is introduced. The linear perturbation approach allows general application of linear analysis techniques in cases where the linear response depends on preloading or on the nonlinear response history of the model. In general, analysis steps and linear perturbation steps, the solution for a single set of applied loads, can be predicted. However, for static analyses covered in this book, it is possible to find solutions for multiple load cases. In this case, the overall analysis procedure can be changed from step to step. This allows the state of the model (stresses, strains, displacements, deformed shapes, etc.) to be updated throughout all general analysis steps. The effects of previous history can be included in the response in each new analysis step by calling the results of a previous history. As an example, after conducting an initial condition analysis step to include residual stresses in cross sections, the initial stresses in the whole cross section will be updated from zero to new applied stresses that accounted for the residual stress effect in metal structures.

It should be noted that linear perturbation steps have no effect on subsequent general analysis steps and can be conducted separately as a whole (overall) analysis procedure. In this case, the data obtained from the linear perturbation steps can be saved in files that can be called into the subsequent general analysis steps. For example, linear eigenvalue buckling analyses, needed for modeling of initial overall and local geometric imperfections, can be conducted initially as a separate overall analysis procedure, and buckling modes can be extracted from the analyses and saved in files. The saved files can be called into subsequent static general analyses and factored to model initial geometric imperfections. The most obvious reason for using several steps in an analysis is to change the analysis procedure type. However, several steps can also be used to change output requests, such as the boundary conditions or loading (any information specified as history or step-dependent data). Sometimes, an analysis may be progressed to a point where the present step definition needs to be modified. ABAQUS [1.29] provides the ability to restart the analysis, whereby a step can be terminated prematurely and a new step can be defined for the problem continuation. History data prescribing the loading, boundary conditions, output, etc., will remain in effect for all subsequent general analysis steps until they are modified or reset. ABAQUS [1.29] will compare all loads and boundary conditions specified in a step with the loads and boundary conditions in effect during the previous step to ensure consistency and continuity. This comparison is expensive if the number of individually specified loads and boundary conditions is very large. Hence, the number of individually specified loads and boundary conditions should be minimized, which can usually be done by using element and node sets instead of individual elements and nodes.

Most current general-purpose finite element computer program divides each step of analysis into multiple increments. In most cases, one can choose either *automatic (direct) time incrementation* or *user-specified fixed time incrementation* to control the solution. Automatic time incrementation is a built-in incrementation scheme that allows the software to judge the increment needed based on equilibrium requirements. On the other hand, user-specified fixed time incrementation forces the software to use a specified fixed increment, which in many cases may be large or small or need updating during the step. This results in the analysis to be stopped and readjusted. Therefore, automatic incrementation is recommended for most cases. The methods for selecting automatic or direct incrementation are always prescribed by all general-purpose software to help modelers. In nonlinear analyses, most general-purpose computer programs will use increment and iterate as necessary to analyze a step, depending on the severity of the nonlinearity. Iterations conducted within an increment can be classified as *regular equilibrium iterations* and *severe discontinuity iterations.* In regular equilibrium iterations, the solution varies smoothly, while in severe discontinuity iterations, abrupt changes in stiffness occur. The analysis will continue to iterate until the severe discontinuities are sufficiently small (or no severe discontinuities occur) and the equilibrium tolerances are satisfied. Modelers can provide parameters to indicate a level of accuracy in the time integration, and the software will choose the time increments to achieve this accuracy. Direct user control is provided because it can sometimes save computational cost in cases where modelers are familiar with the problem and know a suitable incrementation scheme. Modelers can define the upper limit to the number of increments in an analysis. The analysis will stop if this maximum is exceeded before the complete solution for the step has been obtained. To reach a solution, it is often necessary to increase the number of increments allowed by defining a new upper limit.

In nonlinear analyses, general-purpose software uses *extrapolation* to speed up the solution. Extrapolation refers to the method used to determine the first guess to the incremental solution. The guess is determined by the size of the current time increment and by whether *linear, parabolic,* or no extrapolation of the previously attained history of each solution variable is chosen. Linear extrapolation is commonly used with 100% extrapolation of the previous incremental solution being used at the start of each increment to begin the nonlinear equation solution for the next increment. No extrapolation is used in the first increment of a step. Parabolic extrapolation uses two previous incremental solutions to obtain the first guess to the current incremental solution. This type of extrapolation is useful in situations when the local variation of the solution with respect to the time scale of the problem is expected to be quadratic, such as the large rotation of structures. If parabolic extrapolation is used in a step, it begins after the second increment of the step, that is, the first increment employs no extrapolation, and the second increment employs linear extrapolation. Consequently, slower convergence rates may occur during the first two increments of the succeeding steps in a multistep analysis. Nonlinear problems are commonly solved using Newton's method, and linear problems are commonly solved using the stiffness method. Details of the aforementioned solution methods are outside the scope of this book; however, the methods are presented in details in [1.12–1.18].

Most general-purpose computer programs adopt a *convergence criterion* for the solution of nonlinear problems automatically. Convergence criterion is the method used by software to govern the balance equations during the iterative solution. The iterative solution is commonly used to solve the equations of nonlinear problems for unknowns, which are the degrees of freedom at the nodes of the finite element model. Most general-purpose computer programs have control parameters designed to provide reasonably optimal solution of complex problems involving combinations of nonlinearities as well as efficient solution of simpler nonlinear cases. However, the most important consideration in the choice of the control parameters is that any solution accepted as "converged" is a close approximation to the exact solution of the nonlinear equations. Modelers can reset many solution control parameters related to the tolerances used for equilibrium equations. If less strict convergence criterion is used, results may be accepted as converged when they are not sufficiently close to the exact solution of the nonlinear equations. Caution should be considered when resetting solution control parameters. The lack of convergence is often due to modeling issues, which should be resolved before changing the accuracy controls. The solution can be terminated if the balance equations failed to converge. It should be noted that linear cases do not require more than one equilibrium iteration per increment, which is easy to converge. Each increment of a nonlinear solution will usually be solved by multiple equilibrium iterations. The number of iterations may become excessive, in which case, the increment size should be reduced and the increment will be attempted again. On the other hand, if successive increments are solved with a minimum number of iterations, the increment size may be increased. Modelers can specify a number of time incrementation control parameters. Most general-purpose computer programs may have trouble with the element calculations because of excessive distortion in large-displacement problems or because of very large plastic-strain increments. If this occurs and automatic time incrementation has been chosen, the increment will be attempted again with smaller time increments.

Steel and steel-concrete composite bridges may be checked for safety against dynamic loads, as an example bridges constructed in regions that are subjected to earthquakes. In this case, dynamic finite element analyses should be performed. Although this book is not specifically written for detailing dynamic analyses, brief highlights of the different dynamic analyses are highlighted in this book. ABAQUS [1.29] offers several methods for performing dynamic analysis of problems in which inertia effects are considered. Direct integration of the system must be used when nonlinear dynamic response is being studied. Implicit direct integration is provided in ABAQUS (Standard), while explicit direct integration is provided in ABAQUS (Explicit). Modal methods are usually chosen for linear analyses because in direct-integration dynamics, the global equations of motion of the system must be integrated through time, which makes direct-integration methods significantly more expensive than modal methods. Subspace-based methods are provided in ABAQUS/Standard and offer cost-effective approaches to the analysis of systems that are mildly nonlinear.

In ABAQUS (Standard) [1.29], dynamic studies of linear problems are generally performed by using the eigenmodes of the system as a basis for calculating the response. In such cases, the necessary modes and frequencies are calculated first in a frequency extraction step. The mode-based procedures are generally simple to use and the dynamic response analysis itself is usually not expensive computationally, although the eigenmode extraction can become computationally intensive if many modes are required for a large model. The eigenvalues can be extracted in a prestressed system with the "stress-stiffening" effect included (the initial stress matrix is included if the base state step definition included nonlinear geometric effects), which may be necessary in the dynamic study of preloaded systems. It is not possible to prescribe nonzero displacements and rotations directly in mode-based procedures. The method for prescribing motion in mode-based procedures is included in transient modal dynamic analysis. Density must be defined for all materials used in any dynamic analysis, and damping (both viscous and structural) can be specified at the either material or step level.

5.5.2 Linear Eigenvalue Buckling Analysis

Eigenvalue buckling analysis provided by ABAQUS [1.29] is generally used to estimate the critical buckling (bifurcation) load of structures. The analysis is a linear perturbation procedure. The analysis can be the first step in a global analysis of an unloaded structure, or it can be performed after the structure has been preloaded. It can be used to model measured initial overall and local geometric imperfections or in the investigation of the imperfection sensitivity of a structure in case of lack of measurements. Eigenvalue buckling is generally used to estimate the critical buckling loads of *stiff* structures (classical eigenvalue buckling). Stiff structures carry their design loads primarily by axial or membrane action, rather than by bending action. Their response usually involves very little deformation prior to buckling. However, even when the response of a structure is nonlinear before collapse, a general eigenvalue buckling analysis can provide useful estimates of collapse mode shapes.

The buckling loads are calculated relative to the original state of the structure. If the eigenvalue buckling procedure is the first step in an analysis, the buckled (deformed) state of the model at the end of the eigenvalue buckling analysis step will be the updated original state of the structure. The eigenvalue buckling can include preloads such as dead load and other loads. The preloads are often zero in classical eigenvalue buckling analyses. An incremental loading pattern is defined in the eigenvalue buckling prediction step. The magnitude of this loading is not important; it will be scaled by the load multipliers that are predicted by the eigenvalue buckling analysis. The buckling mode shapes (eigenvectors) are also predicted by the eigenvalue buckling analysis. The critical buckling loads are then equal to the preloads plus the scaled incremental load. Normally, the lowest load multiplier and buckling mode is of interest. The buckling mode shapes are normalized vectors and do not represent actual magnitudes of deformation at critical load. They are normalized so that the maximum displacement component has a magnitude of 1.0. If all displacement components are zero, the maximum rotation component is normalized to 1.0. These buckling mode shapes are often the most useful outcome of the eigenvalue buckling analysis, since they predict the likely failure modes of the structure.

Some structures have many buckling modes with closely spaced eigenvalues, which can cause numerical problems. In these cases, it is recommended to apply enough preload to load the structure to just below the buckling load before performing the eigenvalue analysis. In many cases, a series of closely spaced eigenvalues indicate that the structure is imperfection-sensitive. An eigenvalue buckling analysis will not give accurate predictions of the buckling load for imperfection-sensitive structures. In this case, the static Riks procedure, used by ABAQUS [1.29], which will be highlighted in this chapter, should be used instead. Negative eigenvalues may be predicted from an eigenvalue buckling analysis. The negative eigenvalues indicate that the structure would buckle if the loads were applied in the opposite direction. Negative eigenvalues may correspond to buckling modes that cannot be understood readily in terms of physical behavior, particularly if a preload is applied that causes significant geometric nonlinearity. In this case, a geometrically nonlinear load–displacement analysis should be performed. Because buckling analysis is usually done for stiff structures, it is not usually necessary to include the effects of geometry change in establishing equilibrium for the original state. However, if significant geometry change is involved in the original state and this effect is considered to be important, it can be included by specifying that geometric nonlinearity should be considered for the original state step. In such cases, it is probably more realistic to perform a geometrically nonlinear load–displacement analysis (Riks analysis) to determine the collapse loads, especially for imperfection-sensitive structures as mentioned previously. While large deformation can be included in the preload, the eigenvalue buckling theory relies on the fact that there is little geometry change due to the *live* (scaled incremental load) buckling load. If the live load produces significant geometry changes, a nonlinear collapse (Riks) analysis must be used (Figure 5.30).

The initial conditions such as residual stresses can be specified for an eigenvalue buckling analysis. If the buckling step is the first step in the analysis, these initial conditions form the original state of the structure. Boundary

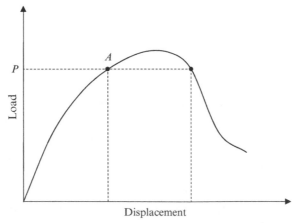

Figure 5.30 Load-displacement behavior that could be predicted by the Riks method available in ABAQUS [1.29].

conditions can be applied to any of the displacement or rotation degrees of freedom (6 degrees of freedom). Boundary conditions are treated as constraints during the eigenvalue buckling analysis. Therefore, the buckling mode shapes are affected by these boundary conditions. The buckling mode shapes of symmetric structures subjected to symmetric loadings are either symmetric or antisymmetric. In such cases, it is more efficient to use symmetry to reduce the finite element mesh of the model. Axisymmetric structures subjected to compressive loading often collapse in non-axisymmetric modes. Therefore, these structures must be modeled as a whole. The loads prescribed in an eigenvalue buckling analysis can be concentrated nodal forces applied to the displacement degrees of freedom or can be distributed loads applied to finite element faces. The load stiffness can be of a significant effect on the critical buckling load. It is important that the structure is not preloaded above the critical buckling load. During an eigenvalue buckling analysis, the model's response is defined by its linear elastic stiffness in the original state. All nonlinear or inelastic material properties are ignored during an eigenvalue buckling analysis. Any structural finite elements can be used in an eigenvalue buckling analysis. The values of the eigenvalue load multiplier (buckling loads) will be printed in the data files after the eigenvalue buckling analysis. The buckling mode shapes can be visualized using the software. Any other information such as values of stresses, strains, or displacements can be saved in files at the end of the analysis. Further details regarding the eigenvalue buckling analysis can be found in [5.1].

5.5.3 Materially and Geometrically Nonlinear Analyses

Materially nonlinear analysis of a structure is a general nonlinear analysis step. The analysis can be also called as *load-displacement nonlinear material analysis*. All required information regarding the behavior of bridges is predicted from the materially nonlinear analysis. The information comprised the ultimate loads, failure modes, and load-displacement relationships as well as any other required data that can be obtained from materially nonlinear analysis. The initial overall and local geometric imperfections, residual stresses, prestressing, and nonlinear stress–strain curves of the construction material are included in the load-displacement nonlinear material analysis. Since most, if not all, bridge components have nonlinear stress–strain curves or linear-nonlinear stress–strain curves, therefore, most of the general nonlinear analysis steps associated with bridges are materially nonlinear analyses.

Materially nonlinear analysis (with or without consideration of geometric nonlinearity) of metal structures is done to determine the overall response of the bridges and bridge components. From a numerical viewpoint, the implementation of a nonlinear stress–strain curve of a construction metal material involves the integration of the state of the material at an integration point over a time increment during a materially nonlinear analysis. The implementation of a nonlinear stress–strain curve must provide an accurate material stiffness matrix for use in forming the nonlinear equilibrium equations of the finite element formulation. The mechanical constitutive models associated with bridge components consider elastic and inelastic response of the material. In the inelastic response models that are provided in ABAQUS [1.29], the elastic and inelastic responses are distinguished by separating the deformation into *recoverable* (elastic) and *nonrecoverable* (inelastic) parts. This separation is based on the assumption that there is an additive relationship between strain rates of the elastic and inelastic parts. The constitutive material models used in most available general-purpose finite element computer programs are commonly accessed by any of the solid or structural elements previously highlighted in Section 5.2. This access is made independently at each constitutive calculation point. These points are the numerical integration points in the elements. The constitutive models obtain the state at the point under consideration at the start of the increment from the material database specified in the step. The state variables include the stresses and strains used in the constitutive models. The constitutive models update the state of the material response to the end of the increment.

Geometrically nonlinear analysis is a general nonlinear analysis step. The analysis can be also called as *load-displacement nonlinear geometry analysis*. The initial overall and local geometric imperfections, prestressing, and residual stresses can be included in the load-displacement nonlinear geometry analysis. If the stress–strain curve of the construction metal material is nonlinear, the analysis will be called as *combined materially and geometrically nonlinear analysis* or *load-displacement nonlinear material and geometry analysis*. All required information regarding the behavior of metal structures is predicted from the combined materially and geometrically nonlinear analysis. The information comprised the ultimate loads, failure modes, and load–displacement relationships as well as any other required data that can be obtained from the combined materially and geometrically nonlinear analysis. Further details regarding the materially and geometrically nonlinear analyses can be found in [5.1].

5.6 RIKS METHOD

The Riks method provided by ABAQUS [1.29] is an efficient method that is generally used to predict unstable, geometrically nonlinear collapse of a structure. The method can include nonlinear materials and boundary conditions. The method can provide complete information about a structure's collapse. The Riks method can be used to speed convergence of unstable collapse of structures. Geometrically nonlinear analysis of bridges may involve buckling or collapse behavior. Several approaches are possible for modeling such behavior. One of the approaches is to treat the buckling or collapse response dynamically, thus actually modeling the response with inertia effects included as the structure snaps. This approach is easily accomplished by restarting the terminated static procedure and switching to a dynamic procedure when the static solution becomes unstable. In some simple cases, displacement control can provide a solution, even when the conjugate load (the reaction force) is decreasing as the displacement increases. Alternatively, static equilibrium states during the unstable phase of the response can be found by using the *modified Riks method* supported by ABAQUS [1.29]. This method is used for cases where the loading is proportional, where the load magnitudes are governed by a single scalar parameter. The method can provide solutions even in cases of complex, unstable response of bridges.

In complex structures involving material nonlinearity, geometric nonlinearity prior to buckling, or unstable postbuckling behavior, a load-displacement (Riks) analysis must be performed to investigate the structures accurately. The Riks method treats the load magnitude as an additional unknown and solves loads and displacements simultaneously. Therefore, another quantity must be used to measure the progress of the solution. ABAQUS [1.29] uses the arc length along the static equilibrium path in load-displacement domain. This approach provides solutions regardless of whether the response is stable or unstable. If the Riks step is a continuation of a previous history, any loads that exist at the beginning of the step are treated as dead loads with constant magnitude. A load whose magnitude is defined in the Riks step is referred to as a reference load. All prescribed loads are ramped from the initial (dead load) value to the reference values specified. ABAQUS [1.29] uses Newton's method to solve the nonlinear equilibrium equations. The Riks procedure uses very small extrapolation of the strain increment. Modelers can provide an initial increment in arc length along the static equilibrium path when defining the step. After that, the software computes subsequent steps automatically. Since the loading magnitude is part of the solution, modelers need a method to specify when the step is completed. It is common that one can specify a maximum displacement value at a specified degree of freedom. The step will terminate once the maximum value is reached. Otherwise, the analysis will continue until the maximum number of increments specified in the step definition is reached.

The Riks method works well with structures having a smooth equilibrium path in load-displacement domain. The Riks method can be used to solve postbuckling problems, with both stable and unstable postbuckling behaviors. In this way, the Riks method can be used to perform postbuckling analyses of structures that show linear behavior prior to (bifurcation) buckling. When performing a load-displacement analysis using the Riks method, important nonlinear effects can be included. Imperfections based on linear buckling modes can be also included in the analysis of structures using the Riks method. It should be noted that the Riks method cannot obtain a solution at a given load or displacement value since these are treated as unknowns. Termination of the analysis using the Riks method occurs at the first solution that satisfies the step termination criterion. To obtain solutions at exact values of load or displacement, the analysis must be restarted at the desired point in the step and a new, non-Riks step must be defined. Since the subsequent step is a continuation of the Riks analysis, the load magnitude in that step must be given appropriately so that the step begins with the loading continuing to increase or decrease according to its behavior at the point of restart. Initial values of stresses such as residual stresses can be inserted in the analysis using the Riks method. Also, boundary conditions can be applied to any of the displacement or rotation degrees of freedom (6 degrees of freedom). Concentrated nodal forces and moments applied to associated displacement or rotation degrees of freedom (6 degrees of freedom) as well as distributed loads at finite element faces can be inserted in the analysis using the Riks method. Nonlinear material models that describe mechanical behavior of the bridges and bridge components can be incorporated in the analysis using the Riks method.

5.6.1 Dynamic Analyses

As mentioned previously, steel and steel-concrete composite bridges may be analyzed for dynamic loads, particularly if the bridges are constructed in regions attacked by earthquakes. In this section, main dynamic analyses supported by ABAQUS [1.29] are highlighted in order to help researchers, designers, academics, and practitioners involved in the dynamic analyses of steel and steel-concrete composite bridges. The dynamic analyses provided in ABAQUS over most analyses that may be needed for the dynamic analysis of steel and steel-concrete composite bridges. The direct-integration dynamic procedure provided in ABAQUS (Standard) offers a choice of implicit operators for the integration of the equations of motion, while ABAQUS (Explicit) uses the central-difference operator. In an implicit dynamic analysis, the integration operator matrix must be inverted and a set of nonlinear equilibrium equations must be solved at each time increment. In an explicit dynamic analysis, displacements and velocities are calculated in terms of quantities that are known at the beginning of an increment; therefore, the global mass and stiffness matrices need not be formed and inverted, which means that each increment is relatively inexpensive compared to the increments in an implicit integration scheme. The size of the time increment in an explicit dynamic analysis is limited, however, because the central-difference operator is only conditionally stable, whereas the implicit operator options available in ABAQUS (Standard) are unconditionally stable and, thus, there is no such limit on the size of the time increment that can be used for most analyses in ABAQUS (Standard), that is, accuracy governs the time increment. The stability limit for the central-difference method (the largest time increment that can be taken without the method generating large, rapidly growing errors) is closely related to the time required for a stress wave to cross the smallest element dimension in the model; thus, the time increment in an explicit dynamic analysis can be very short if the mesh contains small elements or if the stress wave speed in the material is very high. The method is, therefore, computationally attractive for problems in which the total dynamic response time that must be modeled is only a few orders of magnitude longer than this stability limit. Many of the advantages of the explicit procedure also apply to slower (quasistatic) processes for cases in which it is appropriate to use mass scaling to reduce the wave speed.

ABAQUS (Explicit) [1.29] offers fewer element types than ABAQUS (Standard) [1.29]. For example, only first-order, displacement method elements (four-node quadrilaterals, eight-node bricks, etc.) and modified second-order elements are used, and each degree of freedom in the model must have mass or rotary inertia associated with it. However, the method provided in ABAQUS (Explicit) has some important advantages including the following:

(1) The analysis cost rises only linearly with problem size, whereas the cost of solving the nonlinear equations associated with implicit integration rises more rapidly than linearly with problem size. Therefore, ABAQUS (Explicit) is attractive for very large problems.

(2) The explicit integration method is often more efficient than the implicit integration method for solving extremely discontinuous short-term events or processes.

(3) Problems involving stress wave propagation can be far more efficient computationally in ABAQUS (Explicit) than in ABAQUS (Standard).

(4) In choosing an approach to a nonlinear dynamic problem, modelers must consider the length of time for which the response is sought compared to the stability limit of the explicit method; the size of the problem; and the restriction of the explicit method to first-order, pure displacement method or modified second-order elements. In some cases, the choice is obvious, but in many problems of practical interest, the choice depends on details of the specific case.

Direct-solution procedures must be used for dynamic analyses that involve a nonlinear response. Modal superposition procedures are a cost-effective option for performing linear or mildly nonlinear dynamic analyses. The direct-solution dynamic analyses procedures available in ABAQUS include the following:

(1) Implicit dynamic analysis, in which implicit direct-integration dynamic analysis is used to study (strongly) nonlinear transient dynamic response in ABAQUS (Standard).

(2) Subspace-based explicit dynamic analysis, in which the subspace projection method in ABAQUS (Standard) uses direct, explicit integration of the dynamic equations of equilibrium written in terms of a vector space spanned by a number of eigenvectors. The eigenmodes of the system extracted in a frequency extraction step are used as the global basis vectors. This method can be very effective for systems with mild nonlinearities that do not substantially change the mode shapes. However, it cannot be used in contact analyses.

(3) Explicit dynamic analysis, in which explicit direct-integration dynamic analysis is performed in ABAQUS (Explicit).

(4) Direct-solution steady-state harmonic response analysis, in which the steady-state harmonic response of a system can be calculated in ABAQUS (Standard) directly in terms of the physical degrees of

freedom of the model. The solution is given as in-phase (real) and out-of-phase (imaginary) components of the solution variables (displacement, stress, etc.) as functions of frequency. The main advantage of this method is that frequency-dependent effects (such as frequency-dependent damping) can be modeled. The direct method is not only the most accurate but also the most expensive steady-state harmonic response procedure. The direct method can also be used if nonsymmetric terms in the stiffness are important or if model parameters depend on frequency. ABAQUS [1.29] includes a full range of modal superposition procedures. Modal superposition procedures can be run using a high-performance linear dynamics software architecture called SIM. The SIM architecture offers advantages over the traditional linear dynamics architecture for some large-scale analyses. Prior to any modal superposition procedure, the natural frequencies of a system must be extracted using the eigenvalue analysis procedure. Frequency extraction can be performed using the SIM architecture.

ABAQUS [1.29] provides different modal superposition procedures comprising the following:

(1) Mode-based steady-state harmonic response analysis, which is a steady-state dynamic analysis based on the natural modes of the system, can be used to calculate a system's linearized response to harmonic excitation. This mode-based method is typically less expensive than the direct method. The solution is given as in-phase (real) and out-of-phase (imaginary) components of the solution variables (displacement, stress, etc.) as functions of frequency. Mode-based steady-state harmonic analysis can be performed using the SIM architecture.

(2) Subspace-based steady-state harmonic response analysis. In this analysis, the steady-state dynamic equations are written in terms of a vector space spanned by a number of eigenvectors. The eigenmodes of the system extracted in a frequency extraction step are used as the global basis vectors. The method is attractive because it allows frequency-dependent effects to be modeled and is much cheaper than the direct analysis method. Subspace-based steady-state harmonic response analysis can be used if the stiffness is nonsymmetric and can be performed using the SIM architecture.

(3) Mode-based transient response analysis, which provides transient response for linear problems using modal superposition. Mode-based transient analysis can be performed using the SIM architecture.

(4) Response spectrum analysis, which is a linear response spectrum analysis that is often used to obtain an approximate upper bound of the peak significant response of a system to a user-supplied input spectrum (such as earthquake data) as a function of frequency. The method has a very low computational cost and provides useful information about the spectral behavior of a system. Response spectrum analysis can be performed using the SIM architecture.

(5) Random response analysis, in which the linearized response of a model to random excitation can be calculated based on the natural modes of the system. This procedure is used when the structure is excited continuously and the loading can be expressed statistically in terms of a "power spectral density" function. The response is calculated in terms of statistical quantities such as the mean value and the standard deviation of nodal and element variables. Random response analysis can be performed using the SIM architecture. SIM is a high-performance software architecture available in ABAQUS [1.29] that can be used to perform modal superposition dynamic analyses. The SIM architecture is much more efficient than the traditional architecture for large-scale linear dynamic analyses (both model size and number of modes) with minimal output requests. SIM-based analyses can be used to efficiently handle nondiagonal damping generated from element or material contributions. Therefore, SIM-based procedures are an efficient alternative to subspace-based linear dynamic procedures for models with element damping or frequency-independent materials.

ABAQUS [1.29] relies on user-supplied model data and assumes that the material's physical properties reflect experimental results. Examples of meaningful material properties are a positive mass density per volume, a positive Young's modulus, and a positive value for any available damping coefficients. However, in special cases, modelers may want to "adjust" a value of density, mass, stiffness, or damping in a region or a part of the model to bring the overall mass, stiffness, or damping to the expected required levels. Certain material options in ABAQUS allow modelers to introduce nonphysical material properties to achieve this adjustment. Every nonconservative system exhibits some energy loss that is attributed to material nonlinearity, internal material friction, or external (mostly joint) frictional behavior. Conventional engineering materials like steel and high-strength aluminum alloys provide small amounts of internal material damping, not enough to prevent large amplification at near-resonant frequencies. Damping properties increase in modern composite fiber-reinforced materials, where the energy loss occurs through plastic or viscoelastic phenomena as well as from friction at the interfaces between the matrix and the reinforcement. Still, larger material damping is exhibited by thermoplastics. Mechanical dampers may be added to models to introduce damping forces to the system. In general, it is difficult to quantify the source of a system's damping. It usually comes from several sources simultaneously, for example, from energy loss during hysteretic loading, viscoelastic material properties, and external joint friction.

Four categories of damping sources are available in ABAQUS [1.29] comprising material and element damping, global damping, modal damping, and damping associated with time integration. Material and element damping specifies damping as part of a material definition that is assigned to a model. In addition, the software has elements such as dashpots, springs with their complex stiffness matrix, and connectors that serve as dampers, all with viscous and structural damping factors. Viscous damping can be included in mass, beam, pipe, and shell elements with general section properties. Global damping can be used in situations where material or element damping is not appropriate or sufficient. Modelers can apply abstract damping factors to an entire model using global damping. ABAQUS allows modelers to specify global damping factors for both viscous damping (Rayleigh damping) and structural damping (imaginary stiffness matrix). Modal damping applies only to mode-based linear dynamic analyses. This technique allows modelers to apply damping directly to the modes of the system. By definition, modal damping contributes only diagonal entries to the modal system of equations. Finally, damping associated with time integration, which results from marching through a simulation with a finite time increment size. This type of damping applies only to analyses using direct time integration. ABAQUS also applies damping to a linear dynamic system in two forms, which are velocity proportional viscous damping and displacement proportional structural damping, which is for use in frequency domain dynamics. An additional type of damping known as composite damping serves as a means to calculate a model average critical damping with the material density as the weight factor and is intended for use in mode-based dynamics.

5.6.2 Thermal (Heat Transfer) and Thermal-Stress Analyses
5.6.2.1 General
Steel and steel-concrete composite bridges may be analyzed to evaluate temperature-induced thermal stresses. Due to variations of temperatures throughout the year, especially in hot regions, thermal stresses are induced

in steel and steel-concrete composite bridges resulting in additional membrane and bending stresses in cross sections of the bridges as well as longitudinal and lateral thermal expansions. ABAQUS [1.29] provides three main analyses dealing with temperature effects in different cross sections, which are uncoupled heat transfer analysis, sequentially coupled thermal-stress analysis, and fully coupled thermal-stress analysis. Uncoupled heat transfer analysis deals with heat transfer problems involving conduction, forced convection, and boundary radiation analyzed in ABAQUS (Standard). In these analyses, the temperature field is calculated without the knowledge of the stress/deformation state in the structures being studied. Pure heat transfer problems can be transient or steady state and linear or nonlinear. Sequentially coupled thermal-stress analysis can be conducted in ABAQUS (Standard) if the stress/displacement solution is dependent on a temperature field, but there is no inverse dependency. Sequentially coupled thermal-stress analysis is performed by first solving the pure heat transfer problem and then reading the temperature solution into a stress analysis as a predefined field. In the stress analysis, the temperature can vary with time and position but is not changed by the stress analysis solution. ABAQUS allows for dissimilar meshes between the heat transfer analysis model and the thermal-stress analysis model. Temperature values will be interpolated based on element interpolators evaluated at nodes of the thermal-stress model. Finally, in fully coupled thermal-stress analysis, a coupled temperature-displacement procedure is used to solve simultaneously for the stress/displacement and the temperature fields. A coupled analysis is used when the thermal and mechanical solutions affect each other strongly. Both ABAQUS (Standard) and ABAQUS (Explicit) provide coupled temperature-displacement analysis procedures, but the algorithms used by each program differ considerably. In ABAQUS (Standard), the heat transfer equations are integrated using a backward-difference scheme, and the coupled system is solved using Newton's method. These problems can be transient or steady state and linear or nonlinear. In ABAQUS (Explicit), the heat transfer equations are integrated using an explicit forward-difference time integration rule, and the mechanical solution response is obtained using an explicit central-difference integration rule. Fully coupled thermal-stress analysis in ABAQUS (Explicit) is always transient.

5.6.2.2 Uncoupled Heat Transfer Analyses

Uncoupled heat transfer analyses are those in which the temperature field is calculated without consideration of the stress/deformation or the electrical field in the structures being studied. The analysis can include conduction, boundary convection, and boundary radiation. It can also include cavity radiation effects. In addition, the analysis can include forced convection through the mesh if forced convection/diffusion heat transfer elements are used. Uncoupled heat transfer analyses can include thermal interactions such as gap radiation, conductance, and heat generation between contact surfaces. The analyses can be transient or steady state and can be linear or nonlinear. The analyses require the use of heat transfer elements. Uncoupled heat transfer analysis is used to model solid body heat conduction with general, temperature-dependent conductivity; internal energy (including latent heat effects); and quite general convection and radiation boundary conditions, including cavity radiation. Forced convection of a fluid through the mesh can be modeled by using forced convection/diffusion elements. Heat transfer problems can be nonlinear because the material properties are temperature-dependent or because the boundary conditions are nonlinear. Usually, the nonlinearity associated with temperature-dependent material properties is mild because the properties do not change rapidly with temperature. However, when latent heat effects are included, the analysis may be severely nonlinear.

Boundary conditions are very often nonlinear; for example, film coefficients can be functions of surface temperature. Again, the nonlinearities are often mild and cause little difficulty. A rapidly changing film condition (within a step or from one step to another) can be modeled easily using temperature-dependent and field-variable-dependent film coefficients. Radiation effects always make heat transfer problems nonlinear. Nonlinearities in radiation grow as temperatures increase. ABAQUS (Standard) uses an iterative scheme to solve nonlinear heat transfer problems. The scheme uses the Newton's method with some modification to improve stability of the iteration process in the presence of highly nonlinear latent heat effects. Steady-state cases involving severe nonlinearities are sometimes more effectively solved as transient cases because of the stabilizing influence of the heat capacity terms. The required steady state solution can be obtained as the very long transient time response; the transient will simply stabilize the solution for that long time response.

Steady-state analysis means that the internal energy term (the specific heat term) in the governing heat transfer equation is omitted. The problem then has no intrinsic physically meaningful time scale. Nevertheless, you can assign an initial time increment, a total time period, and maximum and minimum allowed time increments to the analysis step, which is often convenient for output identification and for specifying prescribed temperatures and fluxes with varying magnitudes. Any fluxes or boundary condition changes to be applied during a steady-state heat transfer step should be given within the step, using appropriate amplitude references to specify their "time" variations. If fluxes and boundary conditions are specified for the step without amplitude references, they are assumed to change linearly with "time" during the step, from their magnitudes at the end of the previous step (or zero, if this is the beginning of the analysis) to their newly specified magnitudes at the end of the heat transfer step.

Time integration in transient problems is done with the backward Euler method (sometimes also referred to as the modified Crank-Nicolson operator) in the pure conduction elements. This method is unconditionally stable for linear problems. The forced convection/diffusion elements use the trapezoidal rule for time integration. They include numerical diffusion control and, optionally, numerical dispersion control. The elements with dispersion control offer improved solution accuracy in cases where the transient response of the fluid is important. The velocity of a fluid moving through the mesh can be prescribed if forced convection/diffusion heat transfer elements are used. Conduction between the fluid and the adjacent forced convection/diffusion heat transfer elements will be affected by the mass flow rate of the fluid. Natural convection occurs when differences in fluid density created by thermal gradients cause motion of the fluid. The forced convection/diffusion elements are not designed to handle this phenomenon; the flow must be prescribed. Modelers can specify the mass flow rates per unit area (or through the entire section for 1D elements) at the nodes. ABAQUS (Standard) interpolates the mass flow rates to the material points. The numerical solution of the transient heat transfer equation including convection becomes increasingly difficult as convection dominates diffusion. Cavity radiation can be activated in a heat transfer step. This feature involves interacting heat transfer between all of the facets of the cavity surface, dependent on the facet temperatures, facet emissivities, and the geometric view factors between each facet pair. When the thermal emissivity is a function of temperature or field variables, modelers can specify the maximum allowable emissivity change during an increment in addition to the maximum temperature change to control the time incrementation. It should be noted that, by default, the initial temperature of all nodes is zero. Modelers can specify nonzero initial temperatures.

Boundary conditions can be used to prescribe temperatures (degree of freedom 11) at nodes in a heat transfer analysis. Shell elements have additional temperature degrees of freedom 12, 13, etc., through the thickness. Boundary conditions can be specified as functions of time by referring to amplitude curves. For purely diffusive heat transfer elements, a boundary without any prescribed boundary conditions (natural boundary condition) corresponds to an insulated surface. For forced convection/diffusion elements, only the flux associated with conduction is zero; energy is free to convect across an unconstrained surface. This natural boundary condition correctly models areas where fluid is crossing a surface (as, e.g., at the upstream and downstream boundaries of the mesh) and prevents spurious reflections of energy back into the mesh.

Thermal loading in a heat transfer analysis comprises concentrated heat fluxes, body fluxes, and distributed surface fluxes; average-temperature radiation conditions; convective film conditions; and radiation conditions (film properties can be made a function of temperature) as well as cavity radiation effects. Predefined temperature fields are not allowed in heat transfer analyses. Boundary conditions should be used instead to specify temperatures, as described earlier. The thermal conductivity of the materials in a heat transfer analysis must be defined. The specific heat and density of the materials must also be defined for transient heat transfer problems. Latent heat can be defined for diffusive heat transfer elements if changes in internal energy due to phase changes are important. Thermal expansion coefficients are not meaningful in an uncoupled heat transfer analysis problem since the deformation of the structure is not considered.

The heat transfer element library in ABAQUS (Standard) includes (1) diffusive heat transfer elements, which allow for heat storage (specific heat and latent heat effects) and heat conduction; (2) forced convection/diffusion heat transfer elements; (3) shell heat transfer elements; and (4) the first-order heat transfer elements (such as the two-node link, four-node quadrilateral, and eight-node brick), which use a numerical integration rule with the integration stations located at the corners of the element for the heat capacitance terms and for the calculations of the distributed surface fluxes. First-order diffusive elements are preferred in cases involving latent heat effects since they use such a special integration technique to provide accurate solutions with large latent heats. The second-order heat transfer elements use conventional Gaussian integration. Thus, the second-order elements are to be preferred for problems when the solution will be smooth (without latent heat effects) and usually give more accurate results for the same number of nodes in the mesh.

5.6.2.3 Sequentially Coupled Thermal-Stress Analysis

A sequentially coupled heat transfer analysis available in ABAQUS [1.29] is used when the stress/deformation field in a structure depends on the temperature field in that structure, but the temperature field can be found without knowledge of the stress/deformation response. The analysis is usually performed by first conducting an uncoupled heat transfer analysis and then a stress/deformation analysis. The analysis is a thermal-stress analysis in which the temperature field does not depend on the stress field. Nodal temperatures are stored in ABAQUS as a function of time in the heat transfer results (.fil) file or output database (.odb) file. The temperatures are read into the stress analysis as a predefined field; the temperature varies with position and is usually time-dependent. It is predefined because it is not changed by the stress analysis solution. Such predefined fields are always read into ABAQUS (Standard) at the nodes. They are then interpolated to the calculation points within elements as needed. The temperature interpolation in the stress elements is usually approximate and one order lower than the displacement interpolation to obtain a compatible variation of thermal and mechanical strain. Any number of predefined fields can be read in, and material properties can be defined to depend on them.

Appropriate initial conditions for the thermal and stress analysis problems are described in the heat transfer and stress analysis sections. Appropriate boundary conditions for the thermal and stress analysis problems are described in the heat transfer and stress analysis sections. Also, appropriate loading for the thermal and stress analysis problems is described in the heat transfer and stress analysis sections. In addition to the temperatures read in from the heat transfer analysis, user-defined field variables can be specified; these values only affect field-variable-dependent material properties. The materials in the thermal analysis must have thermal properties such as conductivity defined. Any mechanical properties such as elasticity will be ignored in the thermal analysis, but they must be defined for the stress analysis procedure. Thermal strain will arise in the stress analysis if thermal expansion is included in the material property definition. Any of the heat transfer elements in ABAQUS (Standard) can be used in the thermal analysis. In the stress analysis, the corresponding continuum or structural elements must be chosen. For continuum elements, heat transfer results from a mesh using first-order elements can be transferred to a stress analysis with a mesh using second-order elements.

5.6.2.4 Fully Coupled Thermal-Stress Analysis

A fully coupled thermal-stress analysis is performed when the mechanical and thermal solutions affect each other strongly and, therefore, must be obtained simultaneously. The analysis requires the existence of elements with both temperature and displacement degrees of freedom in the model and can be used to analyze time-dependent material response. The analysis cannot include cavity radiation effects but may include average-temperature radiation conditions and takes into account temperature dependence of material properties only for the properties that are assigned to elements with temperature degrees of freedom. In ABAQUS (Standard), a fully coupled thermal-stress analysis neglects inertia effects and can be transient or steady state. On the other hand, in ABAQUS (Explicit), a fully coupled thermal-stress analysis includes inertia effects and models transient thermal response. Fully coupled thermal-stress analysis is needed when the stress analysis is dependent on the temperature distribution and the temperature distribution depends on the stress solution. In ABAQUS (Standard), the temperatures are integrated using a backward-difference scheme, and the nonlinear coupled system is solved using Newton's method. ABAQUS (Standard) offers an exact as well as an approximate implementation of Newton's method for fully coupled temperature-displacement analysis.

A steady-state coupled temperature-displacement analysis can be performed in ABAQUS (Standard). In steady-state cases, modelers should assign an arbitrary "time" scale to the step. This time scale is convenient for changing loads and boundary conditions through the step and for obtaining solutions to highly nonlinear (but steady-state) cases; however, for the latter purpose, transient analysis often provides a natural way of coping with the nonlinearity. Alternatively, modelers can perform a transient coupled temperature-displacement analysis. By default, the initial temperature of all nodes is zero. Modelers can specify nonzero initial temperatures. Boundary conditions can be used to prescribe both temperatures (degree of freedom 11) and displacements/rotations (degrees of freedom 1-6) at nodes in fully coupled thermal-stress analysis. Shell elements in ABAQUS (Standard) have additional temperature degrees of freedom 12, 13, etc., through the thickness. Boundary conditions applied during a dynamic coupled temperature-displacement response step should use appropriate amplitude references. If boundary conditions are specified for the step without amplitude references, they are applied instantaneously at the beginning of the step.

Thermal loads that can be prescribed in a fully coupled thermal-stress analysis comprise concentrated heat fluxes, body fluxes, and distributed surface fluxes; node-based film and radiation conditions; average-temperature radiation conditions; and element- and surface-based film and radiation conditions. In addition, mechanical loads that can be prescribed to the analysis comprise concentrated nodal forces, which can be applied to the displacement degrees of freedom (1-6), as well as distributed pressure forces or body forces. Predefined temperature fields are not allowed in a fully coupled thermal-stress analysis. Boundary conditions should be used instead to prescribe temperature degrees of freedom 11, 12, 13, etc., in ABAQUS (Standard) shell elements. Other predefined field variables can be specified in a fully coupled thermal-stress analysis. These values will affect only field-variable-dependent material properties. The materials in a fully coupled thermal-stress analysis must have both thermal properties, such as conductivity, and mechanical properties, such as elasticity, defined. Thermal strain will arise if thermal expansion is included in the material property definition.

Coupled temperature-displacement elements that have both displacements and temperatures as nodal variables are available in both ABAQUS (Standard) and ABAQUS (Explicit). In ABAQUS (Standard), simultaneous temperature/displacement solution requires the use of such elements; pure displacement elements can be used in part of the model in the fully coupled thermal-stress procedure, but pure heat transfer elements cannot be used. In ABAQUS (Explicit), any of the available elements, except Eulerian elements, can be used in the fully coupled thermal-stress procedure; however, the thermal solution will be obtained only at nodes where the temperature degree of freedom has been activated. The first-order coupled temperature-displacement elements in ABAQUS use a constant temperature over the element to calculate thermal expansion. The second-order coupled temperature-displacement elements in ABAQUS (Standard) use a lower-order interpolation for temperature than for displacement (parabolic variation of displacements and linear variation of temperature) to obtain a compatible variation of thermal and mechanical strain.

5.7 MODELING OF INITIAL IMPERFECTIONS AND RESIDUAL STRESSES

Most structural steel members have initial geometric imperfections as a result of the manufacturing, transporting, and handling processes. Initial geometric imperfections can be classified into two main categories, which are local and overall (bow, global, or out-of-straightness) imperfections. Initial local geometric imperfections can be found in any region of the outer or inner surfaces of metal structural members and are in the perpendicular directions to the structural member surfaces. On the other hand, initial overall geometric imperfections are global profiles for the whole structural member along the member length in any direction. Initial local and overall geometric imperfections can be predicted from finite element models by conducting eigenvalue buckling analysis to obtain the worst cases of local and overall buckling modes. These local and overall buckling modes can be then factored by measured magnitudes in the tests. Superposition can be used to predict final combined local and overall buckling modes. The resulting combined buckling modes can be then added to the initial coordinates of the structural member. The final coordinates can be used in any subsequent nonlinear analysis. The details of the eigenvalue buckling analysis were highlighted in Section 5.5.2. Accurate finite element models should incorporate initial local and overall geometric imperfections in the analysis; otherwise, the results will not be accurate. Efficient test programs must include the measurement of initial local and overall geometric imperfections.

Residual stresses are initial stresses existing in cross sections without the application of an external load such as stresses resulting from manufacturing processes of structural steel members. Residual stresses produce internal membrane forces and bending moments, which are in equilibrium inside the cross sections. The force and the moment resulting from residual stresses in the cross sections must be zero. Residual stresses in structural cross sections are attributed to the uneven cooling of parts of cross sections after hot rolling. Uneven cooling of cross-section parts subjects to internal stresses. The parts that cool quicker have residual compressive stresses, while parts that cool lower have residual tensile stresses. Residual stresses cannot be avoided and in most cases are not desirable. The measurement of residual stresses is therefore important for accurate understanding of the performance of metal structural members.

Extensive experimental investigations were conducted in the literature to determine the distribution and magnitude of residual stresses inside cross sections. The experimental investigations can be classified into two main categories, which are nondestructive and destructive methods. Examples of nondestructive methods are X-ray diffraction and Neutron diffraction. Nondestructive methods are suitable for measuring stresses close to the outside surface of cross sections. On the other hand, destructive methods involve machining/cutting of the cross section to release internal stresses and measure resulting change of strains. Destructive methods are based on the destruction of the state of equilibrium of the residual stresses in the cross section. In this way, the residual stresses can be measured by relaxing these stresses. However, it is only possible to measure the consequences of the stress relaxation rather than the relaxation itself. One of the main destructive methods is to cut the cross section into slices and measure the change in strains before and after cutting. After measuring the strains, some simple analytical approaches can be used to evaluate resultant membrane forces and bending moments in the cross sections. Although the testing procedures to determine residual stresses are outside the scope of this book, it is important to detail how to incorporate residual stresses in finite element models. It should be noted that in some cases, incorporating residual stresses can result in small effect on the structural performance of metals. However, in some other cases, it may result in considerable effect. Since the main objective of this book is to accurately model all parameters affecting the behavior and design of metal structures, the way to model residual stresses is highlighted in this book.

Residual stresses and their distribution are very important factors affecting the strength of axially loaded structural steel members. These stresses are of particular importance for slender columns, with slenderness ratio varying from approximately 40 to 120. As a column load is increased, some parts of the column will quickly reach the yield stress and go into the plastic range because of the presence of residual compression stresses. The stiffness will reduce and become a function of the part of the cross section that is still inelastic. A column with residual stresses will behave as though it has a reduced cross section. This reduced cross section or elastic portion of the column will change as the applied load changes. The buckling analysis and postbuckling calculation can be carried out theoretically or numerically by using an effective moment of inertia of the elastic portion of the cross section or by using the tangent modulus. ABAQUS [1.29] is a popular package that can be used for the postbuckling analysis, which gives the history of deflection versus loading. The ultimate strength of the column could be then obtained from this history.

To ensure accurate modeling of the behavior of metal structures, the residual stresses should be included in the finite element models. Measured residual stresses were implemented in the finite element model as initial stresses using ABAQUS [1.29] software. It should be noted that the slices cut from the cross section to measure the residual stresses can be used to form tensile coupon test specimens. In this case, the effect of bending stresses on the stress–strain curve of the metal material will be considered since the tensile coupon specimen will be tested in the actual bending condition. Therefore, only the membrane residual stresses have to be incorporated in the finite element model. Initial conditions can be specified for particular nodes or elements, as appropriate. The data can be provided directly in an external input file or in some cases by a user subroutine or by the results or output database file from a previous analysis. If initial conditions are not specified, all initial conditions are considered zero in the model. Various types of initial conditions can be specified, depending on the analysis to be performed; however, the type highlighted here is the initial conditions (stresses). The option can be used to apply stresses in different directions. When initial stresses are given, the initial stress state may not be an exact equilibrium state for the finite element model. Therefore, an initial step should be included to check for equilibrium and iterate, if necessary, to achieve equilibrium. Further details regarding incorporating initial geometric imperfections and residual stresses in finite element modeling of steel structural members are found in [5.1].

5.8 MODELING OF SHEAR CONNECTION FOR STEEL-CONCRETE COMPOSITE BRIDGES

Previous investigations by the author have proposed detailed finite element models simulating the behavior of shear connections with headed studs in solid concrete slabs, precast hollow-core concrete slabs, and composite concrete slabs with profiled steel sheeting, as reported in [2.68–2.71]. The finite element models simulated the behavior of headed studs in pushout tests and resulted in extensive data regarding the shear resistance of the studs, failure modes, and load–slip characteristic curves of the studs. The finite element models can be used to represent shear connections in steel–concrete composite bridges. This is attributed to the fact that the finite element models were used to perform extensive parametric studies. The parametric studies investigated shear connections having studs with different diameters, which are used in bridges, as well as investigated shear connections having different concrete strengths, which are also used in bridges.

As an example, the investigation reported in [2.71] investigated the performance of headed stud shear connectors in composite concrete slabs with profiled steel sheeting. A nonlinear 3D finite element model was developed and validated against the pushout tests conducted by Lloyd and Wright [2.57] and Kim et al. [2.58, 2.59]. The pushout tests carried out by Kim et al. [2.58, 2.59] provided the shear connection capacity of 13×65 mm headed

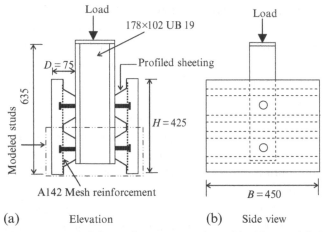

Figure 5.31 Arrangement of the pushout test conducted by Kim *et al.* [2.58, 2.59].

stud welded through deck in composite slabs with profiled steel sheeting. The pushout test specimen arrangement is shown in Figure 5.31. The steel beam used was a $178 \times 102 \times 19 \, \text{kg/m}$ UB section having two $13 \times 65 \, \text{mm}$ headed studs welded on each flange of the steel beam through the profiled steel sheeting. The profiled steel sheeting had a depth (h_p) of 40 mm, average width (b_o) of 136.5 mm, and plate thickness (t) of 0.68 mm. The composite concrete slab had a depth (D) of 75 mm, width (B) of 450 mm, and height (H) of 425 mm. Reinforcement bar mesh of 6 mm diameter and 200 mm spacing between two bars was placed on the top of the profiled sheeting. The concrete slabs of the pushout tests conducted by Kim *et al.* [2.58, 2.59] had the average measured concrete cube strengths of 34.5 MPa, average tensile strength of 2.42 MPa, and Young's modulus of 21.7 GPa. The steel beam had the measured yield stress of 288 MPa and Young's modulus of 189 GPa. The profiled steel sheeting had the measured yield stress of 308 MPa and Young's modulus of 184 GPa. The headed shear studs had the measured yield stress of 435 MPa. The load was applied on the upper part of the steel beam. The movement of the composite concrete slabs relative to the steel beam was measured using six dial gauges attached to either the profiled steel sheeting near the studs or the concrete top surface. The testing arrangements and procedures as well as the specimen dimensions are detailed in Kim *et al.* [2.58, 2.59]. On the other hand, pushout tests conducted by Lloyd and Wright [2.57] provided the shear connection capacity of a $19 \times 100 \, \text{mm}$ headed stud welded through deck in a composite slab with profiled steel sheeting. The profiled steel sheeting had a depth of 50 mm, average width of 150 mm, and plate thickness of 1.2 mm. The pushout tests had the general arrangement as shown in Figure 5.31. The widths of the composite concrete slabs varied from 675 to 1350 mm. The height of the composite concrete slab varied from 600 to 900 mm. The depth of the concrete slab was 115 mm for all pushout test specimens. Different reinforcement areas that were used in the tests varied from A98 to A193. The spacing between two headed studs was 300 mm. The load was applied on the upper part of the steel beam, and the movement of the composite concrete slabs relative to the steel beam was measured.

To show an example of how to simulate the behavior of shear connections using the finite element method, let us present how the pushout tests [2.57–2.59] have been modeled. The finite element program ABAQUS [1.29] was used to investigate the behavior of shear connection in composite beams with profiled steel sheeting tested in [2.57–2.59]. In order to obtain accurate results from the finite element analysis, all components associated with the shear connection must be properly modeled. The main components affecting the behavior of shear connection in composite beams with profiled steel sheeting are concrete slab, steel beam, profiled steel sheeting, reinforcement bars, and shear connector. Both geometric nonlinearity and material nonlinearity were included in the finite element analysis. Combinations of 3D eight-node (C3D8) and six-node (C3D6) solid elements are used to model the pushout test specimens. Assuming that the load is transferred equally from the steel beam to each shear connector, it is decided to model only a single stud welded to each flange of the composite beam as highlighted in Figure 5.31. The predicted shear capacity would be independent of the number of shear connectors used in the experimental investigation, and it can be obtained for different stud diameters by adjusting the finite element mesh. Due to symmetry of the specimens, only a quarter of the pushout test arrangement is modeled. Two finite element models (Model (A) and Model (B)) were developed, as shown in Figures 5.32 and 5.33, respectively. Model (A) presented the actual trapezoidal geometry of the profiled steel sheeting. This model is suitable to investigate the behavior of headed studs welded through profiled steel sheeting with mild side slopes. In this case, the concrete within the ribs of the profiled steel sheeting can be modeled properly. Model (B) simulated the trapezoidal shape of the rib by an equivalent rectangular shape. This model can be used to investigate the behavior of headed studs welded through profiled steel sheeting with stiff side slopes. The two models can be used to study the shear connection in composite beams with different types of profiled steel sheeting. The width

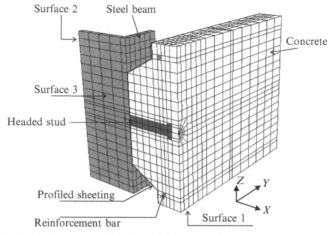

Figure 5.32 Finite element mesh of Model (A) as detailed by Ellobody and Young [2.71].

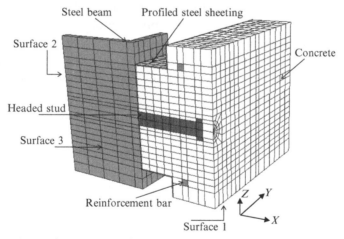

Figure 5.33 Finite element mesh of Model (B) as detailed by Ellobody and Young [2.71].

of the head of the stud is taken $1.5 \times$ the stud diameter, and its thickness is $0.5 \times$ the diameter. The circular cross-sectional area of the reinforcement bar was simulated by the equivalent rectangular cross-sectional area in the finite element modeling. It is assumed that the effect of separation of the profiled steel sheeting from the concrete slab at certain load level has little effect on the concrete slab. Hence, the nodes of the concrete elements are attached to the nodes of the profiled steel sheeting elements. Jayas and Hosain [2.47] observed that the separation of the concrete behind the shear connector occurred even at a low load level. According to this observation, the nodes behind the stud, in the direction of loading, are detached from the surrounding concrete nodes with the other nodes of the stud connected with the surrounding concrete. All nodes of the concrete slab and profiled steel sheeting in the opposite direction of loading (surface 1 in Figures 5.32 and 5.33) are restricted from moving in the Z-direction to resist the applied compression load. All nodes along the middle surface of the steel beam (surface 2) are restricted from moving in the X-direction due to symmetry. All concrete nodes, profiled steel sheeting nodes, reinforcement bar nodes, steel beam flange nodes, steel beam web nodes, and headed stud nodes that lie on the other symmetry surface (surface 3) are restricted from moving in the Y-direction because of symmetry. Following the testing procedures conducted in [2.57–2.59], the load was applied in increments as static uniform load using the RIKS method available in the ABAQUS library. To model the nonlinear behavior of the concrete slab, the yielding part of the concrete stress–strain curve, which is the part after the proportional limit stress, is treated by the Drucker-Prager yield criterion model available in the ABAQUS [1.29] material library. The measured stress–strain curve by Ellobody [2.68] for the 19 × 100 mm headed stud, shown in Figure 5.34, was simulated to a bilinear stress–strain model. The stud material behaved as linear elastic material with Young's modulus (E_s) up to the yield stress of the stud (f_{ys}), and after this stage, it becomes fully plastic. The Young's modulus and yield stress of the stud are taken as 200 GPa and 470.8 MPa. The steel beam and profiled steel sheeting were modeled with yield stresses of 288 and 308 MPa and initial Young's modulus of 189 and 184 GPa, respectively, as measured by Kim *et al.* [2.58, 2.59] using the same bilinear curve as shown in Figure 5.34.

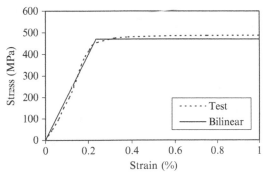

Figure 5.34 Measured and simulated stress-strain curves of stud presented by Ellobody [2.68].

The reinforcement bars were modeled with a yield stress of 460 MPa and initial Young's modulus of 200 GPa, as measured by Ellobody [2.68] using the same bilinear curve as shown in Figure 5.34.

The developed finite element models [2.71] were verified against the pushout tests [2.57–2.59]. The shear connection capacity per stud obtained from the tests (P_{Test}) and finite element analysis (P_{FE}) and the load-slip behavior of the headed shear stud and failure modes have been investigated. It was shown that good agreement has been achieved between both results for most of the pushout tests. A maximum difference of 7% was observed between experimental and numerical results. The experimental load-slip curve obtained for pushout test specimens [2.58, 2.59] was compared with the numerical curve obtained from the finite element analysis, as shown in Figure 5.35. Generally, good agreement has been achieved between experimental and numerical load-slip curves. It is shown that the finite element models successfully predicted the shear connection capacity and stiffness as well as load-slip behavior of the headed shear stud. The maximum load per stud recorded experimentally was 39.2 kN at a slip of 2.1 mm compared with 40.9 kN and 1.3 mm, respectively, obtained from the finite element analysis. The failure mode observed experimentally for pushout specimen [2.58, 2.59] was compared with that predicted numerically. The failure mode was a combination of concrete conical failure and stud shearing as observed experimentally and confirmed numerically. Figure 5.36 showed the stress contour at failure for pushout specimen tested in [2.58, 2.59] and modeled in [2.71]. It should be noted that the maximum stresses in

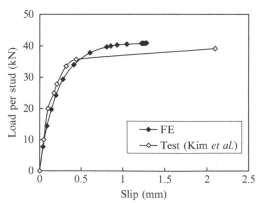

Figure 5.35 Load per stud versus slip for pushout specimen tested in [2.58, 2.59] and modeled in [2.71].

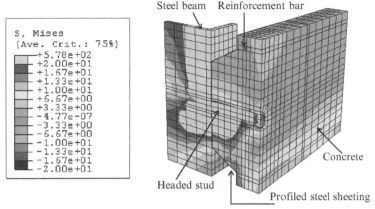

Figure 5.36 Stress contours at failure for pushout specimen tested in [2.58, 2.59] and modeled in [2.71]. (For the color version of this figure, the reader is referred to the online version of this chapter.)

the concrete are in the regions around the stud forming a conical shape. This conical failure mode of concrete was explained in detail both experimentally and numerically by Ellobody [2.68–2.70] for the investigation of pushout tests with solid slabs and precast hollow-core slabs. The conical concrete failure is also known as concrete pullout failure since the tensile force acting on the stud forces the slab to move up and leave a cone of concrete around the stud. The concrete conical failure (or concrete pullout) was also observed experimentally and discussed theoretically in the previous studies on pushout tests with profiled steel sheeting conducted in [2.57–2.59]. The concrete conical failure occurred, and the stud reached its yield stress near the collar. Figure 5.37 showed the deformed shape obtained from the finite element analysis for the 13 × 65 mm headed stud shear connector tested in [2.58, 2.59] and modeled in [2.71]. The verified finite element models [2.71] were used to perform extensive parametric studies. The finite element strengths were compared with design strengths calculated using current codes of practice.

The aforementioned finite element modeling of shear connection can provide a good insight into the local behavior of headed shear studs in the connection. However, to model a full-scale composite girder having many shear studs, we can benefit now from the special-purpose elements provided in ABAQUS [1.29] element library. The shear connectors can be modeled using a nonlinear spring element (using SPRING option). The spring element is of zero length that bears only shear force and obeys the load-slip characteristic of the shear connector used. The positions of

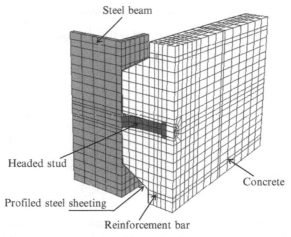

Figure 5.37 Deformed shape at failure for pushout specimen tested in [2.58, 2.59] and modeled in [2.71]. (For the color version of this figure, the reader is referred to the online version of this chapter.)

the spring elements coincide with the positions of the shear connectors used in the composite beam. Because the load–slip characteristic of the shear connector is nonlinear, the force is assumed to be a function of relative displacement in the spring and is defined by giving a table of force values in ascending values of relative displacement. The load–slip characteristic of the shear connector is obtained from the corresponding finite element modeling of the local shear connection. It should be noted that, for a given composite beam, loading, and design method, complete shear interaction is defined as the least number of the connectors such that the bending resistance of the beam would not be increased if more connectors were provided. Partial shear interaction is assumed when the number of connectors used in the composite beam is less than the number of shear connectors that cause full shear interaction. In the design with complete shear connection, it is normally assumed that the failure of shear connectors does not occur and the influence of connector deformation on the structural behavior was neglected. With partial shear connection, the ultimate resistance of the beam depends on the ultimate resistance of the shear connector and its ductility. In these cases, it is important to use the correct load–slip behavior of the connector since it can cause significant redistribution of stresses between the connectors in both serviceability and ultimate limit states.

5.9 APPLICATION OF LOADS AND BOUNDARY CONDITIONS ON THE BRIDGES

Chapter 3 has previously detailed different loads applied on steel and steel–concrete composite bridges as well as highlighted different supports of the bridges. Different loads applied on the bridges must be simulated accurately in finite element models. Any assumptions or simplifications in actual loads could affect the accuracy of results. The loads detailed in Chapter 3 comprised dead, live, wind, static, dynamic, thermal loads, etc., in addition to any loads that designers would like to check the safety of the bridges against them. However, most of these loads applied to steel and steel-concrete composite bridges are either *concentrated loads* or *distributed loads*. Concentrated forces and moments can be applied to any node in the finite element model. Concentrated forces and moments are incorporated in the finite element model by specifying nodes, associated degrees of freedom, magnitude, and direction of applied concentrated forces and moments. The concentrated forces and moments could be fixed in direction or alternatively can rotate as the node rotates. On the other hand, distributed loads can be prescribed on element faces to simulate surface distributed loads. The application of distributed loads must be incorporated in the finite element model very carefully using appropriate distributed load type that is suitable to each element type. Most computer programs specify different distributed load types associated with the different element types included in the software element library. For example, solid brick elements C3D8 can accept distributed loads on eight surfaces, while shell elements are commonly loaded in planes perpendicular to the shell element midsurface. Distributed loads can be defined as *element-based* or *surface-based*. Element-based distributed loads can be prescribed on element bodies, element surfaces, or element edges. The surface-based distributed loads can be prescribed directly on geometric surfaces or geometric edges.

Three types of distributed loads can be defined in ABAQUS [1.29], which are body, surface, and edge loads. Distributed body loads are always element-based. Distributed surface loads and distributed edge loads can be element-based or surface-based. Body loads, such as gravity, are applied as element-based loads. The unit of body forces is force per unit volume. Body forces can be specified on any elements in the global X-, Y-, or Z-direction. Also, body forces can be specified on axisymmetric elements in the radial or axial direction. General or shear surface tractions and pressure loads can be applied as element-based or surface-based distributed loads.

The unit of these loads is force per unit area. Distributed edge tractions (general, shear, normal, or transverse) and edge moments can be applied to shell elements as element-based or surface-based distributed loads. The unit of edge tractions is force per unit length. The unit of edge moments is torque per unit length. Distributed line loads can be applied to beam elements as element-based distributed loads. The unit of line loads is force per unit length. It should be noted that in some cases, distributed surface loads can be transferred to equivalent concentrated nodal loads and can provide reasonable accuracy provided that a fine mesh has been used.

The application of boundary conditions is very important in finite element modeling. The application must be identical to the actual situation in the investigated steel or steel-concrete composite bridges. Otherwise, the finite element model will never produce accurate results. Boundary conditions are used in finite element models to specify the values of all basic solution variables such as displacements and rotations at nodes. Boundary conditions can be given as model input data to define zero-valued boundary conditions and can be given as history input data to add, modify, or remove zero-valued or nonzero boundary conditions. Boundary conditions can be specified using either *direct format* or *type format*. The type format is a way of conveniently specifying common types of boundary conditions in stress/displacement analyses. Direct format must be used in all other analysis types. For both direct format and type format, the region of the model to which the boundary conditions apply and the degrees of freedom to be restrained must be specified. Boundary conditions prescribed as model data can be modified or removed during analysis steps. In the direct format, the degrees of freedom can be constrained directly in the finite element model by specifying the node number or node set and the degree of freedom to be constrained. As an example in ABAQUS [1.29], when modelers specify that (CORNER, 1), this means that the node set named (CORNER) is constrained to displace in direction 1 (u_x). While specifying that (CORNER, 1, 4), this means that the node set CORNER is constrained to displace in directions 1-4 (u_x, u_y, u_z, and θ_x). The type of boundary condition can be specified instead of degrees of freedom. As examples in ABAQUS [1.29], specifying "XSYMM" means symmetry about a plane X=constant, which implies that the degrees of freedom 1, 5, and 6 equal to 0. Similarly, specifying "YSYMM" means symmetry about a plane Y=constant, which implies that the degrees of freedom 2, 4, and 6 equal to 0, and specifying "ZSYMM" means symmetry about a plane Z=constant, which implies that the degrees of freedom 3, 4, and 5 equal to 0. Also, specifying

"ENCASTRE" means fully built-in (fixed case), which implies that the degrees of freedom 1, 2, 3, 4, 5, and 6 equal to 0. Finally, specifying "PINNED" means pin-ended case, which implies that the degrees of freedom 1, 2, and 3 equal to 0. It should be noted that once a degree of freedom has been constrained using a type boundary condition as model data, the constraint cannot be modified by using a boundary condition in direct format as model data. Also, a *displacement-type* boundary condition can be used to apply a prescribed displacement magnitude to a degree of freedom.

All boundary conditions related to the bridge must be carefully applied and checked that the model has not been overconstrained. Symmetry surfaces also require careful treatment to adjust the boundary conditions at the surface. It should be also noted that applying a boundary condition at a node to constrain this node from displacing or rotating will totally stop this node to displace or rotate. When the displacement or rotation is not completely constrained (partial constraint), springs must be used to apply the boundary conditions with constraint values depending on the stiffness related to the degrees of freedom. Different steel bearings briefed in Chapter 3 can be simulated by either roller or hinged boundary conditions, or they can be modeled using solid elements depending on the loads applied. Further details regarding the application of loads and boundary conditions in finite element modeling are found in [5.1].

REFERENCES

[5.1] E. Ellobody, F. Ran, B. Young, Finite Element Analysis and Design of Metal Structures, first ed., Elsevier, 2014, 224 pages, ISBN: 978-0-12-416561-8.

[5.2] E. Ellobody, D. Lam, Finite element analysis of steel-concrete composite girders, Adv. Struct. Eng. 6 (4) (2003) 267–281.

[5.3] W.H. Bowes, L.T. Russell, Stress Analysis by the Finite Element Method for Practicing Engineers, Lexington Books, Toronto, 1975.

[5.4] E. Ellobody, C.G. Bailey, Behaviour of unbonded post-tensioned concrete slabs, Adv. Struct. Eng. 11 (1) (2008) 107–120.

[5.5] A. Hillerborg, M. Modeer, P.E. Petersson, Analysis of crack formation and crack growth in concrete by means of fracture mechanics and finite elements, Cem. Concr. Res. 6 (1976) 773–782.

[5.6] J. Lubliner, J. Oliver, S. Oller, E. Oñate, A plastic-damage model for concrete, Int. J. Solids Struct. 25 (1989) 299–329.

[5.7] J. Lee, G.L. Fenves, Plastic-damage model for cyclic loading of concrete structures, J. Eng. Mech. 124 (8) (1998) 892–900.

[5.8] D.W. Murray, E.L. Wilson, Finite element large deflection analysis of plates, Am. Soc. Civ. Eng. 95 (1) (1969) 143–165.

[5.9] P.G. Bergan, R.W. Clough, Large deflection analysis of plates and shallow shells using the finite element method, Int. J. Numer. Meth. Eng. 5 (1973) 543–556.

[5.10] R.S. Srinivasan, W. Bobby, Nonlinear analysis of skew plates using the finite element method, Comp. Struct. 6 (1976) 199–202.

[5.11] J.H. Argyris, P.C. Dunne, D.W. Scharpf, On large displacement-small strain analysis of structures with rotational freedoms, Comput. Meth. Appl. Mech. Eng. 14 (1978) 401–451.

[5.12] K. Kondoh, K. Tanaka, S.N. Atluri, An explicit expression for the tangent-stiffness of a finitely deformed 3-D beam and its use in the analysis of space frames, Comp. Struct. 24 (1986) 253–271.

[5.13] Y.B. Yang, W. McGuire, Joint rotation and geometric nonlinear analysis, J. Struct. Eng. ASCE 112 (1986) 879–905.

[5.14] J.L. Meek, S. Loganathan, Geometrically non-linear behaviour of space frame structures, Comp. Struct. 31 (1989) 35–45.

[5.15] K.S. Surana, R.M. Sorem, Geometrically non-linear formulation for three dimensional curved beam elements with large rotations, Int. J. Numer. Meth. Eng. 28 (1989) 43–73.

CHAPTER 6

Examples of Finite Element Models of Steel Bridges

6.1 GENERAL REMARKS

Chapters 1–4 have provided the necessary background regarding the general layout, loading, and design of steel and steel-concrete composite bridges. On the other hand, Chapter 5 has provided the required background regarding the issues related to finite element modeling of the bridges. Therefore, it is now possible in this chapter to present examples of finite element models for steel bridges based on the background of finite element analysis detailed in Chapter 5. This chapter presents illustrative examples of finite element models developed to understand the structural behavior of steel bridges. The chapter starts by a brief introduction and a review of recent investigations reported in the literature concerning the modeling of steel bridges. The chapter details how the finite element models were developed and the results obtained. The presented examples show the effectiveness of finite element models in providing detailed data that complement experimental data in the field. The results are discussed to show the significance of the finite element models in predicting the structural response of the different bridges investigated. In overall, it is aimed to show that finite element analysis not only can assess the accuracy of the design rules specified in current codes of practice but also can improve and propose more accurate design rules. The author hopes that the review of recent finite element models reported in the literature together with the illustrative finite element models developed by the author in this chapter can provide readers with a complete piece of work regarding the finite element analysis of steel bridges.

6.2 PREVIOUS WORK

Numerous numerical investigations were reported in the literature highlighting the structural performance of different types of steel bridges subjected to different loadings. The numerical investigations proposed finite element models for the bridges and the bridge components. It should be noted that detailed state-of-the-art review of these investigations is out of the scope of this book. However, in this section, the author provides recent

555

examples showing how other researchers modeled the steel bridges and the bridge components. Earls and Shah [6.1] presented a combined experimental and numerical investigation on high-performance steel I-shaped bridge girders. The investigations were assessed against the American bridge specification (AASHTO) provisions for cross-sectional compactness and adequate bracing. The study showed that the specifications may be inadequate owing to intense interactions between local and global buckling modes in the high-performance steel I-shaped bridge girders. An alternate bracing requirement was proposed by the authors for use with high-performance steel bridge girders. The proposed bracing scenario did not require any additional costs in fabrication or materials. The authors used ABAQUS [1.29] in the numerical investigations, which considered both geometric and material nonlinearities using the modified Riks method. The authors performed an eigenvalue buckling to predict the first buckling mode, which was factored by an assumed maximum initial displacement of the girder (L/1000). The models of the bridge girders considered in this study are constructed from a dense mesh of four-node nonlinear shell finite elements (S4R). The loading in the finite element modeling was imposed as a concentrated load at the midspan of a simply supported (SS) I-shaped girder assembly. The concentrated force simulates the pier reaction of the investigated bridge and the simple supports are placed at the approximate points of inflection on either side of the pier. In the finite element models, an additional length of girder was present beyond the support locations to help simulate the torsional-warping restraint provided by the adjacent beam segments in the actual bridge. The length of the additional beam segments was chosen to be 7.625 m, which corresponds to the distance to the next diaphragm member occurring after the point of inflection, as measured along the longitudinal axis of the girder. The top compression flange had a width of 406 mm and a thickness of 45 mm thickness. The bottom tension flange had the same width but with a modified thickness of 84 mm. A uniaxial representation of the A709 steel grade was used.

Shanmugam *et al.* [6.2] presented a combined experimental and numerical study on the ultimate load behavior of plate girders curved in plan. The investigated girders were medium-sized girders built using rolled steel plates and were tested to failure. The girders were supported at the ends and subjected to a concentrated load applied at the midspan. The behavior of web panels was closely studied in order to investigate the tension-field action. The numerical investigation employed the elastoplastic finite element method and the results were compared with that measured experimentally.

The study indicated that the load-carrying capacity decreases with the increase in curvature. The authors developed a 3D finite element model using ABAQUS [1.29]. Eight-node doubly curved thin shell element with reduced integration points using 5 degrees of freedom per node (S8R5) was used. Riks method in conjunction with the modified Newton-Raphson method was employed. Residual stresses were not considered in the analysis. The authors mentioned that although it is understood [6.3, 6.4] that the shell elements provided by ABAQUS at the plate midthickness could not pick up Saint Venant torsional stresses, this effect was not of main concern in the study since the focus was only on ultimate load-carrying capacity. The secondary girders are represented in terms of appropriate boundary conditions at the support. In the same way, the effect of tie rods used in the experiments to prevent the lateral buckling of the girder at the midspan was taken into account by adopting relevant boundary conditions at the midspan. Geometric imperfections were imposed in terms of the buckled shape of the web plates at the elastic stage. The buckled shape was obtained from ABAQUS analyses in which the girder was loaded without any imperfection. The lateral displacements and the buckling mode thus obtained at the elastic stage were imposed in the final analyses of the girder. Convergence studies were performed to determine the suitable finite element model for the analysis. Three different meshes with 552, 1152, and 1506 elements were considered in the studies. The difference between the ultimate strengths corresponding to models with 552 elements and 1152 elements was about 9% and that between the values corresponding to the models with 1152 and 1506 elements was around 1.8%. Therefore, finite element analysis based on 1152 elements was adopted in the finite element modeling for all the girders curved in plan.

Floor beams of orthotropic plated bridge decks were investigated by Corte and Bogaert [6.5]. The beams have generally elements with a low slenderness, especially in the case of railway bridges. This is attributed to combined flexural and shear deformations. The shear deformations can be considerably large to be neglected. The authors discussed that in a design according to the Pelikan-Esslinger method [6.6, 6.7], this deformation is taken into account in the second stage of the calculation of the orthotropic deck. At this stage, the additional bending moments, shear forces, and floor beam reactions due to the floor beam flexibility are evaluated. The deflection of a directly loaded floor beam creates a distribution of the load to adjacent nonloaded floor beams. In addition, the deflection will affect the longitudinal ribs, increasing the sagging moments at midspan and decreasing the bending moments at the supports of the ribs provided by the floor beams. In the study [6.5], the authors proposed a method accounting for the shear deformation in the floor beam. The validity of the proposed method was checked by full finite element calculation using shell elements that inherently comprise shear deformation. The authors developed a finite element model using four-node Mindlin shell elements. The floor beam model was subjected to a 100 kN concentrated load at midspan. Felkel *et al.* [6.8] evaluated the behavior of bridge girders made of high-performance steel (HPS 70W). The basis for the study was a three-span replacement bridge utilizing HPS 70W girders within all negative moment regions. The study consisted of in situ measurements, experimental tests, and analytic investigations. Three half-scale specimens were tested under monotonic and cyclic loading conditions. The study [6.8] presented results from analytic and experimental investigations highlighting the strength performance of the girders. Data obtained from laboratory tests were used to validate computer models for design evaluations. Parametric studies were performed using the models. The findings of the study indicated that improved structural performance may be obtained when location of bracing was optimized and fabrication imperfection tolerances were minimized. The measured nonlinear material model was adopted in the finite element analysis. Small fabrication and geometric imperfections within tolerances observed in the tests were not initially simulated. Subsequently, imperfections were simulated by applying a small lateral pressure along the entire length of the compression flange.

Galvin and Dominguez [6.9] presented a theoretical and experimental research work on a cable-stayed bridge. Full-scale tests were carried out to measure the bridge dynamic response. The experimental program included the dynamic study for two different live load conditions: the bridge with one-half of its lanes loaded with cars and the bridge unoccupied. Modal parameter estimations were made based on the acquired data. Ten vibration modes were identified in the frequency range of 0-6 Hz by different techniques, two of these modes being very close to each other. The traffic-structure interaction was also studied. Experimental results were compared with those obtained from a 3D finite element model developed in this work. The authors applied a damage identification technique to determine the integrity of the structure. The developed finite element model was a 3D model developed for the numerical analysis of the structure using as-built drawings of the bridge and some double-check in situ measurements. Modal analysis was carried out using ANSYS [6.10]. The arch, supports, and the internal stiffener were represented as two-node beam elements (BEAM44) with 6 degrees of freedom per node. The element permitted the end nodes to be offset from the centroidal axis of the beam. The hangers were modeled as truss elements (LINK10) with 3 degrees of freedom per node. The deck slab was modeled using eight-node shell elements with 6 degrees of freedom per node (SHELL93). The two extreme beams and the vertical supports were connected by spring elements (COMBINE14). A detailed model of all the bridge elements was intended, which resulted in higher number of degrees of freedom. The full model consisted of 10328 beam elements, 17 truss elements, 15672 shell elements, and 8 spring elements, resulting in 26025 elements and 47024 nodes.

Romeijn and Bouras [6.11] developed a finite element model of a tension-tie arch bridge to investigate the in-plane buckling length factor of the arches. The finite element results were compared with the corresponding values given by Eurocode 3. The modeling of the tension-tie arch bridge, the bridges' properties, and the solution procedure were described. Parametric analyses were performed by the authors. The case of one cable missing (broken cable) was also investigated. The finite element software Nastran was used to develop the model. Three different bridge geometries are modeled comprising a bridge with a length of 300 m and a height of 45 m, a bridge with a length of 300 m and a height of 30 m, and a bridge with a length of 300 m and a height of 60 m. For each of the geometries, five different cable configurations were modeled with the number of cables in each arch being $m = 1, 2, 3, 5$, and 11. The center-to-center distance of the crossbeams in the deck was 25 m, while the width of the deck is 30 m. Two different arch inclinations were used comprising the case where the inclination of the arches is 12.5° with the vertical direction and a vertical arch configuration. The different cross sections used were cross section (a), which was the cross section of the main girder having a rectangular box section with a height of 2300 mm, a width of 3000 mm, and a thickness of 25 mm; cross section (b), which was a double-symmetrical I-beam with a height of 2300 mm, flanges 625 mm wide and 25 mm thick, and a web thickness of 20 mm and was used for the crossbeams in the deck; cross section (c), the cross section of the arch, which was a rectangular box section with a height of 4000 mm, a width of 3000 mm, and a thickness of 60 mm; and finally, cross section (d), which was a circular hollow section with a radius of 1000 mm and a thickness of 25 mm. All the cross sections were modeled using beam elements. The cables of the bridge are modeled using rod elements. The rods being used had a circular cross section with a diameter of 120 mm. The concrete deck of the bridge was modeled using plate elements with a thickness of 400 mm. Three different materials have been defined for steel, cables, and concrete. The connections between the deck crossbeams and the main girder, the arch crossbeams and the arch, and the arch and the main girder were assumed rigid. The four corner nodes of the deck were the nodes where the constraints are applied. On one side, only the in-plane rotation is permitted, while on the other side, both the in-plane rotation and the longitudinal translation were allowed.

Eldib [6.12] presented the shear buckling strength and design of curved corrugated steel webs for bridges considering material inelasticity. A finite element analysis was performed to study the geometric parameters affecting the shear buckling strength of curved corrugated steel webs for bridges. Based on the numerical results, a shear buckling parameter formula was proposed. The author presented another formula presented to maximize the shear buckling capacity of curved corrugated web. The proposed formulas agreed well with the published experimental data. It was shown that the curved corrugated webs produced a tremendous increase in the shear buckling strength and considerable weight saving in regard to the corresponding trapezoidal corrugated webs. The corrugation angle had a considerable effect on the behavior of curved corrugated webs, where higher corrugation angles produced a tremendous increase in the shear buckling strength of curved corrugated webs. It was found that the proposed approach provided a good prediction for the shear buckling strength of curved corrugated steel webs of bridges. The general purpose software ANSYS was used in the analysis. The shell element (Shell 63) was used to model the steel web. The finite element has both bending and membrane capabilities. Both in-plane and normal loads were permitted. The element has 6 degrees of freedom at each node. Stress stiffening and large deflection capabilities were included in the element. A mesh sensitivity analysis was performed and six elements per panel were used. A linear elastic buckling analysis was carried out using the models.

Zhang et al. [6.13] investigated a new type of streamlined girder bridge with a thin-walled steel box girder. This bridge had a large width-to-span ratio, which resulted in significant shear lag effects and causes nonuniform stress distribution in the three-cell thin-walled box girder, especially along the flanges of the girder. The authors investigated the effect of shear lag in thin-walled box girder bridges with large width-to-span ratios through both experimental and numerical investigations. A large-scale model was tested under different loading cases. The material parameters were obtained from physical characteristics tests and tensile tests. In addition, a computational model was presented for a comprehensive simulation of a girder bridge including the orthotropic top/bottom/web plates and their ribs. The authors concluded that the finite element analysis can be an effective method to predict properties of this class of bridges. In order to reflect the actual behaviors of the bridge decks accurately, an advanced parametric design language (APDL), in the commercially available software ANSYS [6.10], was used to perform a 3D analysis of the orthotropic steel box girder. The FE model for the steel box girder was built using shell elements (Shell 63). In the 3D bridge model, the total number of shells is 22,496 with a total of 22,574 joints. The authors mentioned that if modeling a larger and more complex bridge is required, it would be difficult to set up and compute an entire shell model; the recommended method would be a combination of a simple entire model with high-fidelity finite elements in the specific local section. Different mesh densities resulted in different computational accuracies to some extent.

Graciano *et al.* [6.14] studied the influence of initial geometric imperfections on the postbuckling behavior of longitudinally stiffened plate girder webs subjected to patch loading. The authors mentioned that upon recognizing the significance of geometric imperfections, a large amount of research has been conducted to develop models of characteristic imperfections for specific structures and then using these models to gain a better estimate of the ultimate load [6.15, 6.16]. Graciano *et al.* [6.14] performed a sensitivity analysis using two approaches (deterministic and probabilistic) in order to investigate the effect of varying imperfections in shape and amplitude on both, the postbuckling response, and ultimate strength of plate girders under patch loading. The sensitivity analysis was performed by means of nonlinear finite element analysis. At first, the initial shape imperfections are modeled using the buckling mode shapes resulting from an eigenvalue buckling analysis. Following the eigenvalue buckling analysis, the amplitude of the buckling shapes for the various modes was factored and then introduced in the nonlinear analysis. The results showed the influence of these modes and amplitudes on the resistance to patch loading. The finite element software ANSYS [6.10] was used. Shell elements (Shell 181) having 4 nodes and 6 degrees of freedom at each node were used to model the web, flanges (top and bottom), and the longitudinal stiffener. Due to symmetry in geometry, loads, and boundary conditions, only one-half of the plate girder was modeled. Transverse stiffeners at the end of the plate girder were taken into account by means of a rigid body kinematical constraint of the degrees of freedom located in the corresponding side. The material herein was considered to have a perfect elastoplastic behavior. Displacement constraints were applied to these loaded nodes in the out-of-plane direction and all rotations were restrained. The finite element analysis was performed using the modified Riks method to properly trace the nonlinear path of the load-displacement response of the girder. In order to model the initial geometric imperfections, the authors performed a linear eigenvalue buckling analysis.

Dynamic and seismic assessment of a double-track railway bridge with four discrete spans located in an earthquake-prone region was presented by Caglayan et al. [6.17]. A 3D computer model of the bridge was generated using a commercial general finite element analysis software COSMOS/M [6.18]. Field measurements such as static and dynamic tests and material tests were conducted on the bridge. The developed 3D finite element model of the bridge structure was used for necessary calculations regarding structural assessment and evaluation according to train loads and seismic loads. Additional members were proposed to transmit seismic loads to supports. The fourth span, which had a permanent imperfection due to truck collision, was studied in detail. The authors considered significant structural irregularities and stiffness changes that existed on the bridge in the finite element. A single span was modeled with beam and spring elements using COSMOS/M [6.18]. Two riveted plate girders functioning as the main girder of each span were simulated with beam elements located on the centroid line and having the same torsional and flexural rigidities of actual main girder. Floor beams were modeled by using beam elements in the transverse direction and rigid bars were used to simulate center of gravity for floor beams and main girders. Since the connections between the rigid bars and crossbeams were semirigid rather than fixed, springs were used to simulate joint rotational rigidities. Connections between the rigid bars and all of the line elements were free to rotate in the longitudinal vertical plane. Also, the connections between the stringers and crossbeams behave semirigidly; therefore, rotational spring elements are used to simulate the rotational rigidities. Vertical, transverse, and longitudinal spring elements for each main girder support were included in the model to simulate the effective stiffness of the combined bearing and pier or abutment structure and the longitudinal restraints at the sliding bearings in each direction.

Altunişik et al. [6.19] presented finite element modeling and operational modal analysis of a full-scale arch-type steel highway bridges. The numerical investigation was performed on a highway bridge, which has arch-type structural system with a total length of 336 m. The 3D finite element model was constructed using project drawings and an analytic modal analysis.

The model was used to generate natural frequencies and mode shapes in the three-orthogonal directions. Ambient vibration tests on the bridge deck under natural excitation such as traffic, human walking, and wind loads were conducted using operational modal analysis. Sensitive seismic accelerometers are used to collect signals obtained from the experimental tests. To obtain experimental dynamic characteristics, two output-only system identification methods were employed, which were enhanced frequency domain decomposition method in the frequency domain and stochastic subspace identification method in time domain. The authors found good agreement between dynamic characteristics in all measurement test setups performed on the bridge deck. It was demonstrated that the ambient vibration measurements using enhanced frequency domain decomposition and stochastic subspace identification methods were enough to identify the most significant modes of steel highway bridges. It was also shown that there were some differences between analytic and experimental natural frequencies, with experimental natural frequencies generally bigger than the analytic frequencies. A 3D finite element model of the bridge was constructed using the SAP2000 software [6.20]. The program can be used for linear and nonlinear, static, and dynamic analyses of 3D models of structures. The program is used to determine the analytic dynamic characteristics based on its physical and mechanical properties. The selected highway bridge was modeled as a space frame structure with 3D prismatic beam elements, which have two end nodes with each end node having 6 degrees of freedom (three translations along the global axes and three rotations about its axes). The key modeling assumptions were as follows:

(1) In the finite element model of the bridge, the fictitious elements were used to determine the torsional and moment effects that are composed of asymmetrical load cases. These elements were massless defined on the axis through the gravity center of uniform and linear loads.

(2) In the finite element model of the bridge deck, diagonal fictitious elements were used to reflect the rigid diaphragm effect of the concrete.

(3) Fictitious elements were modeled as two ends hinged and one end axial sliding.

(4) Rigid link elements were modeled as two higher bending rigidity ends to ensure the torsional moments in the carrier system elements. To determine the length of the rigid element, it was assumed that fictitious elements were located near the gravity center of the loads. For the deck-type arch bridge, the boundary conditions of the side columns connected between the arch and the main girder were fixed in order

to transmit the longitudinal load on the deck. A total of six natural analytic frequencies of the highway bridge were obtained.

Kaliyaperumal *et al.* [6.21] presented finite element analyses for dynamic analysis of steel railway bridges. The analyses of a skew bridge were performed and the results were compared with available field measurements. Initially, eigenvalue analyses of different models were performed in order to obtain the fundamental mode shapes and bridge frequencies and to assess the capability of each model to capture the dynamic behavior of the bridge. Single-span, three-span, and full bridge models were investigated with different elements such as shell, beam, and combinations of these elements. The authors found good agreement between the fundamental dynamic properties of the bridge and empirical results. Following the eigenvalue analyses, time history dynamic analyses were carried out using the full bridge model. The analyses were performed for different train speeds. It was shown that modeling the full bridge using a combination of beam and shell elements was reasonable and computationally efficient in capturing the dynamic behavior of a bridge and estimating the mean stress range for fatigue damage calculations. The finite element models of the bridge were developed using ABAQUS [1.29]. Models with different degrees of complexity, using shell and/or beam elements, were developed in order to investigate the effect of different modeling techniques and computational time on the dynamic behavior of the bridge. Eight-node reduced integration shell elements (S8R) and three-node quadratic beam elements (B32) were used in the FE models. Single-span, three-span, and six-span (full) bridge FE models were developed and analyzed. Both SS and fixed support conditions were assumed in the single-span and three-span models at the two ends of the bridge in order to investigate the effect of boundary conditions. All members were tied to each other, which is equivalent to assuming rigid connections between them. The effect of bracings was included via the single-span FE model by developing a shell model with and a model without bracings. In all the shell element models, the stiffeners in the main girders were also modeled. The intermediate supports were modeled as SS. Eigenvalue analyses were performed for all finite element models of the bridge and the results, in the form of bridge periods (frequencies). The bridge frequencies obtained from the finite element analysis were compared with available empirical formulas suggested by Fryba [6.22] and the International Union of Railways [6.23]. Following the eigenvalue analysis, static and dynamic finite element analyses were carried out to investigate the overall dynamic behavior of the bridge. Two different types of dynamic analyses, that is, modal dynamic and implicit dynamic, were undertaken to investigate the suitability of each to capture the dynamic behavior of the bridge. Explicit dynamic analysis was computationally much more demanding than implicit analyses, and due to the large nature of the finite element model, this type of analysis was excluded from this investigation. A range of different train velocities were employed in the analyses and the results were compared with the available field measurements. The bridge was loaded with the test loco-motive and the axle loads of the train (195 kN) were applied directly to the top flange of the stringers ignoring any load distribution due to the effect of rails and sleepers.

Caglayan et al. [6.24] carried out a series of dynamic tests, acceleration measurements, evaluation, finite element model simulations, and safety index calculations on existing steel railway bridges. Dynamic tests were ful-filled by using a special test train on these bridges to obtain the dynamic parameters, and these parameters were then used to refine the finite element models of the bridges. The updated models were used to represent the actual condition, and safety indexes were calculated for structural components of the bridges for each loading condition. The safety indexes were used to cal-culate failure probabilities of structural members. In addition, the authors performed system reliability of the bridges based on proposed system models of the bridges. It was shown that the study can provide a reliable background for proposed heavier axle loads resulting from new freight trains by realizing the current condition of bridge structures. In employing modal identifica-tion procedures, the authors identified first vibrational mode. Also, in order to define modal parameters, after having preprocessed the collected acceler-ation data using the fast Fourier transform technique, acceleration spectra were obtained for the bridge. The results of the fast Fourier transform anal-ysis and modal identification were used to calibrate the computer models of the bridge. A bridge was modeled with beam elements using the general purpose finite element analysis program COSMOS/M [6.18]. The connec-tions between members were defined by using rotational spring elements to simulate the rotational rigidities. Additionally, the supports were modeled using spring elements and gusset plates were simulated using 1D rigid bar elements. Thai and Choi [6.25] presented a numerical investigation consid-ering both geometric and material nonlinearities for predicting the ultimate strength and behavior of multispan suspension bridges. The geometric non-linearities of the cable members due to sag effects were considered using the catenary element, while the geometric nonlinearities of the beam-column members due to second-order effects were considered using the stability functions. The material nonlinearities of the cable and beam-column members were simulated using elastoplastic hinge and refined plastic hinge models, respectively. A simple initial shape analysis method was presented to determine the deformed shape and initial cable tension of the bridge under dead loads. In addition, the authors presented numerical examples to verify the accuracy and efficiency of the proposed method. Furthermore, a case study on a four-span suspension bridge was carried out to show the capability of the proposed method in estimating the strength and behavior of very large-scale structures.

Recently, Lin et al. [6.26] showed that, due to the increasing aging problems of old railway bridges, structural repair or maintenance technique has been the subject of recent investigations. Rubber-latex mortar, glass fiber-reinforced polymer plates, and rapid hardening concrete can be integrated with the old steel railway bridge to increase its rigidity and reduce both stress levels and structure-borne sound levels of the old steel bridge. The study [6.26] investigated the mechanical performance of the renovated hybrid railway bridge. Material tests on aged structural steel, static loading test on the strengthened bridge, and impact hammer test on the old bridge before and after strengthening were conducted to confirm the effects of present strengthening method. In addition, 3D finite element models were developed to compare between the strengthened and the original steel bridge. It was shown that both experimental and numerical results indicated that the renovation method can greatly enhance the stiffness and reduce the stress levels of steel members, resulting in the extension of the service life of the old steel railway bridge. Furthermore, noise reduction effects by using concrete and rubber-latex mortar were confirmed in the impact test. The finite element modeling of test specimen was carried out in 3D. Three models were developed, solid elements (eight nodes, with 3° at each node) were used to simulate the concrete slab, and shell elements (four nodes, with 5° at each node) were employed to model the steel girder and GFRP plates. Rebar elements (two nodes, with 1° at each node) were used for modeling the reinforcing bars in the concrete slab. Also, in order to account for the slip between concrete slab and longitudinal steel beam, interface elements (eight nodes, with 3° at each node) were employed. The thickness of the interface element was assumed as zero in the numerical analysis. Numerical model of the old steel bridge was built and named as model 1. Cementing agent was not only used between glass fiber-reinforced polymer plate and longitudinal steel girder but also applied between neighboring GFRP plates. In the numerical analysis, perfect bond was assumed between steel girder and

GFRP plate. However, as bond failure between glass fiber-reinforced polymer plates was observed in the experiment but appropriate data about the failure bond stress of the cementing agent were not available, so two numerical models were built on the basis of different assumptions between GFRP plates. Perfect bond between GFRP plates was assumed for model 2 and perfect separation was assumed for model 3. In addition, the authors applied a phase study, in which the first phase was about the dead load and second phase was about the applied load in the experiment. Glass fiber-reinforced polymer was simulated as linear elastic until failure. The elastic module was taken as constant of 18.3 GPa according to material test. In order to account for the effect of rubber-latex and the composite action between steel and concrete, interface element was used in the numerical analysis. The shear force-slip response of interface depended on the slab-steel surface treatment.

6.3 FINITE ELEMENT MODELING AND RESULTS OF EXAMPLE 1

The first example presented in this chapter is for a small-scale built-up I-section plate girder steel bridge tested under bending by Nakamura and Narita [6.27]. The plate girder was a part of an experimental program investigating bending and shear strength of steel and partially encased steel-concrete composite plate I-girders. The plate girder is denoted in this study as (T1). The small-scale plate girder was SS and had a length between supports of 3.6 m as shown in Figure 6.1. The web of plate girder T1 was 900 mm high and 6 mm thick, while the flange of the plate girder was 200 mm wide and 12 mm thick. The web was stiffened by steel plates at the end supports and the loading positions and also stiffened by intermediate stiffeners at intervals of 375 mm. The plate girder was restrained laterally at the end supports to resist lateral-torsional buckling. Tensile coupon tests were conducted to determine the yield and ultimate tensile strength of the steel used, which were 372.3 and 511.4 MPa, respectively. The plate girders were loaded at two points as shown in Figure 6.1 subjecting the plate girder to a pure bending moment zone at midspan with a length of 600 mm. Strain gauges were used to measure the strains in the plate girder section, and displacement transducers were used to measure the midspan deflections during loading. The strain measurements showed that the upper flange was in compression, the lower flange was in tension, and the strain at the web center was nearly zero, indicating that the neutral axis was at the web center. When the strain of the upper flange reached 1000 microstrain, the upper flange

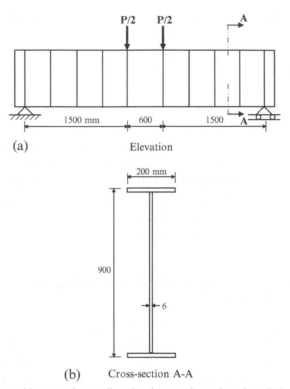

Figure 6.1 General layout of a small-scale plate girder in bending (T1).

started to buckle. The yield stress estimated from the test T1 was 1809 micro-strain, which corresponded to a yield load of 1474 kN. The maximum applied load in the bending test was nearly equal to this yield load. This means that the bending strength of T1 model was almost the same as the yield moment. The load–midspan deflection relationship, load–strain relation-ships, and the deformed shape at failure were observed in the test. The upper flange buckled between the two loading points. The web was also deformed outward so that the web and the flange remained perpendicular. This defor-mation shows a typical torsional buckling shape of the plate girders.

To model the small-scale plate girder (T1) tested by Nakamura and Narita [6.27], the finite element program ABAQUS [1.29] was used. The model has accounted for the measured geometry, initial geometric imper-fections, and measured material properties of the plate girder. Finite element analysis for bucking requires two types of analyses. The first is known as eigenvalue analysis that estimates the buckling modes and loads. Such anal-ysis is linear elastic analysis performed with the load applied within the step.

The buckling analysis provides the factor by which the load must be multiplied to reach the buckling load. For practical purposes, only the lowest buckling mode predicted from the eigenvalue analysis is used. The second is called load-displacement nonlinear analysis and follows the eigenvalue prediction. It is necessary to consider whether the postbuckling response is stable or unstable. The nonlinear material properties and loading conditions are incorporated in the load-displacement nonlinear analysis.

A four-node doubly curved shell element with reduced integration (S4R) was used to model the flanges and web of the small-scale plate girder bridge, as shown in Figure 6.2. The elements are suitable for complex buckling behavior (please refer to Section 5.2 of Chapter 5). The S4R element has 6 degrees of freedom per node and provides accurate solutions to most applications, allows for transverse shear deformation that is important in simulating thick shell elements (thickness is more than about 1/15 of the characteristic length of the shell), allows for the freedom in dealing with further parametric studies on slender and compact sections, and also accounts for finite strain and suitable for large strain analysis as recommended by ABAQUS [1.29]. Since lateral buckling of thin-walled plate girders is very sensitive to large strains, the S4R element was used in this study to ensure the accuracy of the results. In order to choose the finite element mesh that provides accurate results with minimum computational time, convergence studies were conducted. It is found that approximately 75×76 mm

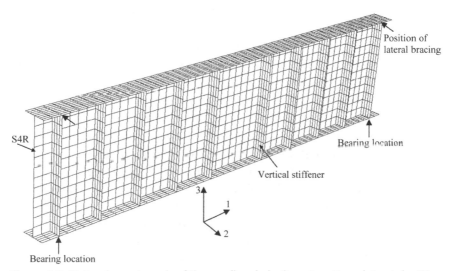

Figure 6.2 Finite element mesh of the small-scale built-up I-section plate girder T1.

(length by width of S4R element) ratio provides adequate accuracy in modeling the web, while a finer mesh of approximately 25×75 mm was used in the flange (see Figure 6.2).

The hinged support of T1, shown in Figure 6.2, was prevented from displacement in the horizontal direction (direction 1-1 in Figure 6.2) and the vertical direction (direction 3-3 in Figure 6.2). On the other hand, the roller support of T1, shown in Figure 6.2, was prevented from displacement in the vertical direction only (direction 3-3 in Figure 6.2). To account for the lateral restraints of the compression flange, the top compression flange was prevented from lateral displacements, in direction 2-2 of Figure 6.2, at the end supports, which is identical to the test T1. The load was applied in increments as concentrated static load, which is also identical to the experimental investigation. The nonlinear geometry was included to deal with the large displacement analysis.

The stress–strain curve for the structural steel given in the EC3 [2.11] was adopted in this study with measured values of the yield stress (f_{ys}) and ultimate stress (f_{us}) used in the tests [6.27]. The material behavior provided by ABAQUS [1.29] (using the PLASTIC option) allows a nonlinear stress-strain curve to be used (see Section 5.4.2 of Chapter 5). The first part of the nonlinear curve represents the elastic part up to the proportional limit stress with Young's modulus of (E) 200 GPa and Poisson's ratio of 0.3 used in the finite element model. Since the buckling analysis involves large inelastic strains, the nominal (engineering) static stress-strain curves were converted to true stress and logarithmic plastic true strain curves as detailed in Section 5.4.2.

Previous investigations by the author have successfully modeled the initial geometric imperfections in steel beams [6.28, 6.29]. Buckling of steel beams depends on the lateral restraint conditions to compression flange and geometry of the beams. Mainly two buckling modes detailed in [6.30, 6.31] could be identified as unrestrained and restrained lateral-distortional buckling modes. The lateral–distortional buckling modes were successfully predicted by the author [6.28, 6.29] by performing eigenvalue buckling analysis (see Section 5.5.2 of Chapter 5) for the investigated steel beams with actual geometry and actual lateral restraint conditions to the compression flange. The same approach [6.28, 6.29] was followed in this book to model initial geometric imperfections of the plate girder investigated T1. Figure 6.3 shows the buckling mode predicted from the eigenvalue buckling analysis detailed in ABAQUS [1.29]. Only the first buckling mode (eigenmode 1) is used in the eigenvalue analysis. Since buckling modes predicted by

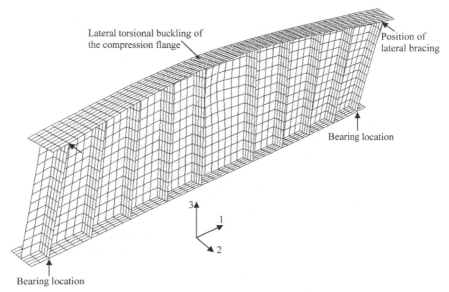

Lateral torsional buckling of
the compression flange

Position of
lateral bracing

Bearing location

Bearing location

Figure 6.3 Elastic lateral buckling mode (eigenmode 1) for the small-scale built-up I-section plate girder T1.

ABAQUS eigenvalue analysis [1.29] are generalized to 1.0, the buckling modes are factored by a magnitude of $L_u/1000$, where L_u is the length between points of effective bracing. The magnitude of $L_u/1000$ is the average measured values in the tests [6.31] and commended in [6.32]. The factored buckling mode is inserted into the load–displacement nonlinear analysis of the plate girder T1 following the eigenvalue prediction. It should be noted that the investigation of plate girders with different slenderness ratios could result in lateral-torsional buckling mode with or without web distortional buckling mode. Hence, to ensure that the correct buckling mode is incorporated in the nonlinear displacement analysis, the eigenvalue buckling analysis must be performed for each plate girder with actual geometry.

The developed finite element model for the plate girder T1 (see Figure 6.2) was verified against the test results detailed in [6.27]. The failure loads, failure modes, and load–midspan deflection curves obtained experimentally and numerically using the finite element model were compared. The deformed shapes of plate girder T1 at failure observed experimentally and numerically were compared as shown in Figure 6.4. It can be seen that the experimental and numerical deformed shapes are in good agreement. The failure mode observed experimentally and confirmed numerically was steel yielding. The data obtained from ABAQUS [1.29] have shown that

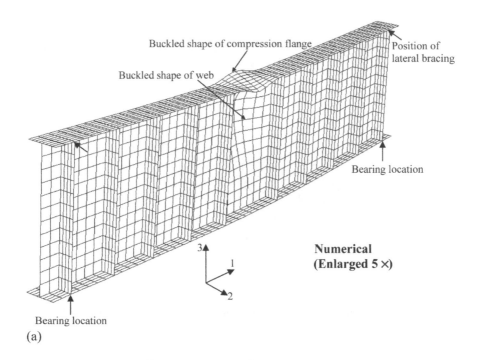

Buckled shape of compression flange

Buckled shape of web

Position of
lateral bracing

Bearing location

Numerical
(Enlarged 5 ×)

Bearing location

(a)

Experimental

(b)

Figure 6.4 Comparison of experimental and numerical deformed shapes at failure of the small-scale built-up I-section plate girder T1.

the von Mises stresses at the maximum stressed fibers at the top and bottom flanges at midspan exceeded the measured yield stresses. In Figure 6.5, the stress (principal stresses in direction 1-1) contours at failure of the small-scale built-up I-section plate girder T1 are plotted. It can be seen that the yield stresses were reached at midspan in the upper (compressive stresses with negative sign) and lower flanges (tensile stresses with positive sign). In addition, in Figure 6.6, the plastic strain (principal strains in direction 1-1) contours at failure of the small-scale built-up I-section plate girder T1 are plotted. It can

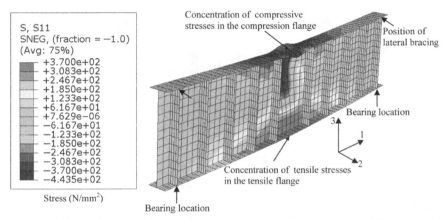

Figure 6.5 Stress (principal in direction 1-1) contours at failure of the small-scale built-up I-section plate girder T1 (enlarged 5 ×). (For the color version of this figure, the reader is referred to the online version of this chapter.)

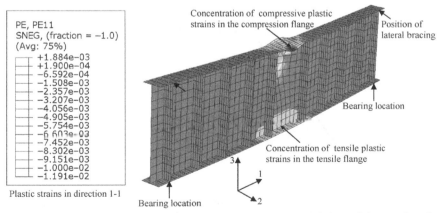

Figure 6.6 Plastic strain (principal in direction 1-1) contours at failure of the small-scale built-up I-section plate girder T1 (enlarged 5 ×). (For the color version of this figure, the reader is referred to the online version of this chapter.)

be seen that the plastic strains were concentrated at midspan in the upper (compressive strains with negative sign) and lower flange (tensile strains with positive sign). Furthermore, in Figure 6.7, the von Mises yield stress contours at failure of the small-scale built-up I-section plate girder T1 are plotted. It can be seen that the yield stresses were reached at midspan in the upper and lower flanges. The load-midspan deflection curves predicted experimentally and numerically were also compared as shown in Figure 6.8. It can

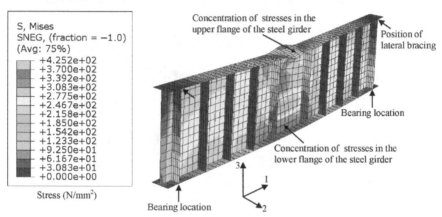

Figure 6.7 Stress (von Mises) contours at failure of the small-scale built-up I-section plate girder T1 (enlarged 5×). (For the color version of this figure, the reader is referred to the online version of this chapter.)

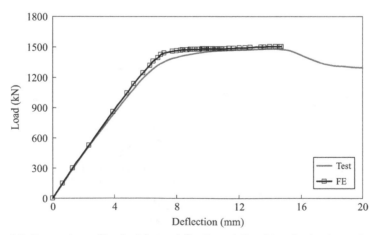

Figure 6.8 Comparison of load-midspan deflection relationships obtained experimentally and numerically for the small-scale plate girder T1. (For the color version of this figure, the reader is referred to the online version of this chapter.)

be shown that generally good agreement was achieved between experimental and numerical relationships. The ultimate failure load observed in the test [6.27] was 1474 kN at a deflection of 15 mm, while the ultimate failure load predicted from the finite element analysis was 1503 kN at a deflection of 14.7 mm. The finite element failure load was 2% higher than that observed in the test.

6.4 FINITE ELEMENT MODELING AND RESULTS OF EXAMPLE 2

The second example presented in this chapter is for a small-scale built-up I-section plate girder steel bridge tested under shear by Nakamura and Narita [6.27]. The plate girder was a part of an experimental program investigating bending and shear strength of steel and partially encased steel-concrete composite plate I-girders. The plate girder is denoted in this study as (T2). The small-scale plate girder was SS with an overhanging cantilever from one end. The plate girder had a length between supports of 2.45 m and the overhanging length from the support to loading was 0.9 m as shown in Figure 6.9. Similar to plate girder T1, the web of plate girder T2 was 900 mm high and 6 mm thick. While the flange of the plate girder T2 was 200 mm wide and 12 mm thick. The web was stiffened by steel plates at the end supports and the loading positions and also stiffened by intermediate stiffeners as shown in Figure 6.9. The plate girder was restrained laterally at the end supports to resist lateral-torsional buckling. Tensile coupon tests were conducted to determine the yield and ultimate tensile strength of the steel used, which were 372.3 and 511.4 MPa, respectively, similar to plate girder T1. The plate girders were loaded at two points, as shown in Figure 6.9, with the loading between supports equals to 0.7 the applied load and loading near the edge of the overhanging cantilever equals to 0.3 the applied load, which subjected the plate girder to high shear forces. Strain gauges were used to measure the strains in the plate girder section, and displacement transducers were used to measure the span deflections at the higher load location between supports during loading. The strain at the web center was greater than other parts and, after the tension field appeared, the stress concentrated around the web center. The diagonal tension-field action was clearly observed in the test T2. No damage was found in the flanges. The main failure mode was buckling owing to shear stresses. The load-deflection relationship, load-strain relationships, and the deformed shape at failure were observed in the test.

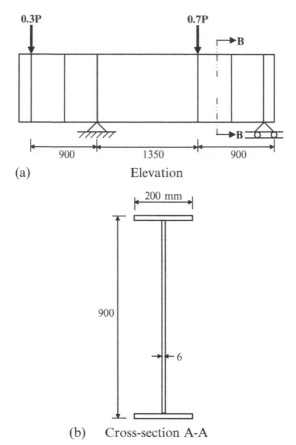

(a)   Elevation

(b)   Cross-section A-A

Figure 6.9 General layout of a small-scale plate girder in bending (T2).

To model the small-scale built-up I-section plate girder (T2) tested by Nakamura and Narita [6.27], the finite element program ABAQUS [1.29] was used. Similar to the modeling of T1, the model has accounted for the measured geometry, initial geometric imperfections, and measured material properties of the plate girder. A four-node doubly curved shell element with reduced integration (S4R) was used to model the flanges and web of the small-scale plate girder bridge, as shown in Figure 6.10. In order to choose the finite element mesh that provides accurate results with minimum computational time, convergence studies were conducted. It is found that approximately 75×76 mm (length by width of S4R element) ratio provides adequate accuracy in modeling the web while a finer mesh of approximately 25×75 mm was used in the flange (see Figure 6.10).

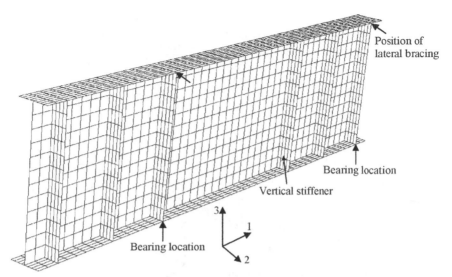

Figure 6.10 Finite element mesh of the small-scale built-up I-section plate girder T2.

The hinged support of T2, shown in Figure 6.9, was prevented from displacement in the horizontal direction (direction 1-1 in Figure 6.10) and the vertical direction (direction 3-3 in Figure 6.10). On the other hand, the roller support of T2, shown in Figure 6.9, was prevented from displacement in the vertical direction only (direction 3-3 in Figure 6.10). To account for the lateral restraints of the compression flange, the top compression flange was prevented from lateral displacements, in direction 2-2 of Figure 6.10, at the end supports, which is identical to the test T2. The load was applied in increments as concentrated static load, which is also identical to the experimental investigation. The nonlinear geometry was included to deal with the large displacement analysis.

The stress-strain curve for the structural steel given in the EC3 [2.11] was adopted in this study with measured values of the yield stress (f_{ys}) and ultimate stress (f_{us}) used in the tests [6.27]. The material behavior provided by ABAQUS [1.29] (using the PLASTIC option) allows a nonlinear stress-strain curve to be used (see Section 5.4.2 of Chapter 5). The first part of the nonlinear curve represents the elastic part up to the proportional limit stress with Young's modulus of (E) 200 GPa and Poisson's ratio of 0.3 used in the finite element model. Since the buckling analysis involves large inelastic strains, the nominal (engineering) static stress-strain curves were converted to true stress and logarithmic plastic true strain curves as detailed in Section 5.4.2.

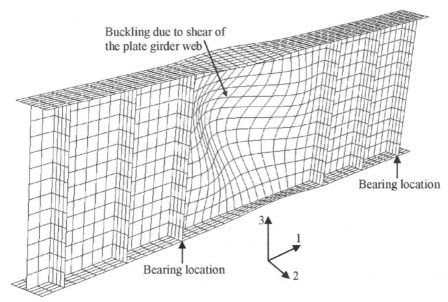

Buckling due to shear of
the plate girder web

Bearing location

3
1
2

Bearing location

Figure 6.11 Elastic lateral buckling mode (eigenmode 1) for the small-scale built-up I-section plate girder T2.

The same modeling approach [6.28, 6.29] was followed in this book to model initial geometric imperfections of the plate girder investigated T2. Figure 6.11 shows the buckling mode predicted from the eigenvalue buckling analysis detailed in ABAQUS [1.29]. Only the first buckling mode (eigenmode 1) is used in the eigenvalue analysis. Since buckling modes predicted by ABAQUS eigenvalue analysis [1.29] are generalized to 1.0, the buckling modes are factored by a magnitude of $L_u/1000$, where L_u is the length between points of effective bracing. The magnitude of $L_u/1000$ is the average measured values in the tests [6.31] and commended in [6.32]. The factored buckling mode is inserted into the load-displacement non-linear analysis of the plate girder T2 following the eigenvalue prediction. It should be noted that the investigation of plate girders with different slenderness ratios could result in different buckling modes. Hence, to ensure that the correct buckling mode is incorporated in the nonlinear displacement analysis, the eigenvalue buckling analysis must be performed for each plate girder with actual geometry.

The developed finite element model for the plate girder T2 (see Figure 6.10) was verified against the test results detailed in [6.27]. The failure loads, failure modes, and load–deflection curves obtained experimentally and numerically using the finite element model were compared. The deformed shapes of plate girder T2 at failure observed experimentally and numerically were compared as shown in Figure 6.12. It can be seen that the experimental and numerical deformed shapes are in good agreement. The failure mode observed experimentally and confirmed numerically

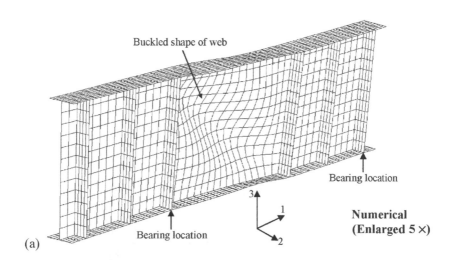

(a)

Buckled shape of web

Bearing location

Bearing location

Numerical (Enlarged 5 ×)

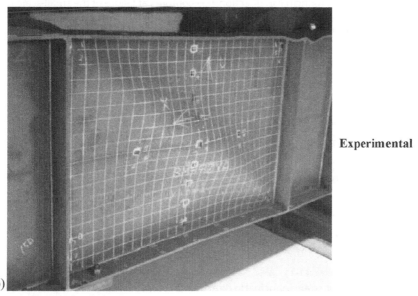

(b)

Experimental

Figure 6.12 Comparison of experimental and numerical deformed shapes at failure of the small-scale built-up I-section plate girder T2.

was buckling owing to shear stresses, which occurred in the wider web panel. The data obtained from ABAQUS [1.29] have shown that the von Mises stresses at the maximum stressed fibers in the web exceeded the measured yield stresses. Figure 6.13 plotted the stress (principal stresses in direction 1-1) contours at failure of the small-scale built-up I-section plate girder T2. It can be seen that the yield stresses were reached at the maximum stresses portions of the web. In addition, in Figure 6.14, the plastic strain

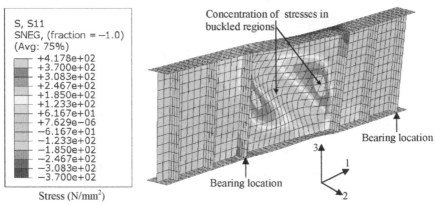

Figure 6.13 Stress (principal in direction 1-1) contours at failure of the small-scale built-up I-section plate girder T2 (enlarged 5 ×). (For the color version of this figure, the reader is referred to the online version of this chapter.)

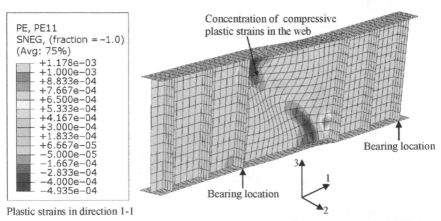

Figure 6.14 Plastic strain (principal in direction 1-1) contours at failure of the small-scale built-up I-section plate girder T2 (enlarged 5 ×). (For the color version of this figure, the reader is referred to the online version of this chapter.)

(principal strains in direction 1-1) contours at failure of the small-scale built-up I-section plate girder T2 are plotted. It can be seen that the plastic strains were concentrated at the maximum stresses portions of the web. Furthermore, in Figure 6.15, the von Mises yield stress contours at failure of the small-scale built-up I-section plate girder T2 are plotted. It can be seen that the yield stresses were reached at midspan in the upper and lower flanges. The load-midspan deflection curves predicted experimentally and numerically were also compared as shown in Figure 6.16. It can be shown that

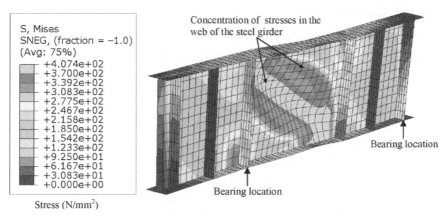

Figure 6.15 Stress (von Mises) contours at failure of the small-scale built-up I-section plate girder T2 (enlarged 5×). (For the color version of this figure, the reader is referred to the online version of this chapter.)

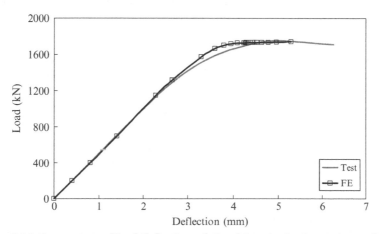

Figure 6.16 Comparison of load-deflection relationships obtained experimentally and numerically for the small-scale built-up I-section plate girder T2. (For the color version of this figure, the reader is referred to the online version of this chapter.)

generally good agreement was achieved between experimental and numerical relationships. The ultimate failure load observed in the test [6.27] was 1752 kN at a deflection of 5 mm, while the ultimate failure load predicted from the finite element analysis was 1742 kN at a deflection of 5.3 mm. The finite element failure load was 1% lower than that observed in the test.

6.5 FINITE ELEMENT MODELING AND RESULTS OF EXAMPLE 3

The third example presented in this chapter is for a full-scale built-up I-section plate girder steel bridge tested under bending by Felkel et al. [6.8]. The plate girder was a part of an experimental program investigating the effect of lateral bracing on the strength and behavior of the bridges. The plate girder is denoted in this study as (T3). The small-scale plate girder was SS and had an overall length of 13.411 m as shown in Figure 6.17. The web of plate girder T3 was 914 mm high and 8 mm thick, while the flange of the plate girder was 229 mm wide and 11-25 mm thick, as shown in Figure 6.17. The web was stiffened by steel plates at the end supports and the loading positions and also stiffened by intermediate stiffeners as shown in Figure 6.17. The plate girder was restrained laterally at the end supports and loading position to resist lateral-torsional buckling. Tensile coupon tests were conducted to determine the yield and ultimate tensile strength of the steel used, which were 558 and 621 MPa, respectively. The plate girders were loaded at midspan using a spreader plate (343 × 229) as shown in Figure 6.17. Strain gauges were used to measure the strains in the plate girder section, and displacement transducers were used to measure the mid-span deflections during loading. The strain measurements showed that no yielding took place and the neutral axis remained stationary until buckling occurred. The failure mode of the specimen was elastic lateral-torsional buckling. The ultimate load resisted by the plate girder was 489.3 kN at a maximum midspan deflection of 40.6 mm. The load-midspan deflection relationship, load-strain relationships, and the deformed shape at failure were observed in the test.

To model the full-scale plate girder (T3) tested by Felkel et al. [6.8], the finite element program ABAQUS [1.29] was used. The model has accounted for the measured geometry, initial geometric imperfections, and measured material properties of the plate girder. A four-node doubly curved shell element with reduced integration (S4R) was used to model

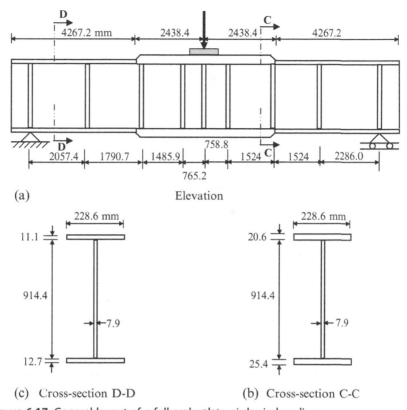

Figure 6.17 General layout of a full-scale plate girder in bending.

the flanges and web of the small-scale plate girder bridge, as shown in Figure 6.18. In order to choose the finite element mesh that provides accurate results with minimum computational time, convergence studies were conducted. It is found that approximately 114×149 mm (length by width of S4R element) ratio provides adequate accuracy in modeling the web, while a finer mesh of approximately 29×149 mm was used in the flange (see Figure 6.18). The hinged support of T3, shown in Figure 6.18, was prevented from displacement in the horizontal direction (direction 1-1 in Figure 6.18) and the vertical direction (direction 3-3 in Figure 6.18). On the other hand, the roller support of T3, shown in Figure 6.18, was prevented from displacement in the vertical direction only (direction 3-3 in Figure 6.18). To account for the lateral restraints of the compression flange, the top compression flange was prevented from lateral displacements, in direction 2-2 of Figure 6.18, at the end supports and the loading position

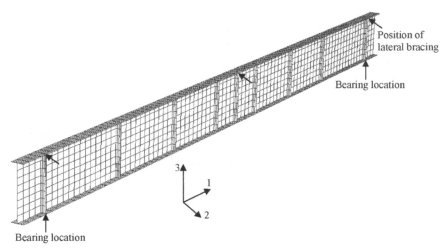

Position of
lateral bracing

Bearing location

Bearing location

Figure 6.18 Finite element mesh of the full-scale built-up I-section plate girder T3.

at midspan, which is identical to the test T3. The load was applied in increments as concentrated static load using the RIKS method (see Section 5.5.4 of
Chapter 5) that is also identical to the experimental investigation. The nonlinear geometry was included to deal with the large displacement analysis.

The stress–strain curve for the structural steel given in the EC3 [2.11] was
adopted in this study with measured values of the yield stress (f_{ys}) and ultimate stress (f_{us}) used in the tests [6.8]. The material behavior provided by
ABAQUS [1.29] (using the PLASTIC option) allows a nonlinear stress–
strain curve to be used (see Section 5.4.2 of Chapter 5). The first part of
the nonlinear curve represents the elastic part up to the proportional limit
stress with Young's modulus of (E) 200 GPa and Poisson's ratio of 0.3 used
in the finite element model. Since the buckling analysis involves large inelastic strains, the nominal (engineering) static stress–strain curves were converted to true stress and logarithmic plastic true strain curves as detailed
in Section 5.4.2.

The same modeling approach [6.28, 6.29] was followed in this book
to model initial geometric imperfections of the plate girder investigated
T3. In Figure 6.19, the buckling mode predicted from the eigenvalue buckling analysis detailed in ABAQUS [1.29] is shown. Only the first buckling
mode (eigenmode 1) is used in the eigenvalue analysis. Since buckling
modes predicted by ABAQUS eigenvalue analysis [1.29] are generalized
to 1.0, the buckling modes are factored by a magnitude of $L_u/1000$, where
L_u is the length between points of effective bracing. The magnitude of

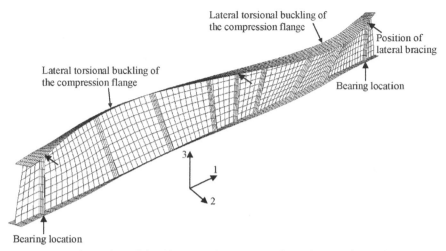

Figure 6.19 Elastic lateral buckling mode (eigenmode 1) for the full-scale built-up I-section plate girder T3.

$L_u/1000$ is the average measured values in the tests [6.8] and commended in [6.32]. The factored buckling mode is inserted into the load–displacement nonlinear analysis of the plate girder T3 following the eigenvalue prediction.

The developed finite element model for the plate girder T3 (see Figure 6.18) was verified against the test results detailed in [6.8]. The failure loads, failure modes, and load–midspan deflection curves obtained experimentally and numerically using the finite element model were compared. The deformed shapes of plate girder T3 at failure observed experimentally and numerically were compared as shown in Figure 6.20. It can be seen that the experimental and numerical deformed shapes are in good agreement. The failure mode observed experimentally and confirmed numerically was lateral-torsional buckling. The data obtained from ABAQUS [1.29] have shown that the von Mises stresses at the maximum stressed fibers at the top and bottom flanges at midspan were not exceeded. In Figure 6.21, the stress (principal stresses in direction 1-1) contours at failure of the full-scale built-up I-section plate girder T3 are plotted. It can be seen that the yield stresses were not reached. In addition, in Figure 6.22, the plastic strain (principal strains in direction 1-1) contours at failure of the small-scale built-up I-section plate girder T3 are plotted. Once again, it can be seen that the plastic strains were not reached. Furthermore, in Figure 6.23, the von Mises yield stress contours at failure of the small-scale built-up I-section plate

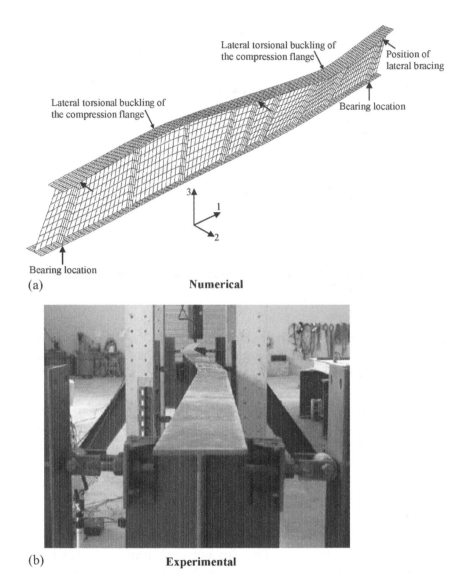

Lateral torsional buckling of
the compression flange

Position of
lateral bracing

Lateral torsional buckling of
the compression flange

Bearing location

Bearing location

(a)   **Numerical**

(b)   **Experimental**

Figure 6.20 Comparison of experimental and numerical deformed shapes at failure of the full-scale built-up I-section plate girder T3. (For the color version of this figure, the reader is referred to the online version of this chapter.)

girder T3 are plotted. It can be seen that the yield stresses were not reached. The load–midspan deflection curves predicted experimentally and numerically were also compared as shown in Figure 6.24. It can be shown that generally good agreement was achieved between experimental and numerical relationships. The ultimate failure load observed in the test [6.8] was

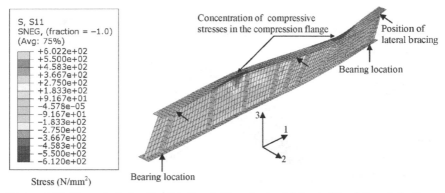

Figure 6.21 Stress (principal in direction 1-1) contours at failure of the full-scale built-up I-section plate girder T3. (For the color version of this figure, the reader is referred to the online version of this chapter.)

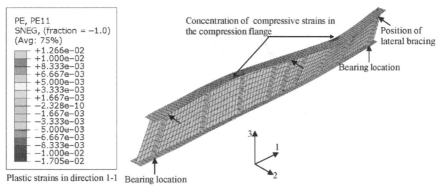

Figure 6.22 Plastic strain (principal in direction 1-1) contours at failure of the full-scale built-up I-section plate girder T3. (For the color version of this figure, the reader is referred to the online version of this chapter.)

489.3 kN at a deflection of 40.6 mm, while the ultimate failure load predicted from the finite element analysis was 490.3 kN at a deflection of 44 mm. The finite element failure load was 0.2% higher than that observed in the test.

6.6 FINITE ELEMENT MODELING AND RESULTS OF EXAMPLE 4

In this example, we can use the finite element modeling approach adopted for the simulation of plate girders T1, T2, and T3 to model a full-scale

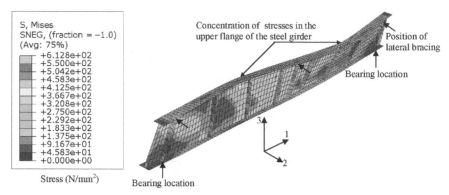

Figure 6.23 Stress (von Mises) contours at failure of the full-scale built-up I-section plate girder T3. (For the color version of this figure, the reader is referred to the online version of this chapter.)

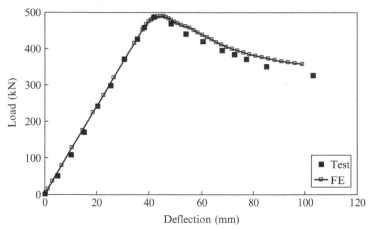

Figure 6.24 Comparison of load-midspan deflection relationships obtained experimentally and numerically for the full-scale plate girder T3.

double-track open-timber floor plate girder railway steel bridge. The SS double-track bridge is similar to that presented in Chapter 1 (see Figure 1.20) having a length of 30 m between supports and an overall length of 31 m. The bridge was designed by the author in Chapter 4 (see Section 4.2) adopting the design rules specified in EC3 [1.27]. In Figure 6.25, the general layout of the bridge and the bridge components comprising main plate girders, cross girders, stringers, bracing members, stiffeners,

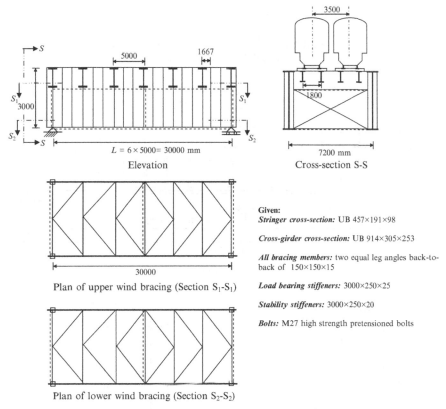

Figure 6.25 General layout of a double-track open-timber floor plate girder railway steel bridge.

etc., is shown. The dimensions and details of the main plate girder are shown in Figure 6.26. The main plate girder web was 3000 mm high and 16 mm width. The flange plate of the main plate girder was 600 mm wide and 30 mm thick, with another plate being added in the middle 14 m region to resist the maximum bending moment at midspan. The stringers were universal beams (UB $457 \times 191 \times 98$), the cross girders were universal beams (UB $914 \times 305 \times 253$), and the bracing members were two equal angles back-to-back ($150 \times 150 \times 15$). The web of the main plate girder was stiffened at supports by load-bearing stiffeners (steel plates of $3000 \times 250 \times 25$ mm) and by intermediate stability stiffeners (steel plates of $3000 \times 250 \times 20$) (see Figures 6.25 and 6.26). The bridge was made of steel having yield and ultimate tensile strengths of 275 and 430 MPa, respectively. The dead loads on each main girder were estimated by 27.6 kN/m,

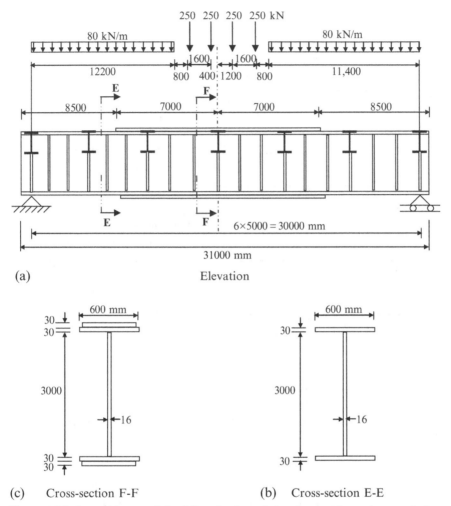

(a)   Elevation

(c)   Cross-section F-F   (b)   Cross-section E-E

Figure 6.26 General layout of the full-scale plate girder of a double-track open-timber railway steel bridge.

while the live loads were Load Model 71, which represents the static effect of vertical loading due to normal rail traffic as specified in EC1 [3.1]. The worst case of loading for bending moment, adopting Load Model 71, is shown in Figure 6.26. Further details regarding the design of the bridge and its components are presented in Section 4.2.

To model the full-scale double-track open-timber floor plate girder railway steel bridge shown in Figures 6.25 and 6.26, the finite element program ABAQUS [1.29] was used. The model has accounted for the bridge geometry, initial geometric imperfections, and nonlinear material properties of the steel used. A four-node doubly curved shell element with reduced integration (S4R) was used to model the flanges and webs of the stringers, cross girders, and main plate girders. The element was also used to model the stiffeners of the web of the main plate girders. The bracing members were modeled using structural 2D truss elements (T2D2) available in the ABA-QUS [1.29] element library. The bolts connecting the cross girders to the main girders and the bolts connecting the stringers to the cross girders were modeled using JOINTC joint elements, available in the ABAQUS [1.29] element library, having stiffnesses in two directions, which simulated the SS end boundary conditions. In order to choose the finite element mesh that provides accurate results with minimum computational time, convergence studies were conducted. It is found that approximately 153×260 mm (length by width of S4R element) ratio provides adequate accuracy in modeling the webs of the main plate girders, while a finer mesh of approximately 75×260 mm was used in the flanges of the main plate girders (see Figure 6.27). Also, it is found that approximately 153×251 mm ratio provides adequate accuracy in modeling the webs of the cross girders, while a finer mesh of approximately 76×251 mm was used in the flanges of the

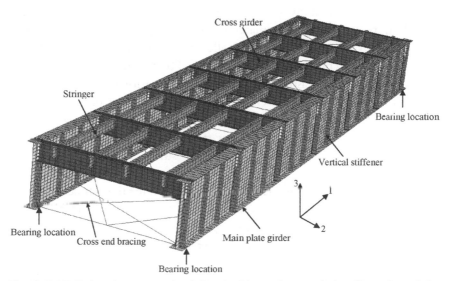

Figure 6.27 Finite element mesh of the double-track open-timber floor plate girder railway steel bridge. (For the color version of this figure, the reader is referred to the online version of this chapter.)

cross girders (see Figure 6.27). Finally, it is found that approximately 153×260 mm ratio provides adequate accuracy in modeling the webs of the stringers, while a finer mesh of approximately 96×260 mm was used in the flanges of the stringers (see Figure 6.27). The hinged supports of the bridge attached to the main plate girders, shown in Figures 6.25 and 6.26, were prevented from displacement in the horizontal direction (direction 1-1) and the vertical direction (direction 3-3). On the other hand, the roller support of the bridge, shown in Figures 6.25 and 6.26, was prevented from displacement in the vertical direction only (direction 3-3).

The developed finite element model shown in Figure 6.27 can be now used to analyze the bridge for any analysis, boundary conditions, geometries, and loadings. As an example in this book, the bridge was analyzed for the unfactored live load case shown in Figure 6.26 and analyzed to predict the ultimate load that can be carried by the bridge up to complete failure. The live load case was applied in increments as concentrated and distributed static loads, which is identical to the Load Model 71 adopted. On the other hand, the ultimate load that can be carried by the bridge was predicted using the RIKS method to cause maximum deflection at midspan. The nonlinear geometry was included to deal with the large displacement analysis.

The stress-strain curve for the structural steel given in the EC3 [2.11] was adopted in this study with the yield and tensile stresses of 275 and 430 MPa, respectively. The material behavior provided by ABAQUS [1.29] (using the PLASTIC option) allows a nonlinear stress-strain curve to be used (see Section 5.4.2 of Chapter 5). The first part of the nonlinear curve represents the elastic part up to the proportional limit stress with Young's modulus of (E) 200 GPa and Poisson's ratio of 0.3 used in the finite element model. Since the buckling analysis involves large inelastic strains, the nominal (engineering) static stress-strain curves were converted to true stress and logarithmic plastic true strain curves as detailed in Section 5.4.2.

An eigenvalue buckling analysis was performed for the whole bridge to model initial geometric imperfections of the bridge. In Figure 6.28, the buckling mode predicted from the eigenvalue buckling analysis detailed in ABAQUS [1.29] is shown. It can be seen from Figure 6.28 that a clear web buckling mode due to bending was predicted at midspan panel for the unfactored live load case shown in Figure 6.26. Only the first buckling mode (eigenmode 1) is used in the eigenvalue analysis. Since buckling modes predicted by ABAQUS eigenvalue analysis [1.29] are generalized

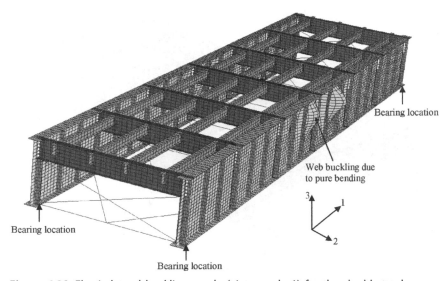

Figure 6.28 Elastic lateral buckling mode (eigenmode 1) for the double-track open-timber floor plate girder railway steel bridge. (For the color version of this figure, the reader is referred to the online version of this chapter.)

to 1.0, the buckling modes are factored by a magnitude of $I_u/1000$, where I_u is the distance between web stiffeners. The factored buckling mode is inserted into the load–displacement nonlinear analysis of the whole bridge following the eigenvalue prediction. The developed finite element model for the whole bridge (see Figure 6.27) was used to perform two analyses as examples in this book. In the first analysis, the unfactored live load case was applied, and the deformed shape after load application, load–displacement relationship, and stress contours were evaluated using the finite element model as shown in Figures 6.29–6.32. On the other hand, in the second analysis, the bridge was taken to overall collapse, and the deformed shape at failure, load–displacement relationship, and stress contours at failure were evaluated using the finite element model as shown in Figures 6.33–6.36.

In Figure 6.29, the deformed shape of the whole bridge after the application of the unfactored live load case is shown. It can be seen that the deformations were in the elastic range. The data obtained from ABAQUS [1.29] have shown that the von Mises stresses at the maximum stressed fibers at the top and bottom flanges at midspan were well below the yield stresses. In Figure 6.30, the stress (principal stresses in direction 1-1) contours after the live load application are plotted. It can be seen that the yield stresses were

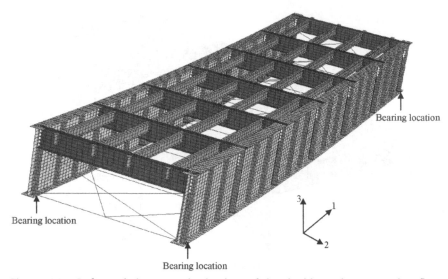

Figure 6.29 Deformed shapes under loading of the double-track open-timber floor plate girder railway steel bridge (enlarged 10 ×). (For the color version of this figure, the reader is referred to the online version of this chapter.)

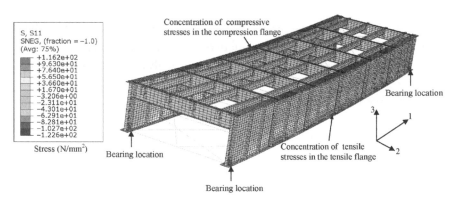

Figure 6.30 Stress (principal in direction 1-1) contours under loading of the double-track open-timber floor plate girder railway steel bridge (enlarged 10 ×). (For the color version of this figure, the reader is referred to the online version of this chapter.)

not reached at midspan in the upper (compressive stresses with negative sign) and lower flanges (tensile stresses with positive sign). In addition, in Figure 6.31, the von Mises yield stress contours after the application of the live load case are plotted. It can be seen that the yield stresses were not reached at midspan in the upper and lower flanges. The load–midspan

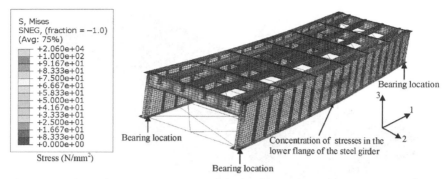

Figure 6.31 Stress (von Mises) contours under loading of the double-track open-timber floor plate girder railway steel bridge (enlarged 10×). (For the color version of this figure, the reader is referred to the online version of this chapter.)

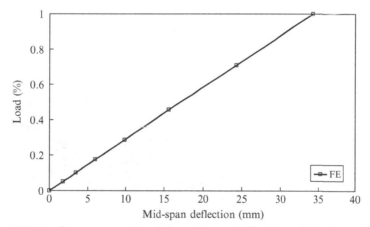

Figure 6.32 Load per one main girder-midspan deflection relationship obtained numerically for the double-track open-timber floor plate girder railway steel bridge under the live loading case.

deflection curve predicted numerically was plotted in Figure 6.32. It can be shown that the relationship is linear, which confirms that the deformations were in the elastic range. The maximum deflection predicted after the application of the live load case was 34.3 mm.

Let us now look at the deformed shape of the whole bridge analyzed to failure and predicted using the RIKS method as shown in Figure 6.33. It can be seen that a clear combined lateral-torsional buckling mode of the upper main plate girder flange and web buckling mode was predicted at midspan

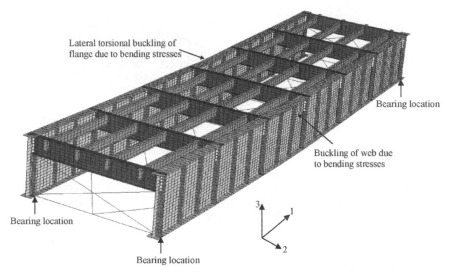

Figure 6.33 Deformed shape of the double-track open-timber floor plate girder railway steel bridge at failure. (For the color version of this figure, the reader is referred to the online version of this chapter.)

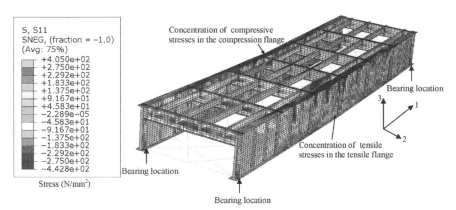

Figure 6.34 Stress (principal in direction 1-1) contours of the double-track open-timber floor plate girder railway steel bridge at failure. (For the color version of this figure, the reader is referred to the online version of this chapter.)

owing to bending stresses as shown in Figure 6.33. The data obtained from ABAQUS [1.29] also showed that the von Mises stresses at the maximum stressed fibers at the top and bottom flanges and web of the main plate girder at midspan exceeded the yield stresses. In Figure 6.34, the stress (principal stresses in direction 1-1) contours at failure are plotted. It can be seen that

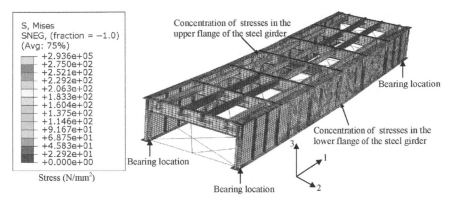

Figure 6.35 Stress (von Mises) contours of the double-track open-timber floor plate girder railway steel bridge at failure. (For the color version of this figure, the reader is referred to the online version of this chapter.)

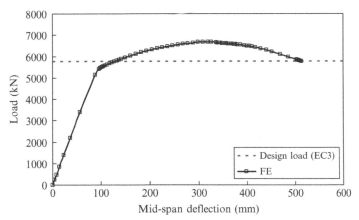

Figure 6.36 Load per one main girder-midspan deflection relationships for the double-track open-timber floor plate girder railway steel bridge at ultimate limit state. (For the color version of this figure, the reader is referred to the online version of this chapter.)

the yield stresses were reached at midspan in the upper (compressive stresses with negative sign) and lower flanges (tensile stresses with positive sign) of the main plate girder. In addition, in Figure 6.35, the von Mises yield stress contours at failure are plotted. It can be seen that the yield stresses were reached at midspan in the upper and lower flanges of the main plate girder. The load–midspan deflection curve predicted numerically was plotted in Figure 6.36. It can be shown that the relationship is nonlinear, which confirms that the deformations exceeded the elastic range. The ultimate load that can be resisted by each main plate girder was 6683.7 kN at a deflection of 318.8 mm. The design load predicted using EC3 [1.27] for each main plate girder was also plotted in Figure 6.36, which was 5763.3 kN. The finite element failure load was 16% higher than the design load, which shows that the design rules specified in the EC3 [1.27] are conservative and accurate.

REFERENCES

[6.1] C.J. Earls, B.J. Shah, High performance steel bridge girder compactness, J. Constr. Steel Res. 58 (2002) 859–880.
[6.2] N.E. Shanmugam, M. Mahendrakumar, V. Thevendran, Ultimate load behaviour of horizontally curved plate girders, J. Constr. Steel Res. 59 (2003) 509–529.
[6.3] Y.L. Pi, N.S. Trahair, Nonlinear elastic behaviour of I-beams curved in plan, J. Struct. Eng. ASCE 123 (9) (1997) 1201–1209.
[6.4] Y.L. Pi, M.A. Bradford, N.S. Trahair, Inelastic analysis and behaviour of steel I-beams curved in plan, J. Struct. Eng. ASCE 126 (7) (2000) 772–779.
[6.5] W.D. Corte, P.V. Bogaert, The effect of shear deformations in floor beams on the moment distribution in orthotropic plated bridge decks, J. Constr. Steel Res. 62 (2006) 1007–1015.
[6.6] R. Wolchuk, Design Manual for Orthotropic Steel Plate Deck Bridges, AISC, Chicago, IL, 1963.
[6.7] M.S. Troitsky, Orthotropic Bridges: Theory and Design, second ed., The James F. Lincoln Arc. Welding Foundation, Cleveland, OH, 1968.
[6.8] J.P. Felkel, D.C. Rizos, P.H. Ziehl, Structural performance and design evaluation of HPS 70W bridge girders, J. Constr. Steel Res. 63 (2007) 909–921.
[6.9] P. Galvin, J. Dominguez, Dynamic analysis of a cable-stayed deck steel arch bridge, J. Constr. Steel Res. 63 (2007) 1024–1035.
[6.10] ANSYS Inc., ANSYS Manual: Theory, Elements and Commands References, ANSYS Inc., Houston, TX, 2005.
[6.11] A. Romeijn, C. Bouras, Investigation of the arch in-plane buckling behaviour in arch bridges, J. Constr. Steel Res. 64 (2008) 1349–1356.
[6.12] M.E.A.H. Eldib, Shear buckling strength and design of curved corrugated steel webs for bridges, J. Constr. Steel Res. 65 (2009) 2129–2139.
[6.13] H. Zhang, R. DesRoches, Z. Yang, S. Liu, Experimental and analytical studies on a streamlined steel box girder, J. Constr. Steel Res. 66 (2010) 906–914.
[6.14] C. Graciano, E. Casanova, J. Martínez, Imperfection sensitivity of plate girder webs subjected to patch loading, J. Constr. Steel Res. 67 (2011) 1128–1133.
[6.15] M. Chryssanthopoulos, M. Baker, P. Dowling, Imperfection modeling for buckling analysis of stiffened cylinders, J. Struct. Eng. ASCE 117 (1) (1991) 1998–2017.
[6.16] M. Chryssanthopoulos, Probabilistic buckling analysis of plates and shells, Thin-Walled Struct. 30 (1–4) (1998) 135–157.
[6.17] O. Caglayan, K. Ozakgul, O. Tezer, E. Uzgider, Evaluation of a steel railway bridge for dynamic and seismic loads, J. Constr. Steel Res. 67 (2011) 1198–1211.
[6.18] COSMOS/M, User's Manual, Solid Works Corporation, Dassault Systémes, Massachusetts, 2008.
[6.19] A.C. Altunişik, A. Bayraktar, B. Sevim, H. Özdemir, Experimental and analytical system identification of Eynel arch type steel highway bridge, J. Constr. Steel Res. 67 (2011) 1912–1921.
[6.20] SAP2000, Integrated Finite Element Analysis and Design of Structures, Computers and Structures Inc, Berkeley, CA, 2008.

[6.21] G. Kaliyaperumal, B. Imam, T. Righiniotis, Advanced dynamic finite element analysis of a skew steel railway bridge, Eng. Struct. 33 (2011) 181–190.

[6.22] L. Fryba, Dynamics of Railway Bridges, Thomas Telford, London, 1996.

[6.23] UIC 776-1 R, Loads to be Considered in the Design of Railway Bridges, International Union of Railways, 1979.

[6.24] O. Caglayan, K. Ozakgul, O. Tezer, Assessment of existing steel railway bridges, J. Constr. Steel Res. 69 (2012) 54–63.

[6.25] H.T. Thai, D.H. Choi, Advanced analysis of multi-span suspension bridges, J. Constr. Steel Res. 90 (2013) 29–41.

[6.26] W. Lin, T. Yoda, N. Taniguchi, M. Hansaka, Performance of strengthened hybrid structures renovated from old railway steel bridges, J. Constr. Steel Res. 85 (2013) 130–139.

[6.27] S. Nakamura, N. Narita, Bending and shear strengths of partially encased composite I-girders, J. Constr. Steel Res. 59 (2003) 1435–1453.

[6.28] E. Ellobody, Interaction of buckling modes in castellated steel beams, J. Constr. Steel Res. 67 (5) (2011) 814–825.

[6.29] E. Ellobody, Nonlinear analysis of cellular steel beams under combined buckling modes, Thin-Walled Struct. 52 (3) (2012) 66–79.

[6.30] M.A. Bradford, Lateral-distortional buckling of steel I-section members, J. Constr. Steel Res. 23 (1992) 97–116.

[6.31] T. Zirakian, H. Showkati, Distortional buckling of castellated beams, J. Constr. Steel Res. 62 (2006) 863–871.

[6.32] S. Chen, Y. Jia, Numerical investigation of inelastic buckling of steel-concrete composite beams prestressed with external tendons, Thin-Walled Struct. 48 (2010) 233–242.

CHAPTER 7

Examples of Finite Element Models of Steel-Concrete Composite Bridges

7.1 GENERAL REMARKS

Similar to Chapter 6, this chapter presents illustrative examples of finite element models developed to understand the structural behavior of steel-concrete composite bridges. The examples are based on the layout, loading, and design background presented in Chapters 1–4 as well as based on the finite element modeling background highlighted in Chapter 5. This chapter is initiated by a brief introduction followed by a review of recent investigations reported in the literature concerning the modeling of steel-concrete bridges. This chapter details how the finite element models were developed and the results obtained. The presented examples show the effectiveness of finite element models in providing detailed data that complement scarce experimental data in the field. The results are discussed to show the significance of the finite element models in predicting the structural response of the different steel-concrete composite bridges investigated. In overall, it is aimed to show that finite element analysis not only can assess the accuracy of the design rules specified in current codes of practice but also can improve and propose more accurate design rules. It should be noted that the main issues that differentiate between the finite element models presented in Chapter 6 and those presented in this chapter are the presence of reinforced concrete slabs and shear connections. The finite element models addressed in this chapter will detail how the aforementioned issues are dealt with in the models based on the material properties highlighted in Chapter 2, the design rules highlighted in Chapter 3, and the finite element modeling variables presented in Chapter 5. The author hopes that the review of recent finite element models reported in the literature together with the illustrative finite element models developed by the author in this chapter can provide readers with a complete piece of work regarding the finite element analysis of steel-concrete composite bridges.

Finite Element Analysis and Design of Steel and Steel–Concrete Composite Bridges
601

7.2 PREVIOUS WORK

Extensive numerical investigations were reported in the literature highlighting the structural performance of different types of steel-concrete composite bridges subjected to different loadings. The numerical investigations proposed finite element models for the composite bridges and the composite bridge components. It should be noted that detailed state-of-the-art review of these investigations is out of the scope of this book. However, in this section, the author provides recent examples showing how other researchers modeled the steel-concrete composite bridges and the composite bridge components. Ålenius [7.1] investigated the stability of a thin-walled box girder steel-concrete composite bridge. The study was proposed a finite element model for the analysis of the bridges. Three different finite element models were analyzed, which were a simply supported rectangular plate uniformly compressed in one direction, a profiled sheeting subjected to shear forces, and finite element model discussing the lateral torsional stability of the bridge. The first analysis provided a simple model for which analytic results were available for comparison with finite element modeling results. It was shown that the finite element analysis provided results well in accordance with the analytic results for the critical buckling load. The analysis of the profiled sheeting in shear studied the attachment techniques of the profiled sheeting. The analysis showed that a substantial reduction of the stresses in the profiled sheeting was obtained with an all-around attachment between the profiled sheeting and box girder, compared to a two-sided attachment. In addition, the analysis showed that the large axial forces that arose at the free edge of the profiled sheeting, when it was attached along two sides, were considerably diminished when an all-around attachment was used. Finally, the third model investigated the lateral torsional stability of the bridge. The analysis studied the overall behavior of the bridge and how it was influenced by initial imperfections and different attachments of the profiled sheeting. The profiled sheeting created a closed cross section, providing a structure that was more rigid than an open cross section. It was concluded that, in order to obtain a closed cross section, the profiled sheeting should be given an all-around attachment and also be strong enough to withstand arisen stresses, mainly at the connections between the profiled sheeting and the box girder. Otherwise, the cross section of the bridge will behave like an open, thin-walled cross section. Shell elements (S4R) available in ABAQUS [1.29] were used in most of the analyses. The analyses comprised linear eigenvalue buckling analysis and nonlinear load-displacement analyses.

Barth and Wu [7.2] performed 3-D nonlinear finite element analyses to predict ultimate load behavior of slab on steel stringer bridge superstructures. The study was accomplished using the general-purpose finite element software ABAQUS [1.29]. Two composite steel girders fabricated from high-performance steel and one four-span continuous composite steel bridge tested to failure have been used to validate the proposed finite element models. In the study, four-node general-purpose shell element with reduced integration (S4R) was used for the steel girders, concrete slab, and stiffeners. The steel reinforcement in the concrete was provided by means of rebar elements. A 3-D two-node beam element (B31) was used to represent cross frames. Full composite action between the RC deck and the steel girder is developed using a beam-type multipoint constraint (MPC beam) between the girder top flange and the deck, which assures nodal compatibility at these locations. The authors mentioned that, to obtain accurate results from the nonlinear finite element analysis, consideration must be given to the element size and mesh density selection. Selection of relatively small elements will eliminate unrealistically low predicted strengths due to the effects of stress concentrations, while incorporating relatively large elements will reduce the need to modify the constitutive model to prevent an overestimation of the energy dissipation capacity [7.3]. The analyses incorporated nonlinear material behavior including a trilinear stress-strain response for structural steel and a complete nonlinear stress-strain curve for steel reinforcement. Concrete was modeled using two concrete models: the smeared crack concrete and concrete damaged plasticity models. The modified Riks method was used in the finite element analyses.

Chung and Sotelino [7.4] investigated finite element modeling of composite steel girder bridges, with the overall flexural behavior of the bridges being the main concern. Four 3-D finite element bridge models were investigated. The finite element models were validated against the results of full-scale tests and a field test conducted by other researchers. In addition, the finite element results were compared with the results of a detailed finite element model that uses solid elements. The numerical simulations were performed using the general-purpose finite element software ABAQUS [1.29]. The authors mentioned that the bridge deck can be modeled either by solid or shell elements. Several shell elements were tested to evaluate their applicability for bridge deck modeling. It was found that a quadrilateral nine-node or eight-node shell element with reduced integration (S9R5/S8R5) and a quadrilateral eight-node thick shell element with reduced integration (S8R) predicted the same response. It was also recommended that the transverse shear deformation may be neglected in typical bridge analyses. In this study, the shear flexible shell element (S8R) was selected to model the concrete bridge deck. The authors also discussed that some models proposed in the literature utilized solid elements to model a concrete bridge deck. The main drawback of solid deck models is the computational cost to predict the correct flexural behavior of the bridge. Multiple layers are required through the thickness direction in order to model the deck with linear solid elements (e.g., eight-node brick elements), since the strain variation of these elements is constant through the thickness. An alternative option is the use of higher-order solid elements, but this may entail an even higher computational cost. In the study [7.4], four different modeling techniques for steel girders, named G1, G2, G3, and G4, respectively, are investigated. The G1 model is a detailed model of a steel girder. The flanges and the web are modeled by shell elements. The next model, G2, was similar to G1 model except that the flange was modeled by beam elements instead of shell elements, which resulted in less computing resources to represent the 3-D nature of the girder structures. G3 model can be used to investigate the possible incompatibility at the element connection between web and flanges found in the previous two models. A typical flat shell element was formulated by superimposing plate bending and membrane action. It was found that the G1 and G2 models shared the drilling rotation of the shell element in the web with the in-plane bending rotation of the shell or beam element in the flanges. Thus, displacement was not compatible along the element boundary of these models. The G3 model placed shell elements at the centroid of girder flanges. Beam elements were placed at the centroid of the girder web. Rigid links, through the constraints of degrees of freedom, were applied to ensure composite action. Finally, G4 model was the simplest model and utilized beam elements with the geometric properties of girder sections. It should be noted that G4 model was not able to represent different material properties of web and flanges. The full composite action between a concrete deck and steel girders was modeled by rigid links through multipoint constraints (MPCs), which are available in ABAQUS [1.29] element library. The authors mentioned that, for bridge models that utilized solid elements in the modeling of the bridge deck with the girder models discussed earlier, care must be taken at the interface between the two different element types, since structural elements and solid elements are incompatible. Unlike structural elements, most solid elements do not have rotational degrees of freedom (DOFs). MPCs must therefore be assigned to relate the shell element rotations to the solid element translations. Sharing nodes with two different types of finite elements may lead to perturbation of element stress and strain at the interface boundary. Rigid links through MPCs connect different elements within the structural model and enforce the kinematics relationships between the degrees of freedom at each node.

The study [7.4] outlined that in the case where shell elements were used to model the bridge deck, the nodes of the girder did not coincide with the nodes of the shell elements in the deck. The shell elements in the bridge deck were connected with the prescribed girder models through an MPC. Typical bending elements, such as the shell element (for the deck and for girder models G1 and G3) and the beam element (for the girder model G2 and G4), should be avoided for the modeling of the composite girder bridge since displacement incompatibility occurs at the interface of two bending elements [7.5]. This incompatibility was noticeable since the axial displacements of the deck and the girder. The incompatibility error disappeared as the mesh was refined and many methods were proposed to eliminate this nonconforming error [7.6–7.8]. In ABAQUS [1.29], the use of S8R elements for the concrete deck and B32 elements for the girder provided full compatibility between the boundaries of two different elements. The applied loading on a bridge deck consisted of pressure loads applied through a tire patch. In the finite element modeling, this requirement imposed the need for a fine mesh in the deck, so that the element is fitted with the patch size. As a part of the research [7.4], the equivalent nodal load algorithm was employed in order to uncouple the patch load from the mesh size. Also, bearings were modeled by assigning boundary conditions to the zero-dimensional elements at their real location. For simply supported beams, rotations in all directions were allowed in order to simulate the simply supported structure. Minimum restraints were assigned for longitudinal and transverse movement, while vertical restraint was placed at the supports. Kinematic constraints were also supplied to nodes between the girders and the deck. Since the main purpose of the study [7.4] was to analyze bridge superstructures, it was assumed that substructures, such as piers and abutments, did not influence the behavior of the superstructure.

The use of ductile end cross frames to reduce the transverse seismic demand in composite steel plate girder bridge superstructures was investigated by Carden et al. [7.9]. The authors showed that the effectiveness of these cross frames was strongly influenced by the transverse flexibility of the superstructure and its capacity for potentially large relative transverse displacements between the deck slab and bearing supports. The study proposed a simplified method for the calculation of these displacements based on the elastic girder stresses and transverse girder stiffnesses, which were shown to compare well with results given by the finite element method. In addition, the proposed method was shown to give results that compared well with experimental data from a 0.4-scale model subject to shake table excitation. Furthermore, parametric studies were described and showed that typical I-girder superstructures were able to accommodate large transverse drifts (up to 17% of the girder height) while remaining in the elastic range. These large drifts were possible without distress to the slab-to-girder connection, by omitting shear studs over a short length of the girder at the support cross-frame locations. Based on the preceding text, a step-by-step procedure was proposed for evaluating the transverse displacement, stiffness, and capacity of the steel girder superstructures in the region of the end and intermediate supports. The developed finite element model was developed for a 9.14 m length of the bridge girder, equal to half of the actual bridge model girder length, with symmetrical boundary conditions at midspan, using elastic shell elements for the flanges and web. The finite element model was generated using SAP2000 [6.20]. Rows of shear studs were placed every 460 mm for the partially composite bridge model except the ends of the bridge. Elements and restraints were used in the finite element model to allow for each shear stud to connect the girder to an assumed transversely rigid deck slab resulting in a constant transverse displacement in the top flange. Contact between the deck slab and top flange was modeled using rigid pin-ended links between the deck slab and edge of the flange. Web stiffeners were modeled using elastic shell elements at the cross-frame locations, with intermediate stiffeners between the cross frames generally neglected except in specific cases where the effect of these was investigated. A rotational spring at the base of the girder was used to model the rotational bearing stiffness.

Bapat [7.10] presented a study to create a database of quantitative information of the long-term performance of selected pilot bridges and to develop a methodology to assess bridge performance. The author discussed that finite element modeling of the pilot bridges was intended not only to assist with instrumentation decisions but also to provide further insight into the behavior of these bridges, which cannot be achieved solely from field testing of the bridges. Finite element models were developed to study the effect of the inclusion of various bridge parameters in the model, such as bridge skew, degree of composite action, thermal gradient, and level of support restraint, on the response of bridges. Initially, the suitability of different modeling techniques and of elements used to model the primary bridge components was assessed using simple models for which analytic solutions are readily available. Based on the studies, it was concluded that shell elements were adequate to model the bridge deck, and beam and shell elements are both satisfactory to model the bridge girders. From the dynamic analyses of two bridges, flexural modes of vibration were found to be highly sensitive to support restraints and to know how the guardrails were modeled and less sensitive to the inclusion of bracing and thermal gradients in the model. The finite element models using extreme boundary conditions were successful in bracketing the field response. The factors identified from these analyses were considered in the analysis of the pilot bridge. Different support restraints and the inclusion of skew and level of composite action in the model had noticeable impact on both the static and dynamic responses of the bridge. The results from these analyses were used to assist with instrumentation decisions prior to field testing. The general-purpose software ABAQUS [1.29] was used to perform the finite element analyses. Two different methods were considered to model the bracing members. In the first method, each part of the bracing assembly was modeled using single linear beam element (B31), available in ABAQUS element library. The author suggested that since no member loads were applied to the bracing, the maximum order of the deflected shape would be cubic and the shape functions assumed in case of linear beam element are cubic. Therefore, it should be able to represent the deflected shape correctly. In the second method, entire bracing assembly was represented by a single beam element modeled at the girder centroid. The effective cross-sectional area was calculated by imposing unit displacement in the horizontal direction to the bracing assemblage. Rigid links were used to connect the bracing with the girder.

Liu and Roe [7.11] discussed the use of headed studs in steel-concrete composite bridges to resist longitudinal shear forces at the interface of steel girder and concrete slab. It was shown that since these studs were subjected to high-cycle fatigue loading due to the growth of traffic and increase in train speed, the study highlighted the dynamic structural behavior of the shear studs during train passages. Different fatigue endurance models were employed for fatigue life estimation. In addition, a parametric study was performed to investigate the effects of different parameters that influence the fatigue life of shear studs. Finally, a fatigue life-cycle design procedure based on the train-bridge interaction analysis and the fatigue endurance model was proposed. A numerical model for a composite bridge with a span of 36 m was developed using ANSYS [6.10]. The concrete slab was connected to the steel girder by headed studs with a diameter of 22 mm. The stud spacing was initially set to 200 mm. The steel girder was represented by shell elements (SHELL63) available in ANSYS element library, the concrete slab was modeled with solid element (SOLID45), and spring elements (COMBIN14) were chosen to represent the headed shear studs. In the study, the corresponding nodes of the concrete slab and steel girder were connected by spring elements in the longitudinal direction and coupled in the other directions. The characteristic of the spring element was derived from load–slip curves obtained by stud push-out tests [7.12]. The change of the stud stiffness during the design life was not taken into account. The train type considered in the study [7.11] was a high-speed train. It was composed of a locomotive followed by eight-passenger cars and another locomotive. The length of the locomotive was 19.7 m, while the length of the passenger car was 26.1 m. The average static axle loads for the locomotives and passenger cars were 176.4 and 112.9 kN, respectively. The train speed was set to 300 km/h. The dynamic response of the bridge was predicted through a moving load model and, alternatively, through a train-bridge interaction model. In the moving load model, the train was simplified as a series of moving loads, while the train-bridge interaction model incorporated subsystems for the train and the bridge. Each vehicle is considered as an independent entity with one car body, two bogies, and four wheel sets. The bogies and the wheel sets were linked by horizontal and vertical springs and dampers. The train subsystem and the bridge subsystem were coupled by the interaction forces and the compatibility of the displacements at the contact points [6.13]. It was shown that during the train passage, variable amplitudes of fatigue loading were generated. Since the train-bridge interaction model predicted the most realistic behavior of the bridge subjected to moving trains, it was recommended for further studies.

Brackus [7.13] discussed that full-depth, precast panel deck systems were becoming common in bridge installation and repair. Therefore, the structural behavior of these systems was the subject of the analyses performed in the study. A steel I-girder bridge containing a precast panel deck system was demolished and provided two full-scale specimens for this project. Destructive testing was performed on the specimens to investigate three failure modes comprising flexural, beam shear, and punching shear. Finite element models were developed using ANSYS [6.10] software to replicate experimental behavior. It was found that the elastic, postelastic, and ultimate behavior of the full-scale bridge sections containing precast panel deck systems can be predicted by the analytic models. The study also investigated changes in dynamic behavior as the system was subjected to flexural yield and failure. Point loads were applied and removed in increments, and dynamic testing was conducted at each load level. It was found that significant damage is somewhat noticeable by monitoring the changes in natural frequencies. The finite element modeling of the bridge specimens was constructed and analyzed using ANSYS. To obtain a comprehensive representation of the bridge specimens and their multiple failure modes, four finite element models were constructed. Each model contained various elements to accurately simulate experimental behavior. The mathematical representation of physical elements was prescribed by four criteria comprising element type, real constants, material association, and key options. The element type designated the element shape, degrees of freedom, and modeling capabilities. The primary elements used in the study were shell elements (SHELL181) and solid elements (SOLID65). In all cases, the girder webs and flanges were modeled with shell elements and the concrete deck was modeled with solid elements. Three connecting elements were used in the study comprising (TARGE170), (CONTA173), and (LINK8) elements available in ANSYS [6.10]. The contact and target elements were used to model the bond at the concrete/steel and concrete/grout interfaces. The SHELL181 element was a quadrilateral planar element with six degrees of freedom at each node (three translational and three rotational). The SOLID65 element was an eight-node solid element with three translational degrees of freedom at each node. Special features of this element include rebar reinforcement and support of a brittle concrete material model, which is capable of compression crushing and tension cracking. Real constants for SOLID65 specify the reinforcement properties by designating a reinforcement material, volumetric ratio of reinforcement to base material, and two angles that describe the orientation of reinforcement. A nonlinear steel material was defined with a multilinear isotropic hardening plasticity model using experimental results. Concrete was modeled using a bilinear stress-strain curve.

Vayas et al. [7.14] presented a modeling technique for simulating steel-concrete composite straight bridges, which was previously detailed in Refs. [7.15,7.16]. The proposed model was based on the representation of steel I-girders through the use of equivalent trusses. The concrete slab was represented by a set of bar elements. Diaphragms and stiffeners could be also taken into account. In contrast to the grillage model, which was used for the analysis of bridges, the recommended 3-D model allowed for a more reliable prediction of deformations and internal forces. The study [7.14] discussed the extension of the model to skewed composite bridges. The presence of skew made the analysis complicated, and for this reason, the grillage analysis was not recommended. The authors showed that phenomenons like differential deflections of the main girders during concreting and lateral displacements of the flanges could be adequately predicted using the proposed model. The proposed models of composite bridges, using a spatial system of beam–like structural elements, could be also used for stability analysis of skewed bridges. Worked examples were provided to illustrate the setup procedure of the proposed modeling and to compare the different ways of analysis. To overcome the difficulties of the grillage and finite element models, a 3-D truss model was proposed where the steel I-girders were modeled by equivalent trusses [7.15,7.16] while the deck slab by a grillage of concrete beams. The main intention was the set up of a global model, which can be used during the erection stages and deck concreting and for the serviceability and ultimate limit states. In the modeling of a composite girder through the use of an equivalent truss, the flanges of the truss were modeled as beam elements with cross section composed of the flange and part (1/3) of the web of the steel girder. The flanges were connected by a hybrid combination of truss and beam elements that represented the web of the steel girder. The concrete section was represented by another beam element connected with the upper flange of the truss through the appropriate offset. In order to verify the validity of the proposed model, numerical investigations for deformations, stresses, buckling, and dynamical modes were performed for a simply supported beam with either steel or composite cross sections. In the grillage analysis of the concrete slab of a composite bridge, concrete slab can be represented by a grillage of interconnected beams. The longitudinal stiffness of the slab was concentrated in the longitudinal beams and the transverse stiffness in the transverse beams. In order to show the differential deflections that occur on a skewed bridge during concreting and to make a first comparison between the two models, a skewed bridge was investigated using a 3-D finite element analysis. The deck of a skewed bridge may be represented through a grillage of beam–like elements. The bridge was simply supported at one edge (hinged supports), while on the other edge, there was free translation along the longitudinal axis.

Adamakos *et al.* [7.17] presented a modeling technique for steel composite bridges, which was previously detailed in Refs. [7.15,7.16]. The proposed modeling technique was based on the representation of steel I-girders by equivalent trusses. The concrete slab was represented by a set of bar elements and the bearings by appropriate springs. Diaphragms and stiffeners may also be taken into account. It was shown that, in comparison to the grillage model, which was usually used for the analysis of bridges, the proposed 3-D model allowed a reliable prediction of deformations, internal forces, and stresses. In addition, it was shown that curved bridges displayed unique behavior characteristics, and for this reason, a grillage analysis was not always suitable. Therefore, the authors concluded that the proposed modeling of composite bridges, using a spatial system of beam-like structural elements, was applied in the study for the modeling of curved composite bridges. Worked examples were provided to illustrate the setup procedure of the proposed modeling and to compare its results with those of the corresponding finite element models. The authors mentioned that a bridge analysis model should be based on the following: (1) The model should reflect the structural response in terms of deformation, strength, and local and global stability; (2) the model should include as many structural elements and parts (cross frames, stiffeners, bearings, etc.) as possible and their possible eccentric connections; (3) the model should cover all construction stages and loading cases; (4) loads should be easily introduced; (5) the model should allow the performance of dynamic analysis and include the most important modes; and finally (6) the model should run with a common analysis and design software. The authors also mentioned that the structural system must reproduce the 3-D behavior of a bridge as accurately as possible. This was achieved through the representation of the steel I-girders by equivalent trusses. The deck slab was idealized by a grillage of concrete beams. The main concept was based on the setup of a global model, which will be easy to modify during the different construction stages, including stages of erection or deck concreting. The study showed that curved composite bridges displayed unique behavior characteristics. The presence of curvature affected the geometry and, as a consequence, the behavior of the structure. Curved bridges were subjected to coupled torsion and bending because of the curvature, and their analysis was more complex than that of straight bridges.

Gara et al. [7.18] proposed a simplified method of analysis for the design of twin girder and single-box steel-concrete composite bridge decks. The method relied on the use of the real width of the slab for the whole bridge length when performing the global analysis, that is, without modifying the deck geometry based on the effective width method, and the ability to evaluate the normal longitudinal stress distribution on the slab by means of a cross-sectional analysis considering the internal actions obtained from the global analysis. The authors considered that in the latter cross-sectional analysis, the properties of the concrete component were based on an effective width calculated using proposed analytic expressions presented in the study. It was shown that the proposed approach was capable of handling different loading conditions, such as constant uniformly distributed loads, envelopes of transverse actions due to traffic loads, support settlements, and concrete shrinkage. These analytic expressions were obtained based on the results of a parametric study performed by means of the finite element formulation described in the first part of the study. The accuracy of the proposed approach was validated for a typical four-span bridge with constant cross section throughout its length against the results obtained based on the finite element method. In addition, a case study of a bridge with varying cross section was considered to show the effectiveness of the proposed methodology.

A new type of beam-to-beam joint used to connect continuously composite beams in small- and medium-span bridges was proposed by Somja et al. [7.19]. The proposed joint was realized by encasing totally the two composite beam ends into a massive composite reinforced concrete block. A direct contact between the ends of the bottom flanges of the steel girders over the support ensured the transfer of the compression forces. The authors designed and fabricated a half-scale joint specimen. The specimen was tested under fatigue loading and monotonically increased loading up to the specimen failure. A numerical finite element model was developed. And the numerical results were compared against the experimental results. In addition, a parametric study was performed to investigate the influence of key parameters governing the joint behavior. The influence of the behavior of this type of joint on the global analysis of a continuous composite beam was studied. Furthermore, a worked example of a two-span continuous railway bridge was presented and effects of intermediate beam-to-beam joint characteristics on the bridge behavior were discussed. The authors developed a 2-D finite element model involving beams and springs. The model reproduced the moment-rotation curve and the slip distribution along the beam. The 2-D model was developed with FineLg software [7.20]. The software is a general nonlinear finite element program first written by Frey [7.21] and mainly developed by de Ville de Goyet [7.22]. Specific concrete beam elements were developed by Boeraeve [7.23]. The beam elements were able to simulate structures undergoing large displacements but small deformations and they were developed using a corotational total description. A 2-D Bernoulli fiber beam element with three nodes and seven degrees of freedom was considered. The total number of degrees of freedom corresponded to one rotational and two translational degrees of freedom for each two nodes located at beam element ends and one relative translational degree of freedom for the node situated at midlength of the beam element. The introduction of the relative translational degrees of freedom for the node at midlength of the beam element was necessary to account for the strong variation of the centroid position when the behavior of the section was not symmetrical. Such situation occurred for concrete sections when cracking propagated. The authors showed that, as usual for fiber element, the section forces at the element nodes were computed using both a longitudinal and transversal integration scheme. The integration along the beam length was performed using four integration points. For each longitudinal integration point, a transversal integration was performed using a multilayer-type scheme. The section was divided into a certain number of layers, each of which was assumed to be in uniaxial stress state. At each transversal integration point, the state of deformation and stress was computed. The connection between the concrete beam element and the steel beam element was introduced by means of specific connection elements comprising two transversal springs and two rotational springs to avoid uplift. Longitudinal springs were uniformly distributed along the element. The resolution of the nonlinear problem was performed using classical algorithms, Newton-Raphson scheme with arc length method.

Zaforteza and Garlock [1.47] investigated numerically the fire response of steel girder bridges by developing a 3-D numerical model for a typical bridge of 12.20 m span length. A parametric study was performed considering different axial restraints of the bridge deck, different types of structural steel for the girders, different constitutive models for carbon steel, different live loads, and different fire loads. The numerical study showed that restraint to deck expansion coming from an adjacent span or abutment should be considered in numerical models. Also, times to collapse were very small when the bridge girders are built with carbon steel (between 8.5 and 18 min), but they can almost double if stainless steel is used for the girders. The authors recommended that stainless steel can be used as a construction material for girder bridges in a high-fire-risk situation. It was also concluded that the methodology developed in the study and the results obtained can be useful for researchers and practitioners interested in developing and applying a performance-based approach for the design of bridges against fire. The numerical study was performed using ABAQUS [1.29]. Due to the symmetry, only half of the bridge was modeled. An uncoupled thermomechanical analysis was used, where in the first phase, a thermal heat transfer analysis provided transient nodal temperatures with respect to time. In the second phase, a structural analysis was performed, and the nodal temperatures were read from the thermal analysis. For the thermal analysis, DC3D8 was employed, which is a 3-D eight-noded linear heat transfer brick element with one degree of freedom per node. For the structural analysis, element C3D8 was used, which is a 3-D eight-noded solid continuum element with three degrees of freedom per node that is compatible with DC3D8 element. Finite element analyses included geometric and material nonlinearities. Since there is no structural connection between the concrete slab and the girder (the bridge was not a composite bridge), the slab was included in the thermal phase of the analysis, but then was deactivated in the structural analysis. In this manner, only the thermal impact of the slab was considered. A finer mesh was used near the supports and the stiffeners because these are areas of high stress and more susceptible to local buckling. The finite element model had 533 nodes and 6560 solid elements. The efficiency of the mesh and finite element model was tested by checking that the difference between the stresses and deflections due to dead loads at ambient temperature given by the beam theory and the FE model was negligible and by checking that an increase of the number of elements in the areas where the mesh was coarser did not have any significant influence in the thermal and structural results. Appropriate boundary conditions were used at the midspan section of the bridge to consider that only half of its structure was modeled. Specifically, midspan section had free vertical displacement, but it was restrained from rotating and from translating on the longitudinal axis. In addition, a vertical support was provided along the surface of the bottom flange beneath the stiffener. Finally, and only for the "fix" analyses, a rigid solid block was created at a distance from the outer cross section of the bridge equal to the width of the expansion joint. This rigid solid block simulated the existence of an adjacent span or abutment, and its goal was to ensure that axial expansion of the nodes of the outer cross section of the bridge was restrained once their horizontal displacement equaled the width of the expansion joint.

Shifferaw and Fanous [1.51] investigated fatigue crack formation in the web gap region of multigirder steel bridges. The authors have shown that the region has been a common occurrence of fatigue crack formation due to differential deflections between girders resulting in diaphragm forces that subject the web gap to out-of-plane distortion. The study investigated the behavior of web gap distortion of a skewed multigirder steel bridge through field testing and finite element analyses. The study also investigated different retrofit methods that include the provision of a connection plate between the stiffener and the girder top flange, loosening of the bolts connecting the cross bracing to the stiffener, and supplementing a stiffener plate opposite to the original stiffener side. The study has shown that the connection plate addition and loosening of bolts alternatives were effective in reducing induced strains and stresses in the web gap region. An inverse relationship between web gap height and induced strains and stresses with the shortest web gap height resulting in the highest strains due to increased bending by diaphragm forces in the web was also shown. The authors concluded that expressions developed to relate vertical stresses and relative out-of-plane displacements combined with measurements of out-of-plane displacements by transducers can be utilized for the prediction of induced stresses at other critical web gap regions of the bridge and at critical locations in the web gaps of similar bridges. The authors used the solid modeling option available in ANSYS [6.10] for node and element generation due to the complexity in geometry and details of the structural members of the bridge. The ANSYS shell element (SHELL63) was adopted for modeling. SHELL63 has four nodes, each with six degrees of freedom, and is capable of modeling bending and membrane behavior. The bridge investigated consisted of four spans with the web gap located near the central pier. Traffic loads acting on the two outer spans, that is, remote from the web gap, had less critical effects on the differential deflection between the exterior and adjacent girders and hence on the out-of-plane distortion of the web gap near the middle pier. Therefore, only the middle two spans were included in the coarse finite element model. The weld connecting the girder flanges to the web was modeled with shell elements that have variable thickness. To connect the nodes corresponding to the plate elements of the bridge deck and girders, rigid link elements were used. This was defined in ANSYS using constraint equations with the nodes along the flange and the deck labeled as master and slave nodes. The ends of the two spans representing the cut sections near piers opposite to the central pier were modeled by imposing fixed boundary conditions at these locations. The support provided by the central pier was modeled as a roller support that restrained the displacement in the direction perpendicular to the plane of the deck. The web gap under investigation was located near the central pier. The size of the elements in the web gap region of the coarse model did not coincide with the spacing of the strain gages that was utilized in the field test and would not allow direct comparison between the analytic and field test strain results. Hence, a submodel near the vicinity of the web gap region was built to ensure better accuracy in capturing the local distortion behavior. The submodel of the web gap region included portions of the bridge deck, bridge girder, stiffener plate, and cross bracing. The effect of the location of the cutoff boundaries on the stress and strain results was also investigated. This was accomplished by comparing the results obtained from the submodel near the cut boundaries with those obtained from the coarse model. The sensitivity study showed that including a portion that is 25 in. (635 mm) long on each side of the stiffener would be sufficient to accurately analyze the web gap region. The common size of the elements used to idealize the components included in the submodel was 0.25 in. (6.35 mm) by 0.25 in. (6.35 mm). The total number of elements in the submodel was about 50,000. Mesh sensitivity by considering smaller element sizes was also examined; however, the differences in the results were negligible.

Zhou *et al.* [7.24,7.25] investigated steel bridges for high-speed trains, which may be vulnerable to excessive fatigue damage owing to stronger dynamic effects induced by the increased train speed. In part I [7.24], the authors conducted dynamic tests on a composite railway bridge for high-speed trains. In addition, a detailed finite element model of the bridge was developed and validated against the dynamic test results. Six types of structural details in the bridge were considered for fatigue evaluation. The stress history of each concerned detail during a single train passage was generated by the validated finite element model. The stress spectrum was used to calculate the fatigue damage of each detail. Among various structural details, the load carrying fillet weld around the gusset plate of the diagonal bracing at the bridge bearing was predicted to be the most fatigue critical detail. In the study, a general methodology for determination of fatigue critical details was presented. In part II [7.25], the authors investigated fatigue assessment based on the dynamic stresses predicted by different approaches, that is, static analysis considering dynamic amplification factor, direct dynamic analysis with a moving load model, or a train-bridge interaction model. Due to the large size of the investigated bridge including seven simply supported spans, the finite element model of the overall bridge would result in a large computational cost. Therefore, a simplified model of a single span was developed with appropriate boundary conditions on the rails and the ballast to simulate the weak coupling between adjacent spans. The second span from the bridge was modeled after the design drawings, by using the general-purpose finite element software ABAQUS [1.29]. Due to symmetry, the span was divided into three types of segments, that is, segment A at the bridge bearing, segment B1 in the external portion near the bearing, and segment B2 in the internal portion close to the midspan. The segments

B1 and B2 are identical in profile except for the plate thickness in webs and bottom flanges. The diagonal and horizontal bracings were modeled by a combination of beam elements in the central part and shell elements at the ends. Couplings of six degrees of freedom were used to join the beam element with the shell elements at a distance of roughly 0.8 m from the bolt connection. The preloaded high-strength bolt connections were assumed to be rigid, realized by TIE constraint between bolt holes and gusset plate. TIE option in ABAQUS is a surface-based constraint used to make all the translational and rotational degrees of freedom equal for a pair of surfaces. Three types of segments were assembled one by one to build up the global finite element model of the second span. The steel box girder was simulated by shell element S4, and the bracings were modeled by beam element B31 in combination with shell element S4. The ballast was assumed as continuum with material properties as found in literature [7.26]. The concrete deck and the safety barrier were modeled by brick element C3D8. The shear studs were not explicitly modeled and TIE constraints were applied to connect the concrete deck to the top flanges of the steel box girder. The FE model contained about 84,000 shell elements, 6400 solid elements, and 1800 beam elements, resulting in more than 85,500 nodes. The symmetrical vibration modes of the span were predominantly excited by the train passages. Therefore, symmetrical boundary conditions were adopted at the ends of the span. The longitudinal translations and rotations of the rails and the ballast were restricted. The bridge bearing system included two fixed bearings, three bidirectional sliding bearings, and one unidirectional sliding bearing. The bearing system allowed relative movements due to thermal expansion and accommodated the bridge to movements due to live loads. The train was composed of a locomotive followed by eight-passenger cars and another locomotive. In the static and dynamic finite element analyses, the moving load model, in which the train axles were represented by a series of moving constant forces, was adopted to simulate the train passage. An ABAQUS user subroutine DLOAD coded by FORTRAN was used to realize the loading scheme of the train.

Finally, Liu et al. [7.27] investigated the performance of composite joints in a truss bridge with double decks. Fatigue tests of three composite joints with different connectors such as headed studs, concrete dowels, and perforated plates under constant repeated loading were carried out. The responses of displacement, strain distribution, crack development, relative slip between concrete, and steel were observed after different loading cycles. The experimental results showed that the deflection increased almost linearly with applied load even after certain repeated loading cycles, but the stiffness reduced gradually with the repeated loading cycles. No serious damage occurred except tiny cracks at the steel-concrete interface caused by slip after 2 million repeated loading cycles, which meant that all three composite joints have good fatigue performance. Based on experimental works, 3-D finite element models of composite joints were developed. The results from finite element analysis were consistent with those from tests in terms of strength and stiffness. In addition, the fatigue details involving reinforcing bars, welding seams, and shear connectors were evaluated according to related specifications. It was concluded that the presented overall investigation may provide reference for design and construction of composite joints in composite truss bridges. The modeling of each composite joint was carried out by using finite element method and software ANSYS [6.10]. The solid elements (SOLID45) were used to simulate both concrete chord and steel structures. As for composite action between concrete and steel trusses, contact elements (TARGE170 and CONTA173) considering the adhesion effects of steel-concrete interface were employed in numerical models. When the surfaces are in contact, normal forces develop between two materials. On the contrary, if the contact element is in tension, the contact surfaces separate from each other resulting in no bonding development. In terms of connectors, three spring elements (COMBIN14) for each stud were applied to simulate shear and axial forces in three directions; gusset plate with concrete dowels and perforated plate connectors were modeled by solid elements (SOLID45) and contact elements, reinforcing bars through the holes, were not included for the reasons of simplification and safety. The free ends of truss members were restrained as hinges like the same condition in the test, and one end of concrete chord was subjected to uniformly distributed load. Because the structure was almost in elastic state under design load that was confirmed in static test, linear elastic analysis was used to investigate the response at different load steps until to design load. More information regarding recent investigations on steel-concrete composite bridges can be found in the state-of-the-art review presented by Ranzi et al. [7.28].

7.3 FINITE ELEMENT MODELING AND RESULTS OF EXAMPLE 1

The first example presented in this chapter is for a simply supported composite steel plate girder tested by Mans [7.29], which is denoted in this study as G1 as shown in Figure 7.1. The main objective of the test was to

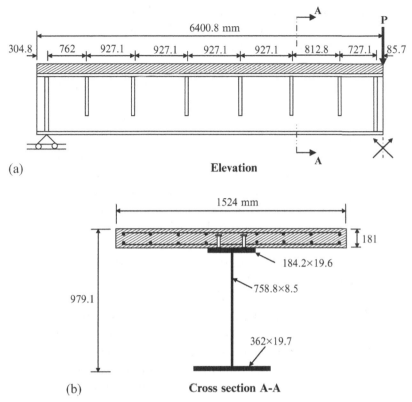

Figure 7.1 General layout and dimensions of composite girder G1.

investigate the ultimate moment resistance and ductility of the composite girder. The structural steel used in the test was a high-strength steel HPS70W having a nominal yield stress of 482 MPa (70 ksi). The general layout and dimensions of composite plate girder G1 are shown in Figure 7.1. The composite plate girder had an overall length of 12,801 mm and a length between supports equal to 12,192 mm. The steel plate girder had a web of 758.8 × 8.5 mm, an upper flange of 84.2 × 19.6 mm, and a lower flange of 362 × 19.7 mm. The flange and web portions of the steel plate girder of G1 had yield and ultimate tensile stresses of 556, 700 MPa and 583, 656 MPa, respectively. The composite plate girder had an overall height of 979 mm. The web of the steel plate girders was strengthened by stiffeners as shown in Figure 7.1 to prevent shear failure. The concrete slab had a width of 1524 mm and a depth of 181 mm. The measured concrete cylinder strength of G2 was 30.5 MPa. The concrete slab had reinforcement steel bars, as shown in Figure 7.1, of Grade 60 having a yield stress of 413 MPa (60 ksi). The reinforcement bars were spaced at 203 mm longitudinally and transversely. The top and bottom reinforcement bars had a cover of 44 mm. The shear connectors were headed studs having a diameter of 19 mm and a height of 114 mm. Eighty pairs of headed studs were used in the composite plate girder G1 as shown in Figure 7.1. The composite plate girders were subjected to a single concentrated load applied at midspan via a spreader beam. The loading was applied in increments using displacement control.

The composite plate girder tested by Mans [7.29] was modeled in this book using ABAQUS [1.29]. In order to obtain accurate results from the finite element analysis, all the composite plate girder components must be properly modeled. The composite plate girder components comprise the steel plate girder, concrete slab, headed stud, and reinforcement bars. The finite element analysis has accounted for the nonlinear material properties and geometry of the components as well as the interfaces between the components that allowed the contact and bond behavior to be modeled and the different components to retain its profile during the deformation of the composite plate girder. The steel-concrete composite plate girder components were modeled using 3-D solid elements (C3D8) available in the ABAQUS [1.29] element library. The elements have three degrees of freedom per node and suit all the strengthened composite girders since lateral torsional buckling of the steel beam compression flange is limited by the surrounding concrete slab properly connected to the top flange via headed stud shear connectors. Only half of the composite plate girder was modeled due to symmetry as shown in Figure 7.2. The total number of elements used in the model was 7958 elements. Different mesh sizes were tried to choose the reasonable mesh that provides both reliable results and less computational time. All the nodes in the middle symmetry surface were prevented to displace in direction 2-2. The roller support nodes were prevented to displace in direction 3-3 only. The load was applied in increments as concentrated static loads at midspan, which is identical to the experimental investigation [7.29]. The nonlinear geometry was included to deal with the large displacement analysis.

The shear forces across the steel plate girder-concrete slab interface of G1 test [7.29] are transferred by the mechanical action of headed stud shear connectors. The load-slip characteristic of headed stud is of great importance in modeling the shear interaction between steel plate girder and concrete slab. The region around the stud is a region of severe and complex stresses. The load-slip characteristic of headed stud depends on many factors including

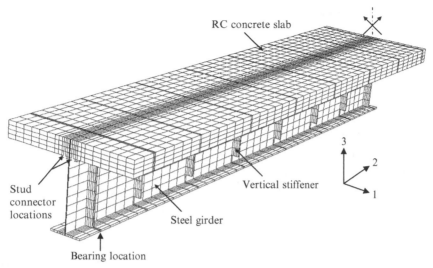

RC concrete slab

Stud connector locations

Steel girder

Vertical stiffener

Bearing location

Figure 7.2 Finite element mesh of composite beam G1.

type of concrete slab, diameter of stud, height of stud, strength of stud, and concrete strength (see Section 5.6 of Chapter 5 in this book). Earlier experimental and numerical investigations reported by the author [2.68,2.69] provided detailed information regarding the capacity and load-slip behavior of headed stud shear connectors in composite girders with solid slabs. In this study, the load-slip characteristic of the studs used in the test [7.29] was predicted based on the detailed experimental and numerical investigations [2.68,2.69]. Following the same approach [2.68,2.69], the load-slip characteristic of the stud was inserted in the finite element model (Figure 7.2) using nonlinear springs in direction 2-2 at the location of the headed studs. On the other hand, the vertical pressure between the concrete slab and the steel beam was simulated by vertical rigid springs with high stiffness in direction 3-3 at the locations of the headed studs.

The steel plate girder-concrete slab interface was modeled by interface elements available within the ABAQUS [1.29] element library. The method requires defining two surfaces that are the master and slave surfaces. In modeling the steel beam-concrete slab interface, the master surface within the model was the top flange of the steel beam upper surface and the slave surface was the bottom surface of the concrete slab. The interface elements are formed between the master and slave surfaces and monitor the displacement of the slave surface in relation to the master surface. When the two surfaces remain in contact, the slave surface can displace relative to the master surface based on the coefficient of friction between the two surfaces. When the two surfaces are in contact, the forces normal to the master surface can be transmitted between the two surfaces. When the two surfaces separate, the relative displacement between the two surfaces can still be monitored but the forces normal to the master surface cannot be transmitted. However, the two surfaces cannot penetrate each other.

Concrete was modeled using the damaged plasticity model implemented in the ABAQUS [1.29] material library. The model provides a general capability for modeling plain and reinforced concrete in all types of structures. The concrete damaged plasticity model uses the concept of isotropic damaged elasticity, in combination with isotropic tensile and compressive plasticity, to represent the inelastic behavior of concrete. The model assumes that the uniaxial tensile and compressive responses of concrete are characterized by damaged plasticity. Under uniaxial compression, the response is linear until the value of proportional limit stress is reached, which is assumed to equal 0.33 times the design compressive strength. Under uniaxial tension, the stress–strain response follows a linear elastic relationship until the value of the failure stress. The tensile failure stress was assumed to be 0.1 times the compressive strength of concrete that is assumed to be equal to 0.67 times the measured concrete cube strength. The concrete cube strength is assumed to be equal to 1.25 the concrete cylinder strength. The softening stress–strain response, past the maximum tensile stress, was represented by a linear line defined by the fracture energy and crack band width. The fracture energy G_f (energy required to open a unit area of crack) was taken as 0.12 N/mm as recommended by the CEB [7.30] and ABAQUS manual [1.29]. The fracture energy divided by the crack band width was used to define the area under the softening branch of the tension part of the stress–strain curve. The crack band width was assumed as the cubic root of the volume between integration points for a solid element, as recommended by CEB [7.30]. The reinforcement bars used in the concrete slab of the composite plate girder test [7.29] were modeled using the (REBAR option) available in the ABAQUS [1.29] element library. Further details regarding the damaged plasticity model and the modeling of reinforcement bars can be found in Section 5.4.3.2 of Chapter 5 in this book.

The stress–strain curves for the structural steel and reinforcement bars given in the EC3 [2.11] and EC2 [2.27], respectively, were adopted for the finite element model of the composite plate girder G1 with measured values of the yield stress and ultimate stresses reported in Ref. [7.29]. The material behavior provided by ABAQUS [1.29] (using the PLASTIC option) allows a nonlinear stress-strain curve to be used (see Section 5.4.2 of Chapter 5 in this book). The first part of the nonlinear curve represents the elastic part up to the proportional limit stress with Young's modulus of (E) 200 GPa and Poisson's ratio of 0.3 that were used in the finite element model.

The developed finite element model for the composite plate girder G1 (see Figure 7.2) was verified against the test results detailed in Ref. [7.29]. The failure loads, failure modes, and load-midspan deflection curves obtained experimentally and numerically using the finite element model were compared. The deformed shapes of composite plate girder G1 at failure observed numerically are shown in Figure 7.3, which is in good agreement with the experimental observations reported in Ref. [7.29]. The failure mode observed experimentally and confirmed numerically was combined steel yielding (SY) in the bottom flange of the steel plate girder and concrete crushing (CC). The data obtained from ABAQUS [1.29] have shown that the von Mises stresses at the maximum stressed fibers at the bottom flanges at midspan exceeded the measured yield stress as well as the stresses in the concrete slab at midspan reached the concrete compressive strength. In Figure 7.4, the stress (principal stresses in direction 2-2) contours at failure of the composite plate girder G1 are plotted. It can be seen that the yield stresses were reached at midspan in the concrete (compressive stresses with negative sign) and lower steel plate girder flanges (tensile stresses with positive sign). In addition, in Figure 7.5, the plastic strain (principal strains in direction 2-2) contours at failure of the composite plate girder G1 are

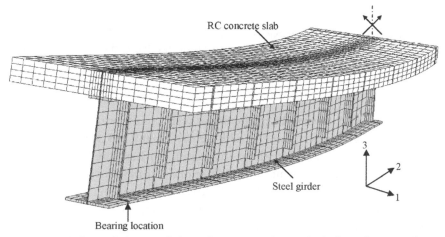

Figure 7.3 Deformed shape at failure of composite beam G1 (enlarged 10 times).

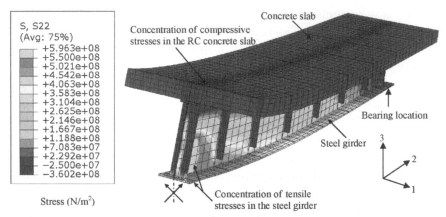

Figure 7.4 Stress (principal in direction 2-2) contours at failure of composite beam G1 (enlarged 10 times). (For color version of this figure, the reader is referred to the online version of this chapter.)

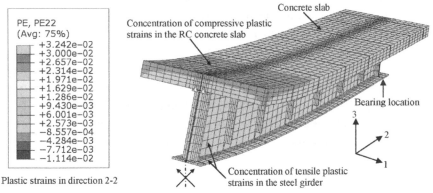

Figure 7.5 Plastic strain (principal in direction 2-2) contours at failure of composite girder G1 (enlarged 10 times). (For color version of this figure, the reader is referred to the online version of this chapter.)

plotted. It can be seen that the plastic strains were concentrated at midspan in the concrete (compressive strains with negative sign) and lower steel plate girder flange (tensile strains with positive sign). Furthermore, in Figure 7.6, the von Mises yield stress contours at failure of the composite plate girder G1 are plotted. It can be seen that the yield stresses were reached at midspan in the lower flange of the steel plate girder. The load–midspan deflection curves predicted experimentally and numerically were compared as shown in

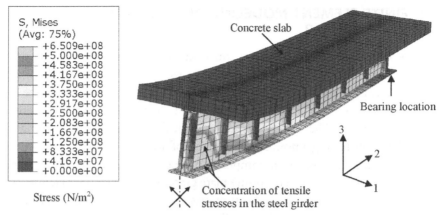

Figure 7.6 Stress (von Mises) contours at failure of composite girder G1 (enlarged 10 times). (For color version of this figure, the reader is referred to the online version of this chapter.)

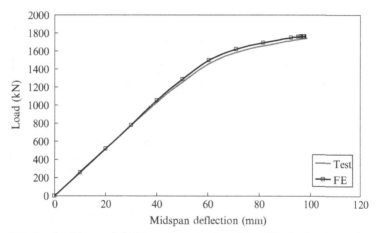

Figure 7.7 Load-midspan deflection of composite girder G1 obtained experimentally and numerically. (For color version of this figure, the reader is referred to the online version of this chapter.)

Figure 7.7. It can be shown that generally good agreement was achieved between experimental and numerical relationships. The ultimate failure load observed in the test [7.29] was 1743.7 kN at a deflection of 95.6 mm, while the ultimate failure load predicted from the finite element analysis was 1765.5 kN at a deflection of 97.7 mm. The finite element failure load was 1.3% higher than that observed in the test.

7.4 FINITE ELEMENT MODELING AND RESULTS OF EXAMPLE 2

The second example presented in this chapter is for another simply supported composite steel plate girder tested by Mans [7.29], which is denoted in this study as G2 as shown in Figure 7.8. Once again, the main objective of the test was to investigate the ultimate moment resistance and ductility of the composite girder. The structural steel used in the test was a high-strength steel HPS70W having a nominal yield stress of 482 MPa (70 ksi). The general layout and dimensions of composite plate girder G2 are shown in Figure 7.8. Similar to G1, the composite plate girder G2 had an overall length of 12,801 mm and a length between supports equal to 12,192 mm. The steel plate girder had a web of 760.4 × 8.9 mm and upper and lower flanges of 182.6 × 19.6 mm. The measured flange and web portions of the steel plate girder of G2 had yield and ultimate tensile stresses of 556, 700 MPa and 583,

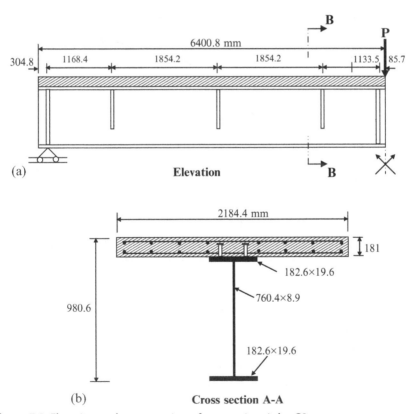

Figure 7.8 Elevation and cross section of composite girder G2.

656 MPa, respectively. The web of the steel plate girders was strengthened by stiffeners as shown in Figure 7.8 to prevent shear failure. The concrete slab had a width of 2184 mm and a depth of 181 mm. The composite plate girder had an overall height of 980 mm. The measured concrete cylinder strength of G2 was 52.5 MPa. The concrete slab had reinforcement steel bars, as shown in Figure 7.8, of Grade 60 having a yield stress of 413 MPa (60 ksi). The reinforcement bars were spaced at 209 mm longitudinally and 356 mm transversely. The top and bottom reinforcement bars had a cover of 44 mm. The shear connectors were headed studs having a diameter of 19 mm and a height of 114 mm. Sixty pairs of headed studs were used in the composite plate girder G2 as shown in Figure 7.8. The composite plate girders were subjected to a single concentrated load applied at midspan via a spreader beam. The loading was applied in increments using displacement control.

The composite plate girder G2 tested by Mans [7.29] was modeled in this book using ABAQUS [1.29]. The finite element analysis has accounted for the nonlinear material properties and geometry of the components as well as the interfaces between the components. The steel–concrete composite plate girder components were modeled using 3-D solid elements (C3D8) available in the ABAQUS [1.29] element library. Only half of the composite plate girder was modeled due to symmetry as shown in Figure 7.9. The total number of elements used in the model shown in Figure 7.9 was 7266. Different mesh sizes were tried to choose the reasonable mesh that provides

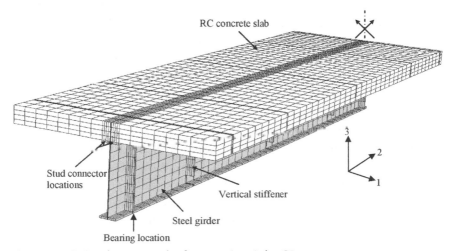

RC concrete slab

Stud connector locations

Vertical stiffener

Steel girder

Bearing location

Figure 7.9 Finite element mesh of composite girder G2.

both reliable results and less computational time. All the nodes in the middle symmetry surface were prevented to displace in direction 2-2. The roller support nodes were prevented to displace in direction 3-3 only. The load was applied in increments as concentrated static loads at midspan, which is identical to the experimental investigation [7.29]. The nonlinear geometry was included to deal with the large displacement analysis.

The shear forces across the steel plate girder-concrete slab interface of G2 test [7.29] are transferred by the mechanical action of headed stud shear connectors. Similar to G1, in this study, the load-slip characteristic of the studs used in the test [7.29] was predicted based on the detailed experimental and numerical investigations [2.68, 2.69].

Under uniaxial compression, the response is linear until the value of proportional limit stress is reached which is assumed to equal 0.33 times the design compressive strength. Under uniaxial tension, the stress-strain response follows a linear elastic relationship until the value of the failure stress. The tensile failure stress was assumed to be 0.1 times the compressive strength of concrete which is assumed to be equal to 0.67 times the measured concrete cube strength. The concrete cube strength is assumed to be equal to 1.25 the concrete cylinder strength. The softening stress-strain response, past the maximum tensile stress, was represented by a linear line defined by the fracture energy and crack band width. The fracture energy G_f (energy required to open a unit area of crack) was taken as 0.12 N/mm as recommended by the CEB [7.30] and ABAQUS manual [1.29]. The fracture energy divided by the crack band width was used to define the area under the softening branch of the tension part of the stress-strain curve. The crack band width was assumed as the cubic root of the volume between integration points for a solid element, as recommended by CEB [7.30]. The reinforcement bars used in the concrete slab of the composite plate girder test [7.29] were modeled using the (REBAR option) available in the ABAQUS [1.29] element library. Further details regarding the damaged plasticity model and the modeling of reinforcement bars can be found in Sections 5.4.3.2 of Chapter 5 in this book.

The stress-strain curves for the structural steel and reinforcement bars given in the EC3 [2.11] and EC2 [2.27], respectively, were adopted for the finite element model of the composite plate girder G1 with measured values of the yield stress and ultimate stresses reported in Ref. [7.29]. The material behavior provided by ABAQUS [1.29] (using the PLASTIC option) allows a nonlinear stress-strain curve to be used (see Section 5.4.2 of Chapter 5 in this book). The first part of the nonlinear curve represents the elastic part up to the proportional limit stress with Young's modulus of (E) 200 GPa and Poisson's ratio of 0.3 that were used in the finite element model. Similar to the composite plate girder G1, the steel plate girder-concrete slab interface was modeled by interface elements available within the ABAQUS [1.29] element library.

The developed finite element model for the composite plate girder G2 (see Fig. 7.9) was verified against the test results detailed in Ref. [7.29]. The failure loads, failure modes, and load–midspan deflection curves obtained experimentally and numerically using the finite element model were compared. The deformed shapes of composite plate girder G2 at failure observed numerically are shown in Figure 7.10, which is in good agreement with the experimental observations reported in Ref [7.29]. The failure mode observed experimentally and confirmed numerically was combined steel yielding (SY) in the bottom flange of the steel plate girder and concrete crushing (CC). The data obtained from ABAQUS [1.29] have shown that the von Mises stresses at the maximum stressed fibers at the bottom flanges at midspan exceeded the measured yield stress as well as the stresses in the concrete slab at midspan reached the concrete compressive strength. In Figure 7.11, the stress (principal stresses in direction 2-2) contours at failure of the composite plate girder G2 are plotted. It can be seen that the yield stresses were reached at midspan in the concrete (compressive stresses with negative sign) and lower steel plate girder flanges (tensile stresses with positive sign). In addition, in Figure 7.12, the plastic strain (principal strains in direction 2-2) contours at failure of the composite plate girder G2 are plotted. It can be seen that the

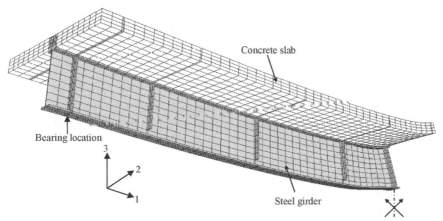

Figure 7.10 Deformed shape at failure of composite girder G2 (enlarged 10 times).

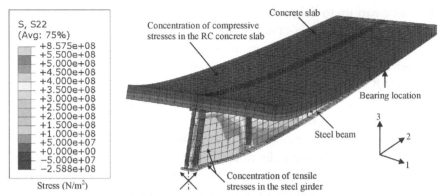

Figure 7.11 Stress (principal in direction 2-2) contours at failure of composite girder G2 (enlarged 10 times). (For color version of this figure, the reader is referred to the online version of this chapter.)

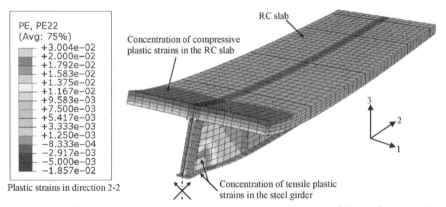

Figure 7.12 Plastic strain (principal in direction 2-2) contours at failure of composite girder G2 (enlarged 10 times). (For color version of this figure, the reader is referred to the online version of this chapter.)

plastic strains were concentrated at midspan in the concrete (compressive strains with negative sign) and lower steel plate girder flange (tensile strains with positive sign). Furthermore, in Figure 7.13, the von Mises yield stress contours at failure of the composite plate girder G2 are plotted. It can be seen that the yield stresses were reached at midspan in the lower flange of the steel plate girder. The load-midspan deflection curves predicted experimentally and numerically were compared as shown in Figure 7.14. It can be shown that generally good agreement was achieved between experimental and numerical

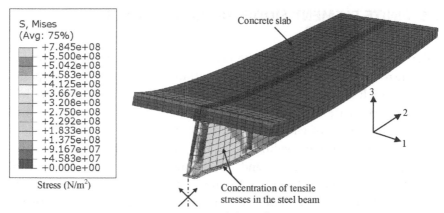

Figure 7.13 Stress (von Mises) contours at failure of composite girder G2 (enlarged 10 times). (For color version of this figure, the reader is referred to the online version of this chapter.)

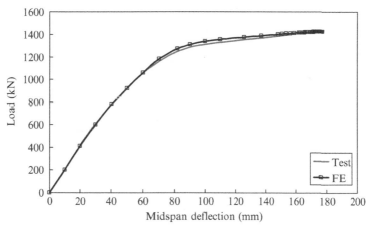

Figure 7.14 Load-midspan deflection of composite girder G2 obtained experimentally and numerically. (For color version of this figure, the reader is referred to the online version of this chapter.)

relationships. The ultimate failure load observed in the test [7.29] was 1432.3 kN at a deflection of 177.8 mm, while the ultimate failure load predicted from the finite element analysis was 1424.9 kN at a deflection of 177.1 mm. The finite element failure load was 0.6% lower than that observed in the test.

7.5 FINITE ELEMENT MODELING AND RESULTS OF EXAMPLE 3

After modeling separate composite plate girders used in bridges, it is now possible to model a full-scale four-span continuous steel-concrete composite bridge, which is detailed in this example 3. The composite bridge was field tested by Burdette and Goodpasture [7.31] to determine ultimate load capacity and mode of failure of the composite bridge. The overall width of the composite bridge width was 10,515.5 mm and the four spans were of 21,336, 27,432, 27,432, and 21,336 mm, with a total length of 97.536 m as shown in Figures 7.15 and 7.15. As shown in Figure 7.15, the bridge had a haunched concrete slab with a constant thickness between haunches of 177.8 mm. The concrete slab rested on four W36 × 170 steel beams spaced at 2540 mm. The measured flange and web portions of the steel beams had yield and ultimate tensile stresses of 275 and 450 MPa, respectively. The concrete slab had a cylinder strength of 47.4 MPa. The composite bridge was loaded at eight locations as shown in Figure 7.16

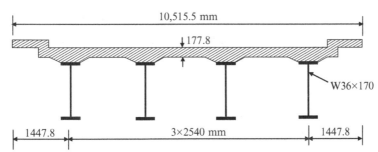

Figure 7.15 Cross section C-C of a four-span continuous steel-concrete composite bridge.

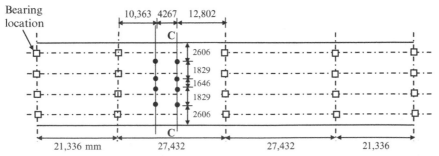

Figure 7.16 Loading configuration of a four-span continuous steel-concrete composite bridge.

simulating the wheel loads of one HS truck in each lane. The point load locations were chosen to produce the maximum positive moment near midspan of the second span. The loads were applied at each of the eight load points through a using 900 kN capacity center hold jack. Since the slab reinforcement and cross-frame details were not reported [7.31], it was calculated according EC2 [2.27] to be T16 spaced at 150 mm transversely and longitudinally. The reinforcement steel bars were Grade 40 having a yield stress of 275 MPa (40 ksi). The top and bottom reinforcement bars had a cover of 40 mm. Cross-frame members were also calculated according to EC3 [2.11] to be two angles back to back of $120 \times 120 \times 12$ spaced at average distances of 7 m.

The full-scale steel–concrete composite plate girder tested by Burdette and Goodpasture [7.31] was modeled in this book using ABAQUS [1.29]. In order to obtain accurate results from the finite element analysis, all the bridge components must be properly modeled. The composite bridge components comprise the steel beams, concrete slab decks, headed stud, reinforcement bars, and support locations. The finite element analysis has accounted for the nonlinear material properties and geometry of the components as well as the interfaces between the components that allowed the contact and bond behavior to be modeled and the different components to retain its profile during the deformation of the composite bridge. The steel–concrete composite bridge were modeled using a combination of 3-D solid elements (C3D8 and C3D6) available in the ABAQUS [1.29] element library. Only half of the composite bridge was modeled due to symmetry as shown in Figure 7.17. The total number of elements used in the model was 9312 elements. Different mesh sizes were tried to choose the reasonable mesh that provides both reliable results and less computational time. All the nodes in the middle symmetry surface were prevented to displace in direction 1-1. The roller support nodes were prevented to displace in direction 3-3 only, while hinged supports were prevented to displace in directions 2-2 and 3-3 only. The load was applied in increments as concentrated static loads at midspan using the Riks method, which is identical to the experimental investigation [7.31]. The steel beams were prevented to displace laterally in direction 2-2 at the locations of the lateral restraints. The nonlinear geometry was included to deal with the large displacement analysis.

The shear forces across the steel beam–concrete slab interface of the bridge test [7.31] were modeled following the same approach [2.68,2.69], and the load-slip characteristic of the stud was inserted in the finite element model (Figure 7.17) using nonlinear springs in direction 2-2 at the location

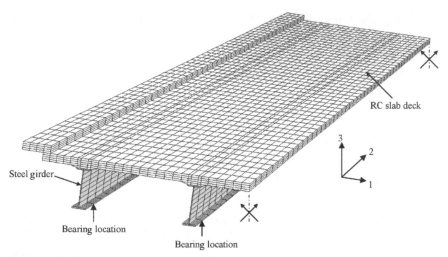

Figure 7.17 Finite element mesh of half of the composite bridge.

of the headed studs. On the other hand, the vertical pressure between the concrete slab and the steel beam was simulated by vertical rigid springs with high stiffness in direction 3-3 at the locations of the headed studs. The steel beam-concrete slab interface was modeled by interface elements available within the ABAQUS [1.29] element library. The stress-strain curves for the structural steel and reinforcement bars given in the EC3 [2.11] and EC2 [2.27], respectively, were adopted for the finite element model of the composite bridge with measured values of the yield stress and ultimate stresses reported in [7.31]. The material behavior provided by ABAQUS [1.29] (using the PLASTIC option) allows a nonlinear stress-strain curve to be used (see Section 5.4.2 of Chapter 5 in this book). The first part of the nonlinear curve represents the elastic part up to the proportional limit stress with Young's modulus of (E) 200 GPa and Poisson's ratio of 0.3 that were used in the finite element model.

Concrete was modeled using the damaged plasticity model implemented in the ABAQUS [1.29] material library. Under uniaxial compression, the response is linear until the value of proportional limit stress is reached which is assumed to equal 0.33 times the design compressive strength. Under uniaxial tension, the stress-strain response follows a linear elastic relationship until the value of the failure stress. The tensile failure stress was assumed to be 0.1 times the compressive strength of concrete which is assumed to be equal to 0.67 times the measured concrete cube strength. The concrete cube strength is assumed to be equal to 1.25 the concrete cylinder strength.

The softening stress-strain response, past the maximum tensile stress, was represented by a linear line defined by the fracture energy and crack band width. The fracture energy G_f (energy required to open a unit area of crack) was taken as 0.12 N/mm as recommended by the CEB [7.30] and ABAQUS manual [1.29]. The fracture energy divided by the crack band width was used to define the area under the softening branch of the tension part of the stress-strain curve. The crack band width was assumed as the cubic root of the volume between integration points for a solid element, as recommended by CEB [7.30]. The reinforcement bars used in the concrete slab of the composite plate girder test [7.29] were modeled using the (REBAR option) available in the ABAQUS [1.29] element library. Further details regarding the damaged plasticity model and the modeling of reinforcement bars can be found in Sections 5.4.3.2 of Chapter 5 in this book.

The developed finite element model for the composite bridge (see Figure 7.17) was verified against the test results detailed in Ref. [7.31]. The failure loads, failure modes, and load-midspan deflection curves obtained experimentally and numerically using the finite element model were compared. The deformed shapes of the composite bridge at failure predicted numerically are shown in Figure 7.18, which is in good agreement with the experimental observations reported in Ref. [7.31]. The failure mode observed experimentally and confirmed numerically was steel yielding (SY) in the bottom flange of the steel beam. The data obtained from ABAQUS [1.29] have shown that the von Mises stresses at the maximum stressed fibers at the bottom flanges of the steel beam at midspan exceeded the measured yield stress. On the other hand, the stresses in the concrete slab at

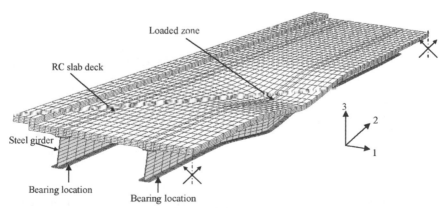

Figure 7.18 Deformed shape at failure of the composite bridge (enlarged 10 times).

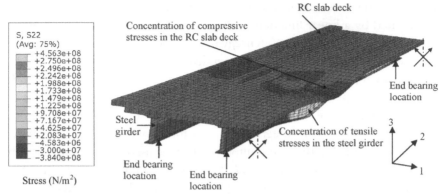

S, S22
(Avg: 75%)
+4.563e+08
+2.750e+08
+2.496e+08
+2.242e+08
+1.988e+08
+1.733e+08
+1.479e+08
+1.225e+08
+9.708e+07
+7.167e+07
+4.625e+07
+2.083e+07
−4.583e+06
−3.000e+07
−3.840e+08

Stress (N/m²)

RC slab deck

Concentration of compressive
stresses in the RC slab deck

End bearing
location

Steel
girder

Concentration of tensile
stresses in the steel girder

End bearing
location

End bearing
location

Figure 7.19 Stress (principal in direction 2-2) contours at failure of the composite bridge (enlarged 10 times). (For color version of this figure, the reader is referred to the online version of this chapter.)

midspan did not reach the concrete compressive strength. In Figure 7.19, the stress (principal stresses in direction 2-2) contours at failure of the composite bridge are plotted. It can be seen that the yield stresses were not reached at midspan in the concrete (compressive stresses with negative sign) and the yield stresses were reached in the lower steel beam flanges (tensile stresses with positive sign). In addition, in Figure 7.20, the plastic strain (principal strains in direction 2-2) contours at failure of the composite bridge are plotted. It can be seen that the plastic strains were concentrated at midspan the lower steel beam flange (tensile strains with positive sign). Furthermore, in Figure 7.21, the von Mises yield stress contours at failure of the composite bridge are plotted. It can be seen that the yield stresses were reached at

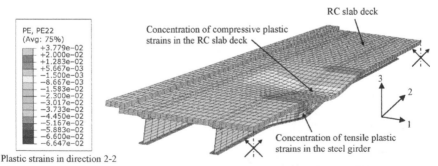

PE, PE22
(Avg: 75%)
+3.779e−02
+2.000e−02
+1.283e−02
+5.667e−03
−1.500e−03
−8.667e−03
−1.583e−02
−2.300e−02
−3.017e−02
−3.733e−02
−4.450e−02
−5.167e−02
−5.883e−02
−6.600e−02
−6.647e−02

Plastic strains in direction 2-2

RC slab deck

Concentration of compressive plastic
strains in the RC slab deck

Concentration of tensile plastic
strains in the steel girder

Figure 7.20 Plastic strain (principal in direction 2-2) contours at failure of the composite bridge (enlarged 10 times). (For color version of this figure, the reader is referred to the online version of this chapter.)

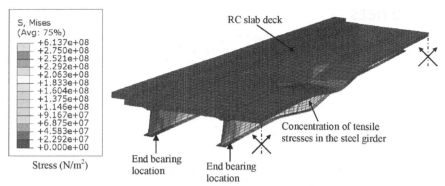

Figure 7.21 Stress (von Mises) contours at failure of the composite bridge (enlarged 10times). (For color version of this figure, the reader is referred to the online ve rsion of this chapter.)

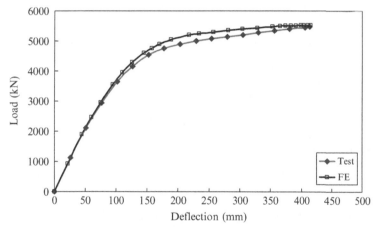

Figure 7.22 Load-deflection relationship of the steel girder at section C-C obtained experimentally and numerically. (For color version of this figure, the reader is referred to the online version of this chapter.)

midspan in the lower flange of the steel beam. The load–midspan deflection curves predicted experimentally and numerically were compared as shown in Figure 7.22. It can be shown that generally good agreement was achieved between experimental and numerical relationships. The ultimate failure load observed in the test [7.31] was 5475.8 kN at a deflection of 414 mm, while the ultimate failure load predicted from the finite element analysis was 5525.5 kN at a deflection of 415.2 mm. The finite element failure load was 0.91% higher than that observed in the test.

REFERENCES

[7.1] M. Ålenius, Finite element modelling of composite bridge stability. MSc Thesis, Department of Mechanics, Royal Institute of Technology, Stockholm, 2003.

[7.2] K.E. Barth, H. Wu, Efficient nonlinear finite element modeling of slab on steel stringer bridges, J. Constr. Steel Res. 42 (2006) 1304–1313.

[7.3] Z.P. Bazant, L. Cedolin, Blunt crack band propagation in finite element analysis, J. Eng. Mech. Div. 105 (1980) 279–315.

[7.4] W. Chung, E.D. Sotelino, Three-dimensional finite element modeling of composite girder bridges, Eng. Struct. 28 (2006) 63–71.

[7.5] A.K. Gupta, P.S. Ma, Error in eccentric beam formulation, Int. J. Numer. Methods Eng. 11 (1977) 1473–1477.

[7.6] R.E. Miller, Reduction of the error in eccentric beam modeling, Int. J. Numer. Methods Eng. 15 (1980) 575–582.

[7.7] A.K. Khaleel, R.Y. Itani, Live-load moments for continuous skew bridges, J. Struct. Eng. ASCE 116 (9) (1990) 2361–2373.

[7.8] E.A. Sadek, S.A. Tawfik, A finite element model for the analysis of stiffened laminated plates, Comput. Struct. 75 (4) (2000) 369–383.

[7.9] L.P. Carden, I.G. Buckle, A.M. Itani, Transverse displacement capacity and stiffness of steel plate girder bridge superstructures for seismic loads, J. Constr. Steel Res. 63 (2007) 1546–1559.

[7.10] A.V. Bapat, Influence of bridge parameters on finite element modelling of slab on girder bridges. MSc Thesis, The Virginia Polytechnic Institute and State University, Blacksburg, VA, 2009.

[7.11] K. Liu, G.D. Roe, Parametric study and fatigue-life-cycle design of shear studs in composite bridges, J. Constr. Steel Res. 65 (2009) 1105–1111.

[7.12] C. Odenbreit, A. Leffer, M. Feldmann, Fatigue Behavior of Shear Studs to Transfer Dynamic Loads Between Steel and Concrete Construction Elements, Sea Tech Week-Fatigue of Maritime Structures, Brest, 2004.

[7.13] T.R. Brackus, Destructive Testing and Finite-Element Modeling of Full-Scale Bridge Sections Containing Precast Deck Panels, Civil and Environmental Engineering, Utah State University, Logan, UT, 2010.

[7.14] I. Vayas, T. Adamakos, A. Iliopoulos, Three dimensional modeling for steel-concrete composite bridges using systems of bar elements—modeling of skewed bridges, Int. J. Steel Struct. 11 (2) (2011) 1–13.

[7.15] I. Vayas, T. Adamakos, A. Iliopoulos, Modeling of steel-composite bridges, spatial systems vs. grillages, in: Proceedings of the 9th International Conference on Steel Concrete Composite and Hybrid Structures, the Organized Institution, UK, 2009.

[7.16] I. Vayas, A. Iliopoulos, T. Adamakos, Spatial systems for modelling steel-concrete composite bridges-comparison of grillage systems and FE models, Steel Constr. Des. Res. 3 (2) (2010) 100–111.

[7.17] T. Adamakos, I. Vayas, S. Petridis, A. Iliopoulos, Modeling of curved composite I-girder bridges using spatial systems of beam elements, J. Constr. Steel Res. 67 (2011) 462–470.

[7.18] F. Gara, G. Ranzi, G. Leoni, Simplified method of analysis accounting for shear-lag effects in composite bridge decks, J. Constr. Steel Res. 67 (2011) 1684–1697.

[7.19] H. Somja, S. Kaing, A. Lachal, New beam-to-beam joint with concrete embedding for composite bridges: experimental study and finite element modelling, J. Constr. Steel Res. 77 (2012) 210–222.

[7.20] FineLg User's Manual, V9.2, Greisch Info, Department ArGEnCo, Ulg, 2011.

[7.21] F. Frey, L'analyse statique non linéaire des structures par laméthode des éléments finis et son application à la constructionmétallique. Doctoral Thesis, Université de Liège, 1977.

[7.22] V. de Ville de Goyet, L'analyse statique non linéaire par la méthode des elements finis des structures spatiales formées de poutres à section non symétrique. Doctoral Thesis, Université de Liège, 1989.

[7.23] P. Boeraeve, Contribution à l'analyse statique non linéaire des structures mixtes planes formées de poutres, avec prise en compte des effets différés et des phases de construction. Doctoral Thesis, Université de Liège, 1991.

[7.24] H. Zhou, K. Liu, G. Shi, Y.Q. Wang, Y.J. Shi, G.D. Roeck, Fatigue assessment of a composite railway bridge for high speed trains. Part I: modeling and fatigue critical details, J. Constr. Steel Res. 82 (2013) 234–245.

[7.25] K. Liu, H. Zhou, G. Shi, Y.Q. Wang, Y.J. Shi, G.D. Roeck, Fatigue assessment of a composite railway bridge for high speed trains. Part II: conditions for which a dynamic analysis is needed, J. Constr. Steel Res. 82 (2013) 246–254.

[7.26] W.M. Zhai, K.Y. Wang, J.H. Lin, Modelling and experiment of railway ballast vibrations, J. Sound Vib. 270 (4–5) (2004) 673–683.

[7.27] Y. Liu, H. Xin, J. He, D. Xue, B. Ma, Experimental and analytical study on fatigue behavior of composite truss joints, J. Constr. Steel Res. 83 (2013) 21–36.

[7.28] G. Ranzi, G. Leoni, R. Zandonini, State of the art on the time-dependent behaviour of composite steel-concrete structures, J. Constr. Steel Res. 80 (2013) 252–263.

[7.29] P.H. Mans, Full scale testing of composite plate girder constructed using 70-ksi high performance steel. MSc Thesis, University of Nebraska-Lincoln, USA, 2001.

[7.30] CEB, RC elements under cyclic loading, Comite Euro-International Du Beton (CEB), Thomas Telford, 1996.

[7.31] E.G. Burdette, D.W. Goodpasture, Final Report in Full-Scale Bridge Testing an Evaluation of Bridge Design Criteria, Department of Civil Engineering, The University of Tennessee, USA, 1971.

INDEX

Note: Page numbers followed by *f* indicate figures and *t* indicate tables.

A

AASHTO. *See* American Association of State Highway and Transportation Officials

AASHTO M270, 48–50

ABAQUS, 102–104
  3D shell or membrane element in, 480*f*
  bridge deck shell elements used in, 605
  buckling modes in, 578
  C3D8 and C3D6 elements in, 498–500
  classical plasticity model in, 502
  composite plate girders modeled by, 620, 627–628
  compressive inelastic strain in, 515*f*
  concrete damaged plasticity model in, 509–516
  concrete smeared cracking model in, 503–509
  connector behavior in, 484–485
  contact analysis capabilities of, 485–486
  contact domains defined in, 489–490
  contact elements in, 492–493
  contact pair approach in, 486*f*
  contact surface classifications in, 485–486
  Coulomb friction model in, 494–496
  cracking displacement in, 513–515
  cracking strain in, 512*f*
  damping sources categories in, 533
  deformable bodies making contact in, 486–487
  distributed loads in, 551–552
  for double-track open-timber floor plate girder railway steel bridge, 590–592
  dynamic analyses in, 528–529
  eigenmodes in, 522
  eigenvalue buckling analysis in, 522–523, 570–571, 584–585
  finite element modeling using, 469, 470–471, 564–565, 616–617
  as finite element program, 99
  first-order coupled temperature-displacement elements in, 540

fracture energy cracking model in, 507*f*
  friction models in, 493–494, 494*f*, 495*f*
  general contact approach in, 485*f*
  heat transfer element library in, 537
  interactions between bodies defined in, 490–491
  JOINTC elements in, 484–485, 484*f*
  kinetic friction coefficient in, 494–496
  master contact surfaces in, 491–492
  material behavior in, 570, 584
  material's physical properties and, 532–533
  modified Riks method in, 527
  nonlinear equilibrium equations in, 527–528
  nonlinear spring force-relative displacement relationship in, 483*f*
  plane stress in, 505*f*
  postbuckling analysis in, 542
  postfailure stress-displacement curve in, 514*f*
  postfailure stress-fracture energy curve in, 514*f*
  prestressing tendons in, 479
  rebar force output in, 481–482
  rebar location in, 480*f*
  rebar orientation in, 480*f*
  reinforcement bars in, 479
  sequentially coupled thermal-stress analysis in, 538
  spring elements available in, 482
  static friction coefficient in, 494–496
  static Riks procedure used in, 523–524, 524*f*
  steady-state coupled temperature-displacement analysis in, 539
  temperature analyses provided in, 533–534
  tension-stiffening model in, 506*f*, 512*f*
  true stress and plastic true strain in, 501–502
  uniaxial behavior of concrete in, 504*f*, 511*f*
  yield and failure surfaces in, 505*f*